D0930864

FAILURE ANALYSIS

Case Histories and Methodology

FAILURE ANALYSIS

Case Histories and Methodology

Dr.-Ing. Friedrich Karl Naumann
Max-Planck-Institut für Eisenforschung

Translators from the German version:

Dr. Claus G. Goetzel
Mrs. Lilo K. Goetzel
Portola Valley, California

Technical Editor:

Dr. Harry Wachob
Senior Metallurgical Engineer
Failure Analysis Associates
Palo Alto, California

Dr. Riederer-Verlag GmbH
Stuttgart, West Germany

American Society for Metals
Metals Park, Ohio 44073

Library of Congress Catalog Card No. 83-71811

ISBN: 0-87170-171-5

SAN 204-7586

Printed in the United States of America

PREFACE TO TRANSLATION

The translation of **Das Buch der Schadensfälle** by Dr. Friedrich Karl Naumann provides a vast collection of case histories that will greatly expand the available literature on failure analysis. Every attempt was made to retain the author's intent, description, and analysis of these studies. Editorial license was only exercised to clarify particular passages or to supplement the German text, e.g. providing equivalent American alloys or general chemical compositions in order to make the book more readily usable to the metallurgical engineering community in the English speaking countries. In several instances, Dr. Naumann has referenced German standards or test procedures. If the reader desires more information on an particular item, clarification or information can be obtained from the following sources:

1. **DIN – English Translations of German Standards,**
   Beuth Verlag GmbH
   P. O. Box 1145
   D-1000 Berlin 30
   Federal Republic of Germany

2. A. **Stahl – Eisen Prüfblatt,** test sheet of
   Verein Deutscher Eisenhüttenleute, 1961,
   Verlag Stahleisen mbH

   B. **Stahl-Eisen-Werkstoffblatt,** materials reference sheet in
   Taschenbuch der Stahl-Eisen-Werkstoffblätter, 3rd. Ed., 1980,
   Verlag Stahleisen mbH

   C. **Werkstoff-Handbuch Stahl und Eisen,** 4th Ed., 1965,
   Verlag Stahleisen mbH
   P. O. Box 8229
   D-4000 Düsseldorf 1
   Federal Republic of Germany

We have the pleasant duty of recognizing the assistance of many people during the preparation of this translation. First, we gratefully acknowledge the encouragement and technical support provided by the American Society for Metals; particularly Messrs. William H. Cubberly and Timothy L. Gall. Likewise, our thanks are extended to the Dr. Riederer-Verlag, in particular Mr. H. Schneider, for the many ways of rendering support and assistance. Additionally, the supplemental descriptions and correspondence with the author, Dr. Friedrich K. Naumann, have clarified technical issues throughout the text and were extremely helpful and greatly appreciated. We would like to thank the management and experts at Failure Analysis Associates at Palo Alto, California, for their support and encouragement throughout the technical review.

Dr. Harry Wachob
Technical Editor
Senior Metallurgical Engineer
Failure Analysis Associates
Palo Alto, California

Dr. Claus G. Goetzel
Mrs. Lilo K. Goetzel
Translators
Portola Valley, California

PREFACE

This book will assist those producers and users of machines, apparatus, or structures who are concerned with an analysis of failures and an elucidation of their causes. It is based primarily upon the experience of the author and therefore must by necessity be limited to failures of iron and steel parts. But even within the framework of this large territory, no complete description of the manifold phenomena can be given. To compensate for this, individual cases or specific areas that may be of technical or economic significance, but not yet of general knowledge, will be described in some detail.

The description will be supplemented by instructions on how an investigation should be prepared and conducted in order to be most successful. The book is organized in such a way that the time sequence proceeds from planning and production to actual service of the parts in which the defects that lead to failure originate.

The book has been prepared from the point of view of the materials scientist and metallographer. It was suggested first by Professor Dr. G. Petzow, the editor-in-chief of the journal **Praktische Metallographie/Practical Metallography.**

The investigations were conducted in large part at the Max-Planck-Institut für Eisenforschung, Düsseldorf. The author is indebted to his colleagues who assisted in the examinations. This is particularly true for Mr. Ferdinand Spies, who not only took part in conducting the various investigations, but also assisted in the selection of failure cases and preparation of illustrations.

Düsseldorf, 1980

Dr.-Ing. Friedrich Karl Naumann
Max-Planck-Institut für Eisenforschung

Contents

1. Examination

Failure analyses are useful not only for adjudications of legal disputes, but also because the elucidation of their causes facilitates the prevention of future failures.

Research has also benefited to a large extent from failure analyses. A fountain of knowledge has been gained from precipitation and corrosion phenomena in research that was based upon investigations of aging and corrosion failures in boiler plates and pipes.

1.1 **Preparation of Examination**

Before we proceed with this analysis, some introductory statements should be made. Often a metallographer or materials testing engineer receives a piece of steel from someone who has sustained damages with the request to determine chemical composition, strength or other properties. The fact that a failure analysis is desired, is often not even mentioned. Therefore, as a rule, the questions asked do not serve those interested in the origin of the failure. They presuppose that the cause is already known to the inquirer, so that all he needs is confirmation. But if this is not the case, the answer may be meaningless. In such a case a cursory look by a materials specialist is more useful than many an irrelevant inquiry.

The following may be helpful to illustrate the point. A **chain** manufacturer sent a piece of 3 mm wire asking whether it was made of open hearth or Bessemer steel. Further questioning elicited that the wire could not be welded in a resistance-butt welding machine. The manufacturer was asked to send a piece of a chain with poorly welded links. No weldable comparison wire could be sent because manufacture had only recently begun and no welds had so far been successful.

It was immediately apparent from the chain links that a welding error had occurred. Metallographic investigation served to confirm this. **Figure 1.1** shows a longitudinal section through the weld of a link. The weld seam was completely open. On either side of the joint zone, craters were visible. Apparently these were current contact points. From the fine structure **(Fig. 1.2)** it can be seen that the steel had elongated grains at the joint. This confirms that this part was not heated to a high enough temperature, whereas high resistance at the current entry points heated these areas to partial fusion and apparently hardened them in spots. In this case the material was erroneously thought to be at fault. Any answer to the original question, even if such were possible, would not have been of any help to the manufacturer. But the findings now made it possible for him to remedy the situation.

Another failure analysis, in which the answer to the question posed could not contribute anything useful was the following. An armature factory sent a bushing with a screwed-on ring from the **safety valve** of a seagoing vessel with the request to establish the chemical composition, yield strength, tensile strength, elongation, and notch impact strength. The parts showed evidence of substantial pitting corrosion and were covered with a black crust. The ring was screwed on improperly and had cracked from the outside to the inside **(Fig. 1.3).** The thread was strongly corroded on both the male and female parts **(Fig. 1.4).** This was reason enough to assume that the failure had occurred due to stress corrosion (see also section 15.3.4.2).

Analysis showed that both parts consisted of a titanium-stabilized austenitic stainless steel with approx. 18 % Cr, 10 % Ni and 2 % Mo. Thus the material selection was correct for the proper function of the valve, as far as corrosion by the seawater was concerned. That made the determination of the mechanical properties superfluous, even if they could have been established in these parts, since they had no bearing on corrosion restistance. Metallographic investigation confirmed that this was a typical case of transgranular stress corrosion cracking **(Fig. 1.5).** The stress and deformation were caused by the improper threading of the ring. It was assumed that seawater acted as corrosive agent. This was confirmed by an analysis of the black crust. It contained 1.02 % C, 0.24 % S, and 0.45 % Cl. Carbon and sulfur may have originated in the lubricant, while the chlorine was introduced by the seawater.

The principal cause of failure in this case was the improper thread engagement which caused high stress in the ring, bushing, and especially the thread. This failure, like the preceding one, could be explained by mere visual inspection of the parts. Analysis served solely to confirm that the right material was selected.

Another example: A messenger of a machine manufacturer arrived with a tooth that had broken out of a **pinion gear.** An inquiry was made to establish whether the material corresponded to a certain steel according to standards. To answer this question would have been of little use. Since it could be assumed that this was not the only tooth that had broken out of the gears, only an inspection of the entire assembly with all gears could show the primary tooth fracture. A single tooth, among many other damaged ones, was found that exhibited an incipient fracture with a typical fatigue failure structure. Only by examining this tooth, which had fractured first, and whose fracture affected all other damage, it was possible to establish the failure cause.

As can be seen from these examples, the materials testing engineer who is to analyze a failure, must be provided with all the information available. This is a prerequisite for any successful failure analysis. A personal discussion with all interested parties may be necessary and where indicated, the manufacturer of the failed part as well as the raw material producer should be consulted. A very important question, that almost always should be asked of the claimant, is what has been done differently in the present situation than before. This is especially important in cases where failure occurs suddenly and repeatedly after many years of satisfactory performance. The search for a deliberate or inadvertent change is often difficult, but it is of greater importance than the best examination. Questions and discussions of this type serve a purpose only if the parties will disclose all pertinent information.

It is also helpful to the investigator if he receives comparison parts together with those that failed and particularly those that have proven satisfactory in operation. The answer to the question at which stage of manufacture the failure occurred can often be answered by an examination of the raw material, if that is still available.

Finally, a visit to the plant of the claimant may be advisable for a better understanding of the function of the failed part and knowledge of the operational conditions to which it was exposed.

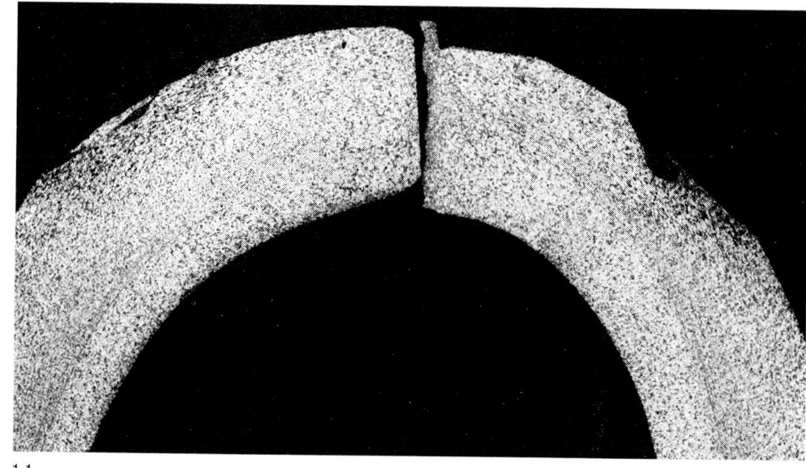

1.1

1.2

Fig. 1.1 and 1.2. Poor welding of a chain link, longitudinal section. Etch: Nital

Fig. 1.1. Overview 10 ×. Overheated and molten spots from current entry at both sides of non-bonded weld

Fig. 1.2. Elongated grain structure through cold deformation at welded joint. 100 ×

1.3

1.4

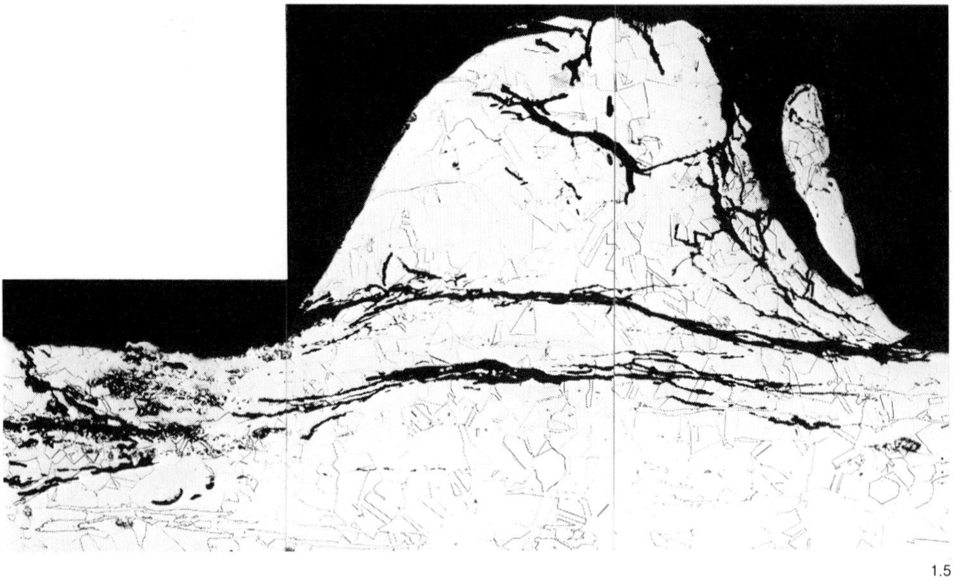

1.5

Fig. 1.3 to 1.5. Ring from safety valve of seagoing vessel destroyed by stress corrosion

Fig. 1.3. External view. 0.7 ×

Fig. 1.4. Internal view after cutting open of the ring. 0.7 ×

Fig. 1.5. Longitudinal section through thread of bushing. Etch: V2A-etchant*. 85 ×

* For composition see Appendix II

FIGURE #4

1.2 **Procedure**

After the history of the failed object has been determined, the part itself should be thoroughly examined. Especially important are the state of the surface and in case of fractures, their origin and nature. If only one single broken specimen is available from which the configuration of the entire piece cannot be reconstructed, a sketch should be procured from which this can be determined.

The cause of many failures can be determined best by careful examination of the surface with either the naked eye, a hand-held magnifying glass, or a binocular microscope. High strength wires are especially sensitive to surface damage. An example:

Non-magnetic stainless steel wires of 0.7 mm diameter with approx. 18 % Cr and 12 % Ni (AISI type 304L) which were cold drawn to an ultimate strength of 1570 to 1770 MPa (228–257 ksi) frequently broke during a not specified "klanken test"* (a dynamic, free-handed loop tensile test) and occasionally during drawing. It should be noted that the wires predrawn to 1.5 to 1.25 mm diameter were subjected to an intermediate anneal in a continuous furnace, protected by an atmosphere of cracked ammonia at 1020 °C, but without removal of the stearate that had been used as lubricant. Initially the fracture occurred after minor cold working. There did not seem to be any relationship to the melting practice, nor was this probable, since the failure occurred sometimes in certain places of the same coil, while in others none occurred. But observation had shown that wires of 7 mm diameter were especially susceptible to such failure, while thicker of thinner wires were less susceptible. Also wires made of stainless steel with approx. 10 % Cr and 9 % Ni (type 302) showed less tendency to break, even though the leaner nickel austenite is less stable, than did wires made of more nickel-rich non-magnetic steel with approx. 12 % Ni (type 305). The wire manufacturer had exhausted all means of determining the cause of failure, and thus the Max-Planck-Institut für Eisenforschung was requested to undertake a special investigation and conduct micro-analytic and magnetic tests, as well as X-ray analyses.

During an interview and an inspection of the wire drawing plant the necessary samples were selected, namely satisfactory as well as defective coils of both steel types and various diameters. In addition, good and bad spots were also collected from coils that showed different behavior during drawing or during klanken testing.

The manufacturer's observation that even within the same coil local differences in tendency for crack formation had occurred, lead to the suspicion that the fractures had some connection with local drawing or surface defects. Before starting with the desired special testing program, all wires were examined over their entire length with a binocular microscope at twenty times magnification. Furthermore, at good and defective sites, klanken tensile tests of the dynamic type used by the manufacturer (designated in the following as klanken tear tests) were conducted and supplemented by static tensile tests using knotted round wires.

Regardless of the type of steel, heat to heat variation, or dimensions, surface defects of various kinds could be determined in those spots displaying poor klanken tear behavior, while the satisfactory sections were smooth **(Fig. 1.6).** Failures were mostly initiated in longitudinal grooves **(Fig. 1.7)** where the wires sometimes developed cracks during drawing **(Fig. 1.8).**

In the transverse section, surface defects showed up as rough spots or laps. **Figures 1.9 to 1.11** show these in comparison with a smooth surface. In one coil, sections were also made of some drawing defects in addition to the surface defects and the slight cracks originating from them. They were associated with surface carburization effects **(Fig. 1.12).** Both satisfactory and unsatisfactory specimens showed the same structure of strongly deformed austenite with a small amount of elongated ferrite **(Fig. 1.13).**

A comparison of all the manufacturer's observations with the results of many hundreds of klanken tear tests and the accompanying examinations showed the existence of a definite con-

* see Appendix I

nection between the surface defects and failure during klanken tear testing. Finally it could be predicted from visual surface inspection whether the klanken tear tests would lead to fracture or not. The knotted round wire tensile tests, conducted according to a German standard procedure*, was less reliable than the klanken tear test.

According to these macroscopic tests, supplemented by only a few microsections, it could be established that the failures observed in these high strength wires after a minor increase in strength were caused by comparatively insignificant surface defects. In addition, crack formation during drawing was apparently promoted by carburization, which probably resulted from the adhering lubricant. Therefore it made no sense to conduct any further tests as per the original request.

* DIN 51214

1.6

1.7

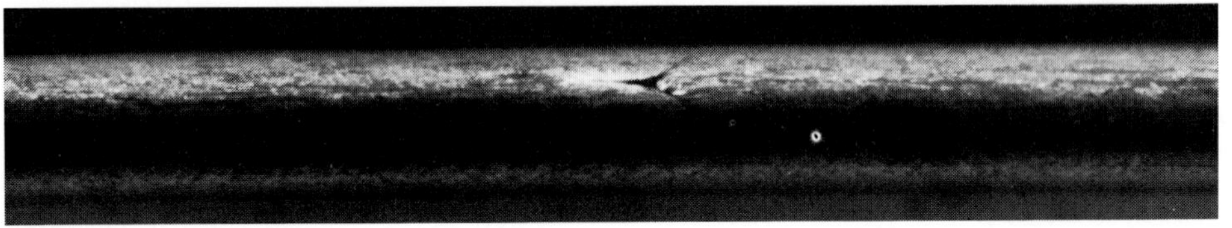

1.8

Fig. 1.6. to 1.13. Austenitic steel wires drawn to high strength that broke in part during **klanken** tear test

Fig. 1.6. to 1.8. Surface view. 20 ×
Fig. 1.6. Smooth area
Fig. 1.7. Spot with groove
Fig. 1.8. Spot with incipient crack

Fig. 1.12. Tensile crack at spot carburized by drawing lubricant. Etch: V2A-etchant. 200 ×
Fig. 1.13. Grain structure in longitudinal section. Etch: V2A-etchant. 500 ×

Fig. 1.9. to 1.11. Unetched cross sections. 500 ×
Fig. 1.9. Smooth area
Fig. 1.10. Average condition
Fig. 1.11. Individual deep grooves

1.9

1.10

1.11

1.12

1.13

The cause of the frequent cracks of high strength tensile wires in a prestressed concrete structure could also be found merely by visual inspection of the wire surface. The cracks and fractures originated in small corrosion pits **(Fig. 1.14 and 1.15).** Such oxide pits act as very sharp notches especially under fatigue stresses. Stress concentrations form at the bases of these notches (see also section 15.3.4.2). The wires were of high quality, free of defects as established by metallographic and mechanical tests.

The previous examples have shown the importance of a preliminary thorough visual inspection of the defective specimens. For this purpose the materials test engineer should allow sufficient time. In almost all cases this effort will be worthwhile. This is particularly true of fractures (see also section 2.1). In addition to the location of the fracture origin and its structure, the surface condition at the origin of the fracture is of interest.

Prior to sectioning of the specimen, a photograph or sketch should be made in which the subsequent sections should be drawn in. All parts should be clearly designated, so that their position in the specimen can be reestablished later. The next failure analysis shows that errors can occur even at this stage, which may later produce defects.

A **bone drill** of stainless tool steel[1]) fractured during drilling of the marrow cavity in the leg of a patient which should be splinted subsequently by driving a pin through the fracture. The fracture propagated from a fatigue crack shown in **Fig. 1.16** and is designated with A2. At the fracture origin, the remains of a number could be seen **(Fig. 1.17)** which had been inscribed with an electrical engraving tool. The larger part of this number was on the unavailable opposite fractured part. The cross section of the drill was weakened at this point by a milled groove from whose edge another fatigue fracture propagated (A1). A longitudinal section through A2 showed that the material had melted and hardened during cooling. It also showed several incipient cracks **(Fig. 1.18).** This crack initiation led to the formation of the fatigue fracture. Any damage to the surface should be meticulously avoided, especially in specimens which are cyclically stressed.

1.14

1.15

Fig. 1.14. and 1.15. Broken hard drawn high strength tensile wire of 4.5 mm thickness and 2160 MPa (313 ksi) strength. Fracture origin: Corrosion pits that were present already prior to assembly

Fig. 1.14. Surface, pickled with ammonium citrate solution

Fig. 1.15. Longitudinal section through corrosion pit, unetched. 80 ×

This may be illustrated by the following failure analysis:

A **rear-axle side shaft** of a truck broke during driving. It was said to consist of chromium-molybdenum alloy steel of AISI type 4140 and heat treated to 460 to 515 HB. The automotive manufacturer wished to find out if quality and heat treatment corresponded to standards. The fracture of the shaft was of a torsional nature that propagated from a small fatigue crack **(Fig. 1.19).** It originated in grinding grooves which in turn were caused by the grinding of a flat plane for a Brinell hardness test **(Fig. 1.20).** Such grinding grooves are deep notches at whose bottoms

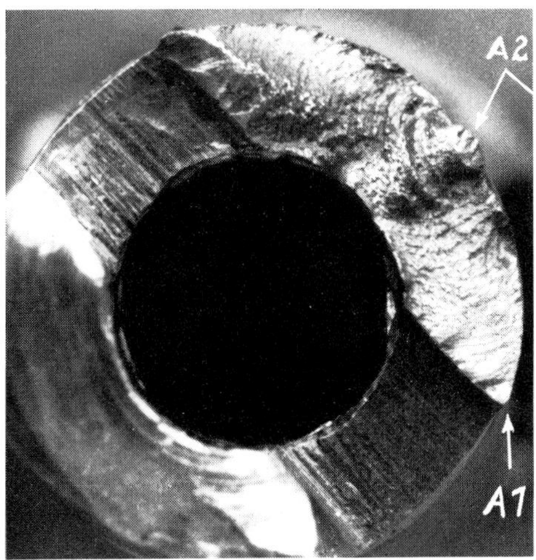

1.16

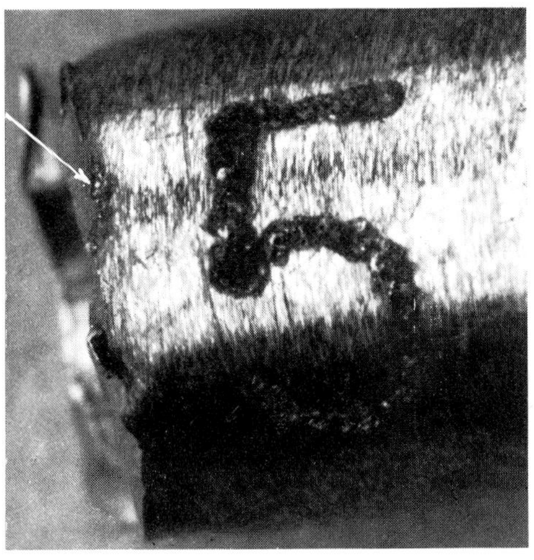

1.17

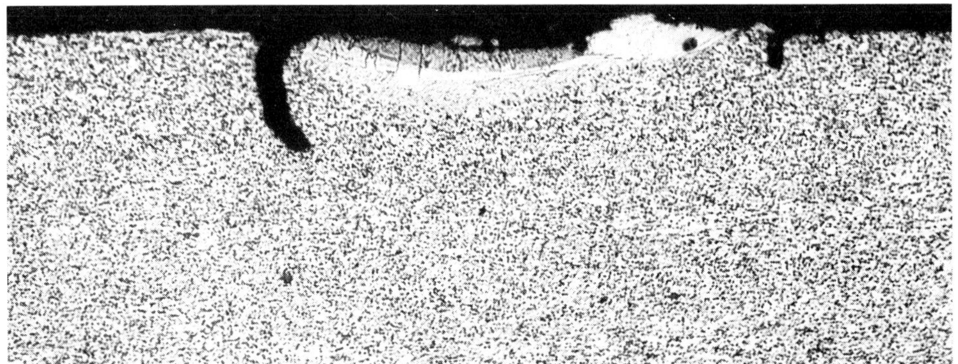

1.18

Fig. 1.16. to 1.18. Bone drill broken by burn of electrical engraving tool

Fig. 1.16. and 1.17. Views 10 ×. Fatigue fracture propagated from A1 and A2

Fig. 1.16. Fracture plane

Fig. 1.17. Surface at fracture

Fig. 1.18. Microstructure at fracture origin A2. Longitudinal section. Etch: V2A-etching solution. 250 ×

high stresses concentrate (compare with section 2.1). Chemical and metallographic tests showed that steel quality and heat treatment were satisfactory. But the martensitic structure indicated that the tempering temperature was fairly low. In consideration of this fact, the hardness which ranged from 572 to 606 HV 10 was extremely high. The high hardness, although not damaging to a smooth surface, here prevented stress relief by deformation in the grinding grooves. The fracture therefore was caused by grinding grooves and the high notch sensitivity of the hard material.

Only after a thorough examination of the specimen from all sides, as well as taking of a photograph or making a sketch, test specimens can be prepared. In some cases it may be useful or necessary to resort to non-destructive testing or pretesting[2]).

The pieces should be inspected thoroughly during sectioning of the specimens for metallographic examination. Jamming of the saw or separating of the cut may indicate high residual stresses that are relieved through the section. Spark formation during grinding may point to the presence of certain alloying elements. During cutting and working of the specimens, slow reduction and adequate cooling are essential, especially in the case of hard or high work-hardening steels. These precautions may prevent formation of cracks whose origin might later be subject to misinterpretation (see also section 8.2)[3]). For the same reason pickling with hydrogenating acids is an unsuitable method for exposing cracks or developing the primary microstructure in these steels (see also section 9.1)[4]). Magnetic particle testing or the dye penetration method and primary etching with copper ammonium chloride solution according to Heyn are preferable in these cases.

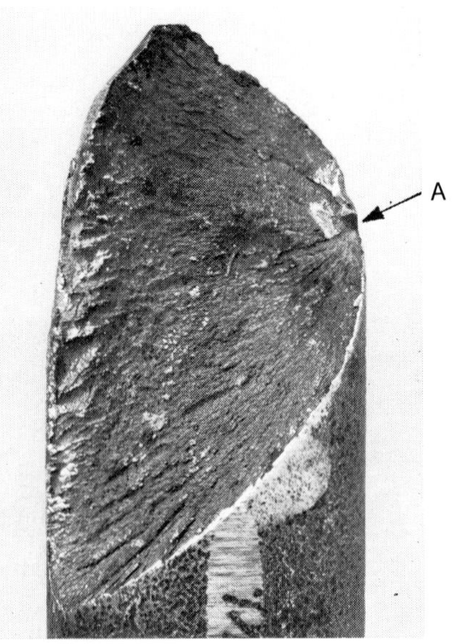

1.19

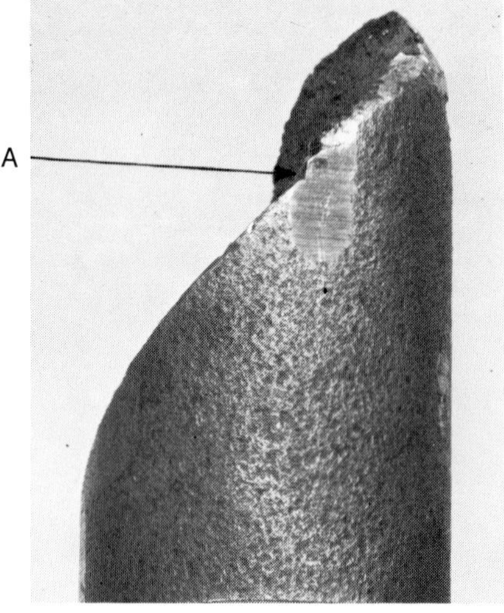

1.20

Fig. 1.19 and 1.20. Rear axle side shaft with torsion fracture originating at grinding mark (A). 1 ×

Fig. 1.19. Fracture

Fig. 1.20. Surface at fracture origin

Metallographic examination is usually preceded by macroscopic observation. This can be performed through a surface etch prior to cutting of the specimens. Regions overheated by friction, for instance, can be shown very easily by this method (see also section 13.4)[5])[6]). When testing for hydrogen sulfide attack of pipe and wire surfaces, the Baumann sulfur print method has proved of great value (see also section 15.3.4.2)[7])[8]).

Macroscopic methods are also best for determining segregations, peripheral blisters, aging phenomena, flaking, melt top crust pieces*, and other inhomogeneities as will be shown in the respective sections. Segregations and fiber orientation may, for instance, indicate whether a specimen was made from sheet or strip and whether it was forged to shape or machined from a large billet. A simple primary etching method answered the important question of a construction engineer for a failure analysis of a constriction at the fracture of a **concrete steel (Fig. 1.21).** Was it caused by local corrosion or by mechanical overloading? The fiber orientation of a longitudinal section etched according to Oberhoffer **(Fig. 1.22)** proved that corrosion induced material loss accurred across the fiber, while in a fracture constriction the fibers would have pulled together.

Metallographic sections should also be observed first with the naked eye or under a magnifying glass. Peculiarities of the microstructure sometimes show up more clearly in the higher concentration of the macrograph than at high magnification. An example for this is provided in the

* see Appendix I

1.21

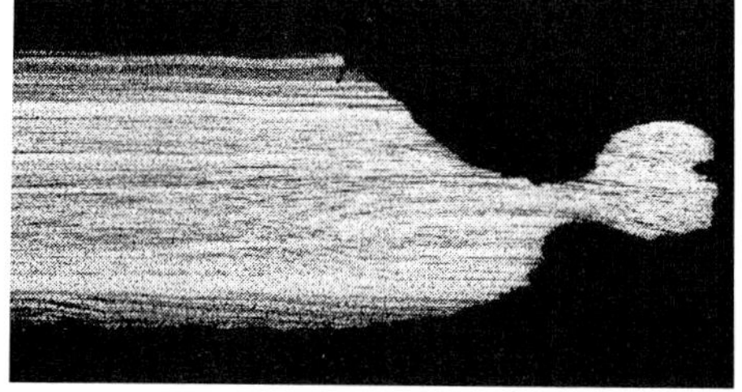

1.22

Fig. 1.21 and 1.22. Corroded concrete steel with constriction like pit

Fig. 1.21. View. 1 ×

Fig. 1.22. Longitudinal section through pit. Etch according to Oberhoffer. 3 ×

following: A **high pressure pipeline** of 108 mm O.D. and 60 mm I.D. made of hydrogen-resistant chromium-molybdenum-vanadium steel, containing approx. 0.2 % C, 3 % Cr, 0.5 % Mo, and 0.5 % V, cracked a few days after being put into operation under an operational stress that did not approach one-half its yield strength. It cracked lengthwise close to a bend **(Fig. 1.23).** The air quenched pipe had been bent at an angle of 90 degrees, after local heating, in two vertical planes that were mounted one upon another, and subsequently tempered. The section containing the crack was cut perpendicular to the crack plane for examination. The crack was broken open and discovered that the crack had propagated from the inner diameter. An annular ring zone with lighter etch toning was observed in the cross section through the crack area by etching with dilute nitric acid. This zone extended around three-fourths of the circumference **(Fig. 1.24)** and attained its greatest depth at the crack origin cross section. The crack was almost exactly opposite the deepest spot of the zone. The microstructure of the two zones could be differentiated by the fact that the inner one was transformed predominantly into bainite, while the outer one was transformed to martensite **(Fig. 1.25a and b).** In the part not heated for bending, the pipe showed a more highly tempered heat treated structure **(Fig. 1.25c).**

The crack therefore was caused by thermal and transformation stresses that in turn were caused by uneven, faster cooling on one side of the pipe. These differences in structure are more evident with macroscopic observation than in the microstructure.

This example should serve to stress again the old rule, that the metallographer should always begin his observations with the naked eye assisted by a magnifying glass, followed by microscopic examination, if necessary, after a preliminary good overview. Even then a low magnification should be tried first before using higher magnifications or the electronmicroscope.

Additional useful information may be gleaned quickly from simple bench tests such as folding, twisting, and impact tests.

Literature Section 1

1) F. K. Naumann u. F. Spies: Bruch eines Knochenbohrers. Prakt. Metallographie 5 (1968) S. 222/24

2) H. Christian, F.-X. Elfinger, P. Löbert u. B. Raible: Die zerstörungsfreie Prüfung im Dienste der Schadensverhütung. Prakt. Metallographie 15 (1978) S. 423/40

3) F. K. Naumann u. F. Spies: Schleifrisse. Prakt. Metallographie 5 (1968) S. 291/98

4) F. K. Naumann u. F. Spies: Beizblasen und Beizbrüchigkeit. Prakt. Metallographie 4 (1967) S. 663/70

5) F. K. Naumann u. F. Spies: Durch Heißlaufen gerissene Drehbankspindel. Prakt. Metallographie 6 (1969) S. 125/28

6) Dieselben: Durch Rutschen des Walzgutes warm gewordene und gerissene Stahlwalzen. Prakt. Metallographie 5 (1968) S. 647/51

7) Dieselben: Schäden an Gestängeverbindern in einer schwefelwasserstoffführenden Erdgasbohrung. Prakt. Metallographie 10 (1973) S. 100/08

8) Dieselben: Untersuchung einer blasigen und rissigen Erdgasleitung. Prakt. Metallographie 10 (1973) S. 475/80

1.23

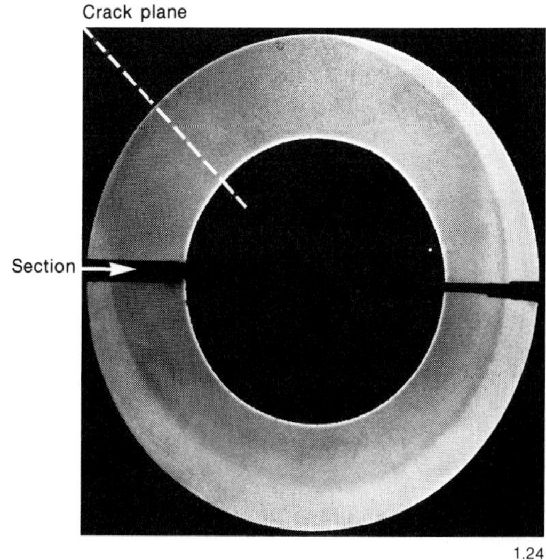

1.24

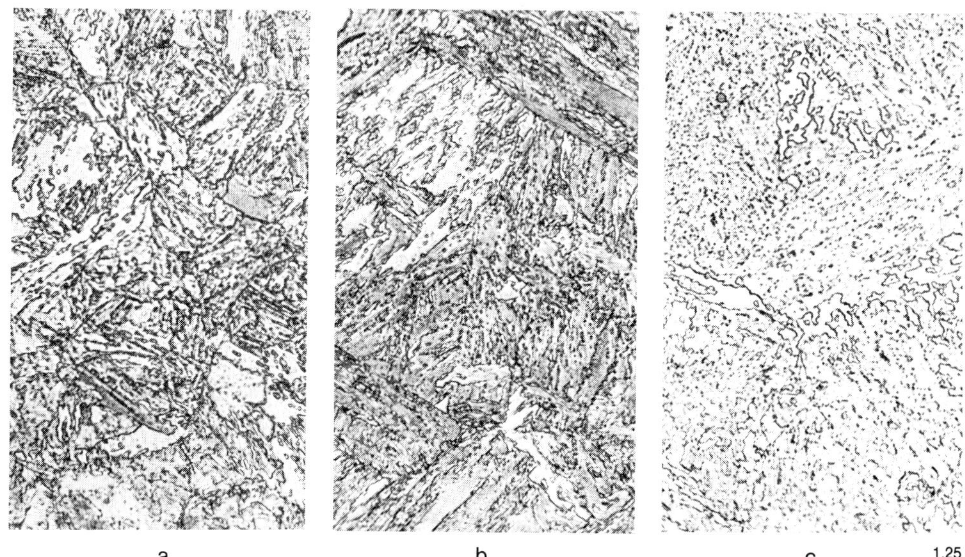

a b c 1.25

Fig. 1.23 to 1.25. High pressure pipe of 108 mm O.D. and 60 mm I.D. that broke shortly after service initiation

Fig. 1.23. External view. 0.33 ×

Fig. 1.25a to c. Microstructure in cross section Q--Q. Etch: Nital. 500 ×
a internal zone, b external zone, c microstructure of straight pipe section

Fig. 1.24. Cross section Q--Q. Etch: Aqueous nitric acid solution. 0.5 ×

2. Types of Failures

2.1 Overview

What types of failures shall be discussed? By far the most cases received by the author for analysis were crack or separation failures. We shall report on them in the following paragraph and again under the individual chapter headings. The latter are arranged according to specific causes for the failures. Often these also occur through wear, abrasion, and erosion. Generally the designer may be held responsible for these due to a construction error or selection of unsuitable material; else they may be the fault of the user because of overloading errors in assembling or insufficient care and maintenance in operation. The same holds true for corrosion failures of all kinds from general material loss to formation of pits and porosity to crack and fracture formation in containers and reaction vessels. In this connection errors should also be mentioned that were found during routine material surveillance and which had not yet caused any defects.

2.2 Crack and Fracture Failures

Cracks are separations that have not yet led to failures, but often are their incipient cause. We shall not deal here with the conditions of formation and propagation in the sense of fracture mechanics. Cracks are often closed so tightly that they must be made visible by suitable means. If a hardened or slightly tempered steel with high residual stresses is to be examined, which is susceptible to pickling cracks[1]) (see also section 9.1), it is preferable to use magnetic particle testing or dye penetration rather than pickling with a hydrogenating acid.

The external appearance of the cracks often indicates their origin or cause. Longer straight cracks are characteristic indications of first order stresses. As a rule, tension cracks, induced by rapid heating or drastic quenching, are of this type.

Short cracks which often appear in large quantities either as straight stringers or as networks, may have been caused by grinding or pickling (see also sections 8.2 and 9.1)[1])[2]).

Crack propagation is very important also with reference to the grain structure of the steel. Cracks as a rule run predominantly transgranular at room temperature, i.e. on certain lattice planes across the grain, while when heated they propagate preferably in an intergranular manner, i.e. along the grain boundaries. Stress cracks with intergranular propagation may occur when the grain boundaries are weakened due to precipitates. Such effects are caused by oxides and silicates in oxidized (burnt) steels (see also section 6.1.2.); by sulfides in the heat-affected zones next to welding seams (see also section 7.2); through carbides and nitrides on the primary grain boundaries of finished steel castings (see also section 4.3.2); and on the austenitic grain boundaries of stainless steels (see also section 15.3.3.). Caustic cracks in structural steels for boilers propagate strictly in intergranular fashion (see also section 15.3.4.1). Stress corrosion cracks in austenitic stainless steels in most cases propagate purely in a transgranular manner (see also section 15.3.4.2), but may be deflected into the grain boundaries by carbide or nitride precipitates (see also section 15.3.4.1).

In order to determine crack origin and type of fracture, the cracks should be opened up if possible.

In examining **the causes leading to failure,** an evaluation of its origin, propagation, and nature is decisive. In failure analysis it is important to protect both fracture surfaces from dirt and corrosion.

The **fracture origin** or initiation can be recognized from a concentration of fracture fibres at that point or their radiating from it, respectively. In order to find the origin, some additional work and much patience is sometimes necessary. If it is not possible to determine the origin of the fracture, its cause may also be elusive. Each fracture starts at the point of highest stress. Such points may be found in sharp corners, edges, or changes in cross section, as well as at bore holes and surface damage where multi-axial stresses, that can not be released through slip, are concentrated.

Fracture propagation generally is determined by stress conditions. Tensile stress fractures propagate perpendicular to the direction of stress, shear stress fractures at an angle of 45° to it. Torsion stresses are subject to the predominant effect of shear stress, and therefore also run at an angle of 45° to the bar axis.

Figures 1.19 and 1.20 had shown a typical torsion fracture. **Helical springs,** too, break if overloaded with torsion fractures. These normally propagate from the inner surface where stresses are highest. **Figure 2.1** shows the fracture of a helical spring originating from a small fatigue crack. In this case stress was increased by a comparatively insignificant surface defect **(Fig. 2.2),** shown in the cross section as a flat groove **(Fig. 2.3).**

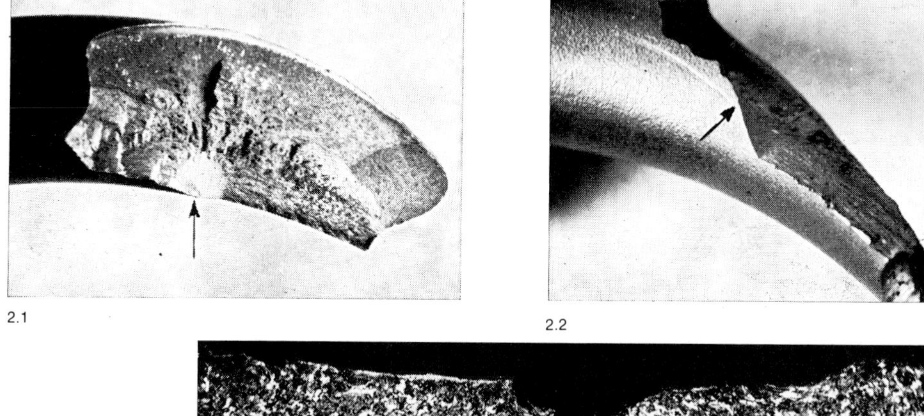

2.1

2.2

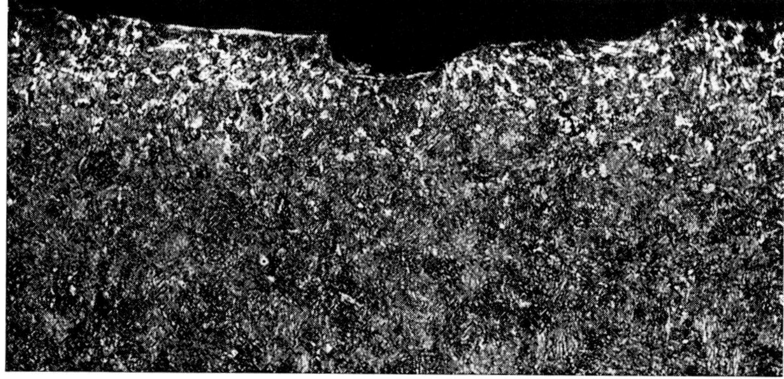

2.3

Fig. 2.1 to 2.3. Fracture of helical spring caused by mechanical surface defect (arrow)

Fig. 2.1. Fracture. 2 ×

Fig. 2.2. Side view. 2 ×

Fig. 2.3. Cross section through defect. Etch: Picral. 200 ×

A second companion **spring** showed a similar fracture **(Fig. 2.4).** But this one deviated somewhat from the former inasmuch as it originated at a spot in the cross section that was offset against the internal fiber by about 90°. In this case the stresses were increased by grinding or friction grooves **(Fig. 2.5).** The spring had become so hot at this site by the grinding or friction that the structure under that surface upon rapid cooling resulted in a martensitic phase transformation **(Fig. 2.6).** This brought about the superimposition of thermal and transformation stresses on those caused by the operation.

A very special type of fatigue fractures can occasionally be found in highly stressed short **journal pins of crankshafts.** A crankshaft that was overloaded on the test stand showed a longitudinal crack in the crank pin. This crack branched out when meeting the fillets at the transition to the two crankarms[3]). The crack consisted of many small cracks, as can be seen in **Fig. 2.7.** All were at a 45° angle to the longitudinal axis and were thus caused by torsion stresses. From the cross

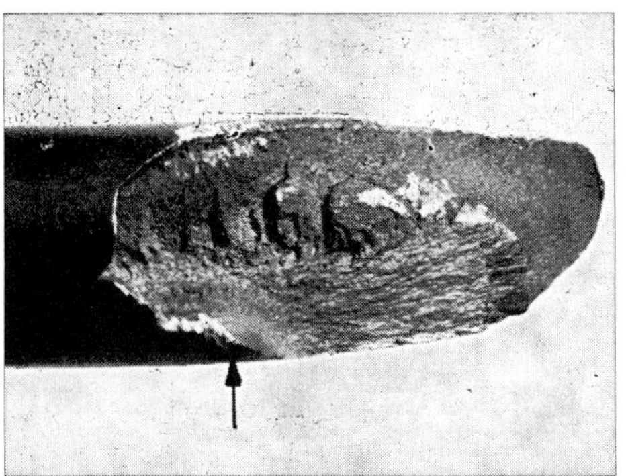

2.4

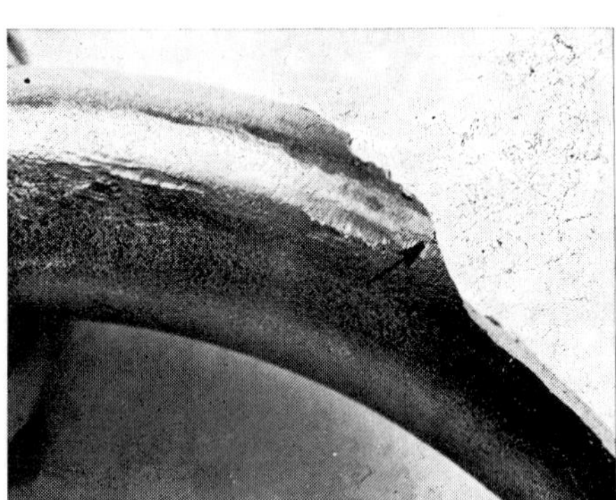

2.5

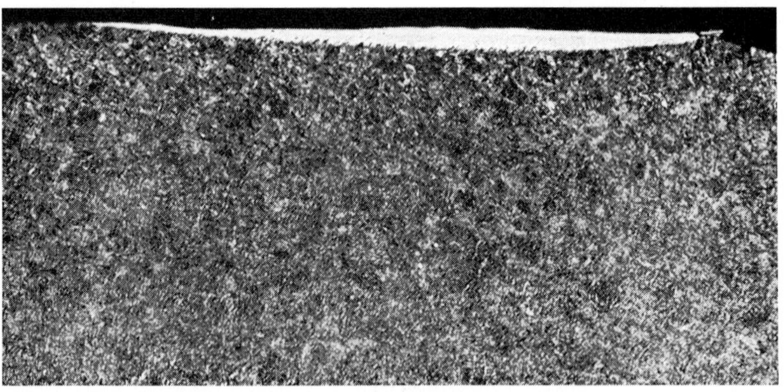

2.6

Fig. 2.4 to 2.6. Fracture of helical spring caused by grinding and chafing (arrow)

Fig. 2.4 Fracture. 2 ×

Fig. 2.5. Side view. 2 ×

Fig. 2.6. Cross section through chafing spot. Etch: Picral. 100 × The featureless light microstructure under the surface is martensite.

section it could be seen that the crack had penetrated up to 14 mm deep into the pin that had been surface hardened up to 2.5 mm depth. The crack had branched out into the martensitic peripheral structure as well as the bainite structure of the core **(Fig. 2.8).** During prying the cracks open, the fracture followed the steps of the small incipient cracks in the bearing surface **(Fig. 2.9),** then penetrated into the pin in a semi-circular way and ended again in steps **(Fig. 2.10).** The small cracks were apparently torsion fatigue fractures. The fracture path **(Fig. 2.11)** as well as the external branching (Fig. 2.8) indicated that their origin was a point below the bearing surface. This was probably the spot at which the sum of the torsion fatigue stress and residual compressive stress first exceeded the fatigue strength. The investigation showed no errors in forming, material, treatment, or machining. Therefore failure must have been caused by overloading.

Brittle materials break with cleavage fracture, tough ones with shear or mixed fracture[4])[5]). Ductile fractures have a fibrous appearance **(Fig. 2.16).** The individual crystallites are deformed by slip in such a way that their outlines can no longer be recognized in the fracture. This can be seen in **Fig. 2.12,** a highly magnified SEM photo[6])[19])[20]). It shows the artificially produced fracture of a heat treated saw blade made of nickel-chromium steel. Brittle fractures may run along either the cleavage planes of the grains or the grain boundaries. **Figure 2.13** shows the fracture of such a saw blade that was embrittled by hydrogen absorption during hard chrome plating and had broken with an intergranular separation. **Figure 2.14** shows a transgranular brittle fracture of a low carbon fine grained steel that had broken at –100° C.

Such fractures rarely occur without prior deformation, and therefore they are designated in the following not as non-deformation, but as low-deformation fractures. Expressions, such as brittle fracture or brittle fracture propensity, should not exclude such small deformations. Grain size has a decisive effect upon deformability and also by definition upon the fracture's surface. Coarse grains promote the propagation of a transgranular cleavage fracture due to fewer restraining grain boundaries. Some drop forged and normalized 1.5 ton **freight hooks** of carbon steel (type 1035), which were required for safety reasons to bend open at three times the highest stress without rupturing, broke prematurely during hoisting trials. In tests encompassing eleven broken and sixteen bent-open hooks, the former could be distinguished from those that had passed this test only in the fracture and grain size. The forced-open hooks had a purely ductile fracture (Fig. 2.16); those that broke showed the grainy-fibrous appearance of a mixed fracture **(Fig. 2.15).** The ferrite grain size in both cases was not really coarse, but significantly more so in the broken hooks **(Fig. 2.17)** than in the bent ones **(Fig. 2.18).** The difference was not very great, but nevertheless proved decisive at the unusually high stress. For such hooks a fine grained steel or a low alloy heat treatable steel should therefore be used (see also section 6.2).

The typical case of an intergranular brittle fracture is the "rock candy"* or **primary grain boundary fracture (Fig. 2.19 and 2.20),** that is caused by the precipitation of difficult-to-dissolve phases, such as nitrides, carbides, oxides and sulfides, onto the grain boundaries or twin boundaries of the primary or austenitic grain (see section 4.3.2). Low-deformation transgranular tensile cracks, however, form flakes (Fig. 4.79) that are caused by precipitation of hydrogen (see also section 4.3.3).

---

* see Appendix I

2.7

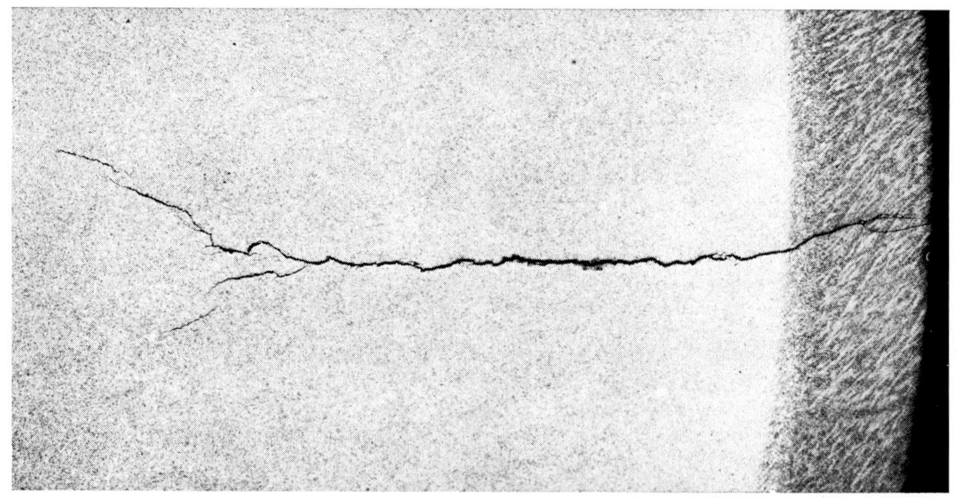

2.8

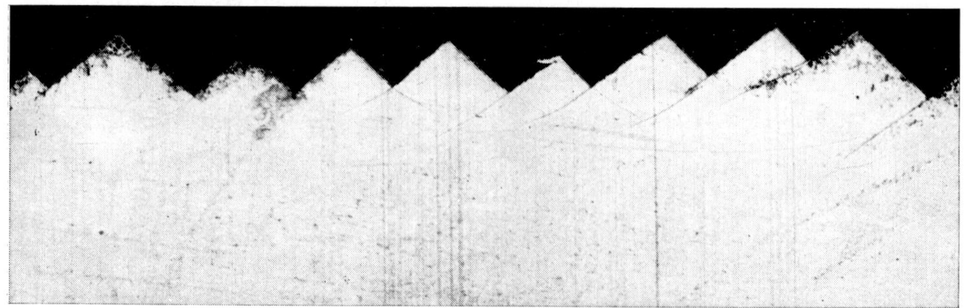

2.9

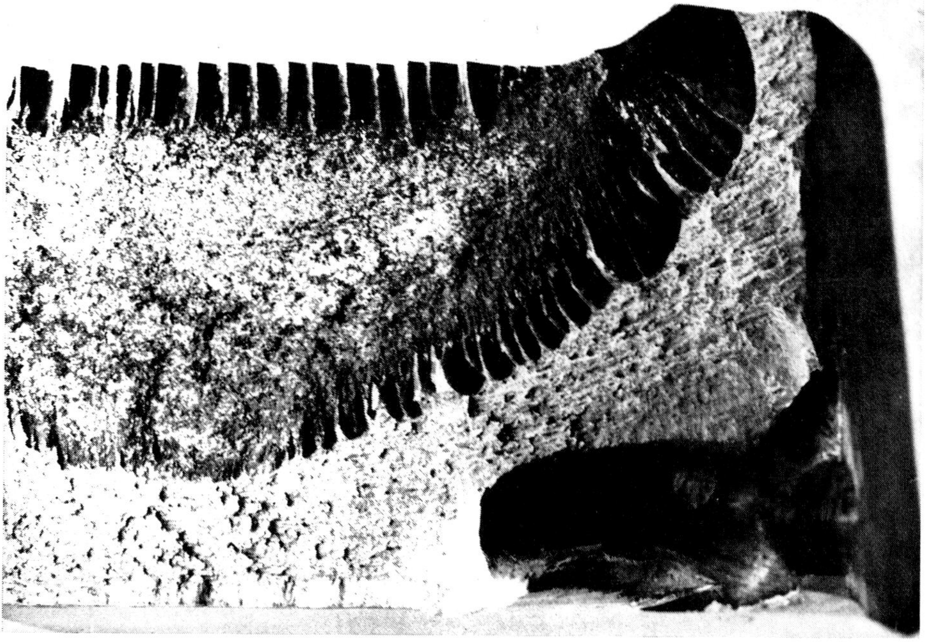

2.10

2.11

Fig. 2.7 to 2.11. Torsion fatigue fractures in bearing pin of highly stressed crankshaft

Fig. 2.7. Bearing surface of crank pin. 10 ×

Fig. 2.8. Crack propagation in cross section. Etch: Nital. 8 ×

Fig. 2.9. Fracture edge of opened crack according to figure 2.7 seen from the bearing surface of opened crack. 10 ×

Fig. 2.10. Fracture surface of opened crack. 3 ×

Fig. 2.11. Torsion fatigue fractures under bearing surface. 10 ×. Fracture origin designated by arrows

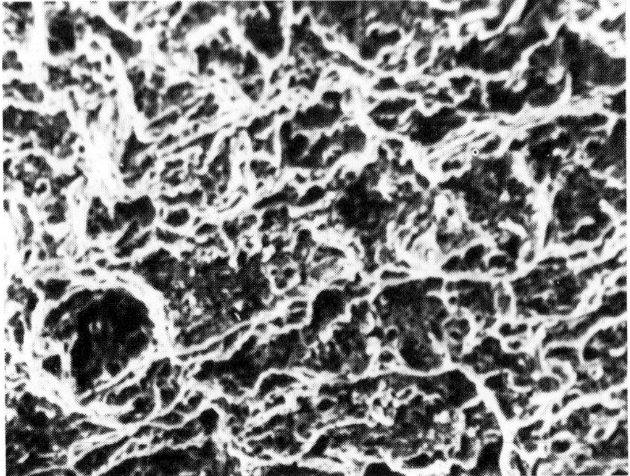

2.12

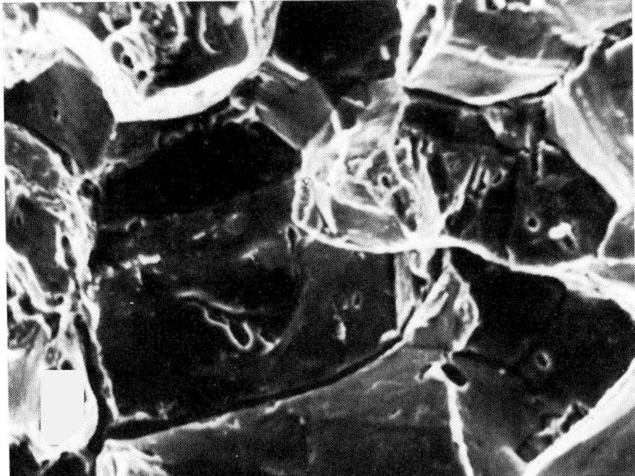

2.13

2.14

Fig. 2.12 to 2.14. Fracture shown by SEM

Fig. 2.12. Ductile fracture (waffle fracture) of heat treated saw blade

Fig. 2.13. Intergranular separation fracture of similar saw blade after hard chrome plating. 500 ×

Fig. 2.14. Transgranular separation (crevice) fracture of low carbon, fine-grained steel broken at –100 °C. 200 ×

2.15

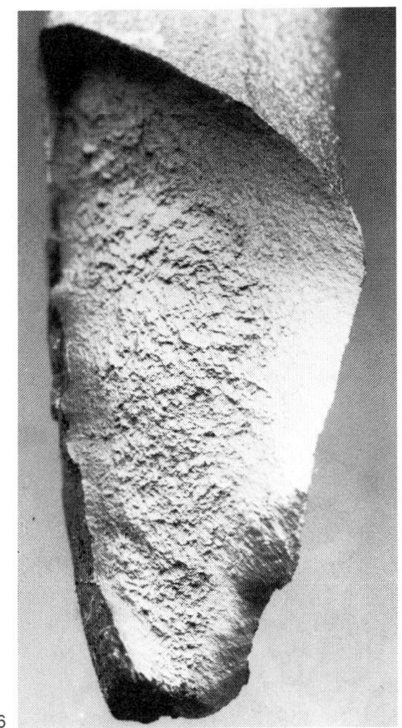

2.16

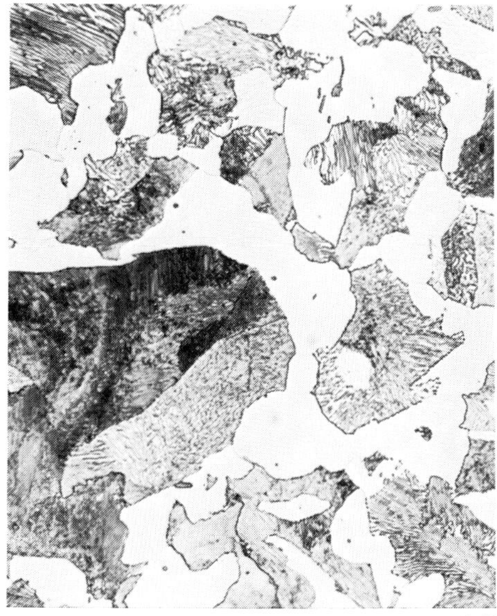

2.17

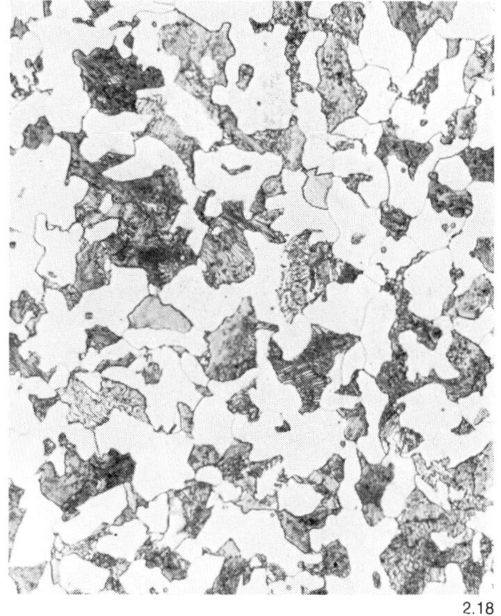

2.18

Fig. 2.15 to 2.18. Freight hooks of 1035 steel with different deformability

Fig. 2.15. and 2.16. Views of fracture. 2 ×

Fig. 2.15. Broken hook

Fig. 2.16. Bent-open hook

Fig. 2.17. and 2.18. Microstructure. Etch: Nital. 500 ×

Fig. 2.17. Broken hook

Fig. 2.18. Bent-open hook

Fig. 2.19. "Rock candy" fracture (primary grain boundary fracture) in steel casting. 1 ×

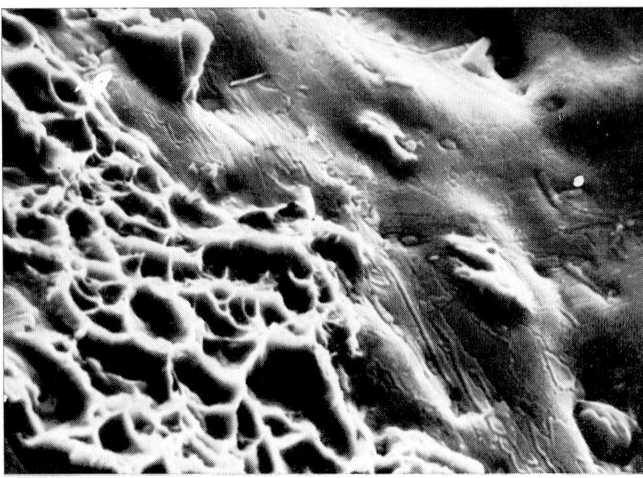

Fig. 2.20. "Rock candy" fracture in forged cast billet, SEM picture. 2000 × Lower left: deformation fracture. Precipitates of aluminum nitride on "rock candy" fracture surface (right). (see Fig. 4.72)

Low deformation intergranular fractures may also occur in the case where stresses are applied for a long time in the temperature range of about 500° C. High temperature screws of a very low elongation material broke in many cases while in use in steam power stations with such intergranular fractures without necking[7]). For instance, in such a power station several **screw bolts** of 35 mm shaft diameter fractured during the reassembly of a turbine following overhaul after approximately 8 years of operation. The bolts consisted of a creep resistant steel with 0.28 % C, 1.61 % Cr, 1.26 % Mo and 0.20 % V. They were exposed to primary steam of 500° C and heated to 470 to 475° C under these conditions. The fractures were composed of a mostly well advanced, oxidized initial part that had formed during operation, and a lightly toned remaining part that apparently was formed only during the tightening of the screws after overhaul of the turbine **(Fig. 2.21).** After dislodging of the oxide layer, observation under a binocular microscope showed that the oxidized part was a brittle intergranular fracture, whereas the remaining part was a ductile fracture. Longitudinal sections through the fracture plane confirmed this observation. They also showed that in the vicinity of the failures further intergranular separations had occurred **(Fig. 2.22).** This proved that the fractures were of a static nature and not due to cyclic loading as had been assumed by others.

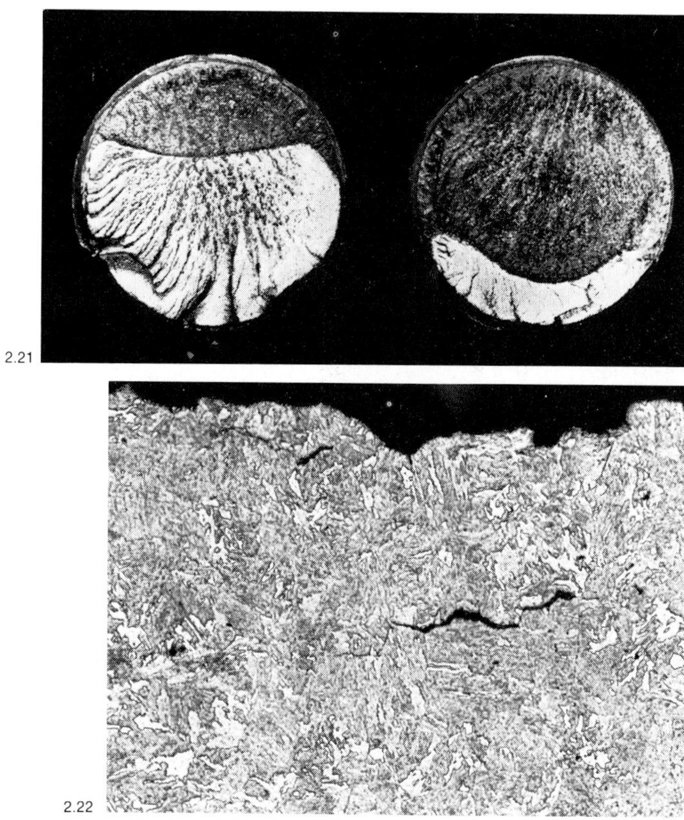

2.21

2.22

Fig.2.21. and 2.22. Screw bolts of power turbine with intergranular quench cracks

Fig. 2.21. Fracture plane of two screw bolts. 1 ×. Top: Oxidized intercrystalline incipient fracture. Bottom: Catastrophic fracture caused by tightening of screw after overhauling

Fig. 2.22. Longitudinal section through incipient fracture with grain boundary fragmentations in its vicinity. Etch: Nital. 200 ×

The cause of these intergranular brittle fractures is as yet unknown. Their formation is probably promoted by precipitates on the austenitic grain boundaries which in turn decrease the cohesive strength of the grain boundaries below the shear strength of the crystallites[8]).

Fatigue fractures, observed macroscopically, may also be assigned to the group of low deformation transgranular fractures[9]). But under the microscope it can be seen that here, too, a slip deformation precedes crack formation in those crystal planes favorably oriented in the direction of principal stress **(Fig. 2.23)**[10])[11])[12]). Fatigue fractures may occur under repeated alternating stresses, often at nominal stresses that may lie far below the yield strength of the material. Fatique fractures proceed in steps which may be either higher or lower depending upon the magnitude of the stress, and the fracture plane shows a characteristic pattern of "arrest lines"* that arrange themselves concentrically around the origin of the fracture. SEM microscopy shows similar, but much finer lines for fatigue fractures (Fig. 3.13). Each fracture line compressed here into a small space corresponds to a change of load[13]) and their number therefore may indicate the timetable of crack formation.

The point or points of origin can usually be found easily in fatigue fractures. Fractures from a reverse bending stress have two points of origin or two groups of origins lying opposite one another in the cross section. But rotational bending usually shows several points of origin distributed around the circumference. Often several fracture origins are grouped closely behind or next to each other and joined in one step later. A fast propagating fatigue fracture may lose its characteristic form and then may no longer be recognized as such with certainty. In this case it may be called a low cycle fatigue failure. G. Henry and D. Horstmann[19]) present a detailed description of fractures and their formation mechanism, as well as their appearance under the optical and scanning electron microscope.

* see Appendix I

Fig. 2.23. Deformation by slip on the surface of a structural steel test bar subjected to reverse bending, with beginning crack formation (slightly etched during electropolishing). 200 ×

Figure 2.24 shows a far advanced bending fatigue fracture in a **crankshaft** made of an alloy steel with approx. 0.35 % C and 1 % Cr (AISI type 5135) which failed after only 150 operating hours as a result of overloading. The rupture propagated from the hollow groove of the main bearing and penetrated across the arm almost to the hollow groove of the neighboring crank bearing. The grooves were well rounded and cleanly machined. No material defects were apparent, and the fiber was well oriented vis-a-vis the direction of stress **(Fig. 2.25).**

Notches, at which multi-axial stresses are formed due to inhibited deformation, also lower resistance against cyclic type loads to the extent that the sharper the notch, the lower the resistance **(Fig. 2.26).** Notches in this sense may consist of abrupt cross section changes, sharp edges, screw threads, grooves or borings, working defects such as turning and grinding grooves, as well as thermal stress-induced cracks due to grinding or welding. Most fatigue fractures start from such sites of high stress concentration. For instance, screw bolts often break in the first supporting thread, shafts at the bearing sites, poorly rounded or machined hollow grooves, sharp-edged keyways, and leaf springs from friction points which occur at the shackle, and axles from friction-induced oxidation points at loose seats. Also oxidized and decarburized surfaces reduce fatigue strength considerably (see curve e in Fig. 2.26). Notches of a special kind are actually corrosion pits (see curves f and g in Fig. 2.26). The combined effect of vibrational stresses and corrosion, a phenomenon that may be designated as corrosion fatigue, is the cause of many defects (see also section 15.3.5).

Fatigue strength may be increased and notch sensitivity reduced by surface hardening or other methods that subject the peripheral zone to a compressive prestress. Fracture in this case does not originate at the surface, but at a point in a zone lying below it, in which the sum of tensile and compressive stresses first exceeds the fatigue strength.

Fatigue failures as a result of surface notches are exemplified in **Fig. 2.27 to 2.29,** showing the oxidized and pitted surface and the peripheral structure of an unmachined **axle journal.** Other examples of the effect of notches are cited in sections 3.1 under design defects, 4.3.3 under hydrogen in steel, 7 under welding defects, 8 under machining errors, 9 under picking errors, 13.3 under wear, and 15.3.5 under corrosion fatigue. Examples of fatigue failures due to case decarburization are cited in section 6.1.3.

Internal defects may also act as notches and be the cause of failures. In section 4.3.3 such failures are reported, in which flake cracks were the cause. The fatigue fracture of a **piston rod** of a drop hammer shown in **Fig. 2.30** was caused by a defect in the interior of the cross section. The primary structure showed a minor defect in the fracture origin **(Fig. 2.31).** Microscopic examination showd a cavity **(Fig. 2.32)** accompanied by non-metallic inclusions. But this defect was so minor in comparison with the cross section of the rod that it could not be considered the cause of failure, but merely the point of lowest resistance at which the fracture started during overloading in service.

A steel is more notch sensitive, the higher its strength (see Fig. 2.26). The purpose for the high strength may therefore be neutralized by unsuitable design or poor machining.

What can fractures indicate about material defects? It has been shown already that certain defects such as inclusions and precipitates find expression in the fracture appearance. Slag streaks of non-metallic inclusions cause the fracture to be fibrous (see also section 4.1). This is even more so if the specimen was hot when it fractured so that the fracture is discolored blue, or if it was heat treated to high toughness before breaking. This is useful to the materials engineer, but should not lead to the wrong conclusions. Thus, for instance, the heat treated boiler plate of **Fig. 2.33** may seem to contain more impurities than the one normalized. In actuality both samples came from the same sheet. Flakey spots are considerably more apparent on the fracture surface of a heat treated steel than on the surface of forged or annealed steel fractures. Precipitates on the grain boundaries often lead to "rock candy" fractures only if such strength has been imparted to the steel through heat treatment that the shear strength of the grains exceeds the cohesive strength of the grain boundaries.

2.24

2.25

Fig. 2.24. and 2.25. Bending fatigue fracture of a crankshaft broken by overloading after 150 h

Fig. 2.24. Fracture.1 ×

Fig. 2.25. Longitudinal section through fracture origin (arrow). Etch according to Heyn. 0.75 ×

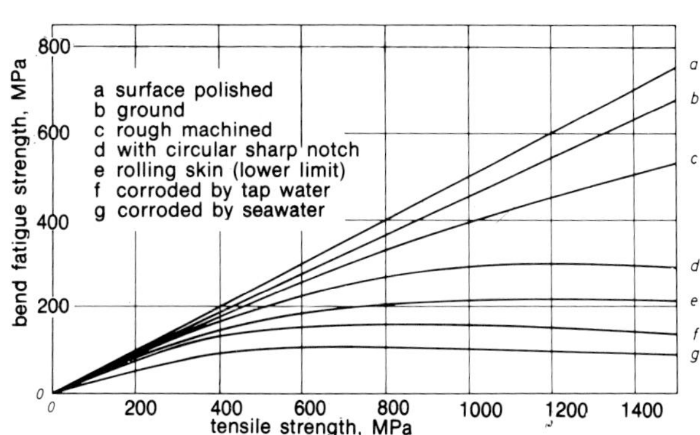

Fig. 2.26. Relation between reverse bending fatigue strength and tensile strength. Data according to **Werkstoff-Handbuch Stahl u. Eisen,** 4th Ed, Sheet D 11

2.27

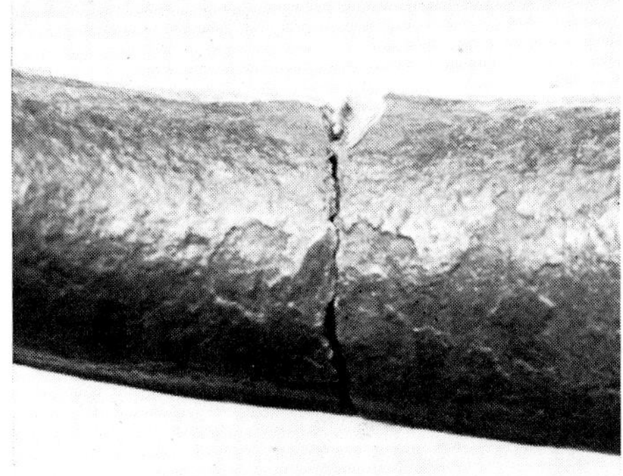

2.28

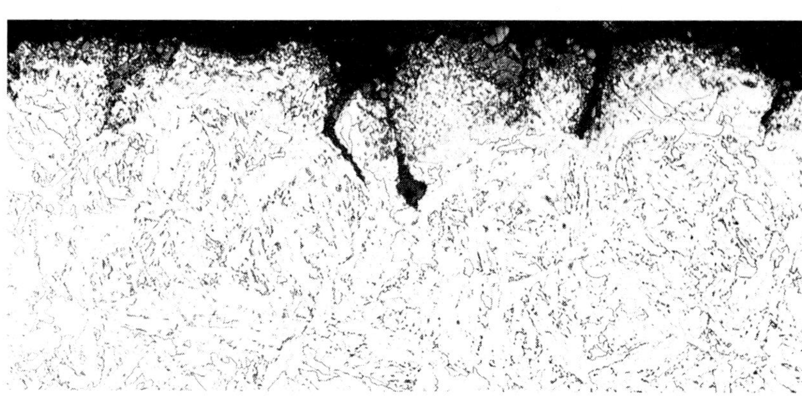

2.29

Fig. 2.27 to 2.29. Broken axle journal with unmachined surface

Fig. 2.27. Fracture. 2 ×

Fig. 2.28. Fracture surface. 2 ×

Fig. 2.29. Peripheral structure. Etch: Nital. 200 ×

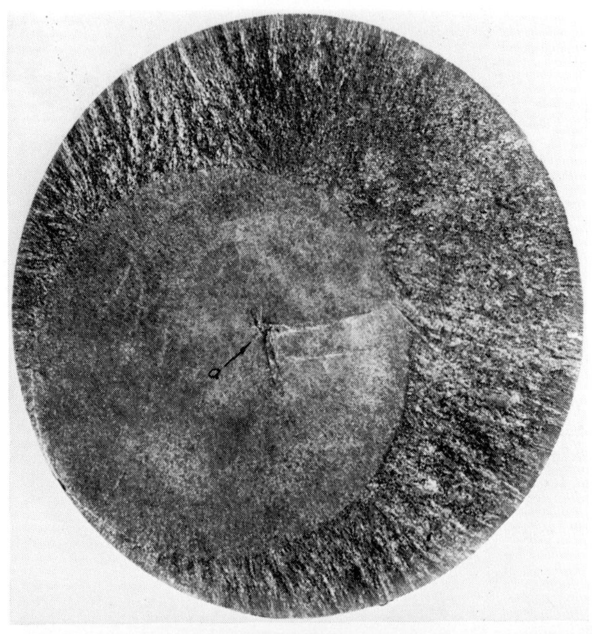

2.30

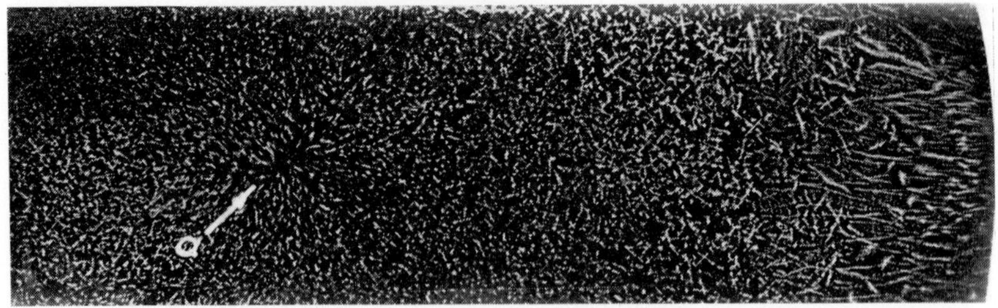

2.31

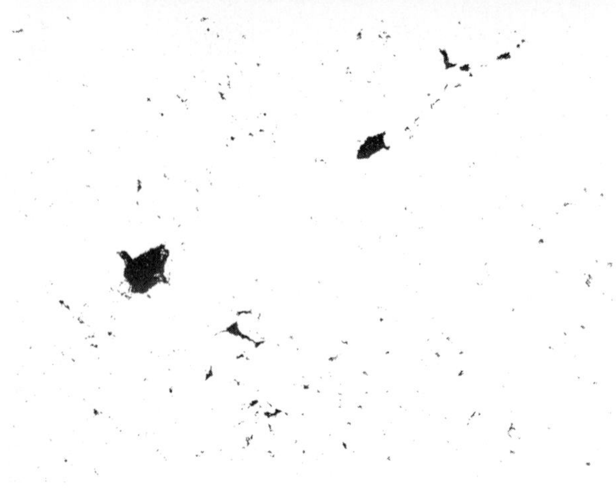

2.32

Fig. 2.33. Fracture specimens of a boiler plate. 1 ×
N = normalized. V = heat treated

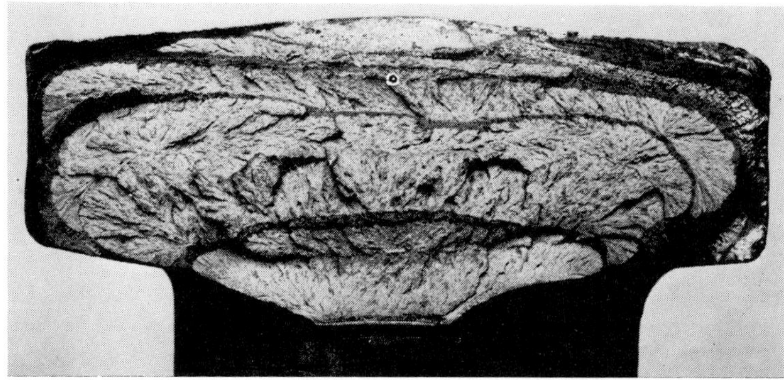

Fig. 2.34. Arm fracture of a crankshaft that was induced by hammering a wedge with several blows. 1 ×

Fig. 2.30 to 2.32. Fatigue fracture propagating from the interior of a piston rod of 190 mm diameter of a drop forge hammer

Fig. 2.30. Fracture. Origin at a. Approx. 0.4 ×

Fig. 2.31. Cross section with fracture origin (a). Etch according to Oberhoffer. 1 ×

Fig. 2.32. Point a in unetched cross section. 100 ×

On the other hand, certain fracture appearances may be mistaken for the presence of inhomogeneities in the material. Such a mistake could be made in the case of the well known fluted fracture in tensile specimens with strong necking[14]). Nor is the steel defective in the case shown in **Fig. 2.34** where an arm fracture of a crankshaft made of AISI type 1045 carbon steel with 810 MPa (117 ksi) tensile strength and 19 % elongation is shown which was deliberately fractured in an automotive factory to test the material. This fracture was apparently procuded in stages by hammering a wedge between the arms of the crankshaft. The fine-grained areas correspond in this case to the high stress induced failure rate at the beginning of each hammer blow, and the fibrous areas correspond to the diminished failure rate at the end.

After this description of the nature of fractures, the cause of brittle failures should be summarized. Brittle fractures or the tendency to embrittlement may be inherent in the material, but need not be so. It has been known for a long time that phosphorus makes steel brittle. In heat

2.35

2.36

Fig. 2.35 and 2.36. Fractured Nimonic 75 pipe bend from cooling vessel of a synthetic ammonia installation

Fig. 2.35. Fracture. 1 ×

Fig. 2.36. Surface after dye penetration crack testing. 1 ×

treated steels alloyed with chromium and manganese, phosphorus also promotes temper embrittlement, i.e. the lowering of notch toughness during slow cooling after tempering, or during long term holding at temperatures between 450 and 600 °C[15]). Nitrogen makes the steel sensitive to strain aging embrittlement (see section 4.3.1). Highly alloyed chromium and chromium-nickel steels tend to embrittle through transformation of the solid solution into a superlattice phase, such as the so-called σ-phase[16]). It has already been shown that steel loses toughness through formation of coarse grains (Fig. 2.15 to 2.18). This may be caused by recrystallization or overheating during working or heat treatment (see sections 5 and 6). The sensitivity to overheating of a steel may be decreased by deoxidation or denitridation.

Hydrogen which has been absorbed by iron in the solid or liquid state may also lead to brittle fractures[1]) (see also sections 4.3.3, 9.1 and 9.2). **Figure 2.35** shows a **pipe of Nimonic 75,** that had torn open in a bend without prior deformation. Several transverse cracks were made visible at the inside of the bend **(Fig. 2.36)** during crack examination by dye penetration. A straight piece of the same pipe split open during transverse bending tests after only minor deformation. It contained 28 ml hydrogen per 100 g metal, while an unused pipe contained only 1.8 ml/100 g metal, i.e. less than one-fifteenth. The fact that nickel can absorb large amounts of hydrogen is well known. But the source of hydrogen could not be established. The typical smell of rotten eggs indicates its reduction from diluted hydrogen sulfide (see also section 15.3.4.2). Often hydrogen cannot be proved to be the cause of brittle fractures because it is very mobile in the iron lattice and may be expelled within a few days at room temperature, or within a few hours at elevated temperatures, at least in part. The hydrogen as well as the embrittlement caused by it may have disappeared by the time the materials test engineer lays his hands on the specimen. Sheets and plates of the same melt have been known to show cracks or blisters if they were machined soon after pickling, but were free of such defects if they were stored for a while.

Therefore a tendency to brittle fracture need not be a material property. In many cases the cause may be found in the design, the operating conditions, or in the type of loading. They are promoted by three factors which are: 1) multi-axial stresses, 2) low temperature, and 3) high stress rate. All three impair plastic flow and therefore lead to brittle type fractures. It has already been mentioned that sharp notches and cross sectional transitions promote the formation of high multi-axial stresses.

The fracture of a **pipe flange** of an oil pipeline may serve as an example for brittle failure as a result of multi-axial stresses. After almost 3 years of operation followed by a period of non-use, the flange ruptured during start-up. The fracture is designated by arrows in **Fig. 2.37.** It is located in the part drawn to 5 mm thickness, and about 10 mm adjacent to the welding seam with which the flange had been welded to the pipe of 7 mm wall thickness. In this case the stress was strongly multi-axial and consisted of a tangential tensile stress produced by an internal pressure of 50 atm, an axial tensile stress increased by the shock wave during start, and residual stresses during welding. Furthermore, elongation was inhibited by the massive flange part. The flange consisted of a killed cast low carbon steel of 360 MPa (152 ksi) tensile strength that was believed to be open hearth steel because of its low nitrogen content. The structure was fairly coarse grained and permeated by narrow and smooth deformation twinning, of the type designated as C-twinning by A. Kochendörfer[17]), in the vicinity of the fracture, particularly on the inside **(Fig. 2.38).** Since the operating temperature was not unusually low, the presence of such twinning in the structure is a confirmation that the stress must have been strongly multi-axial. The stress rate may have been unusually high due to the initial startup shock.

An example of the effect of low temperature ist presented by the **frame of a crane hanger** shown in **Fig. 2.39,** that fractured on a cold winter day during lifting of a casting ladle. A blow from the unraveling of a chain sling may have been a contributing factor. The fracture **(Fig. 2.40)** propagated without substantial deformation from a point on the surface at the inside of the bend. Examination showed the material to consist of a rimmed cast steel with 0.15 % C, 0.021 % P and 0.005 % N. It was free of defects, not aged, and had good mechanical properties with 370 MPa (54 ksi) tensile strength, 20 % elongation, and 117 J notch toughness. The cause of failure therefore was not to be found in the composition or properties of the material.

The SEM picture of the fracture of a low carbon fine-grained steel fractured at –100 °C was shown in Fig. 2.14.

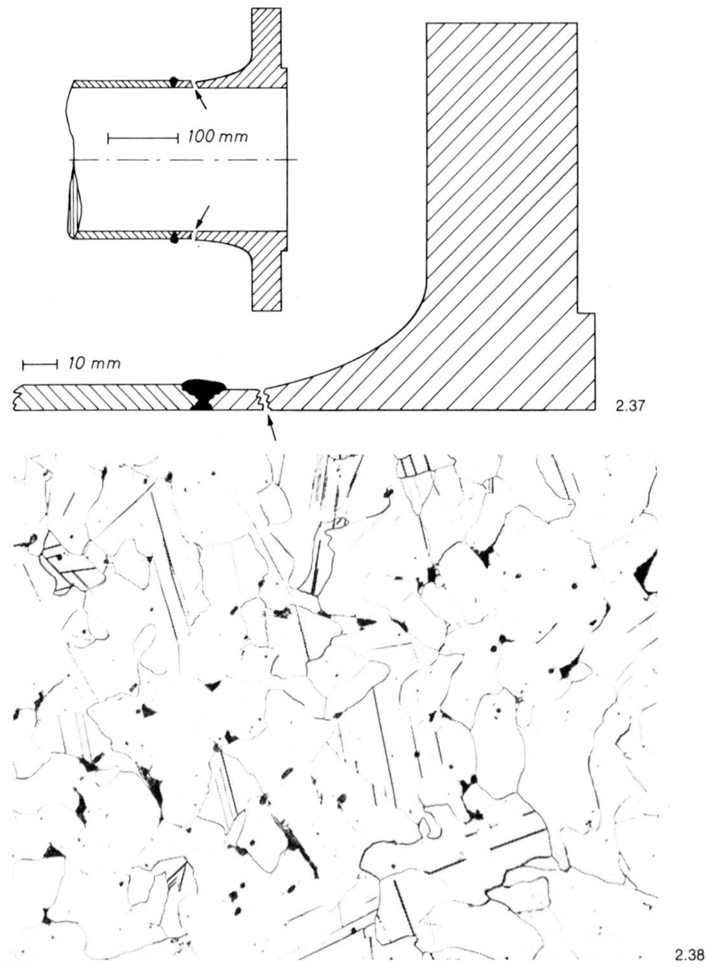

Fig. 2.37 and 2.38. Broken flange of oil pipeline

Fig. 2.37. Longitudinal section

Fig. 2.38. Microstructure adjacent to fracture, inside. Longitudinal section. Etch: Nital. 100 ×

The effect of high initial stress rate ist shown in the fracture of a **ring of a grabbing cable of a jet catapult installation.** It occurred adjacent to a perfect butt welding seam **(Fig. 2.41)** and showed only minor deformation **(Fig. 2.42** left). The ring was free of material defects and had a nominal heat treated bainite structure. Its strength, calculated from hardness, was about 880 MPa (127 ksi). A tensile specimen taken from the ring showed a perfect deformation fracture with good neck formation after tensile testing **(Fig. 2.42,** right). This figure may serve to show how fractures and deformation energy may differ in the same specimen if the stresses that lead to rupture differed in type and rate.

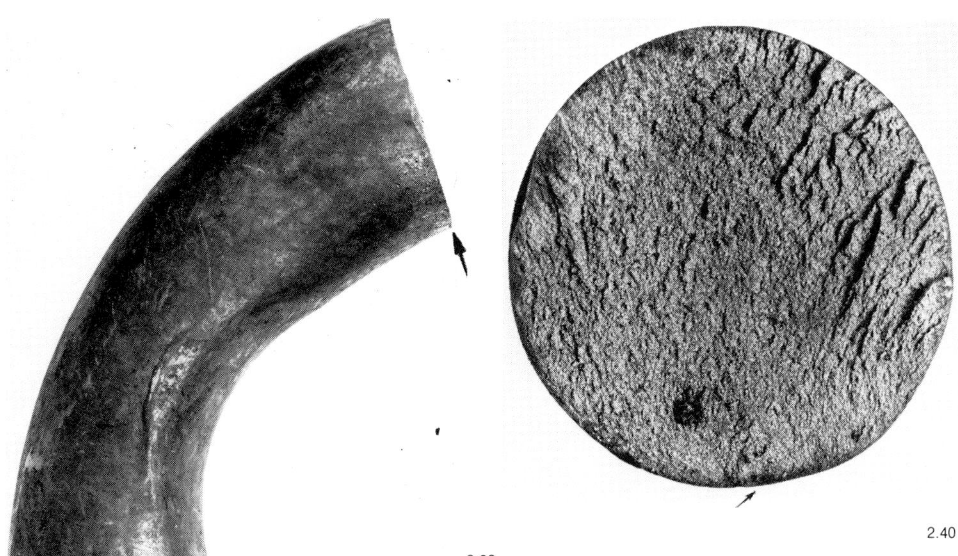

2.39 2.40

Fig. 2.39 and 2.40. Frame of crane hanger made brittle by exposure to cold

Fig. 2.39. Side view. 0.4 ×. Fracture origin designated by arrow

Fig. 2.40. Fracture. 0.75 ×

2.41

a b 2.42

Fig. 2.41 and 2.42. Fracture of a ring of grabbing cable of jet catapult installation

Fig. 2.41. Longitudinal section through fracture end. Etch according to Oberhoffer. 2 ×

Fig. 2.42. Fractures. a = ring fracture, b = tensile bar fracture. 1 ×

Literature Section 2

1) F. K. Naumann u. F. Spies: Beizblasen und Beizbrüchigkeit. Prakt. Metallographie 4 (1967) S. 663/70

2) Dieselben: Schleifrisse. Prakt. Metallographie 5 (1968) S. 291/98

3) Dieselben: Kurbelwelle mit Torsionsdauerbrüchen in einem induktiv oberflächengehärteten Pleuelzapfen. Prakt. Metallographie 10 (1973) S. 234/38

4) E. J. Pohl: Das Gesicht des Bruches metallischer Werkstoffe. Ergebnisse von Untersuchungen an beschädigten Maschinenteilen, ausgeführt durch die Abteilung für Maschinenversicherung der Allianz Versicherungs-Aktiengesellschaft

5) G. Richter: Was sagt die Ausbildung eines Bruches über die Bruchursache aus? Der Maschinenschaden 29 (1956) S. 97/106

6) Engel - Klingele: Rasterelektronenmikroskopische Untersuchungen von Metallschäden. Gerling-Institut für Schadensforschung und Schadensverhütung GmbH, Köln 1974

7) E. Houdremont: Werkstofffragen im neuzeitlichen Kesselbau unter Berücksichtigung der Rohstofflage. Mitt. Ver. Großkesselbes. 63 (1937) S. 229/42; Stahl u. Eisen 58 (1938) S. 1175

8) F. K. Naumann, H. Keller, H. Kudielka u. A. Krisch: Die Änderung des Mikrogefüges von warmfesten Stählen durch Zeitstandbeanspruchung. Arch. Eisenhüttenwes. 42 (1971) S. 439/47

9) F. K. Naumann u. F. Spies: Dauerbrüche. Prakt. Metallographie 6 (1969) S. 447/56

10) M. Hempel u. E. Houdremont: Beitrag zur Kenntnis der Vorgänge bei der Dauerbeanspruchung von Werkstoffen. Stahl u. Eisen 73 (1953) S. 1503/11

11) M. Hempel: Verformungserscheinungen und Rißbildung an biegewechselbeanspruchten Rundproben eines vergüteten Chrom-Molybdän-Stahles. Arch. Eisenhüttenwes. 37 (1966) S. 887/95

12) M. Hempel: Die Entstehung von Mikrorissen in metallischen Werkstoffen unter Wechselbeanspruchung. Arch. Eisenhüttenwes. 38 (1967) S. 446/55

13) F. Forsyth u. D. Ryder: Aircraft Engineering 32 (1960) S. 96

14) H.-F. Klärner u. E. Hougardy: Das Auftreten des Fräserbruchs bei Zerreißproben. Arch. Eisenhüttenwes. 41 (1970) S. 587/93

15) H. Bennek: Einfluß des Phosphors auf die Anlaßsprödigkeit. Arch. Eisenhüttenwes. 9 (1935/36) S. 147/54

16) F. K. Naumann: Beitrag zum Nachweis der σ-Phase und zur Kinetik ihrer Bildung und Auflösung in Eisen-Chrom- und Eisen-Chrom-Nickel-Legierungen. Arch. Eisenhüttenwes. 34 (1963) S. 187/94

17) A. Kochendörfer u. H. Scholl: Zwillingsbildung und Sprödbruchneigung von Stählen. Arch. Eisenhüttenwes. 28 (1957) S. 483/88

18) F. K. Naumann u. F. Spies: Ritzel mit ausgebrochenen Zähnen. Prakt. Metallographie 6 (1969) S. 627/31

19) G. Henry u. D. Horstmann: Fraktographie und Mikrofraktographie. De ferri metallographia, Bd. 5, Verlag Stahleisen mbH., Düsseldorf 1979

20) R. Mitzsche, F. Jeglitsch, H. Scheidl, St. Stanzl u. G. Pfefferkorn: Anwendung des Rasterelektronenmikroskops bei Eisen- und Stahlwerkstoffen. Radex-Rundschau 1978, S. 575/890

3. Failures Caused by Planning Errors

A large number of failures can be ascribed to the planning stage, such as those that occur during construction, material selection or specifications for processing, machining, assembly, etc.

3.1 Design Errors

It has been known for a long time that during construction of machine parts and tools sharp edges and cross sectional changes should be avoided because stress concentrations form at such places. These can result in stresses substantially above the nominal stress. It is surprising how often this well known design axiom is violated even today.

Such unfavorably designed parts may even show cracks or fractures during heat treating. **Figure 3.1** shows the fracture of a **lathe tool bit** made of tool steel with approx. 1.45 % C and 1.4 % Cr that was hardened in oil at 870 °C, which was at least 20° too high. The fracture propagated from a rectangular cross sectional transition that was not properly filleted, and moreover was rough machined as shown by the grooves in **Fig. 3.2** (see also section 8.1). Thereby the notch effect was further aggravated.

3.1

3.2

Fig. 3.1 and 3.2. Lathe tool bit of 1.45 % C, 1.4 % Cr steel with acute-angled and rough machined cross-sectional transition which fractured during hardening

Fig. 3.1. Fracture. 1 ×

Fig. 3.2. View into angle. 2 ×

Many failures in service, especially those caused by shock or cyclic loads, can be caused by such design errors. **Figure 3.3** shows a **bolt of a self service elevator**[1]) which failed as a result of reverse bending fatigue. In this case the fracture also propagated from a sharp-edged cross-sectional transition. In order to avoid further damage and prevent potential accidents, 24 other bolts that had not yet failed were examined metallographically or in bend tests. Eight of these proved to have incipient fatigue cracks in the cross-sectional transitions. The bolts were in part normalized, and in part heat treated. Their strength which was determined from Brinell hardness, was between 440 and 700 MPa (65 and 100 ksi). The cracks had occurred in the annealed as well as the heat treated bolts, i.e. in soft as well as hard bolts. The higher strength of the heat treated bolts was made ineffective by the unfavorable design.

A failed and a bent **chain link** made of unalloyed structural steel and used as elements in a transportation device that was exposed to shocklike overloading **(Fig. 3.4),** had the same strength and microstructure. The only difference was that the failed link had a sharp-edged transition from which the fracture propagated, while the bent one was rounded out. Both links were undoubtedly overloaded. But the overloading led to fracture only in one case, while in the other it was prevented by a more suitable design.

A machine tool factory sent **rock drills** made of a steel with 0.95 % C, 1.2 % Cr, and 0.25 % Mo for examination. They had failed after a short period of service in the hexagonal shaft, while others had proved free of defects[2]). At first glance it could be seen that the broken drills had sharp edges **(Fig. 3.5a)** while those free of defects were well rounded off **(Fig. 3.5b).** Fatigue fractures propagated from the sharp edges. These in turn led to catastrophic failures under the shock loads **(Fig. 3.6).** The design differences could be clearly seen in the cross section **(Fig. 3.7a and b).** Etching showed that the failed drills also were surface decarburized which further reduced the fatigue strength (see also section 6.1.3).

Fig. 3.3. Self service elevator bolt with reverse bending fatigue fracture propagating from sharp-angled cross-sectional transition. 1 ×

The shafts of two **compressor transmissions** broke in the pin on the side of the compressor **(Fig. 3.8),** and several teeth broke off the smaller gear on the drive shaft. The fractures of the shaft were of a torsional fatigue nature that initiated at the sharp edges at the end of the keyway **(Fig. 3.9).** The shape of the keyway should have been changed. The fractures at the **pinions** of the drive shaft were also fatigue fractures **(Fig. 3.10).** They had started at the sharp angular edge of the helical gear. It was recommended to round off the edge or to provide a transition to the straight gear if the stress level did not require a strengthening of the gear cross section in the first place. It could not be determined whether there was any connection between shaft and gear fractures and which of the fractures had occurred first.

An exact calculation of stresses is often difficult with complex shapes and pronounced multi-axial stresses.

The upper part with a latch closure of a **high pressure vessel** of 1600 mm O. D. and 400 mm wall thickness made of chromium-nickel-molybdenum steel of 780 MPa (113 ksi) strength ruptured after approx. 2000 load cycles. The water pressure changed three times hourly between 0 and 2000 atmospheres. The fracture propagated from eight fatigue cracks that were offset by 45 ° against each other. The cracks were located in those places where the cover latches were braced against the seal areas of the vessel wall. These spots are designated by arrows in **Fig. 3.11. Figure 3.12** shows one of the cracks between two closure latches. From the distance of the crack lines in the SEM picture **(Fig. 3.13)** and the given load cycles it may be concluded that the first crack occurred soon after the vessel was put into operation. The depth of the fatigue cracks was uniformly almost exactly 10 mm, while their total area was only about 2 % of the total fracture. The stress therefore must have been uniformly distributed, as well as very high, and probably was predominantly multi-axial. The catastrophic fracture was nevertheless not a brittle fracture. A section through the fillet showed that it had a radius of 5 mm. A second section was made through it which ran parallel to the first fatigue fracture **(Fig. 3.14).**

Fig. 3.4. Broken or bent chain links from transportation device overloaded by impact. 0.8 ×. Left: Broken link with sharp cross-sectional transition. Right: Bent link with rounded-out cross-sectional transition

3.5

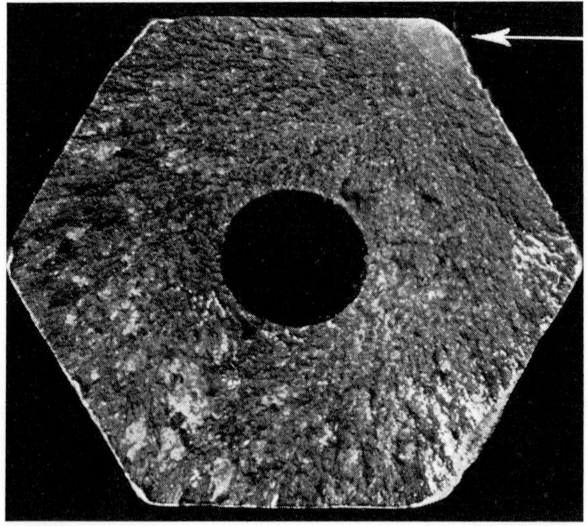

3.6

Fig. 3.5 and 3.6. Rock drills of different durability

Fig. 3.5a and b. Edge formation of broken and satisfactory drills. 2 ×

Fig. 3.6. Fracture of drill shaft. 3 ×

A thorough investigation of the vessel showed that the steel was clean and the heat treating perfect. The mechanical properties corresponded to specifications. It must be concluded that the stress in the fillet was considerably higher than that assumed by the designer.

Yet not only ruputure strenght, but also corrosion resistance can be endangered by unsuitable design. Metals are attacked by corrosion much more in narrow crevices than in smooth planes, as will be discussed later in section 15.3.2. An example below:

Screens of type 316 stainless steel, containing approx. 0.05 % C, 18 % Cr, 12 % Ni and 2 % Mo, that were exposed to brackish water from the mouth of a river, were made unserviceable due to parts of the bars breaking out after a few months[3]). The bars of which the screen was assembled consisted of wires of wedge-shaped cross section and pressed-in ribs **(Fig. 3.15).** They were bent into loops at approx. 100 mm intervals and cold formed into final shape in two operations. With the aid of these loops they were then placed on rods. Short lengths of tubing were employed to maintain a distance of 8 mm between them.

All failures originated at a loop specifically at the top of the contact surface of the turns **(Fig. 3.16).** This surface was strongly attacked by **crevice corrosion (Fig. 3.17).** The cross section showed substantial pitting **(Fig. 3.18)** and in part also stress corrosion cracks **(Fig. 3.19).** The smooth areas of the rods did not corrode. The inherent resistance of the steel against this very aggressive medium was therefore counteracted by a design error.

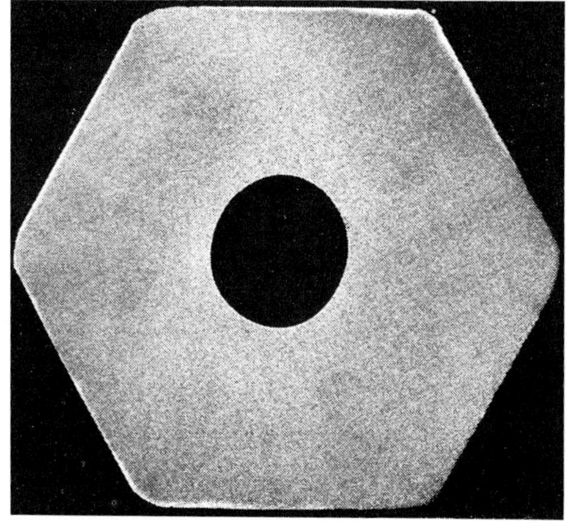

3.7 a

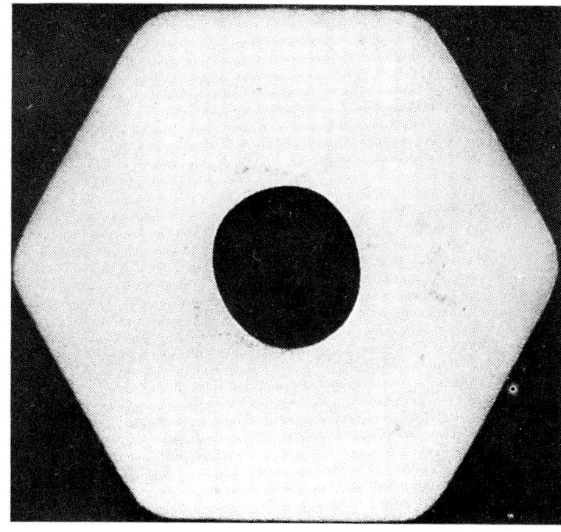

b

Fig. 3.7a and b. Cross section through drill shafts. Etch: Picral. 2.5 ×

Fig. 3.7a. Broken drill, sharp-edged and decarburized

Fig. 3.7b. Undamaged drill with rounded-off edges and no decarburization

Fig. 3.8. Compressor transmission shaft with fracture progagating from acute-angled keyway. Approx. 0.5 ×

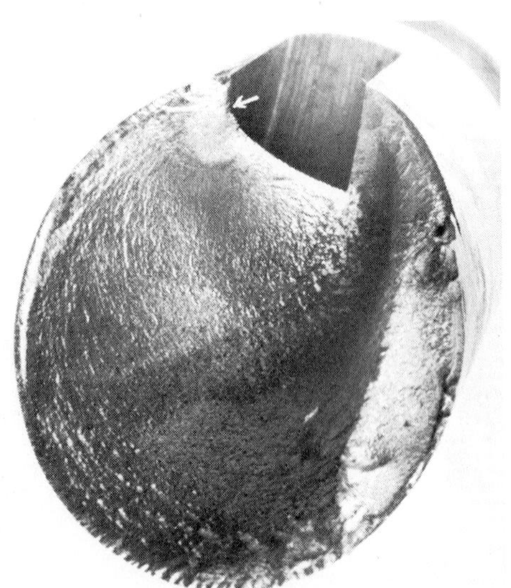

Fig. 3.9. Torsion fatigue fracture of transmission shaft according to Fig. 3.8. 2 ×

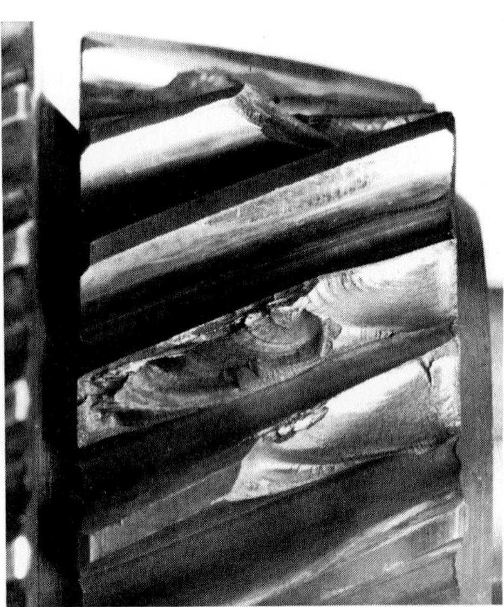

Fig. 3.10. Drive shaft pinions with fatigue fractures propagating from the acute-angular edge of helical gear. 2 ×

3.2 **Selection of an Unsuitable Material, Mistakes**

Ignorance, carelessness, and false economies in the selection of materials cause many errors. Not every steel user is in a position to select the most suitable material for his purpose from the many varieties available. In cases of doubt he should consult the steel manufacturer whose materials specialists have the prerequisite knowledge of mechanical and technological properties, as well as experience in corrosion resistance. In order to make the proper recommendation, the materials expert naturally must find out what conditions the materials will be exposed to during machining and service. The melting process, deoxidation, and alloying must be appropriate for the ultimate purpose. A close cooperation between the materials specialist of the producer and the planners and plant engineers of the user is the best formula for success.

Some applications of unsuitable materials are the result of a mix-up in materials or of poor caretaking of the available stock. The examples cited below will show this.

Grabbing hooks of a jet-catapult installation occasionally fractured under very high impact in the rectangular cross-sectional transition to the hook or the low angle to the eye part. The hooks consisted of a strongly case hardenable alloy steel with approx. 0.18 % C, 2 % Cr and 2 % Ni and were cut out of a flat profile. Two broken hooks (1 and 2) and a less defective one (3, **Fig. 3.20)** were examined. The fractures had a grainy or fibrous structure **(Fig. 3.21 and 3.22).** In hook 3 according to Fig. 3.20 only one piece of the case layer had broken out at the back of the nose.

Oberhoffer etch of the fracture was used to confirm metallographically the unfavorable fiber orientation with respect to the load direction. Microscopic examination also showed a crack in hook 3 which had not yet fractured at the transition to the nose **(Fig. 3.23).** All hooks were case hardened. The carburized outer layer had a perfectly fine acicular hardened structure with finely distributed carbides **(Fig. 3.24).** The grain structure of hook 3 consisted of pure martensite, while in hook 2 it consisted of bainite, and in hook 1 of a mixture of the two (**Fig. 3.25).** The specimen was through hardened, as was to be expected from the type of steel used. Core hardness for the hooks 1, 2 and 3 was approximately 480, 420, and 450 HV 10.

The purpose of surface hardening, namely to combine wear resistance with impact resistance, was counteracted by the high core hardness and corresponding brittleness. Since in this case evidently toughness was of greater importance than wear resistance, it would have been preferable to dispense with maximum surface hardness and to produce the hooks from a heat treatable steel to obtain maximum toughness. Additionally the cross sectional transitions should have been rounded off. To achieve a more favorable fiber orientation, free form or die forging of the individual parts would have been preferable to the machining of the parts from flat steel.

In a **concrete structure** a ceiling was hung on flat bars of 30 x 80 mm cross section. The bars were borne by a slit steel plate and supported by tabs that were welded onto the flat sides **(Fig. 3.26).** One of the bars failed during mounting when it was dropped from a height of about 1 m onto the opposite support[4]). The failure was a low ductility fracture as can be seen from **Fig. 3.27.** It propagated from one of the fillet welds. Macroetching and Baumann replicas of a cross section indicated the existence of a distinct core segregation. Therefore the material consisted of rimmed cast steel. Etchings according to Fry of longitudinal sections proved through flow line analysis that the bar was strain aged after cold working, probably as a result of straightening **(Figs. 3.28 a and b).** The bar had tensile strength values that corresponded to specifications for a low carbon structural steel. Notch impact toughness, determined on longitudinally cut V-notch samples, fell at 40 °C within a wide scatter band between 90 and 230 J/cm^{2}*, whereas at room temperature the impact resistance was only 35 J/cm^{2}** **(Fig. 3.29).** Therefore a rimmed cast steel

* 430 – 1095 ft lb / in^2
** 165 ft lb / in^2

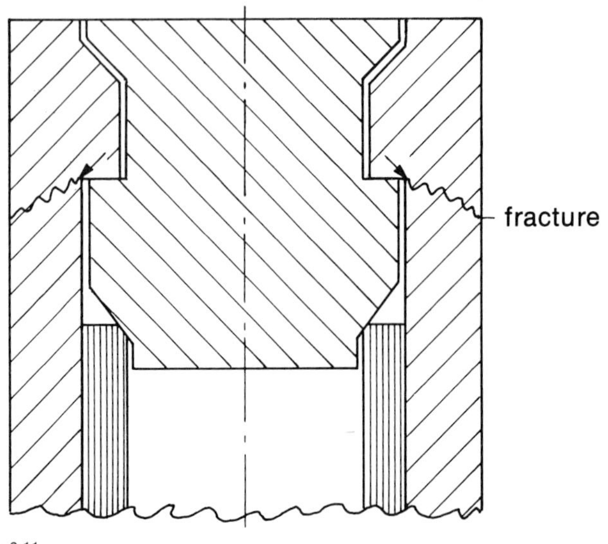

3.11

3.12

3.14

Fig. 3.11 to 3.14. Fracture of a high pressure vessel at latch closure

Fig. 3.11. Longitudinal section through closure. 0.04 × Two of the eight fatigue fractures distributed over the perimeter are designated by arrows

Fig. 3.12. Fatigue fracture between two latches

Fig. 3.14. Section transversely through fillet. 1 ×. Bottom: Fracture. Above: Second incipient fatigue fracture

Fig. 3.13. Fatigue fracture plane as seen through SEM. 2400 ×

had been used in this instance for a welding construction, for which such material is less suitable because of its strain aging tendency than a killed steel. Strain aging failures are discussed in more detail in section 4.3.1.

The following failure is also due to the use of an unsuitable steel for welding. During welding of **saddle sockets** onto pipes of 150 mm O.D. it happened that the sockets failed repeatedly during cooling along the circumferential seam **(Fig. 3.30)**[5]). The fractures were composed of a plurality of incipient cracks all of which propagated from the weld seam into the pipe material. According to chemical analysis, the pipe consisted of a killed cast steel with 0.51 % C and 0.75 % Mn; the socket consisted of a rimmed steel with less than 0.1 % C, according to metallographic examination. Sections through the weld showed that the seam itself was free of defects, but that the material adjacent to the seam had cracked in many places **(Fig. 3.31).** This area had a purely martensitic structure **(Fig. 3.32)** and a hardness of 700 to 800 HV 0.1. Accordingly, the pipe consisted not only of an unsuitable steel, but also was cooled uncommonly fast after welding. The failure of the pipeline therefore was caused by thermal and transformation stresses that could not be reduced by deformation because of their multi-axial nature and brittleness of the hardened zone.

A material that was not only unsuitable but expensive as well, was used in the following case. This case also represents an example of the above mentioned embrittlement through precipitation of σ-phase, an intermetallic compound with approximately 50 at.% Cr. A **recuperator** for blast heating of a cupola furnace became unserviceable due to the brittle fracture of several finned tubes made of heat resistant cast steel with 1.4 % C, 2.3 % Si and 28 % Cr[6]). The service temperature was reported to be 850°C. This already led to the suspicion that the breaking had something to do with the formation of sigma phase which not only embrittles the alloy, but also causes stress due to a volume reduction associated with the formation of the precipitated phase[7]).

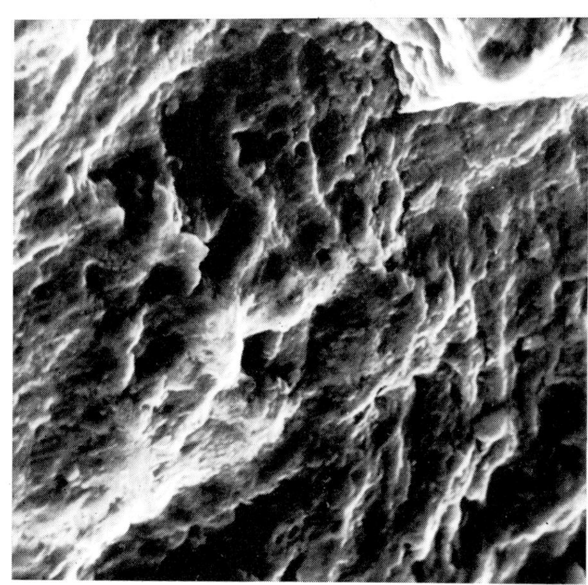

3.13

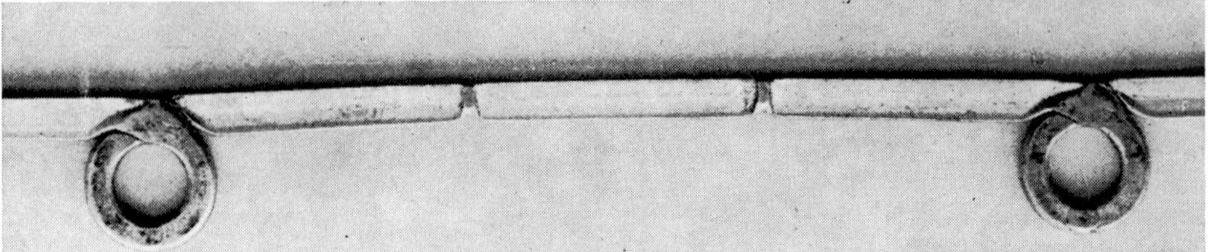

3.15

3.16

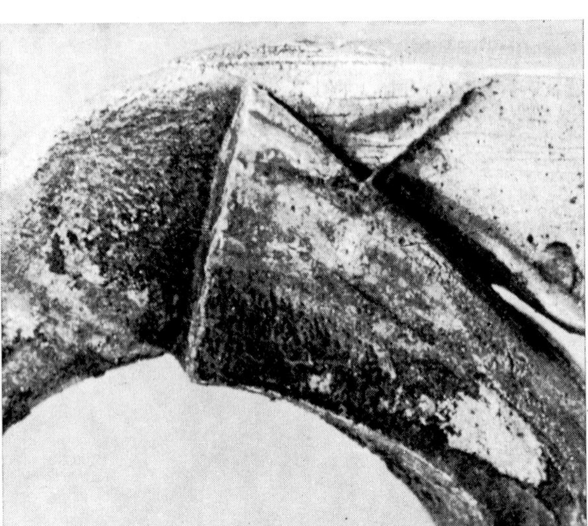

3.17

Fig. 3.15 to 3.19. Stainless steel screen bars fractured by crevice and stress corrosion while exposed to brackish water

Fig. 3.15. Unused bar. 1 ×

Fig. 3.16. Fracture in loop. 10 ×

Fig. 3.17. Crevice corrosion in loop. 10 ×

Fig. 3.18. Pitting corrosion at interface of loop. Unetched cross section. 100 ×

Fig. 3.19. Pitting and stress corrosion cracks adjacent to fracture. Longitudinal section. 200 ×

The fractures of the finned tubes were partly coarse-, partly fine-grained and not deformed which is not uncommon for this casting alloy **(Fig. 3.33).** For metallurgical analysis in which an unused recuperator was also included for purposes of comparison, the sections were first etched electrolytically in a 1:10 dilution of saturated ammonia water at a terminal voltage of 1.5 V after a routine etching with V2A-etching solution* which only attacks the ferritic matrix. This served to dissolve the chromium carbide and stain it by means of a brown precipitate. Immediately thereafter and without intermediate polishing, the section was etched at 2 V terminal potential in a 10N solution of caustic soda which additionally etched and colored the sigma phase. The photomicrographs of the fractured recuperator are shown in **Fig. 3.34a, b, c** and of the unused one in **Fig. 3.35a, b, c.** It should be noted that the finned tubes of the unused heat exchanger contained in addition to the chromium-silicon alloyed solid solution only chromium carbide as a precipitate; the caustic soda etch had no effect. On the other hand, in the fractured finned tubes, a considerable amount of sigma phase had formed as a result of the prolonged annealing at 850 °C. In the photomicrograph this phase appears attacked and discolored by the second etch. The previously formed carbides served as nuclei for the precipitation of the sigma phase.

The multi-axial stresses caused by sigma phase formation and the related embrittlement must be viewed as the cause of failure of the recuperator. In this case a steel of lower chromium content with no or little tendency for sigma phase formation would have had adequate corrosion resistance at the relatively low service temperature.

In the following failure analysis things were somewhat different. In a petrochemical installation, several **hooks** made of heat resistant alloy steel castings, containing approx. 0.4 % C, 1.8 % Si, 26 % Cr, and 14 % Ni, that also had a tendency to sigma phase formation, broke at temperatures of 650 to 800 °C **(Fig. 3.36)** with a coarse-grained cleavage type fracture **(Fig. 3.37)**[8]). The fracture showed at its starting point at the inner surface of the bend a zone that was stained green

* For composition see Appendix II

3.18

3.19

3.20

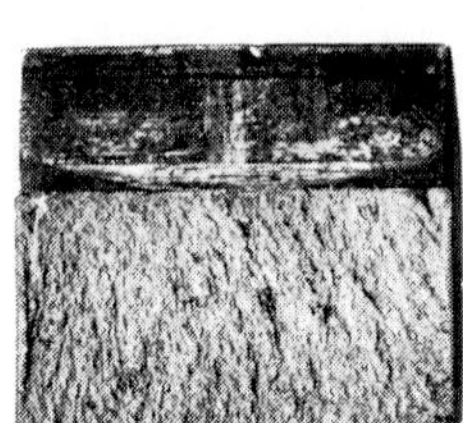

3.21

3.22

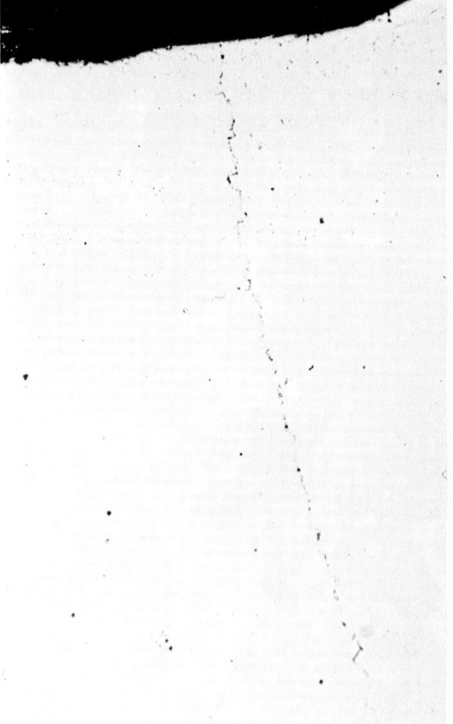

3.23

Fig. 3.20 to 3.25. Broken grabbing hooks of jet-catapult installation

Fig. 3.20. Grabbing hook 3 with fragmentation at nose. 0.8 ×

Fig. 3.21. Fracture at nose of hook 1. 1 ×

Fig. 3.22. Fracture in obtuse angle of hook 2. 1 ×

Fig. 3.23. Incipient fracture in transition to nose of hook 3. Unetched section. 100 ×

Fig. 3.24. Outer microstructure of hook 3. Etch: Picral. 500 ×

Fig. 3.25. Core microstructure of hook 1. Etch: Picral. 500 ×

by chromium oxide and was of a dendritic structure. A longitudinal section showed that it consisted of a high density of microcavities and intergranular thermal cracks **(Fig. 3.38 and 3.39).** The fine structure consisted of austenite with much carbide mostly precipitated out of the melt, and sigma phase **(Fig. 3.40).** An unused hook that was examined for comparison was free of sigma formation **(Fig. 3.41)** and coarse porosities, but also had a coarsely grained solidification structure, strongly transcrystalline up to the center. During tensile testing, it fractured at barely one-half the normal service load.

In this case, therefore, the sigma formation was not primarily responsible for the fracturing of the hooks; instead the coarse, partly porous cast structure was the primary reason for failure. Structural parts of this type should preferably be bent from rolled rods or forged in a drop forge. Furthermore, a steel with less tendency to coarse grain or sigma formation and lower chromium content would have been preferable.

The following failure occurred because of a mix-up in materials. A **steel socket pipe line** of 150 mm nominal bore (NB) failed during pressure testing; the failure occurred adjacent to the welding seam over almost the entire perimeter. The crack propagated through one of the pipes in part within the weld undercut* in part immediately adjacent to it **(Fig. 3.42).** A longitudinal section through an area not yet broken showed, after Heyn etching, that the welded pipes consisted of different materials, of which the undamaged one had a lighter etch toning while the one fractured at the transition to the weld seam had a darker tone **(Fig. 3.43).** The first one consisted of a fine structure of ferrite and pearlite which had coarsened in the vicinity of the weld seam. In the failed pipe the darker region was composed of martensite next to the weld seam, and in the unheated part the microstructure consisted of pearlite and bainite. The failed pipe therefore probably contained more carbon and also alloying elements to increase hardenability. **Table 1** shows a chemical analysis of the pipes. The non-fractured pipe consisted of a soft steel that probably was intended for this type of application, while the failed one was composed of a special chromium-manganese alloy steel.

The failure therefore could be attributed to stresses that occurred due to delayed austenite transformation into the martensite.

* see Appendix I

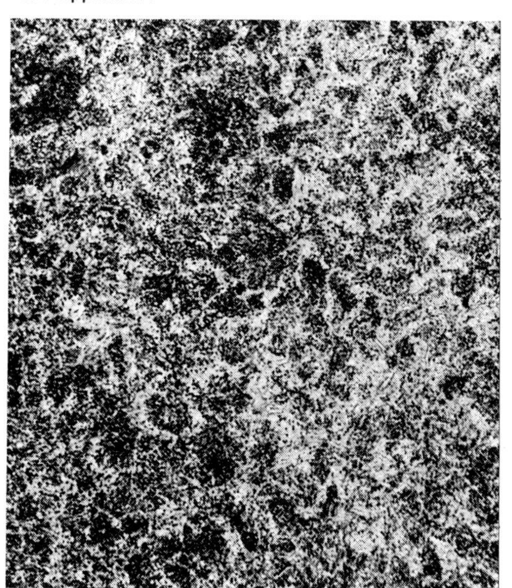

3.24

3.25

3.26

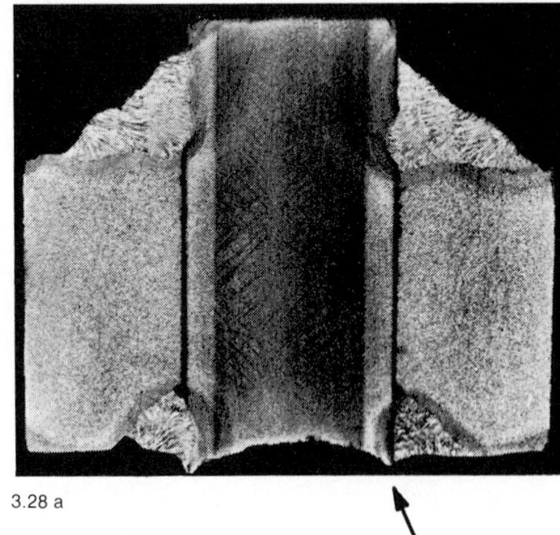

3.28 a

3.27

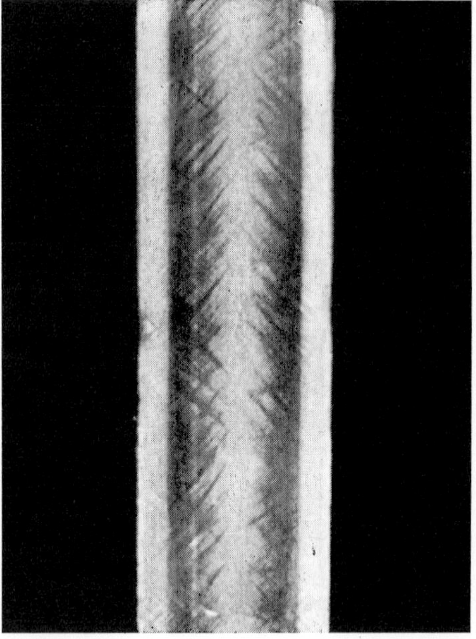

3.28 b

Fig. 3.26 to 3.29. Broken suspension bar from concrete structure

Fig. 3.26. Upper end of suspension bar. Side view with welded-on tab. Approx. 0.75 ×

Fig. 3.27. Fracture view. Approx. 0.75 ×. Fracture origin designated by arrow

Fig. 3.28a and b. Longitudinal sections. Etch according to Fry. Approx. 0.75 ×

Fig. 3.28a. Region of fracture. Origin designated by arrow

Fig. 3.28b. At distance of approx. 200 mm from fracture

Fig. 3.29. Notch impact toughness of steel as function of impact temperature

Table 1. Chemical Composition of Welded Socket Pipes

	C %	Si %	Mn %	Cr %
Non-fractured pipe	0.11	0.07	0.33	0
Fractured pipe	0.35	0.11	0.99	2.08

As a deterrent example for the careless use of unsuitable materials the following failure may be cited[9]). Two bolts from the roof structure or ceiling of a **church** had failed during stressing. They were said to consist of a high strength steel with at least 590 MPa (85 ksi) yield point and 880–980 MPa (127–142 ksi) tensile strength. Visual examination of the bolts showed that the double-V-groove weld of the shaft was not completely welded, and that the fracture of the bolt that probably failed first had originated in this spot **(Fig. 3.44 and 3.45).** The machining grooves at the end faces of the head piece could still be seen in the root. Tinting of the fracture proved that the seam had cracked while exposed to heat. The weld seams were not attacked by copper ammonium chloride solution during etching of the bolt surface according to Heyn. Therefore a highly alloyed welding rod material had been used.

Table 2. Chemical Composition of A Fractured Bolt

	C %	Si %	Mn %	P %	S %	Cr %	Mo %
Head	0.19	0.50	1.50	0.026	0.034	0	0
Shaft	0.41	0.20	0.58	0.012	0.012	1.10	0.19

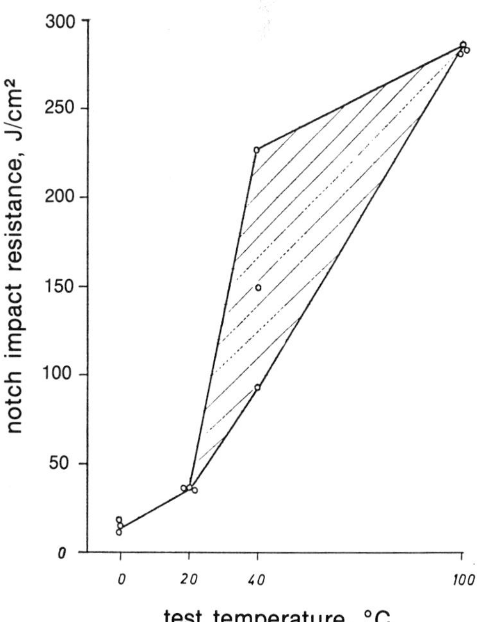

3.29

3.30

3.32

3.31

Fig. 3.30 to 3.32. Pipe of 150 mm diameter made of steel with 0.51 % C, cracked during welding of a saddle socket

Fig. 3.30. Fracture region. 0.5 ×

Fig. 3.31. Cross section through weld seam. Etch: Picral. 10 ×

Fig. 3.32. Microstructure of cracked region in pipe material adjacent to weld seam. Etch: Picral. 500 ×

Spectroscopic examination showed that the head pieces consisted of manganese steel and the shafts of a chromium-molybdenum steel. Chemical analysis of a bolt showed the data presented in **Table 2.** Thus the composition of the head piece corresponded approximately to the low carbon, high manganese steel AISI No. 1522, a weldable construction steel with higher yield point and strength, while the shaft consisted of a chromium-molybdenum steel of AISI type 4140.

Longitudinal sections **(Fig. 3.46)** confirmed that the root had remained completely open. The orientation of the fiber showed that the head pieces were cut out of billets or plates in such a way that the fiber ran transversely to the principal stress. Judging by the fine structure, the head pieces were either untreated or normalized **(Fig. 3.47),** while the shafts were heat treated **(Fig. 3.48).** In the secondary banded structure of the heads, the unfavorable direction of the fibers could again be discerned. The welding seams consisted of an austenitic steel. They were characterized by intergranular hot tears, as could be seen from heat tinting of the fractures **(Fig. 3.49).** The transition zones were hardened martensitically on both sides and had cracked in some places due to transformation stresses **(Fig. 3.50).** The fractures of these bolts were therefore due to the mistaken selection of unsuitable materials as well as to processing and welding errors of all types.

Many mistakes are also being made in the selection of materials for good corrosion resistance. The varying conditions of exposure make a selection in these cases difficult. But mix-up of materials during storage or assembly may also lead to failures.

An example of the latter may be cited here. **Bolts** said to be made from stainless low carbon austenitic chromium-nickel steel of type 304L corroded rapidly against all expectations when in contact with water from the Mediterranean Sea **(Fig. 3.51).** The gap between bolt and nut threads played apparently a contributory part in corrosion (crevice corrosion, see also section 15.3.2). Micrographic examination showed that the steel contained large amounts of sulfide inclusions **(Fig. 3.52).** The sulfur content was analyzed and found to be 0.153%. In this case then a stainless free cutting screw machine steel had been used instead of the one originally intended. Free cutting steel is often used for bolts because of its easy machinability, but it is less corrosion resistant.

Several methods, such as spark testing or spot reactions with specially constituted solutions[10]) can be utilized for a quick determination of **material mix-ups** or to sort out mixed-up storage lots. For instance, some wires of molybdenum-free chromium-nickel steel were erroneously woven into a **200 mesh screen** consisting of wires of an acid-resisting austenitic steel containing approx. 0.05 % C, 18 % Cr, 12 % Ni and 2 % Mo (AISI type 316). The screen was etched with hydrochloric acid with an auxiliary indicator of potassium ferricyanide[11]). The potassium ferricyanide in the solution serves only as an indicator for the dissolved Fe^{3+}-ions. The ferricyanide-ion $(Fe(CN)_6)^{4-}$forms a deep blue tinted complex compound with the iron ion. This compound is called Berlin blue.* The reaction is utilized frequently in metallographic etching. **Figures 3.53 and 3.54** shows sections of this screen in natural size and at 15-fold magnification after etching. Sixty of the approximately 2000 wires that had a diameter of but 20 to 30 μm were attacked and covered with a bluish-green layer. Analysis confirmed that they consisted of a molybdenum-free stainless steel of type 304 with approx. 18 % Cr and 10 % Ni.

The etching method ist based upon the fact that the molybdenum-containing steel of otherwise the same composition is more resistant against reducing acids than the molybdenum-free steel. **Figure 3.55** shows the potential for specimens of both steels in hydrochloric acid diluted to a 1:1 ratio as a function of time. For the same initial passive potential, the specimen of the molybdenum-free steel became active immediately, whereas the steel alloyed with molybdenum remained passive for ten minutes.

* see Appendix I

3.3 **Unsuitable Methods**

Errors in the selection of processing methods may also be designated as planning errors. They occur during manufacture, machining, assembly, shipping or installation.

If for reasons of economics a steel susceptible to strain aging is to be used, it should be remembered that cold working of such a steel can lead to failures. **Figure 3.56,** for instance, shows a **screw** from the guide for a pulley rope of a hoist, that cracked in the first supporting thread with a low deformation fracture **(Fig. 3.57).** The bolt consisted of a free cutting steel with high phosphorus, sulfur, and nitrogen contents and as a result showed a corresponding inclusion-rich structure. For a component that is subject to impact stresses during unraveling of the rope slings this was certainly not a suitable material. Additionally the thread had been cold rolled as proven by the etching of flow figures with Fry's solution **(Fig. 3.58).** Consequently an unsuitable method had been used for the processing of an unsuitable material.

Impact tests on **railway track screws** made of basic converter steel of an old heat, whose threads were partially cut and partially cold rolled, showed that the bolts with the cut thread retained their toughness during artificial strain aging at 220 °C, while the bolts with rolled thread were in part strongly embrittled. **Figures 3.59 and 3.60** show longitudinal sections of bolts with cut and rolled threads, respectively, which had been etched with Fry's solution. In the latter, strain aging can readily be discerned by the etch of the deformed areas. In the production of

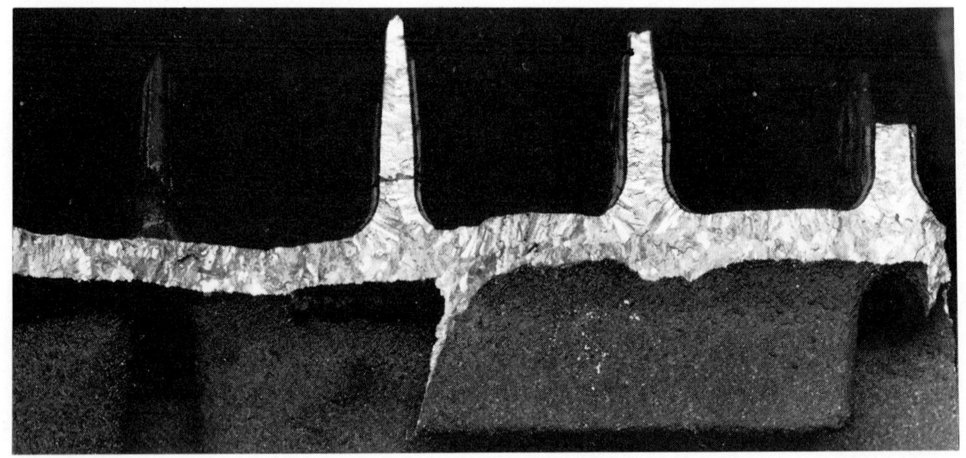

3.33

Fig. 3.33 to 3.35. Broken recuperator of heat resistant cast steel

Fig. 3.33. Fracture of finned tube. 1 ×

Fig. 3.34a and 3.35a. Etch: V2A-etching solution

Fig. 3.34b and 3.35b. Etch: ammonia water 1.5 V (carbide etch)

Fig. 3.34c and 3.35c. Etch: ammonium hxdroxide + 10 N caustic soda 1.5 V (carbide and sigma etch)

Fig. 3.34a to c. Microstructure of broken recuperator. 500 ×

Fig. 3.35a to c. Microstructure of an unused recuperator. 500 ×

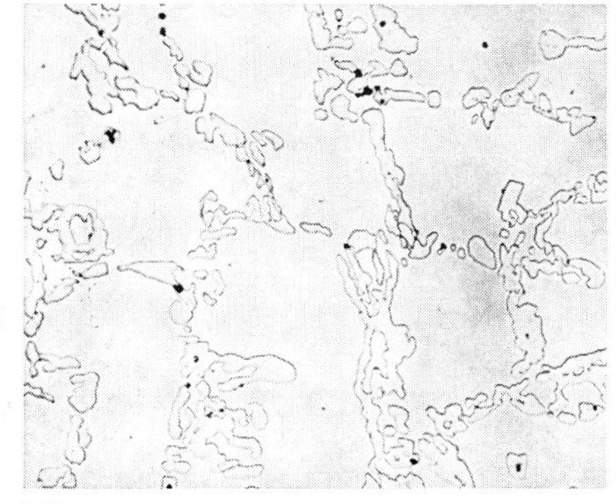

3.34a

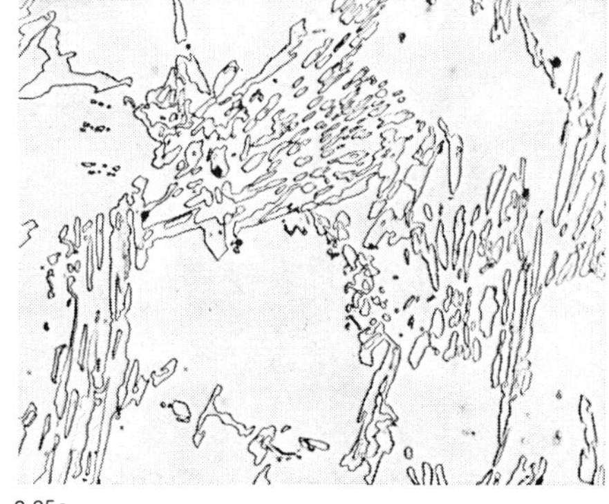

3.35a

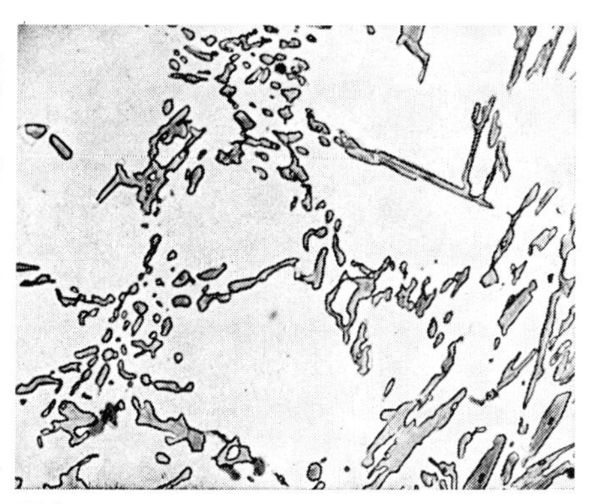

3.34b

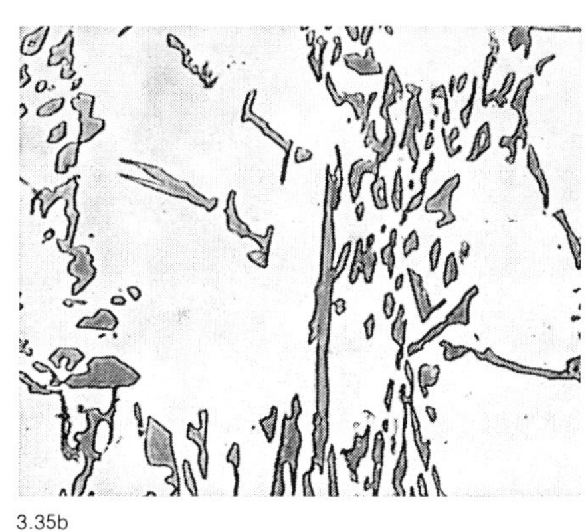

3.35b

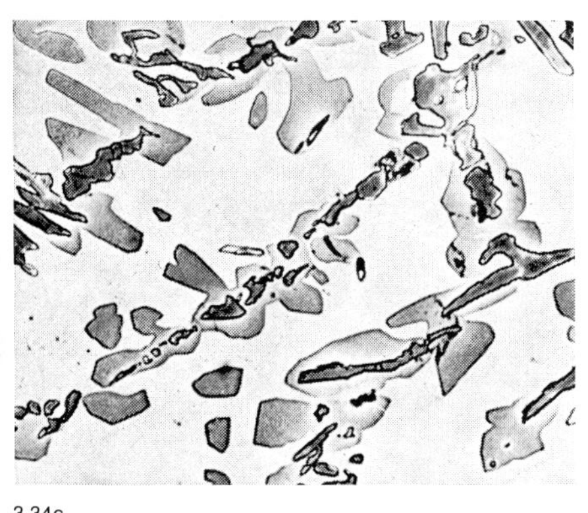

3.34c

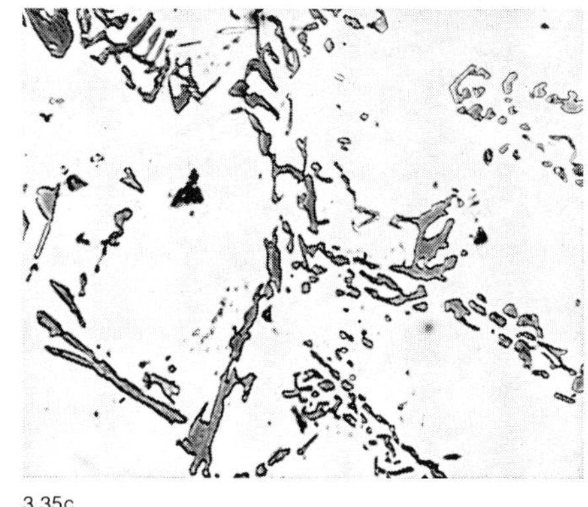

3.35c

3.36

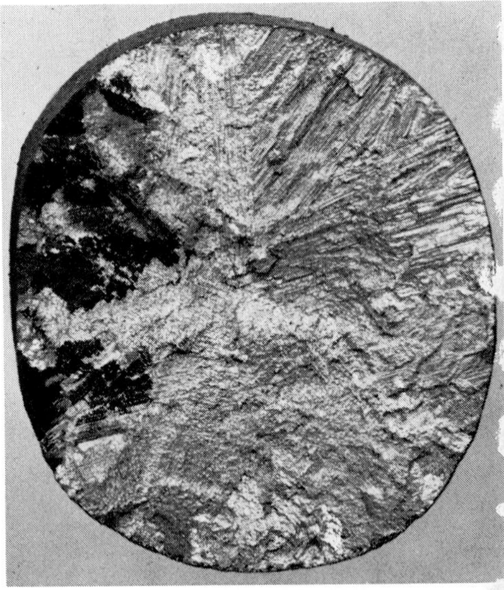

3.37

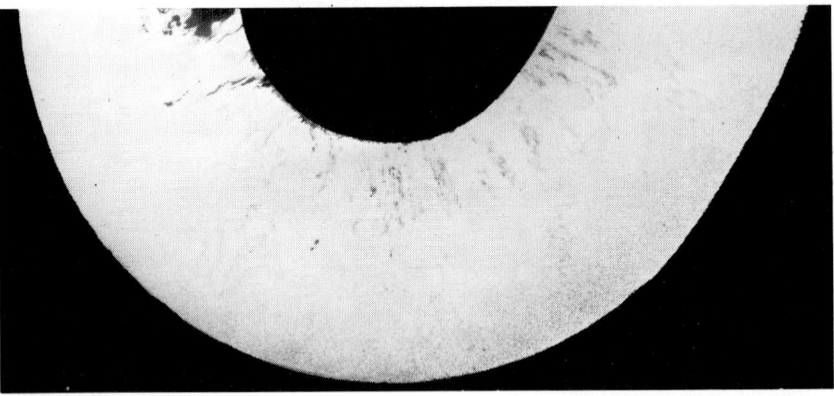

3.38

Fig. 3.36 to 3.41. Broken hook of heat resistant cast steel

Fig. 3.36. Broken hook. 1 ×

Fig. 3.37. Fracture, origin stained green by chromium oxide. 2 ×

Fig. 3.38. Longitudinal section. 1 ×. Shrinkage cavities and hot tear zone inside the bend made visible by dye penetration method

Fig. 3.39. Microcavities and thermal cracks at interior of bend. Longitudinal section, unetched. 20 ×

Fig. 3.40 and 3.41. Microstructure of hooks. Etch: 10 caustic soda 2 V. 500 ×

Fig. 3.40. Broken hook

Fig. 3.41. Unbroken hook

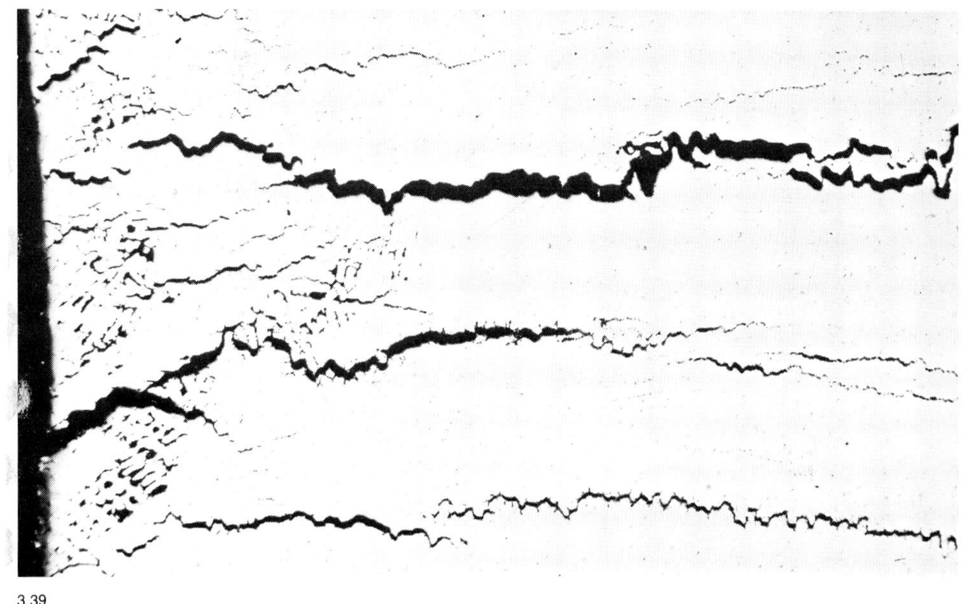

3.39

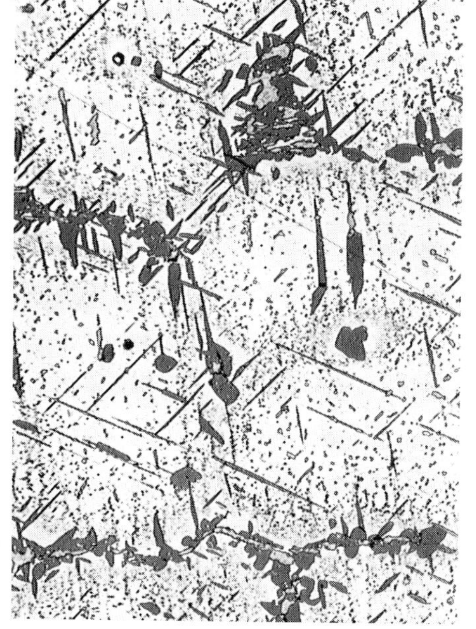

3.40

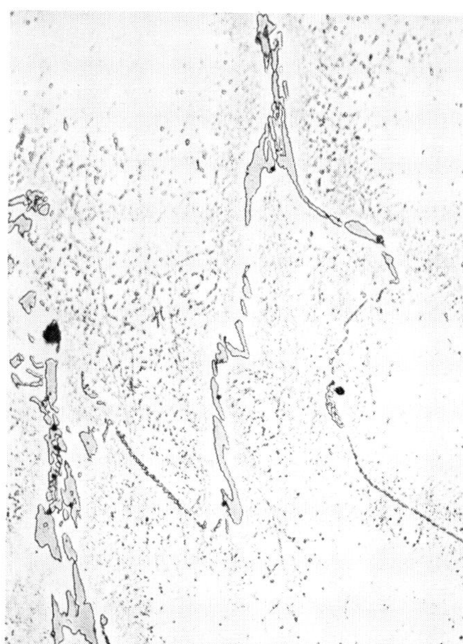

3.41

highly stressed parts, the principal stress should not be directed transversely to the fiber, if possible (see sections 4.1 and 4.2). Die or drop forging methods are often more suitable for this purpose than machining from a billet. An example of unsuitable fiber structure was previously discussed in section 3.2 (Figure 3.22).

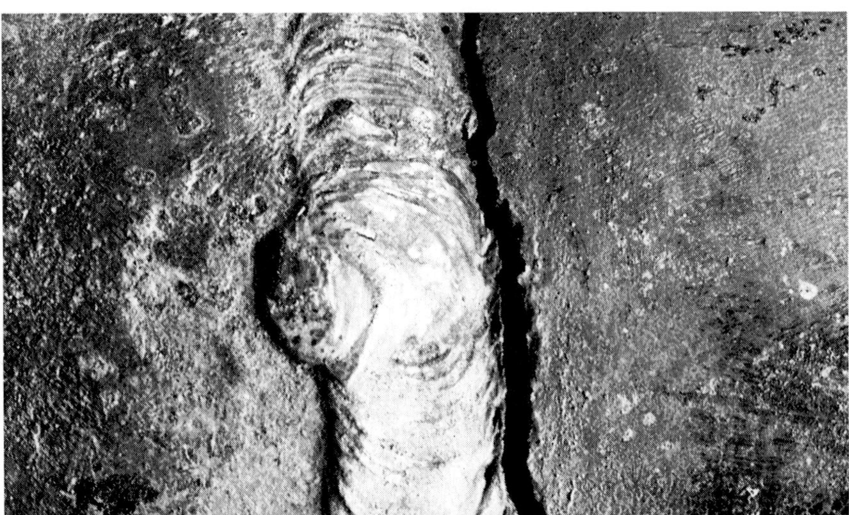

3.42

3.43

Fig. 3.42 and 3.43. Steel socket pipeline of 150 mm NB that failed next to weld seam

Fig. 3.42. View of fracture. Approx. 1 ×

Fig. 3.43. Longitudinal section through weld opposite crack. Etch according to Heyn. 1 ×

Wheel bolts of AISI type 4140 steel were produced by cold upsetting of a collar and were heat treated to 980 MPa (142 ksi) tensile strength. The zone under the collar exhibited widely scattered impact energy values that were too low. As can be seen from **Fig. 3.61,** the fiber was so strongly deflected by the upsetting operation that the microstructure under the collar was oriented almost transversely to the longitudinal axis of the bolt. A shape modification in the sense, that the collar was blended over a cone section into the cylindrical shaft of the bolt in a less abrupt way, led to a minor orientation change of the fiber and thus to higher impact resistance **(Fig. 3.62).** The amount and magnitude of inclusions was minor in both cases. In addition, the heat treated components exhibited a fine acicular structure throughout. Therefore it was quite remarkable that the fiber reorientation had such a strong effect upon toughness.

During machining from a billet, especially of rimmed cast steels, segregation areas, that are harmless in the interior, are often cut open and then form stress concentrations. This was the case, for instance, in a **pipe connection sleeve** of a hydraulic forging press, as shown in **Fig. 3.63.** It had fractured without deformation in the sharp-edged cross-sectional transition between pipe and collar, whereby coarse turning grooves also may have played some part. A longitudinal section through the head part showed, after Heyn etching **(Fig. 3.64),** that the sleeve was machined out of an ingot of rimmed cast steel in such a way that the thin pipe wall was situated almost entirely in the segregation zone and that the cross sectional transition which was exposed to high multi-axial stresses was located in the most impure boundary areas between the pure ingot skin and the segregated core. The brittle fracture of the sleeve may have been decisively facilitated by this manufacturing error.

Welding should be avoided, whenever possible, for the joining of strongly hardenable steels. If this is not possible, special precautionary measures must be taken (see section 7.3). That this is not always the case, is shown in the following example.

Spindles made of hardenable 13 % chromium steel with approx. 0.4 % C (AISI type 420), **Fig. 3.65,** selected because of its good antifriction properties, were fastened to a superheated steam push rod made of high temperature structural steel with 0.13 % C, 0.85 % Cr and 0.45 % Mo by means of screwing and welding[13]). one of the spindles failed immediately adjacent to the weld seam during the first operation of the rod. Metallographic examination of a longitudinal section through the fracture origin **(Fig. 3.66)** showed that the low-alloy welding rod material was alloyed so highly by melting of the spindle steel that it was no longer attacked by 1 % nital solution in the transition zone, and displayed a martensitic structure. The area of the spindle steel subject to thermal effects had hardened also. Originally it had an annealed structure of ferrite and carbide. Short cracks had formed in the two adjacent areas under the effects of transformation stresses **(Fig. 3.67).** The fracture apparently originated in one of these cracks.

During examination of the other spindles that had not failed, cracks could be seen with the naked eye without application of nondestructive testing methods **(Fig. 3.68).** The cracks also propagated from the heataffected zone of the weld seam or the spindle material and would likewise have led sooner or later to the failure of the spindles.

The errors in this case could have been avoided if the additional welding to secure the joining of spindles and rods had been omitted; alternatively, a spindle material with lower hardenability should have been selected. In any case the weld should have been thoroughly preheated to approximately 200 °C prior to applying the first welding bead, thus delaying cooling.

A similar case occurred with **bridge bolts,** with which the concrete roadbed of a river bridge was connected to the upper flange of an I-beam. Several of these bolts broke during lowering of the bed. The fracture origin in all bolts was determined to be at the root of the thread **(Fig. 3.69).** In the fracture, this spot was characterized by a narrow, very fine-grained zone. Weld spatter could be seen on the thread **(Fig. 3.70).** In the etched longitudinal section through the fracture origin, the zone under the fine-grained fracture was very light as compared to the strongly etched basic structure **(Fig. 3.71).** The structure in this area was martensitic, while it was purely pearlitic in the remaining cross section. The bolts therefore were locally overheated and quickly cooled. The

3.44

3.45

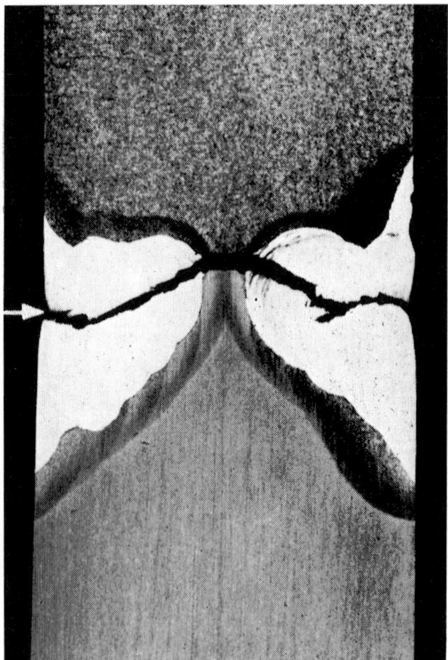

3.46

Fig. 3.44 to 3.50. Failed stress bolts of church roof

Fig. 3.44. Fracture of a bolt. 2 ×

Fig. 3.46. Longitudinal section through the fracture region. Etch according to Heyn. 2 ×

Fig. 3.45. Side view after etch according to Heyn. 1 ×

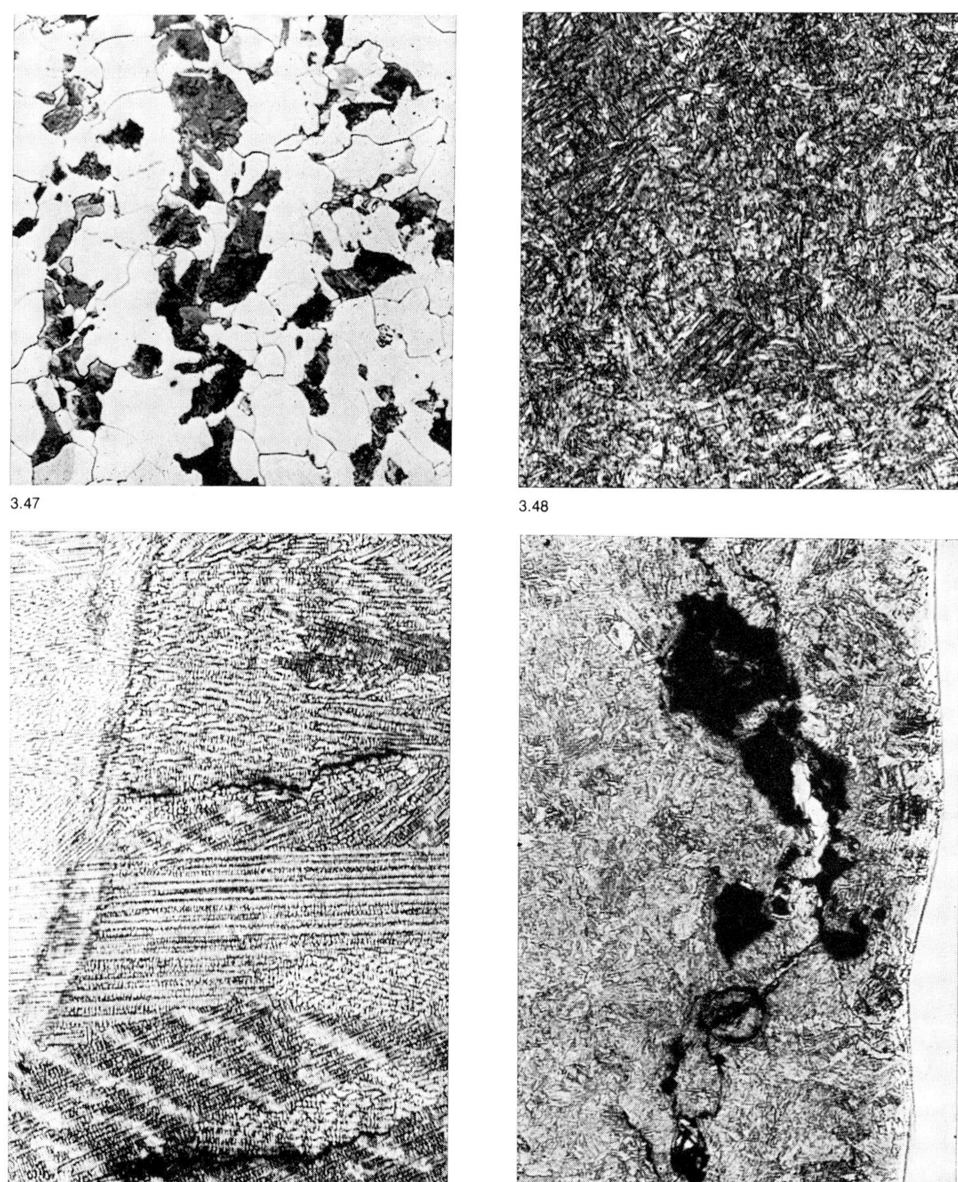

3.47

3.48

3.49

3.50

Fig. 3.47 to 3.50. Microstructure of bolts and weld seams. Longitudinal sections. Etch: 3.47, 3.48 and 3.50 nital, 3.49 V2A-etching solution

Fig. 3.47. Head piece. 500 ×

Fig. 3.49. Weld seams. 50 ×

Fig. 3.48. Shafts. 500 ×

Fig. 3.50. Transition zone with stress crack in bolt material. 100 ×

3.51

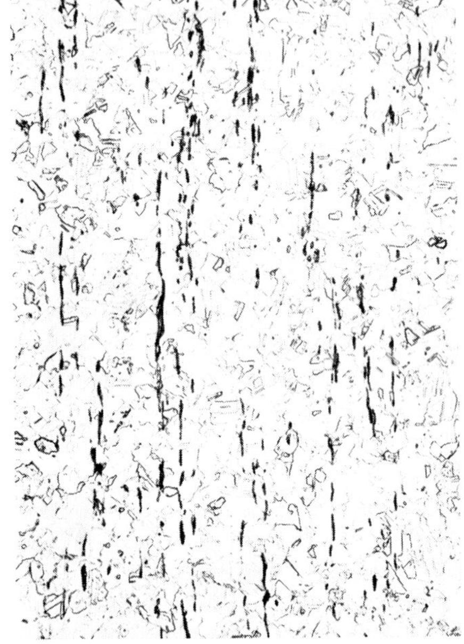

3.52

Fig. 3.51 and 3.52. Free cutting stainless steel screw bolts corroded by seawater

Fig. 3.51. View. 1 ×

Fig. 3.52. Microstructure, longitudinal section. Etch: V2A-etching solution. 200 ×

3.53

melted beads at the fracture origin indicate that this was caused by a welding torch. Presumably welding was used to secure the screws or nuts against turning. The fracture of the screws therefore was caused by quenching stresses and perhaps even quench cracks. In a screw steel of such high carbon content, welding was quite unsuitable regardless of the purpose for which it may have been intended.

Hardened parts are unable to reduce overloading by plastic flow, and therefore do not tolerate stresses that exceed the yield point. Such overloading is caused easily through impact. In surface hardened parts the tough core may well prevent the spontaneous fracture of the entire part by blocking the expansion of surface cracks. But these cracks consist of sharp notches that later may lead to failures, especially during cyclic stresses. Such was the case with a **nitrided piston rod** made of a steel with approx. 0.35 % C, 1.7 % Cr, 1 % Ni, 1 % Al, and 0.2 % Mo, that broke in the keyway after almost 10,000 h of service **(Fig. 3.72)**[14]. The fracture propagated from both sides of the keyway at the point where the flat side faces change into the semi-cylindrical front faces. The failure was caused by well advanced fatigue fractures **(Fig. 3.73).** A section through

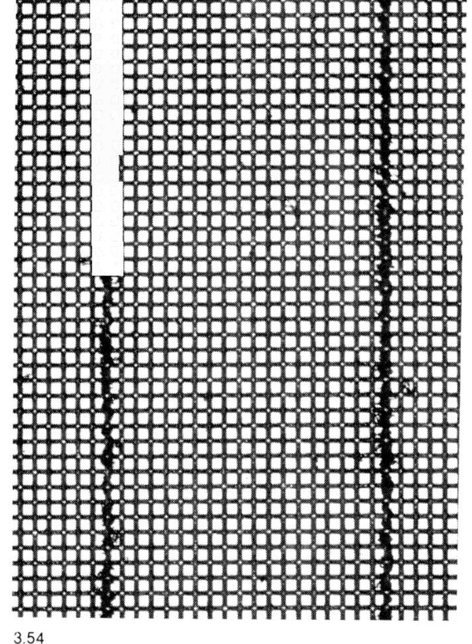

3.54

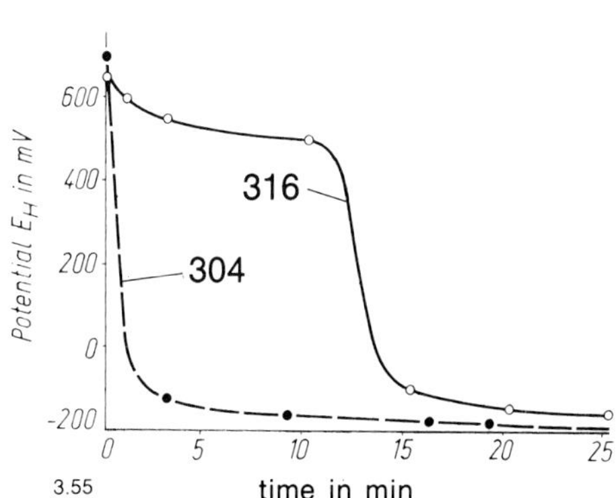

3.55

Fig. 3.53 to 3.55. 200-mesh screen of stainless chromium-nickel-molybdenum steel with woven-in wires of molybdenum-free steel

Fig. 3.53 and 3.54. Views. The molybdenum-free wires stained blue-green after special etching in hydrochloric acid with potassium ferricyanide

Fig. 3.53. 1 ×

Fig. 3.54. 15 ×

Fig. 3.55. Change of potential with time for molybdenum-containing and molybdenum-free 18-8 stainless steel submerged in 50 % aqueous hydrochloric acid

3.56

3.57

3.58

Fig. 3.56 to 3.58. Broken screw from pulley rope guide of a hoist

Fig. 3.56. View. 1 ×

Fig. 3.57. Fracture. 2 ×

Fig. 3.58. Longitudinal section. Etch according to Fry. 2 ×

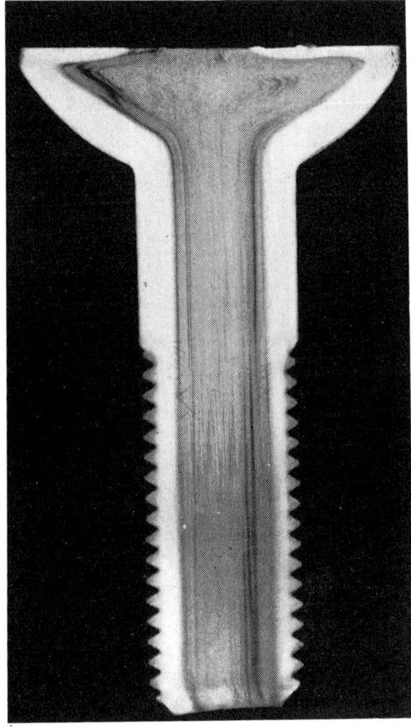

3.59

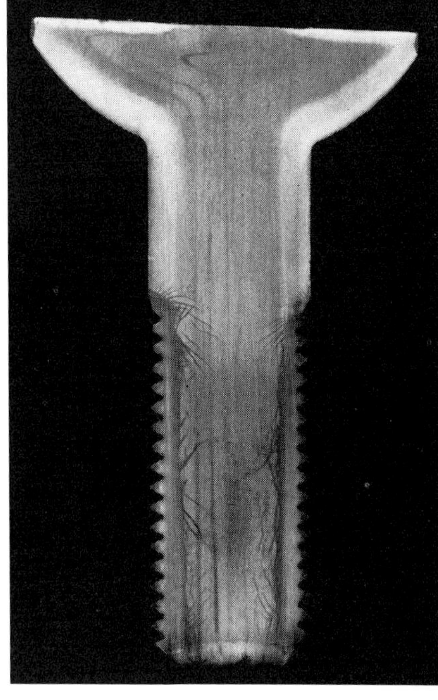

3.60

Fig. 3.59 and 3.60. Railway track screws of basic converter steel of an old heat. Longitudinal section. Etch according to Fry. 1 ×

Fig. 3.59. Cut thread. Impact energy measured on four other longitudinally sectioned bolts, > 206 J

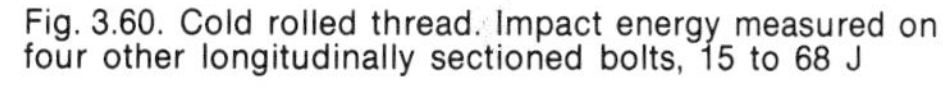

Fig. 3.60. Cold rolled thread. Impact energy measured on four other longitudinally sectioned bolts, 15 to 68 J

3.61

3.62

Fig. 3.61 and 3.62. Wheel bolts with low (Fig. 3.61) and high (Fig. 3.62) impact resistance. Longitudinal sections. Etch according to Heyn. 1 ×

Fig. 3.61. Old shape

Fig. 3.62. New shape

3.63

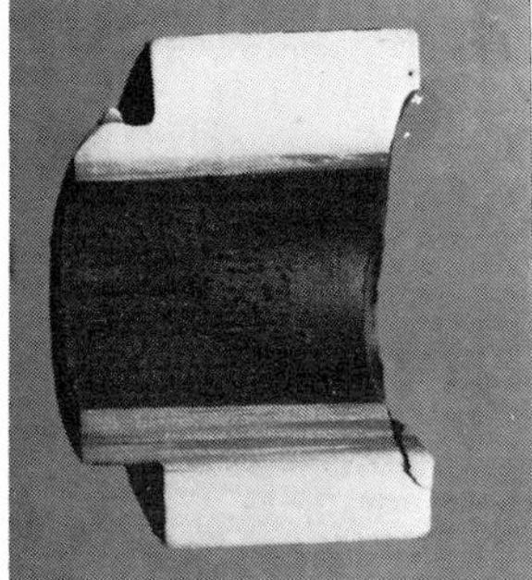

3.64

Fig. 3.63 and 3.64. Broken pipe connecting sleeve of a hydraulic forging press

Fig. 3.63. Fracture in sharp edged and rough machined cross sectional transition.1 ×

Fig. 3.64. Longitudinal section through head part. Etch according to Heyn. 1 ×. Pipe wall consisted almost exclusively of segregated core material

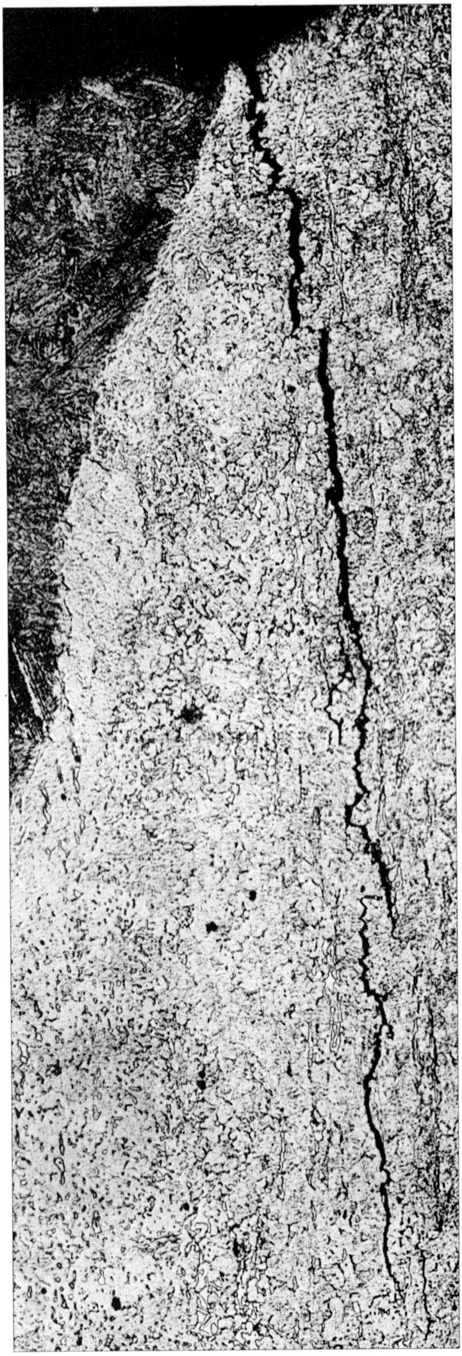

3.67

3.65

3.66

3.68

Fig. 3.65 to 3.68. Failed superheated steam pushrod spindle of stainless steel, cracked at weld

Fig. 3.65. Pushrod with unbroken spindle. Approx. 0.25 ×

Fig. 3.66. Longitudinal section through fracture origin of another spindle. Etch: Nital. 3 ×. Incipient fracture and cracks designated by arrows

Fig. 3.67. Crack with martensitic transition structure of spindle material. Etch: V2A-etching solution. 100 ×

Fig. 3.68. Weld region of a spindle with crack (Fig. 3.65). 1 ×

3.69

3.70

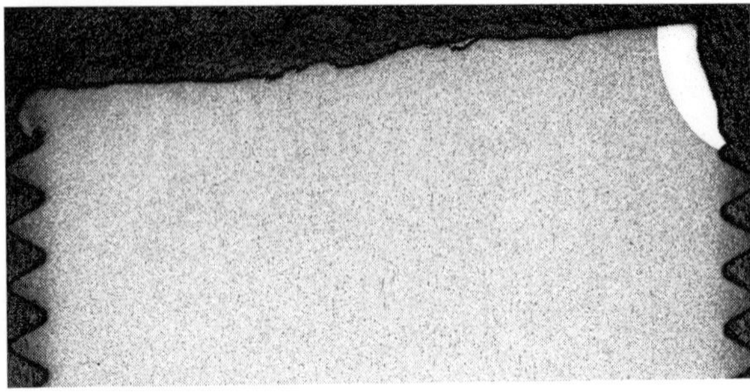

3.71

the keyway showed that its surface was also nitrided **(Fig. 3.74).** The nitrided layer had cracked in many places probably during the pounding in of the key **(Fig. 3.75).** The fatigue cracks started from such cracks. The nitrided layer had been removed at both ends of the keyway at the bearing surface. This was done by grinding flat faces at right angles to the keyway (F in Fig. 3.74). These smooth faces were intended to facilitate subsequent disassembly of the joint. Without this precaution, protuberant metal deformed during pounding of the key would have caused cold welding. The designer had therefore assumed a heavy loading of the key. It would have been proper for him to note on the design that the surface of the keyway was to be kept soft during nitriding.

A designer who plans the application of materials should be fully familiar with the properties and characteristics of these materials during machining and use and should consult specialists as necessary. For instance, the designation "stainless steel" does not necessarily mean that the alloy is corrosion-resistant under all circumstances. Its stability is based upon the formation of a passive oxide layer, that is resistant only to oxidizing atmospheres and oxidizing acids. If this passive layer is penetrated locally, corrosion elements are formed under high current density and a correspondingly high rate of corrosion results in deep pitting. The passivity may be locally counteracted by the effect of chlorides, for instance, or through the formation of electrochemical couples at dust particles (see also section 15.3.2). An example of this is given in the following.

An architect had intended a casing of fine sheets of 0.35 mm thick stainless steel of type 304L for a **facade** of an administration building in an industrial area located between coking plants and chemical factories. The sheets were cold pressed to a narrowly folded oaktree bark-like profile, whose ribs were located horizontally in the structure and consequently facilitated the deposition of construction dust and soot **(Fig. 3.76).** Due to corrosion of the sheets this facade looked unsightly even before construction was finished **(Fig. 3.77).** At the western side, where the dust deposits were rinsed off regularly by rain, the sheets were in considerably better condition than at the eastern side. The dust which had been deposited mainly in the upward turned surface in the profile fold, was of a light grey color, such as of lime or cement, and therefore originated in the construction. Under the dust particles oxide pustules and pinpoint cavities had formed which were filled with water. The pitting character of this type of corrosion is shown better in detail at higher magnification in Fig. 15.80.

The question, whether or not it made any sense to case a facade with sheets of stainless steel in an industrial area, is immaterial. In any event it would have required constant cleaning. But it certainly was a mistake to use sheets with such a distinct profile. In any case the surface should have been protected from dust during the construction period.

The following failure analysis constitutes an example for the application of **unsuitable straightening methods.** The analysis concerns an **extruder worm** used in a plastic fiber processing plant[12]). The worm consisted of stainless steel with 13 % Cr and approx. 0.1 % C (AISI type 410) and showed in several places of the surface some bright, round or oval, somewhat deeper patches with pits or ruptures **(Fig. 3.78).** In the etched cross section it could be seen that the heat treated structure in these areas was transformed into martensite up to almost 2 mm depth

Fig. 3.69 to 3.71. Bridge bolt failure at weld

Fig. 3.69. Fracture with origin in sickle-shaped hardened zone. 3 ×

Fig. 3.70. Side view with weld spatter in fracture origin. 3 ×

Fig. 3.71. Longitudinal section through fracture with martensitic (light) zone in fracture origin. 3 ×

3.72

3.73

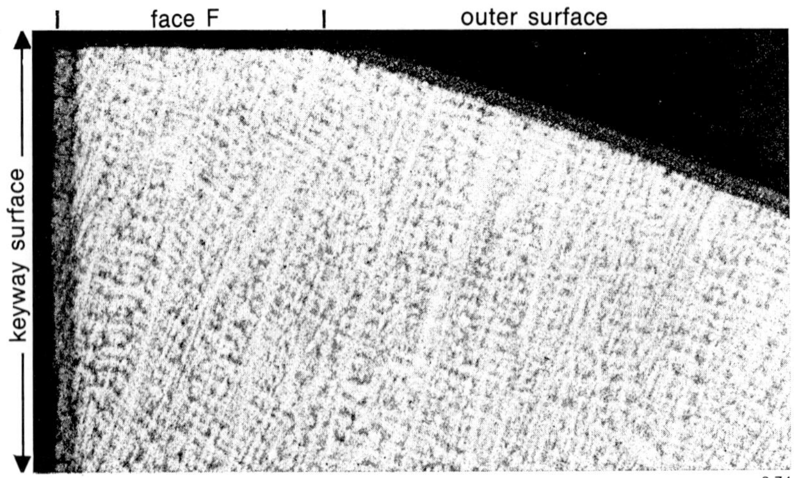

3.74

(Fig. 3.79). The hardened spots had cracks in several places and some of these had splinters broken out. The worm therefore had been heated in part within a narrow space into the austenitic zone. Later it was shown that this had occurred to compensate for distortions through straightening. Such high local heating would have been inadvisable in any other steel as well because of the thermal stresses that had to be expected, but in this high hardenability steel, additional high transformation stresses had to lead to failures of the type described.

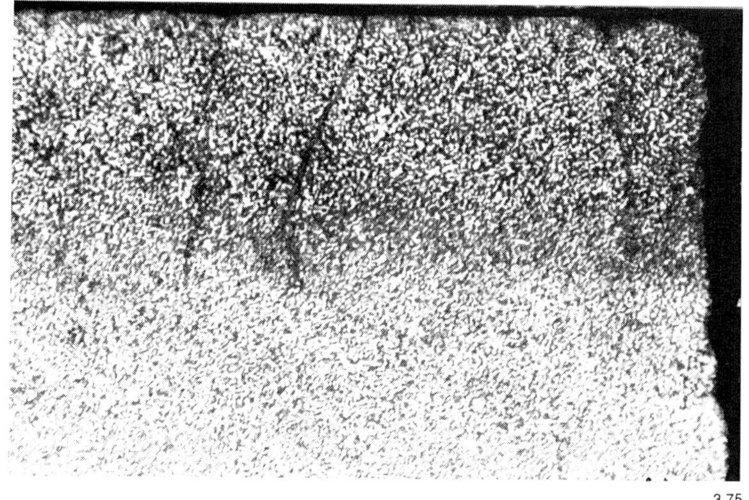

3.75

Fig. 3.72 to 3.75. Broken pivot of piston rod with incipient fracture in nitrided layer of keyway

Fig. 3.72. View of pivot. 0.4 ×. Fracture origin designated by arrows

Fig. 3.73. Fracture surface. 0.5 ×

Fig. 3.74. Cross section adjacent to fracture origin. Etch: Picral. 10 ×

Fig. 3.75. Surface microstructure at keyway. 100 ×

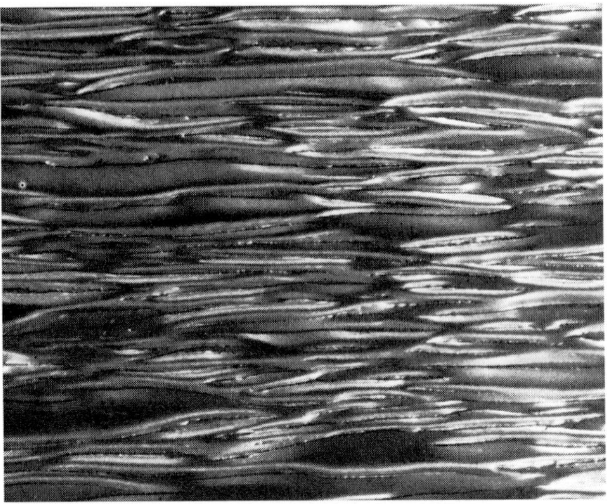

3.76

3.77

Fig. 3.76 and 3.77. Facade sheets of stainless steel corroded by dust deposits during construction

Fig. 3.76. New sheet. 1 ×
Fig. 3.77. Corroded facade. 0.05 ×

In order to show that during **assembly and disassembly** unsuitable methods are often used, the following examples may be cited. Cracks were found in the **lid of a centrifuge[15])** that was overhauled after several years of service. The lid consisted of tempered chromium-nickel-molybdenum steel of 1010 MPa (146 ksi) tensile strength and 16 % elongation. The cracks were situated on the cylindrical, threaded part of the lid and were for the most part tightly closed so that they could not be detected without the aid of magnetic particle inspection **(Fig. 3.80).** They occurred in groups and were located parallel to one another and aligned preferentially with the axis of the centrifuge. In a longitudinal section parallel to the cracks, crescent-shaped zones were found after Heyn etching, that showed a stronger etch attack **(Fig. 3.81).** These cracks were located in zones which were only 1 to 2 mm deep and in most cases did not extend beyond these regions, or only slightly so, as can be seen from the cross section in **Fig. 3.82.** The lid therefore had been heated in this place. The heated zones showed a martensitic structure that could be distinguished clearly from the heat treated structure of the lid by its finer austenitic grain size. The temperature therefore reached the austenitic range. The crescent shape of the zones points to heating with a welding torch. The reason for the heating of the lid could not be discovered. The lid may possibly have jammed in the thread during disassembly and an attempt may have been made to loosen it by heating. The cracks had formed due to thermal and transformation stresses induced by high heating and fast cooling.

An assembly error was the cause for the following failure[16]). During construction of a **prestressed concrete viaduct,** several wires of 12.2 mm diameter ruptured after applying the tension, but before the channels were grouted. They were made of a heat treated prestressed alloy steel of at least 1420 MPa (205 ksi) yield point and 1570 MPa (228 ksi) tensile strength. While the wire bundles, each containing over 100 wires, were being drawn into the channels, they were repeatedly pulled over the sharp edges of square section guide blocks. The wires showed at the surface locally shiny or bluish abrasion zones, that had either cracked or could be made to crack by bending **(Fig. 3.83).** The wire fractures initiated at these abrasion zones. The fracture surfaces showed a kidney-shaped crack darkened by corrosion and therefore old, as well as a large catastrophic fracture with deformed border **(Fig. 3.84). Figure 3.85** shows a longitudinal section of the peripheral structure of a wire through an abrasion zone. Strong deformation and tongue-like overlaps were visible. The material had gotten so hot through friction and deformation that martensite formed during fast cooling. Another cracked wire as well as the wires that had not yet cracked, showed similar structural changes.

The martensite formation may have led to cracks in the structure as a result of transformation stresses, or else the brittle martensite may have been cracked by localized plastic deformation of the wires. In short term tensile tests the wires still showed the specified strength, but elongation and reduction of area were distinctly lower in specimens that had failed in the abrasion zones. Sharp notches, as do occur through crack formation in the hardened zone, may lead to fractures under fatigue tensile stress, especially if there is additional corrosion present; this is true even if the nominal stress lies far below the yield point[17]).

During **overhauling** too, errors often occur that are inherent in the method used. We will report on this in section 11.

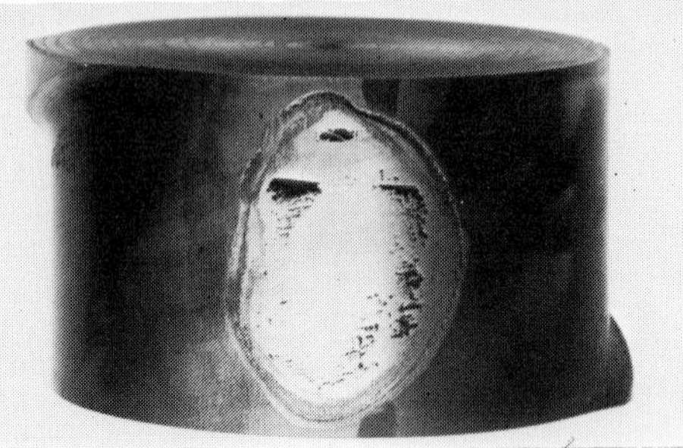

3.78

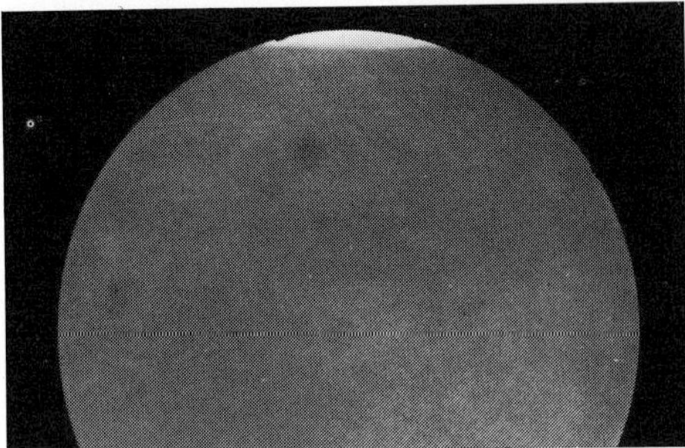

3.79

Fig. 3.78 and 3.79. Stainless chromium steel extruder worm section locally heated for straightening

Fig. 3.78. Surface. 1 ×

Fig. 3.79. Cross section. Etch: V2A-etching solution. 1 ×

Fig. 3.80 to 3.82. Chromium-nickel-molybdenum steel centrifuge cover, locally heated for disassembly

Fig. 3.80. Outer surface. Magnetic particle inspected. 1 ×

Fig. 3.81. Longitudinal section through cylindrical part. Etch according to Heyn. 1 ×

Fig. 3.82. Cross section through crack zone. Etch: Nital. 8 ×

3.80

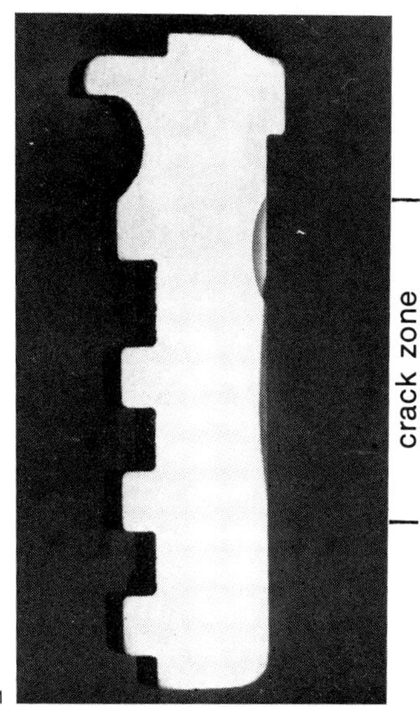

3.81

3.82

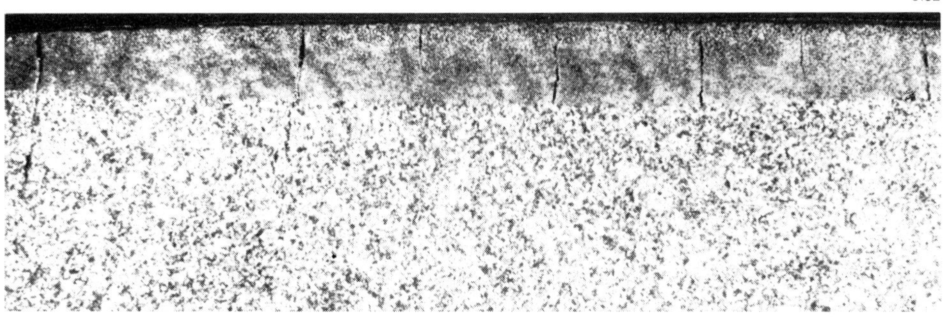

3.83

3.84

3.85

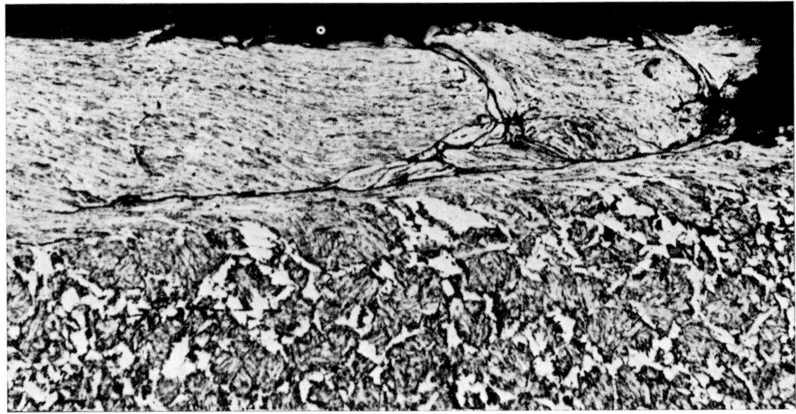

Literature Section 3

1) F. K. Naumann u. F. Spies: Untersuchung von Kabinenbolzen. Prakt. Metallographie 5 (1968) S. 689/91

2) Dieselben: Gesteinsbohrer mit unterschiedlicher Haltbarkeit. Prakt. Metallographie 6 (1969) S. 579/84

3) Dieselben: Zerstörte Siebstäbe aus nichtrostendem Stahl. Prakt. Metallographie 9 (1972) S. 592/96

4) Dieselben: Gebrochene Hängestange. Prakt. Metallographie 11 (1974) S. 418/21

5) Dieselben: Beim Schweißen gerissenes Rohr. Prakt. Metallographie 12 (1975) S. 329/32

6) Dieselben: Gebrochener Rekuperator aus hitzebeständigem Stahlguß. Prakt. Metallographie 9 (1972) S. 708/10

7) F. K. Naumann: Beitrag zum Nachweis der σ-Phase und zur Kenntnis ihrer Bildung und Auflösung in Eisen-Chrom- und Eisen-Chrom-Nickel-Legierungen. Arch. Eisenhüttenwes. 34 (1963) S. 187/94

8) F. K. Naumann u. F. Spies: Gebrochene Haken aus hitzebeständigem Stahlguß. Prakt. Metallographie 15 (1978) S. 460/66

9) Dieselben: Gebrochene Spannbolzen. Prakt. Metallographie 11 (1974) S. 616/20

10) E. Baerlecken: Schnellprüfverfahren zur Erkennung von Stahlverwechslungen. Stahl u. Eisen 73 (1953) S. 30/34

11) F. K. Naumann u. W. Carius: Ätzverfahren zur Unterscheidung von chemisch beständigen Chrom-Nickel- und Chrom-Nickel-Molybdän-Stählen. Arch. Eisenhüttenwes. 28 (1957) S. 641/44

12) F. K. Naumann u. F. Spies: Gebrochener Anker eines Schweißgenerators. Prakt. Metallographie 7 (1970) S. 285/87

13) Dieselben: Untersuchung von Heißdampfschieberspindeln. Prakt. Metallographie 11 (1974) S. 621/27

14) Dieselben: Bruch einer nitrierten Kolbenstange. Prakt. Metallographie 7 (1970) S. 47/51

15) Dieselben: Rissiger Zentrifugendeckel und Extruderschnecke mit Oberflächenausbrechungen. Prakt. Metallographie 5 (1968) S. 397/401

16) Dieselben: Gerissene Spanndrähte von einer Talbrücke. Prakt. Metallographie 10 (1973) S. 592/95

17) W. Jäniche, W. Puzicha u. H. Litzke: Untersuchungen zum zeitabhängigen Bruch vergüteter hochfester Stähle. Arch. Eisenhüttenwes. 36 (1965) S. 887/96

Fig. 3.83 to 3.85. Fracture of tension wire at abrasion zone that was caused by tightening of the wire bundles in the tension channel

Fig. 3.83. Cracks occurring during bending in an abrasion zone. 20 ×

Fig. 3.84. Wire fracture. 4 ×

Fig. 3.85. Longitudinal section through abrasion zone at fracture origin. Etch: Picral. 500 ×

4.1 Metallic Additions and Non-Metallic Inclusions

Errors that are based upon melting and composition of steels will be discussed first. Primarily unintentional, and in part unavoidable, additions of phosphorus, sulfur, oxygen, copper and tin are considered damaging to steel. Copper and tin are more noble than iron, and therefore cannot be removed from the steel by oxidation, but are concentrated and deposited at the steel surface. Consequently a firmly adhering dense oxide scale is formed that is hard to remove by pickling. This results in surface defects in thin sheets. If metal is precipitated onto the iron surface below the scale, the metal may penetrate the grain boundaries and cause "hot shortness" during hot working[1][2]. Only a few hundreds of a per cent of tin are needed to substantially enhance the hot shortness caused by the copper[3]. According to K. Born[4], copper hot shortness is possible above 950 °C.

Figure 4.1 shows a section of **bar steel** that cracked at the edges during rolling. Spectroscopic examination of the steel showed substantial amounts of copper and tin; wet chemical analysis also showed 0.16 % Cu and 0.81 % Sn. During the search for the origin of such extraordinary high amount of tin it was determined that the melt feed stock consisted predominantly of tin sheet scrap, which had been insufficiently detinned, or not at all. The scrap contained from 0.19 to 0.24 % Cu and 0.1–2.5 % Sn.

Figure 4.2 illustrates the pickled surface of a low-carbon open hearth steel sheet that is marred by adhesive scale. **Figure 4.3** shows the unetched section of **adhesive scale** on the surface of a strip containing 0.23 % Cu. After nickel plating, cold rolled steel strip was found to have defective areas where the nickel coating had defoliated **(Fig. 4.4).** Examination of the strip and of one previously delivered that was considered acceptable, as well as the starting material in variòus stages of working, disclosed the defects to have been caused by **rolled-in scale.** This could not be completely removed by normal pickling. Analysis of eight acceptable and five defective strips indicated a copper content between 0.05 and 0.09 % for the former and between 0.20 and 0.22 % for the latter. This supported the failure analysis.

In order to avoid such failures, the scrap used for steel melting should contain no more than 0.15 % Cu and 0.02 % Sn[5]. These are values that cannot be reliably maintained in commercially purchased scrap, especially from automotive sources, so that steel plants should rely on their own pure recycled scrap supply for sensitive parts, or alternatively reduce the scrap fraction substantially.

Phosphorus causes low temperature embrittlement in steel, just like nitrogen which will be discussed later. Since the old type basic converter steel is rarely produced anymore, failures caused by phosphorus have become rare. In heat treated steels, phosphorus and tin promote a tendency toward temper embrittlement during slow cooling from tempering or stress relief annealing.

Sulfur and oxygen in iron and steel appear in the form of non-metallic sulfide and oxide inclusions. Iron sulfide forms a low-melting eutectic with the iron and therefore causes hot shortness during hot working. In manganese-containing steels it occurs only in highly segregated regions. Hot shortness failures by sulfur are described in section 4.2

Fig. 4.1. Edge cracks caused by hot shortness in bar steel with 0.16 % Cu and 0.81 % Sn. 1 ×

Fig. 4.2. Surface defect caused by scale adhesion in pickled sheet surface. 1 ×

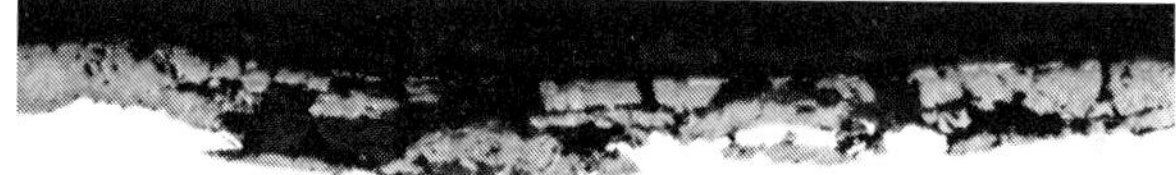

Fig. 4.3. Unetched section of scale adhesion on surface of strip with 0.23 % Cu. 200 ×

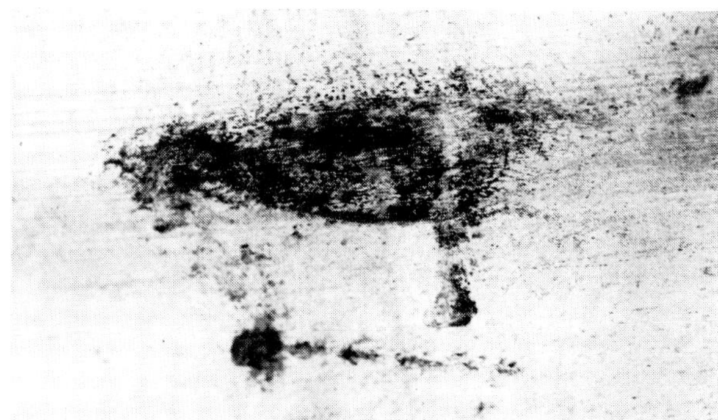

Fig. 4.4. Surface defect through rolled-in scale in cold rolled and nickelplated strip. 1 ×

As a rule, sulfur and manganese combine to form high melting **manganese sulfide.** It has a bluish color in contrast to the yellow iron oxide and appears in cast iron often in the shape of cubic crystals, while in cast steels it appears as round droplets that can be stretched plastically and widened by rolling or forging into lenticular shapes. **Silica,** resulting from deoxidation with silicon, and **silicates** also solidify as droplets. The silicates are either easier or harder to deform according to their silica and alumina content. Silicate inclusions, consisting of furnace slag, are usually stretched during rolling just like manganese sulfides, but are somewhat darker than the latter. Oxides such as **aluminium oxide,** resulting from the deoxidation by aluminium, and spinels solidify in a crystalline form and are not deformable. The same holds true for aluminium oxide-rich silicates which are the result of deoxidation, or slagged refractories and fireclays of the ladles, troughs, and channels. They appear in the steel in the form of crystalline agglomerates or as stringers of crumbly inclusions that have been fragmented during deformation.

After tapping, the steel is left in the pouring ladle for a while in order to allow time for the nonmetallic inclusions that have formed in part as deoxidation products in the ladle to conglomerate and segregate. But since this period cannot be extended indefinitely, it is unavoidable that some of the smaller inclusions remain in the steel. Therefore a certain inclusion content has to be tolerated. Smaller, uniformly distributed inclusions do not affect the quality of the steel materially. They may even act as nuclei during solidification of the steel that lead to the formation of a fine-grained structure, if they have been precipitated prior to the beginning of solidification of the iron.

Coarser inclusions can lead to fracture initiation sites on a bearing surface which experiences higher surface stresses, such as occur in roller bearings. Macroscopically visible inclusions may result in fatigue initiation sites in highly stressed machine parts such as crankshafts. If they are found during machining, the particular part ist usually scrapped.

The precipitation of inclusions onto the primary grain boundaries may cause intergranular fractures with brittle fracture modes (see also section 4.3.2).

During pickling, galvanizing, or enameling, hydrogen may be precipitated at the inclusions and cause blistering (see also sections 9.1 und 9.2).

Inclusion stringers decrease transverse toughness. During stressing solely in the rolling direction, they do not lead to failure. During torsional stressing the inclusion stringers lying directly below the surface may tear open.

By binding sulfur to rare earths such as cerium, sulfides of lower deformability are obtained. This inhibits the formation of connected slag stringers and increases transverse toughness[6]). The same effect is obtained by the addition of titanium which results in the formation of mixed sulfides of manganese and titanium or titanium carbo-sulfides[7]).

If required, the sulfur and oxygen content and thus the amount and magnitude of inclusions can be substantially decreased by special control of the melting practice or by its subsequent treatment. In certain cases these inclusions may also be useful. For instance, additions of 0.2 to 0.3 % S, with or without the addition of 0.15 % to 0.3 % Pb, facilitates automatic screw machining and improves surface quality because the chip breaks off at the inclusions because smearing is prevented. Overheating sensitivity may be decreased substantially through fine precipitates of oxide, carbide, and nitride phases (see section 6.2).

Fig. 4.5 to 4.8. Stub axle with slag inclusions

Fig. 4.5. Surface defects shown by magnetic particle inspection. 3 ×

Fig. 4.6. Slag stringer cut during machining, in unetched cross section. 100 ×

Fig. 4.7 and 4.8. Unetched longitudinal section through aluminum oxide stringer located directly below surface

Fig. 4.7. 100 ×

Fig. 4.8. 500 ×

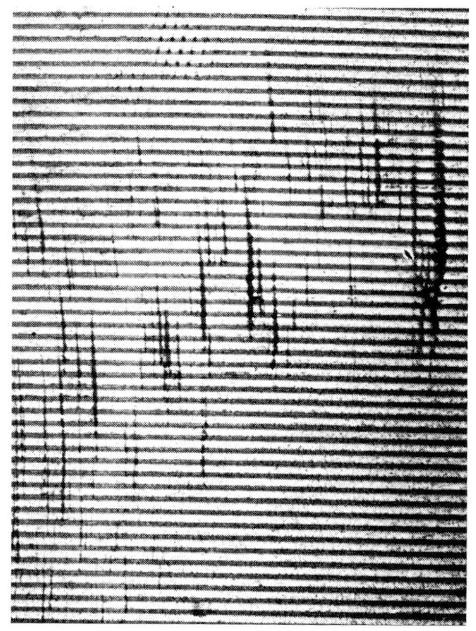

4.5

4.6

4.7

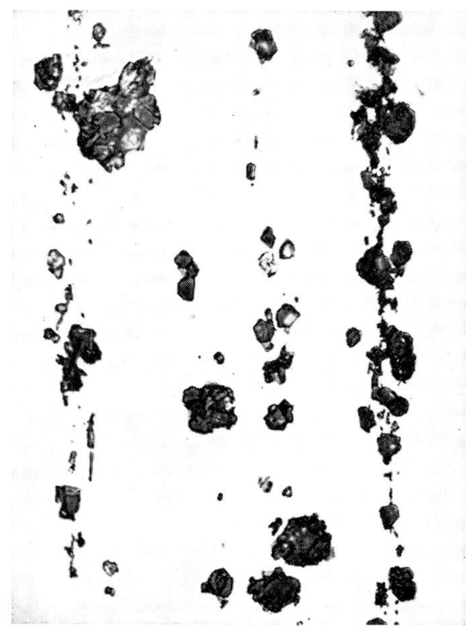

4.8

A
B
C
D
E
F
G
1
2
3
4
5
6
200
×
200

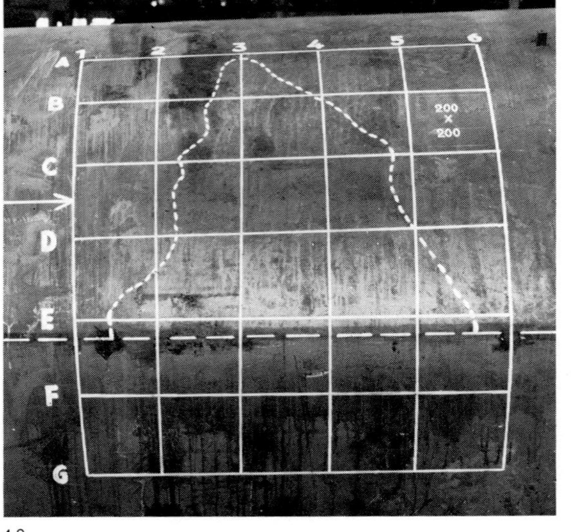

4.9

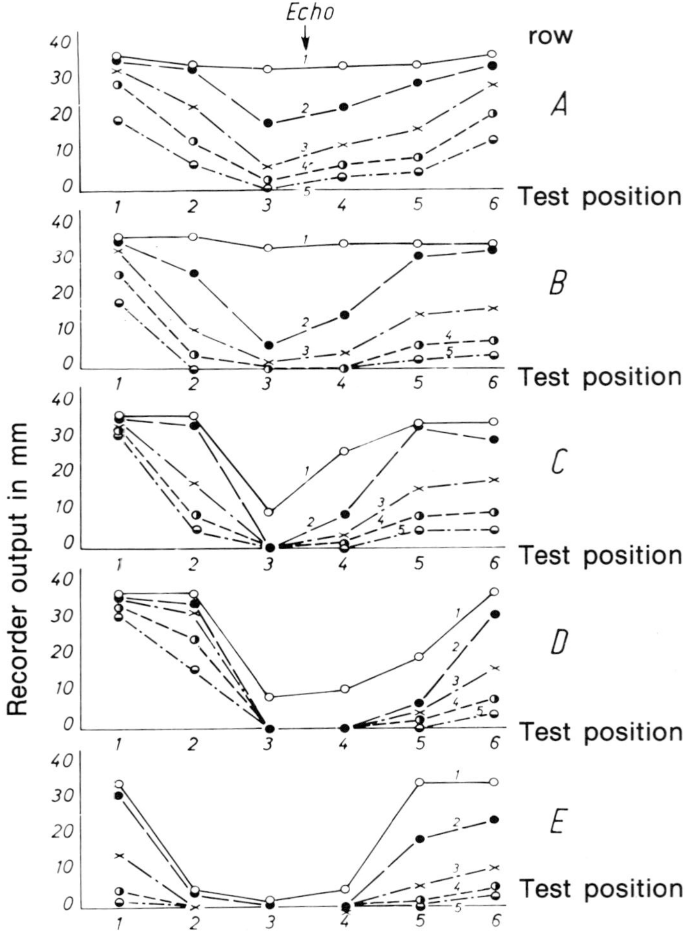

Echo
row
A
B
C
D
E
Test position
Recorder output in mm
0
10
20
30
40
1
2
3
4
5
6

4.11

Fig. 4.9 to 4.11. Defective section of unfired locomotive boiler

Fig. 4.9. External view. Coordinate grid to establish test site. Defective area based on plant ultrasonic inspection marked by broken lines

Fig. 4.10. Tests at points of row D according to Fig. 4.9

Fig. 4.11. Results of ultrasonic inspection

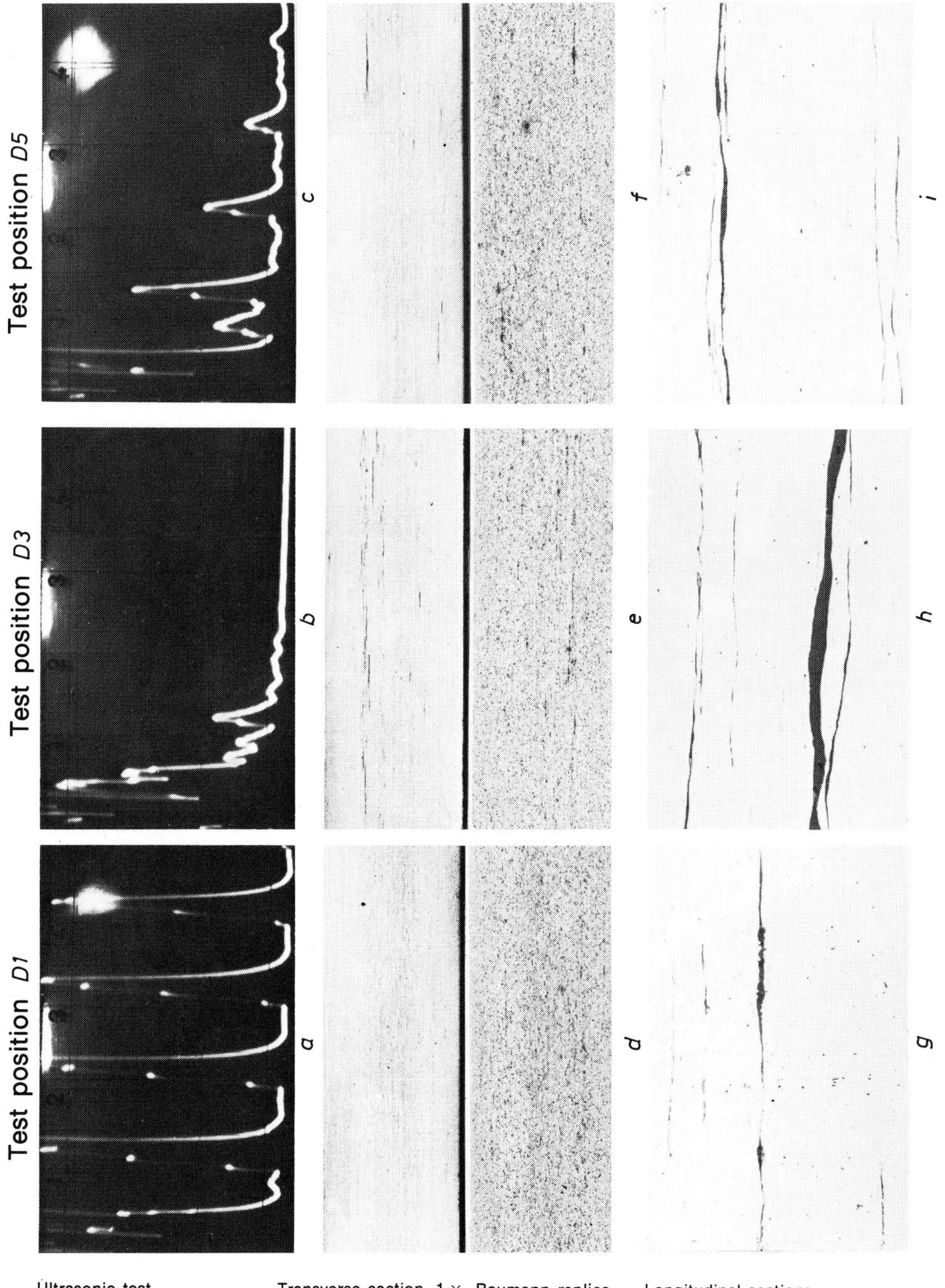

Ultrasonic test

Transverse section. 1 × Etch according to Heyn

Baumann replica

Longitudinal sections. Unetched. 100 ×

4.10

Identification of inclusions can be made metallographically according to their color, shape, and distribution, as well as their sensitivity to attack by acid or alkaline etch media. Micro- and scratch hardness testing may be used in addition. Futhermore, inclusions can be analysed directly by electronmicroprobe or microchemically after electrolytic isolation. Their composition is often complex.

Coarse, elongated inclusions in sheets can be clearly distinguished during ultrasonic testing according to the pulse echo method by the indication of an intermediate echo, as will be shown in a later example.

The amount and distribution of inclusions can be counted and measured electronically, by quantitative metallography according to specified procedures*, or by deep etching. Visual inspection of fresh surfaces produced by machining cylindrical specimens to progressively smaller diameters can serve to illustrate macroscopic inclusions**.

The origin of the inclusions is more difficult to determine and often cannot be established since they usually no longer have their original shape and composition, but have reacted with other components of the steel and the slag.

"Scabs"*** may serve as an example[8]). During magnetic particle inspection of drop forged **stub axles** in an automotive plant, longitudinal, crack-life defects were found at the surface **(Fig. 4.5).** Metallographic examination showed that these were not cracks but coarse slag stringers that were located in part at the surface where they were cut open during machining, and in part located directly below the surface **(Fig. 4.6 and 4.7).**

The inclusions were multi-phased and consisted of crystals or crystal fragments in a glassy transparent matrix. They had crumbled during forging, but were not elongated **(Fig. 4.8).** It could be concluded that they consisted predominantly of aluminium oxide or aluminium oxide-rich silicates. It probably was refractory slag material from the fireclay bricks of the pouring channels. Based upon this investigation it had to be recommended that the defective stub axles should be scrapped since the anticipated high loads would result in a fatigue failure.

In the following, an example is given for the appearance, proof, and effect of **plastically deformed silicate and sulfide inclusions.** During the construction of an **unfired locomotive boiler,** defects appeared during weld tensile testing. The boiler consisted of two sections connected with a round weld seam, and each section in turn consisted of a longitudinally welded 20 mm thick plate having a length of 5800 mm and a width of 2500 mm. During in-plant pulse-echo ultrasonic inspection the defects were found to be concentrated in a tongue-like area in the rolling direction. These had extended from the longitudinal seam into the sheet.

Figure 4.9 shows the defective section. A rectangular grid of coordinates of 200 mm mesh width has been superimposed on it to establish the test sites. The ultrasonic inspection took place at the intersections; transverse sections for metallographic examination were subsequently cut through them. The defective area is delineated by a broken line in Fig. 4.9.

The defects became apparent during ultrasonic testing by an attenuation of the back face ultrasonic image and the faster damping of the oscillograph peaks. Within the delineated area they increased from outside to inside and from top to bottom.

Figures 4.10a, b, and c show representative diagrams of the test data points D 1, D 3 and D 5. D 1 lies outside of the defective area. The fifth echo is still clearly discernible and only slightly weaker than the first. But at the data point D 3 which is located in the middle of the defective field, the first echo is much weaker, while the second one is barely recognizable. At data point D 5 that lies at the border of the delineated area, the first amplitude is notably lower, but three others are still discernible.

---

* **Stahl-Eisen Prüfblatt** 1570. ASTM E 45–63
** **Stahl-Eisen Prüfblatt** 1580
*** see Appendix I

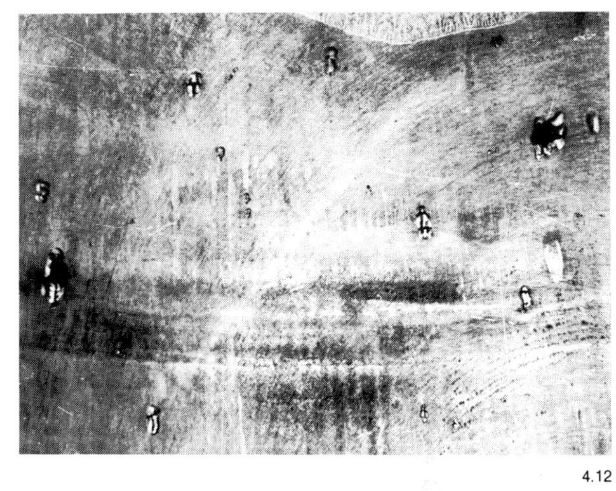

4.12

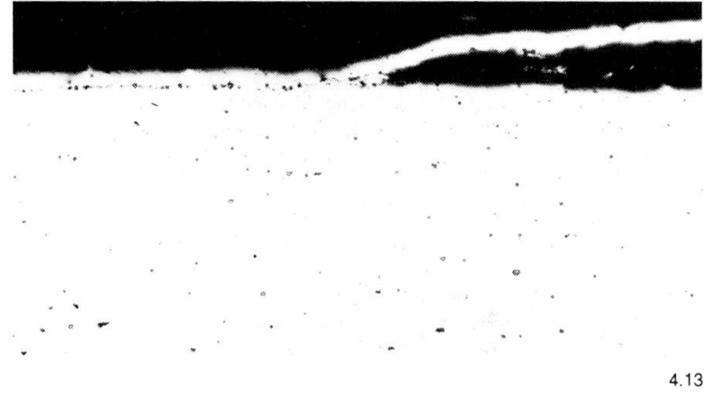

4.13

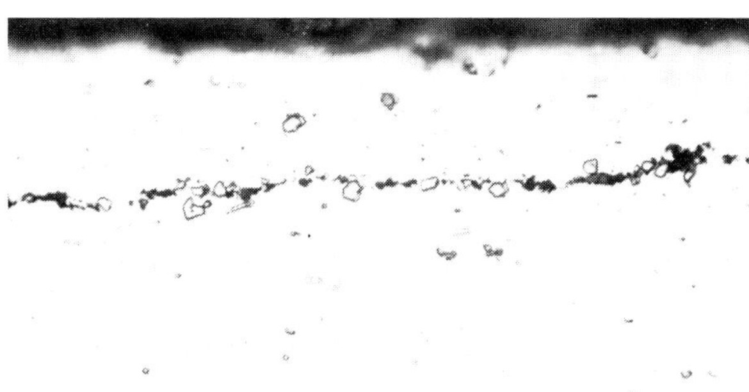

4.14

Fig. 4.12 to 4.14. Deep drawn steel pot with pickling blister

Fig. 4.12. Surface. 0.33 ×

Fig. 4.13. Unetched longitudinal section through blister. 100 ×

Fig. 4.14. Section of Fig. 4.13. 500 ×

4.15

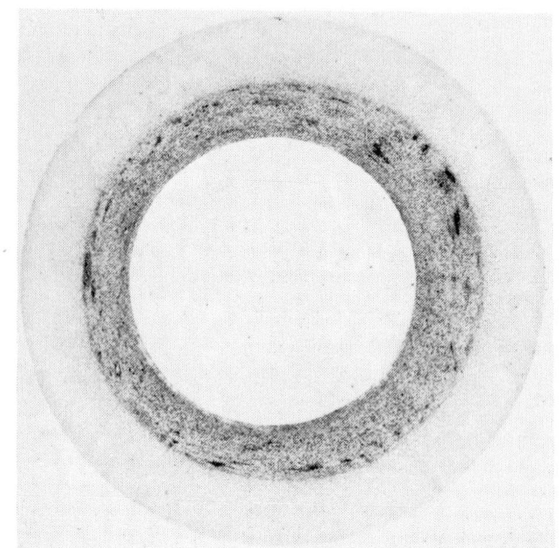

4.16

4.17

4.18

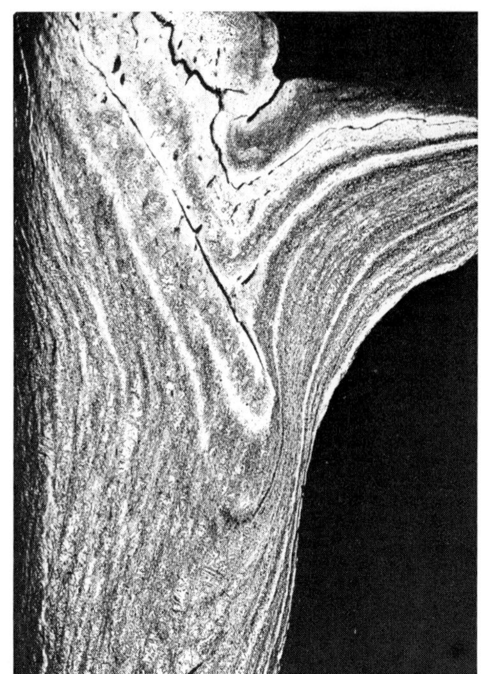

Figure 4.11 summarizes all the ultrasonic inspection results in such a way that the amplitudes for the data points 1 to 5 are shown for the horizontal rows A to E. From this presentation it can be readily seen that the defects evident from the attenuation of the impulses of the back face ultrasonic image of row 1 increase in the direction to the right, achieve a maximum in row 3, and decrease in intensity again from rows 4 to 6.

The figures 4.10d, e, f are macroetchings and Baumann replicas of transverse sections at the intersects D 1, D 3, and D 5. At the point D 1, located outside the defective zone, the sheet was so pure that no inclusions or segregations were visible macroscopically (Fig. 4.10d). The middle of the defective area was strongly permeated by elongated inclusions (Fig. 4.10e). In the transition zone (D 5) a few long stringers were still present (Fig. 4.10f).

Microscopic examination of longitudinal sections showed that the inclusions consisted principally of stretched-out silicate slags as well as of manganese sulfides (Figs. 4.10g, h, i) which could be recognized by their lighter shading and diminished thickness. The inclusions in the delineated region were of 4 to 25 mm length and 0.025 to 0.09 mm thickness, while outside of this zone they had a length of 1 to 3 mm and a thickness of 0.002 to 0.02 mm.

The results of the examination indicated that the defective region was connected with the segregation zone at the end of the cast ingot. Therefore the rejection of the boiler by the locomotive manufacturer was justified.

However, it was of interest to know to what extent these defects would actually have affected the usability of the boiler. Since it was predominantly subject to tangential tension, i.e. in the rolling direction of the sheet, these defects would not necessarily reduce its utilization. For a clarification of this question, two tensile specimens were cut from each of two regions – one lying outside the defective zone, square A–B x 1–2, and the other inside the defective zone, square C–D x 3–4. They had a width of 10 mm in the circumferential direction of the vessel and were of the full gage of the sheet. They were cut from the section along the circumference, i.e. in the rolling direction of the sheet (L) and in the direction of the boiler axis, i.e. transverse to the sheet rolling direction (Q). **Table 1** shows the tensile test data.

Table 1: Results of Tensile Tests

Square	Specimen	Tensile Strength MPa (ksi)	Elongation (gage length L_o = 40 mm) %
A–B x 1–2	L1	451 (65.4)	42.7
	L2	435 (63.1)	42.5
	Q1	411 (59.6)	38.3
	Q2	411 (59.6)	37.5
C–D x 3–4	L1	413 (59.9)	40.8
	L2	424 (61.5)	41.0
	Q1	419 (60.7)	35.3
	Q2	422 (61.2)	36.0

Fig. 4.15 and 4.16. Hot short pipe

Fig. 4.15. View of I.D. surface. 1 ×

Fig. 4.16. Baumann sulfur print of transverse section. 0.5 ×

Fig. 4.17 and 4.18. Hot press nut with splitting tears in phosphorus and sulfur segregations

Fig. 4.17. View. 0.8 ×

Fig. 4.18. Longitudinal section. Etch according to Oberhoffer, illuminated vertically. 6 ×

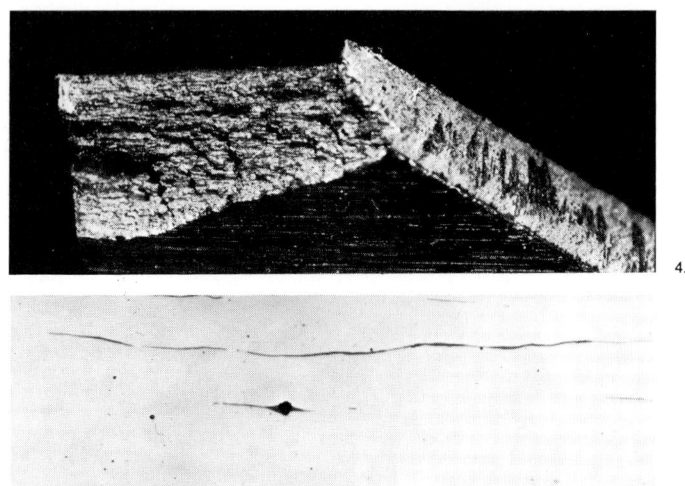

4.19

4.20

4.21

4.22

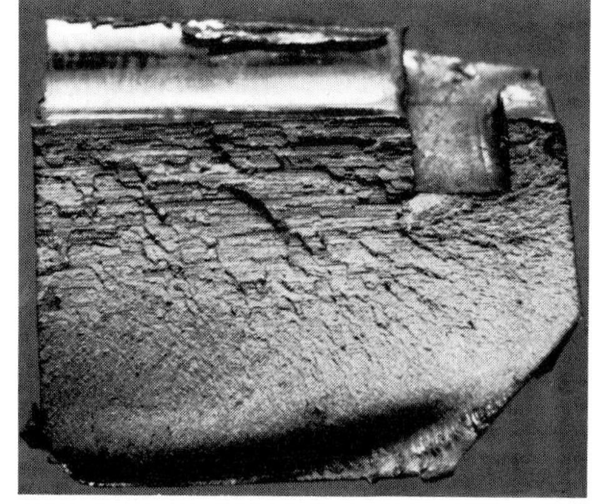

4.23

4.24

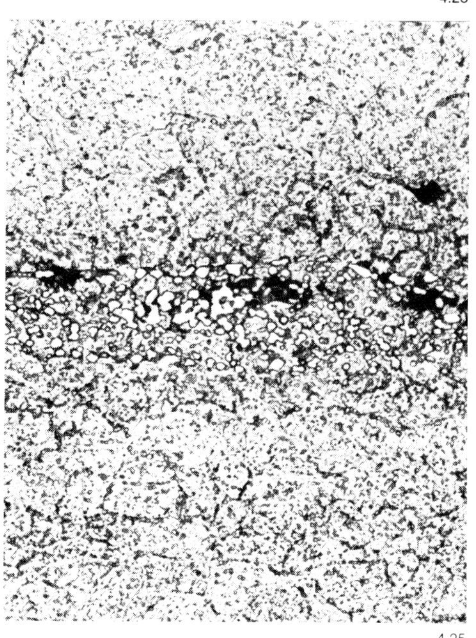

4.25

Fig. 4.23 to 4.25. Broken centrifuge finger of stainless knife steel

Fig. 4.23. Fracture. 0.33 ×

Fig. 4.24 and 4.25. Microstructure in segregation zone. Longitudinal section. Etch: V2A-etching solution

Fig. 4.24. 100 × Fig. 4.25. 500 ×

Fig. 4.19 to 4.22. Sheet of wood saw split open during stamping and setting of teeth

Fig. 4.19. Fracture surface of tooth. 8 ×

Fig. 4.20. Unetched longitudinal section. 500 ×

Fig. 4.21. Longitudinal section, etched according to Oberhoffer. 100 ×

Fig. 4.22. Section through split open stamped edge. Etch: Nital. 100 ×

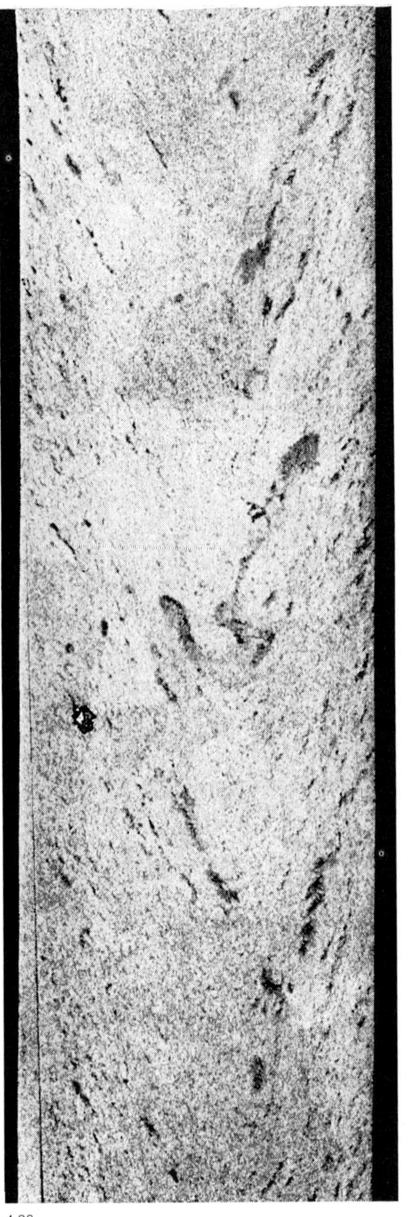

4.26

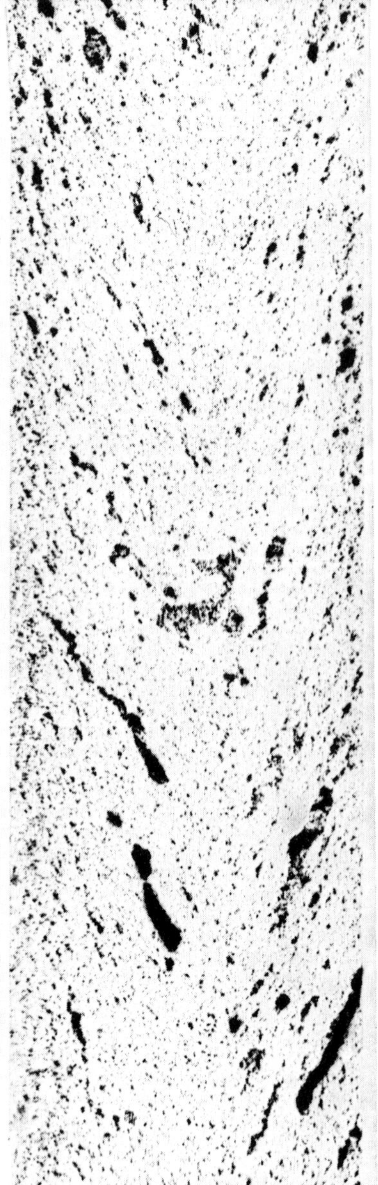

4.28

All specimens had pronounced reductions in area. The longitudinal specimens from the defective area were subject to longitudinal cracks during tensile testing. Ultimate strength and elongation were only affected slightly by the slag inclusions. Accordingly the boiler would probably have performed well in service for many years after having survived the welding.

The following is an example of **blister formation through hydrogen precipitation** at slag inclusions. A deep drawn pot of stainless steel of the titanium stabilized grade containing approx. 0.1 % C, 18 % Cr, 10 % Ni and 2 % Mo had become unusable because defects arranged in streaks appeared at the surface. As can be seen from **Fig. 4.12** they are blisters that had in part torn open. A section through such areas showed that separation had occurred at inclusions located directly below the surface **(Fig. 4.13).** These consisted primarily of yellow cubes of titanium nitride. In the cast ingot they were apparently arranged in clusters and rolled out into stringers during forming **(Fig. 4.14).** During pickling, hydrogen precipitated at these inclusions, and the thin surface layer had bulged, probably during intermediate annealing. The failure may have been caused or aggravated by aggressive pickling or by the use of an unsuitable pickling solution. Section 9.1 will describe pickling blister formation in detail.

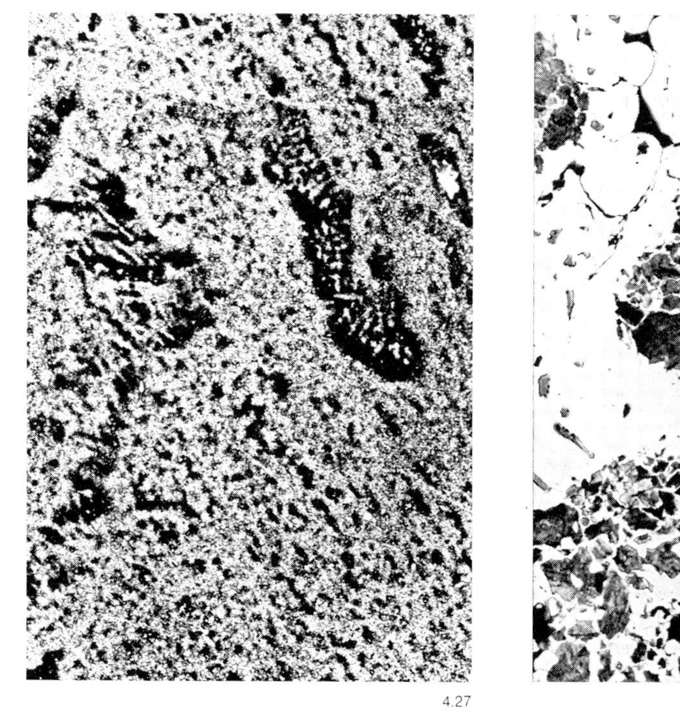

4.27

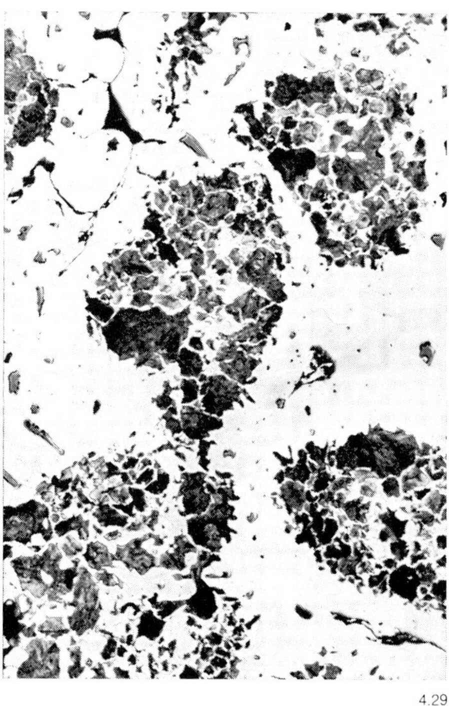

4.29

Fig. 4.26 to 4.29. Longitudinal section through wall of blistered hollow cylinder of die cast steel

Fig. 4.26. Heyn etch. 0.5 ×

Fig. 4.27. Oberhoffer etch. Illuminated obliquely. 2 ×

Fig. 4.28. Baumann sulfur print. 0.5 ×

Fig. 4.29. Etch: Picral. 50 ×

In larger parts, such as forgings, defects are rarely caused by inclusions. The Forging Committee of the Verein Deutscher Eisenhüttenleute tested the effect of large inhomogeneities upon the failure susceptibility of large free-form forged shafts. Torsion and bend cyclic fatigue tests were performed on crankshafts that contained large defects; they were predominantly longitudinal, such as slag inclusions and segregations[9]). When overloading during the torsional cyclic tests, eleven of thirteen crankshafts (130 to 215 mm pin diameter) broke, irrespective of some very coarse defects, in the fillet between crank pin and arm. One defect originated at a lubricating hole and another in a turning groove of an oil bore. That means that in 12 cases the fractures propagated from areas of stress concentrations caused by design while in one case a machining error caused the failure, i.e. by forming a notch that also introduced a stress concentration. No failures were due to material defects. During the fatigue bend tests all three crankshafts fractured at the fillet through the crank arm. In no case could any effect of the defects be found on the limiting stress for the fatigue fracture.

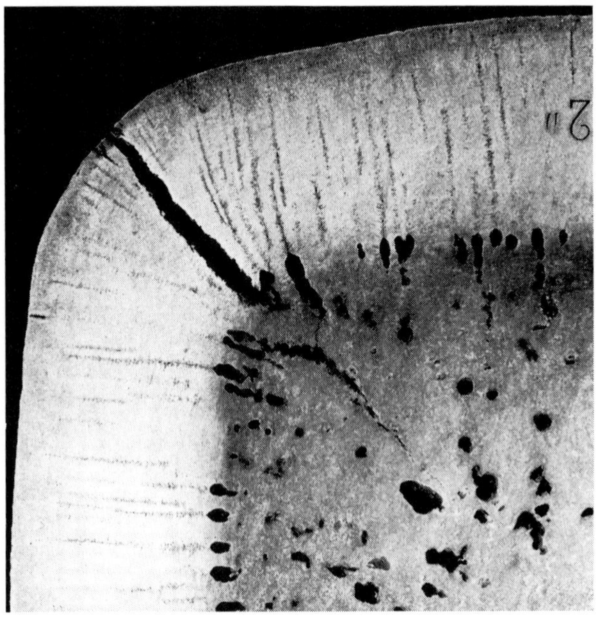

Fig. 4.30. Square ingot of rimmed cast screw steel with gas blister segregations. Cross section through ingot edge. Etch according to Heyn. 0.4 ×

4.2 **Segregations**

Segregations are regions of chemical phase separations that occur during the solidification interval. The regions that solidify first, i.e. at the outside and the bottom of a cast ingot and at the cores or axes of a dendrite, then consist of a high melting pure metal. The zones solidifying last, i.e. the interior and upper parts of the ingot and the outer shells of the dendrite or the interstices of the dendrites are enriched by lower melting foreign materials and alloying metals. The larger the temperature difference between the liquidus and solidus line, the greater the chemical segregation becomes. Concentration gradients are equilibriated at a slower rate, the lower the diffusion rate of the foreign material or alloying metal. The phenomena occurring during continuous castings are described by H. Jacobi and K. Wünnenberg*.

Phosphorus, for instance, is a strongly segregating element. Phosphorus segregations are confirmed by etching with acid solutions according to Oberhoffer, in which the phosphorus-rich areas – in contrast to etching according to Heyn – have a more noble potential than the phosphorus-deficient regions. In contrast to the latter they are not attacked[10]).
Ingot segregation in rimmed steel castings is shown by the formation of a carbon- and impurity-deficient case, as well as a sharply delineated segregated core. In this instance segregations may be advantageous because the pure ingot skin facilitates cold working. e.g. by deep drawing. But if the segregation zone and in particular the strongly impure transition zone comes to the surface or is cut during machining, failures may occur (see also Fig. 3.63 and 3.64).

In large ingots considerable accumulations of segregated elements may occur even during solidification of killed steel. Such segregations constitute areas of weakness that can cause failures or at least play a contributory part.

The structural inhomogeneities caused by segregation can be compensated either completely or in part by long time annealing. Usually very high temperatures are necessary as well, which makes this application of diffusion annealing impractical. Attaining equilibrium takes longer, the thicker the ingot and the coarser the primary grain structure. Forging or rolling shortens the diffusion paths which in turn facilitates concentration equilibrium.

Segregations in ingots as well as in grains are stretched into stringers by forging or rolling. Such stringer formation decreases the toughness of steel during transverse loading. Segregation furthermore promotes **hot shortness** during warm forging. Iron sulfide forms a low melting eutectic with iron that is segregated in the form of thin films between the grains and may lead to fracture of the steel during hot working. Generally, manganese is added to the steel in sufficient quantities to combine with the sulfur to form manganese sulfide with a high melting point. However, if sulfur is segregated locally in larger amounts, the manganese content may be insufficient to prevent rupture or cracking of the steel in the segregation zone during deformation. The following case is an example of this type of hot shortness.

A large number of **pipe blooms** fractured during passage through the first rolls. The blooms consisted of steel with 0.02 % C, trace of silicon, 0.19 % Mn, 0.013 % P and 0.052 % S. Their inner wall surfaces were characterized by numerous gaping cracks **(Fig. 4.15).** Metallographic examination (using a sulfur print) showed the existence of strong core segregations in the blooms **(Fig. 4.16).** In this zone the sulfur content had risen to 0.106 %. The cracks were limited in depth to the segregation zone. Uncommonly large wustite inclusions indicated that the steel was overdoxydized during melting. Rupture of the pipe blooms through hot shortness during drawing was caused by two contributing factors, (1) the unusually high sulfur content, doubled locally by segregation, and (2) the low manganese content, further diminished by partial combination with oxygen.

Figures 4.17 and 4.18 demonstrate the connection between fractures of **hot press nuts** with phosphorus (and sulfur) segregations.

---

* **Stahl u. Eisen,** Vol. 97, 1977, p 1075-81

The following failure analysis is an example for the reduction in the transverse deformability as a result of segregation. A sheet for **saw blades** of 2 mm thickness made of oil quenched, heat treated steel with 0.9 % C, 0.4 % Si and 0.8 % Mn split open at the stamped edges, or else the teeth broke off, during stamping and tooth setting. The fracture surfaces had a fibrous appearance **(Fig. 4.19).** In the longitudinal section it could be seen that the sheet was strongly contaminated by non-metallic inclusions **(Fig. 4.20).** These were predominantly manganese sulfides, as confirmed by Baumann replicas. By etching of the surface according to Oberhoffer, the primary banded structure became visible **(Fig. 4.21).** The stronger etched bands appear darker under vertical illumination applied in this case, and correspond to the branches of the original dendritic structure. Therefore they contain more iron, while the lighter ones in which the inclusions are located, correspond to the phosphorus-rich and alloy-rich dendritic interstices. The darker bands had a microhardness of 426 to 445, while the lighter ones had a hardness of 495 to 538 HV 0.03. **Figure 4.22** shows a longitudinal section through a stamped edge. The cracks ran in part along the light, inclusion-rich and alloy-rich bands of the secondary structure[11]) and occasionally jumped from one row to the next. Therefore the bands are the cause of crack formation and fracture.

Carbon also segregates. Large castings have a markedly higher carbon content on the inside and at the top than on the outside and bottom of the ingots. Grain segregation generally is equalized quickly by diffusion. Only ledeburitic and hypereutectoid steels (especially those with hard-to-dissolve special carbides) tend toward segregations. In the worked state these steels also tend toward banding, which may cause trouble during working and may affect service properties.

A **centrifuge finger** of martensitic stainless steel with approx. 0.45 % C and 13 % Cr broke during a test run in an ultracentrifuge at 84,000 r.p.m., damaging the device considerably. The finger was examined together with one that was found free of defects under similar test conditions. The fingers were produced from rolled bars and heat treated to achieve a high strength.

The broken finger showed a fibrous fracture around the inner bore **(Fig. 4.23),** while the good finger had a non-fibrous deformation after breaking it in a bench test. Chemical analysis and hardness testing showed that the broken finger consisted of the specified steel, while the good one consisted of the same 13 % Cr steel but with a lower carbon content (approx. 0.2 %) with correspondingly lower hardness. In the longitudinal section the broken finger showed (in the segregation area) more carbide and secondary banding in addition to non-metallic inclusions **(Fig. 4.24 and 4.25).** In contrast, the good steel had fewer inclusions and no carbide bands.

Inclusions and segregations of this type would hardly lead to failure under purely axial stress, but under the very high tangential tensile stress acting on these fingers, they apparently contributed measurably to fracture.

The lower carbon steel proved superior to the prescribed knife steel in spite of its inferior strength because of its lower tendency toward carbide segregation. The improved deformability associated with the lower strength probably enabled the steel to eliminate stress concentrations by slip and thus had a favorable effect. Therefore this steel should be more suitable for the highly stressed fingers than the high-strength knife steel. It is also more corrosion-resistant than the latter. In order to obtain a more favorable fiber orientation the fingers should be forged individually rather than machined out of bars.

Fig. 4.31 to 4.34. Rivets torn open during cold upsetting of heads due to peripheral blister formation

Fig. 4.31. View after magnetic particle inspection. 1 ×

Fig. 4.32. Steel bar with surface defect, etched with 50 % hydrochloric acid. 1 ×

Fig. 4.33. Transverse section through shaft of rivet. Etch according to Oberhoffer. 20 ×

Fig. 4.34. As Fig. 4.33. Etch: Nital. 100 ×

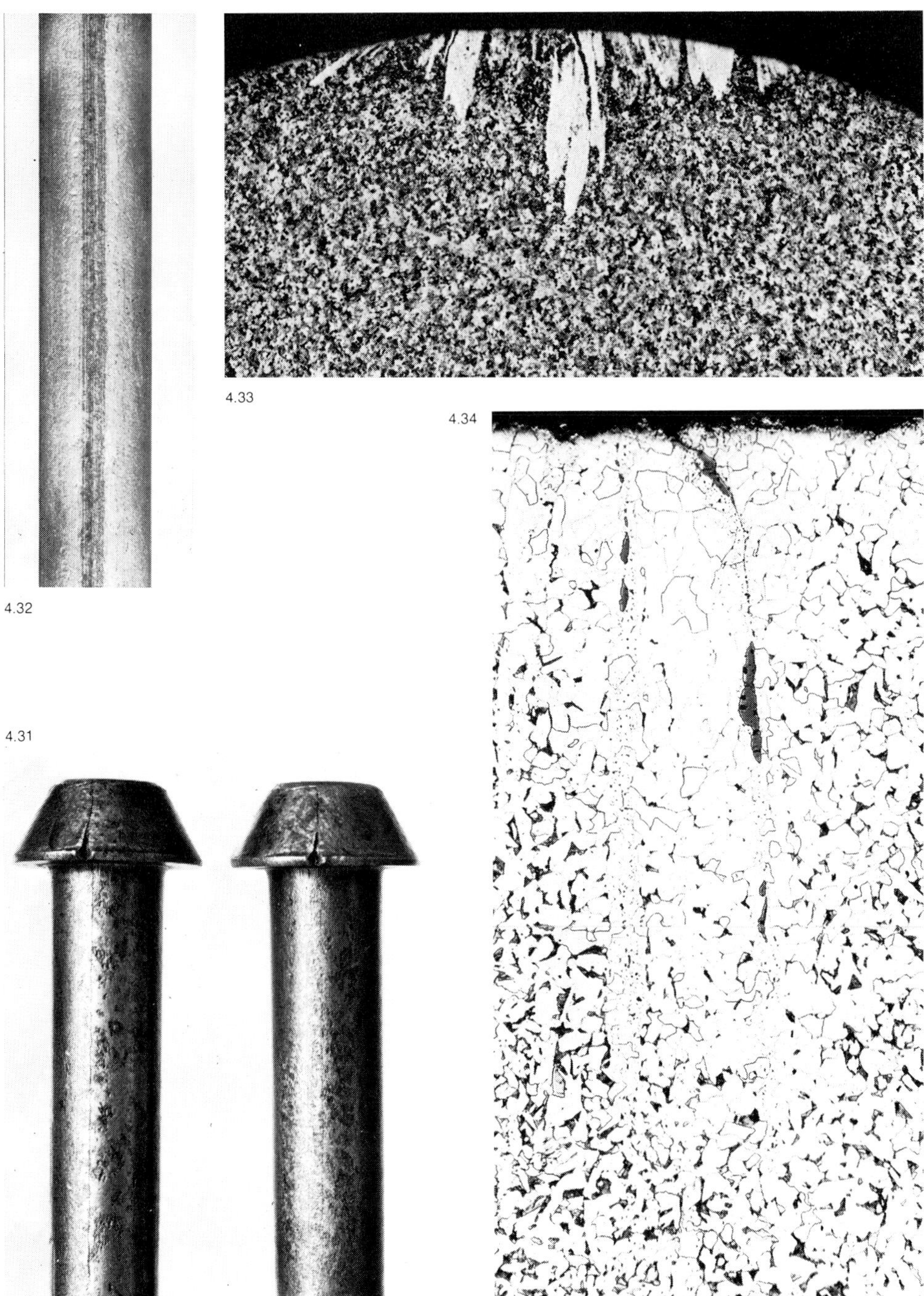

4.32

4.33

4.34

4.31

4.35

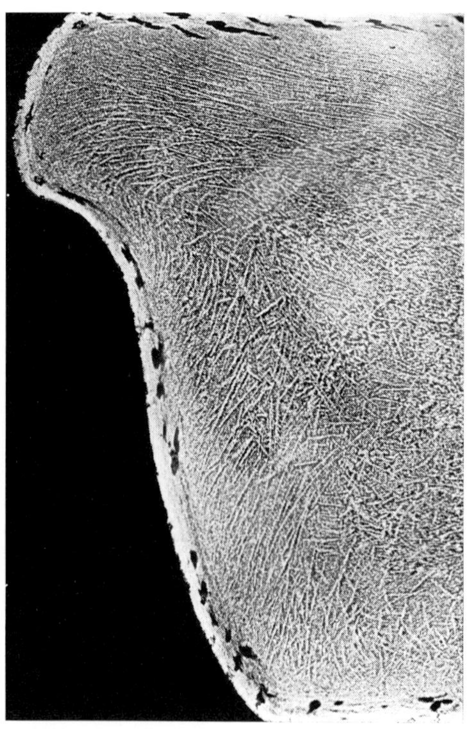

4.36

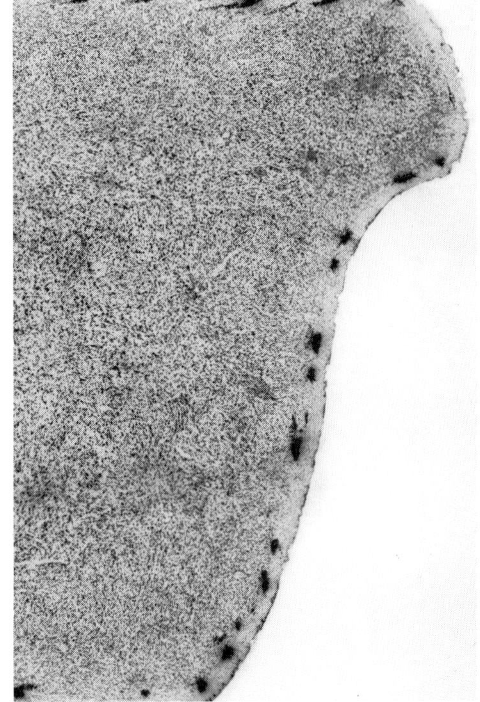

4.37

4.38

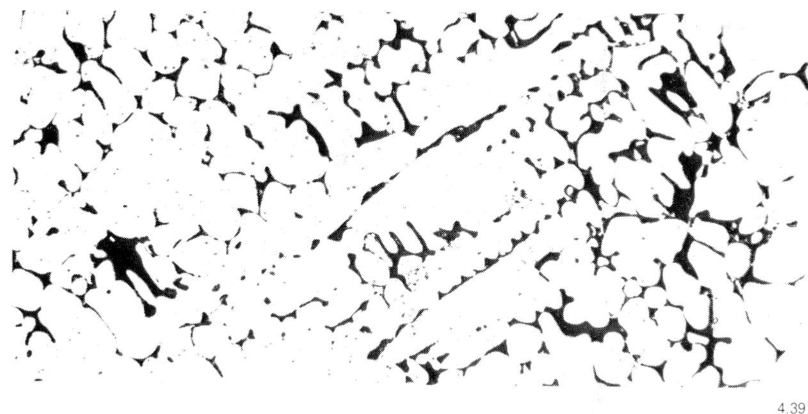

4.39

Fig. 4.38 and 4.39. Interdendritic voids in cast ingot disks of 80-20 Ni-Cr heating element alloy caused by evolution of gases during solidification

Fig. 4.38. Fracture of disk. 1 ×

Fig. 4.39. Microstructure of disk, unetched. 100 ×

Fig. 4.35 to 4.37. Surface defect caused by opened-up external blow holes in wheel rim blank. 1 ×

Fig. 4.35. Defective running surface

Fig. 4.36. Transverse section. Etch according to Oberhoffer. 0.67 ×

Fig. 4.37. Baumann sulfur print of transverse section. 0.67 ×

A special type of segregation is **gas bubble segregation.** Steel comes in contact with the gases from the atmosphere as well as with oxides and humidity in the raw materials during melting. Nitrogen from the air is in part soluble in liquid iron. Oxygen and oxides react with the carbon from the steel to form carbon monoxide gas. Water vapor (from the atmosphere, wet raw materials, and additions) reacts with iron to form hydrogen, which in turn dissolves in the iron. During the solidification of the steel, the solubility of gases changes in steps (Fig. 4.80); most of the gases precipitate and escape. The effect of hydrogen remaining in solution will be discussed later. The bubbles which can no longer rise to the surface because they are either too small or the temperature of the steel is too low, remain at first as gas-filled cavities in the solidified steel.

They may suck in residual melt enriched by impurities if the gas volume and the pressure in the interior decrease with decreasing temperature. This phenomenon is called gas bubble segregation[12]).

Rimmed steel, in which the reaction of oxygen with the carbon of the steel resulting in the formation of carbon monoxide still takes place in the mold, is particularly permeated by bubbles. If they are located far enough from the surface, which can be achieved by maintaining a suitable pouring temperature and rate, they are welded shut during rolling or forging without leaving any defects. Surface failures of the product may be caused by either the oxidation by bubbles below the surface during casting, or the oxidation by blisters that break open under the effect of segregation products during shaping.

Hydrogen blisters also appear in fully deoxidized and killed steels. The hydrogen may originate in the lacquer used for coating of the mold or in the wet molding sand.

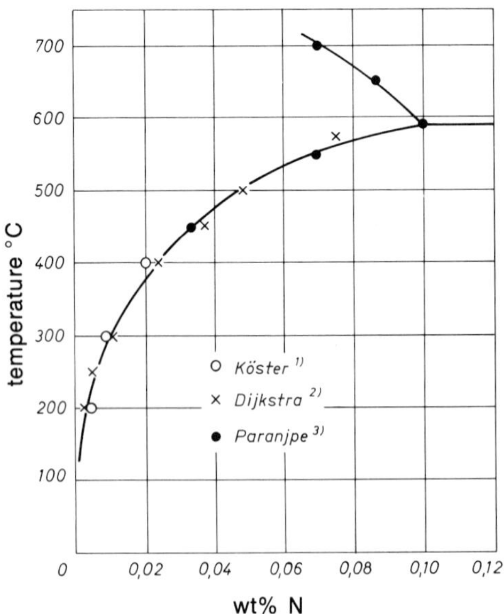

Fig. 4.40. Solubility of nitrogen in α-iron as function of temperature

1) W. Köster: Arch. Eisenhüttenwes. 3 (1929/30) S. 637/58
2) L. J. Dijkstra: J. Metals, Trans. 1 (1949) S. 252/60
3) V. G. Paranjpe u. a.: J. Metals, Trans. 2 (1950) S. 261/67

Rising gas bubbles that are filled with segregation products sometimes pull a train of segregation products with them. If such areas are cut during machining, they can be discerned as dark stripes during finish machining and are called "**shadow stripes**". **Figure 4.26** shows a longitudinal section etched according to Heyn, of a 95 mm thick wall of a **hollow cast steel cylinder.** During lapping of the outer surface these shadow stripes could be seen[13]). The segregated regions are attacked by copper ammonium cloride solution, thus are phosphorus-rich. During etching with Oberhoffer's reagent they remained free of attack[10]), as can be seen from **Fig. 4.27.** The Baumann prints according to **Fig. 4.28** shows that the blisters and their paths are also rich in sulfides. In the metallographic section of **Fig. 4.29** the ferrite of the blisters appears lightly etched by picral, therefore it is rich in phosphorus[10]), while the high sulfur contents is expressed by the presence of numerous manganese sulfide inclusions. Voids are also visible in the micrograph.

Since the steel was strongly deoxidized with silicon and aluminum, the gas forming the blisters could not have been carbon monoxide. It was probably hydrogen.

A **square ingot** of 570 mm per side made of rimmed screw steel showed longitudinal cracks at the edges upon arrival at the rolling mill. Melt analysis showed it contained 0.42 % Mn and 0.065 % P. Material analyses of the outer and core zone resulted in the data shown in **Table 2.** Considering the intended application, phosphorus and sulfur contents are very high, especially in the segregation zone. **Figure 4.30** is an illustration of a transverse section through an edge of the ingot. The crack follows an interdendritic segregation band of the transgranular zone which ends in each case in a gas blister. This anomalic segregation phenomenon is considered to be the failure cause. Probably a casting error was involved.

Table 2. Chemical Composition of a Screw Steel Ingot

	C %	Si %	Mn %	P %	S %
Outer zone	0.03	trace	0.37	0.046	0.042
Core	0.07	trace	0.38	0.100	0.110

Rivets made from a cast killed open hearth steel with a tensile strength of 330 to 410 MPa (48-60 ksi) cracked open during cold upsetting of the heads. Two rivets and a specimen of the steel bar from which they were made, were investigated. Both rivets showed an open crack at a location of the head. Magnetic particle inspection was used to make the straight longitudinal defects and the extensions of the cracks into the rivet shafts visible **(Fig. 4.31).** Smaller defects of the same type appeared after pickling with hydrochloric acid in other areas around the circumference of the rivets, as well as on the steel bar section **(Fig. 4.32).** Baumann replicas and Oberhoffer etching proved that there were external blowhole segregations on transverse sections of the rivet shafts and steel bar **(Fig. 4.33).** During upsetting of the heads, the bar tore open in such an external blowhole that had stretched during rolling. **Figure 4.34** shows a micrograph of such a defect. During etching for secondary structure development the defect could be mistaken for a rolling lap that had been decarburized and welded shut. But primary etching for phosphorus segregation presents an unequivocal finding of an external blowhole as the failure cause.

Figure 4.35 shows surface cracks on the bearing surface of **a wheel rim blank.** These cracks occurred when segregated external blowholes burst during rolling. The segregations could also be confirmed by primary etching **(Fig. 4.36)** and Baumann sulfur print **(Fig. 4.37).**

4.3 **Gases in Steel**

In connection with gas blister segregations we have already reported on the absorption of gases by liquid steel and the failures that may result during solidification of the steel by liberated gases, for instance by the formation of voids in the solid metal.

A special case is the following dealing with a porous cast ingot of the **heating element alloy** type NiCr 80 20. This alloy is particularly sensitive to blister formation because nickel dissolves large amounts of hydrogen. **Figure 4.38** shows the fracture of a transversely sectioned disk of this ingot. It has a dendritic fracture in the core as well as in the broad transgranular outer zones. The metallographic section of the ingot showed that the zone had numerous interdendri-

4.41

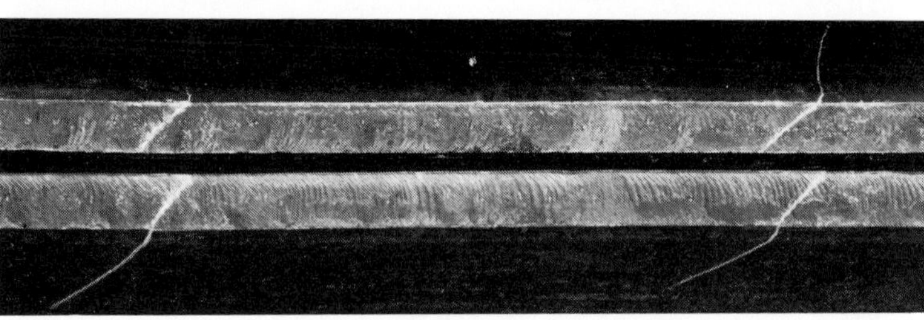

4.42

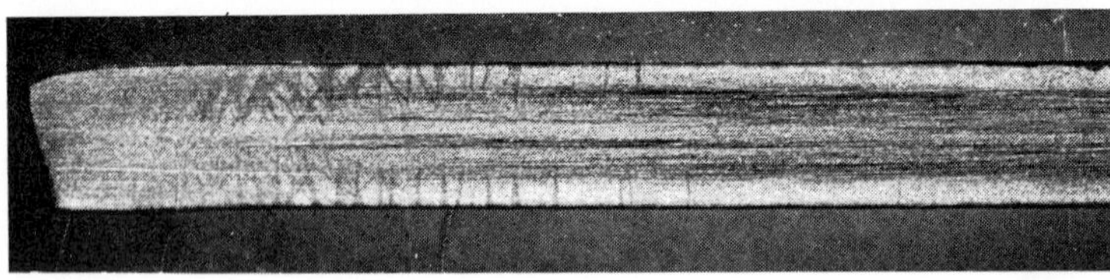

4.43

Fig. 4.41 to 4.45. 20 mm plates of rimmed basic converter steel trimmed by circular knife edge shears

Fig. 4.41. Edge crack in plate trimmed at 250 °C. Crack made visible by magnetic particle inspection

Fig. 4.42. Edge crack in plate trimmed at 150 °C after drop test. Cracks made visible by magnetic particle inspection

Fig. 4.43. Transverse section through cut edge (left). Etch according to Fry

Fig. 4.44. Plane section through edge crack. Etch: Nital. 100 ×

Fig. 4.45. Schematic presentation of temperature-dependent effects upon impact resistance of edge zone

tic voids in the vicinity of the surface **(Fig. 4.39).** Therefore it was improbable that these were shrinkage cavities. It was more to the point to interpret these voids as the consequence of gas liberation. Nickel used for melting of alloys may have absorbed large amounts of gas such as carbon monoxide or hydrogen during its production by electrolysis or by the carbonyl process. For the melting of such high nickel-containing alloys, degassed nickel (if necessary vacuum remelted) should be used.

The largest part of the dissolved gases in liquid iron escapes, as previously mentioned, during solidification. But a part also remains dissolved in the solid iron. In the following we will report on failures that were caused by dissolved gases or those precipitated from a solid solution.

The solubility of **oxygen** in solid iron is very low. Apart from the fact that the participation of oxygen in the aging of low carbon steel is still open to dispute, failures due to dissolved oxygen are unknown. The effect of bound oxygen in oxides or silicates was already described in section 4.1. But nitrogen and hydrogen are soluble up to considerable concentrations in gamma- and alpha-iron.

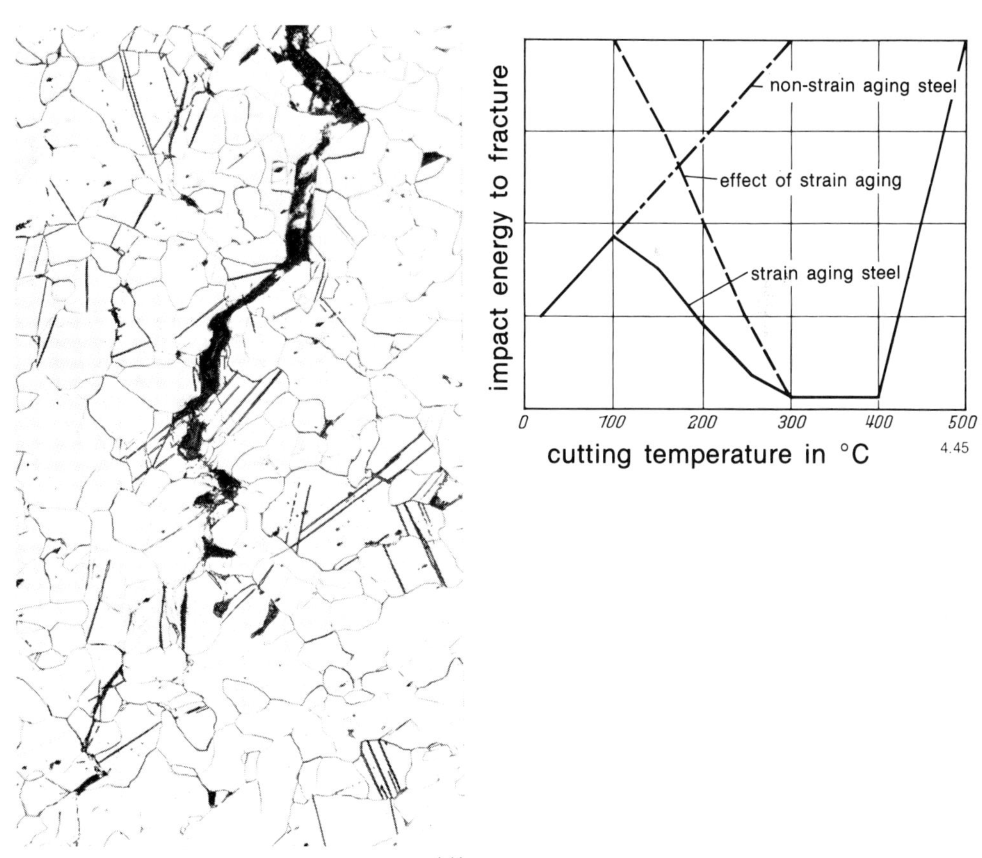

4.44

4.45

4.3.1 **Failures Caused by Nitrogen. Aging**

Nitrogen is soluble in alpha-iron up to 0.1 wt % at 590 °C. Its solubility decreases considerably with decreasing temperature just like that of carbon **(Fig. 4.40).** The phenomena connected with the precipitation of nitrogen and carbon from the lattice at low temperatures are known as aging[14]). They are promoted by deformation and lead to strengthening and embrittlement. Such aging may appear during deformation at temperatures of about 260 °C and is called **blue brittleness.** If the steel is cooled so fast that nitrogen and carbon remain in solution, these elements may also be precipitated during storage or even faster during tempering. These phenomena are known as natural or artificial **quench aging.** The strongest effect occurs during cold deformation and subsequent storage or tempering of steel that contains carbon or nitrogen in alpha solution. In this case it is designated as natural or artificial **strain aging.** It is assumed that carbon plays a larger role in quench aging, while nitrogen plays a larger part in strain aging. Cast low carbon and high nitrogen steels are particularly sensitive to the latter.

In the following a failure analysis is reported that occurred during working in the blue brittle range[15]). Edge cracks occurred on 20 mm thick **plates of structural steels** with an ultimate strength of 410, 460, and 550 MPa (60, 67, and 80 ksi), respectively, during a cold winter after trimming with circular knife edge shears. It could not be established when they had occurred. In order to determine the conditions and cause of their formation, plant tests were conducted in which cutting temperatures ranged from 65 to 250 °C. They are temperatures that were measured in the plant over a period of time. Since blue brittleness was assumed to be instrumental in the crack formation, plates made of basic converter steel were used for these tests. They showed edge cracks when trimmed at 220 to 250 °C **(Fig. 4.41).** The cracks initiated at the lower edge of the sheared face and had not yet propagated through the entire plate thickness. When opened, they showed a non-deformation type cleavage fracture. The sheets trimmed at 65 and 150 °C remained crack free. This indicates that the cracks may have occurred during the cutting operation. But it remained to be established whether the cracks had occurred afterwards during piling when the plates were bounced on each other. To test this assumption the plates trimmed at different temperatures were dropped flat from a 6 m high crane rail bed onto a pile located below it. This produced edge cracks also in the plates trimmed at 150 °C similar to those rolled

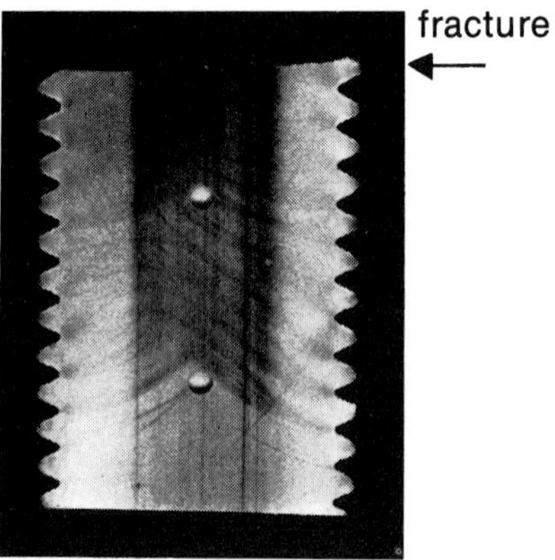

Fig. 4.46. Longitudinal section through screw bolt embrittled by strain aging and ruptured deformationless at overloading. Etch according to Fry. 1 ×

a

b

Fig. 4.47. Welded and strain aged mold support fractured when dropped from scaffolding a) approx. 0.1 × b) approx. 0.5×

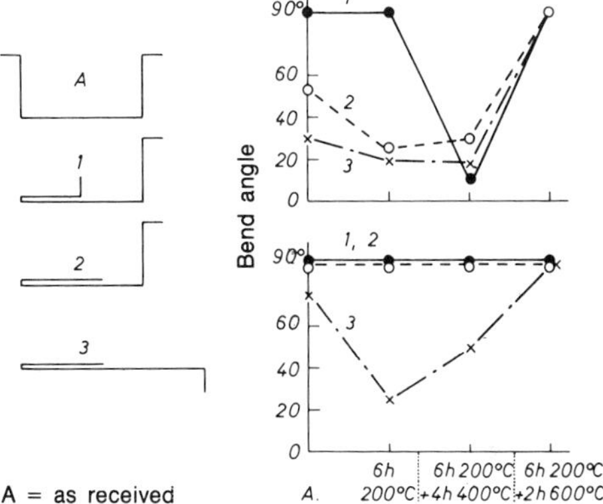

Fig. 4.48. Impact fold and bend test with aged cap profile Top: Specimen from fractured profile. Below: Specimen from unbroken profile

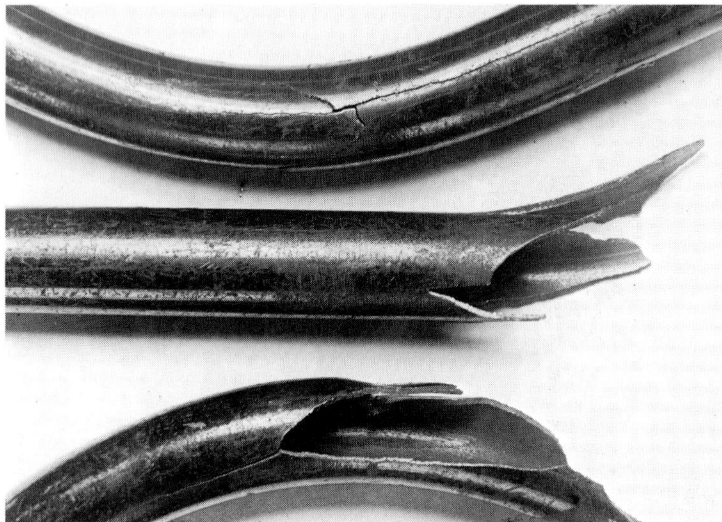

Fig. 4.49. Cold drawn pipes embrittled by strain aging during lacquering. About 0.5 ×

4.50

4.51

during the winter **(Fig. 4.42).** Only those trimmed below 65 °C, i.e. below the blue fracture range, remained crack-free. The results therefore proved that crack formation occurred only in sheets that were trimmed in the brittle temperature range; the cracks formed either directly upon cutting or later during piling. Plates made of open hearth steel were not subject to cracking in these tests.

Hardness and tensile tests showed that the cut edge and especially its lower part were severely hardened and that the outer zone contained a complicated system of high multi-axial stresses. In sheets made of basic converter steel, slip lines were developed in the bent zone in metallographic sections perpendicular to the edge by etching according to Fry **(Fig. 4.43).** Deformation twins of the C type according to A. Kochendörfer and H. Scholl[16]) appeared in metallographic sections across the cracks **(Fig. 4.44),** i.e. parallel to the direction of rolling. These twins appeared in addition to the cracks and were an indication of multi-axial stresses.

It was confirmed by impact bending tests on specimens of 30 mm width, machined along the cutting edges, that this zone was embrittled **(Fig. 4.45).** In open hearth steel plates, tested in the same manner, brittleness decreases with rising cutting temperature due to progressive stress relief. In the basic converter sheets this tendency was reversed at cutting temperatures above 100 °C, i.e. at the beginning of the brittle temperature range, due to aging. Impact energy first

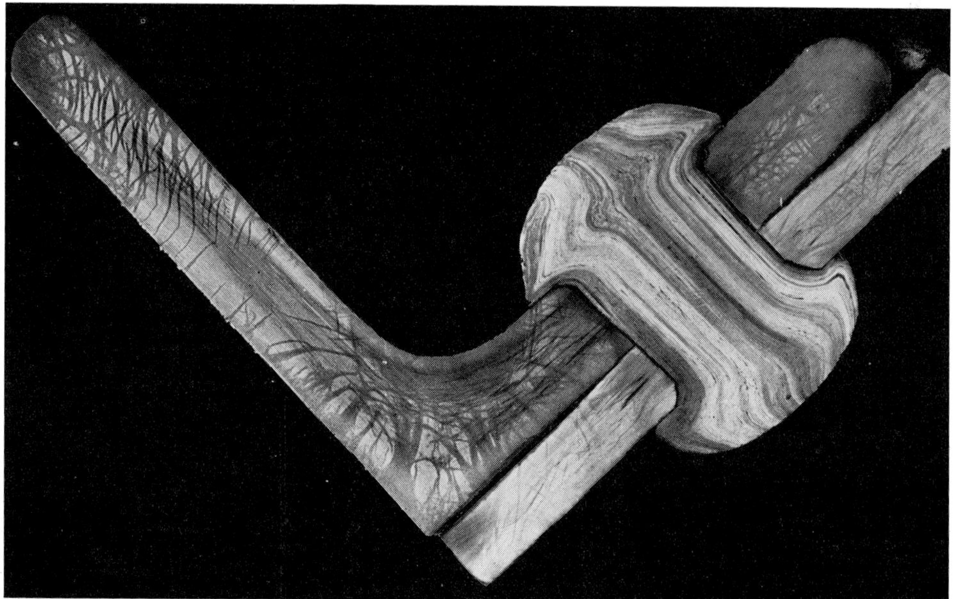

Fig. 4.52. Section through riveted connection in upper top girder joint of an old bridge made of basic converter steel, tempered for 5 h at 200 °C and etched according to Fry. 1 ×

Fig. 4.50 and 4.51. Pipe of basic converter steel embrittled by strain aging during galvanizing and fragmented during thread cutting

Fig. 4.50. Fragments. 1 ×

Fig. 4.51. Fracture. 1 ×

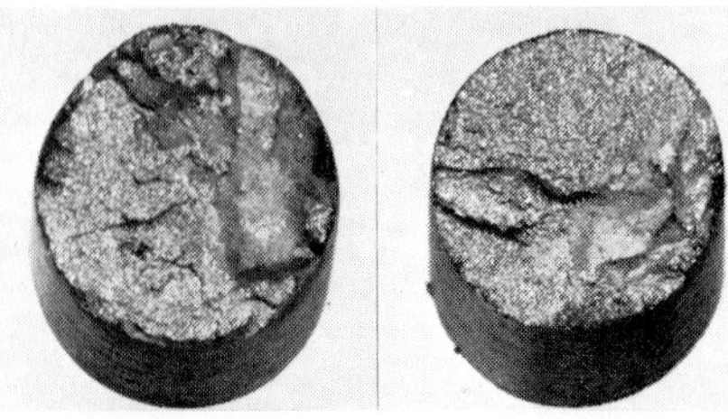

4.53

4.54

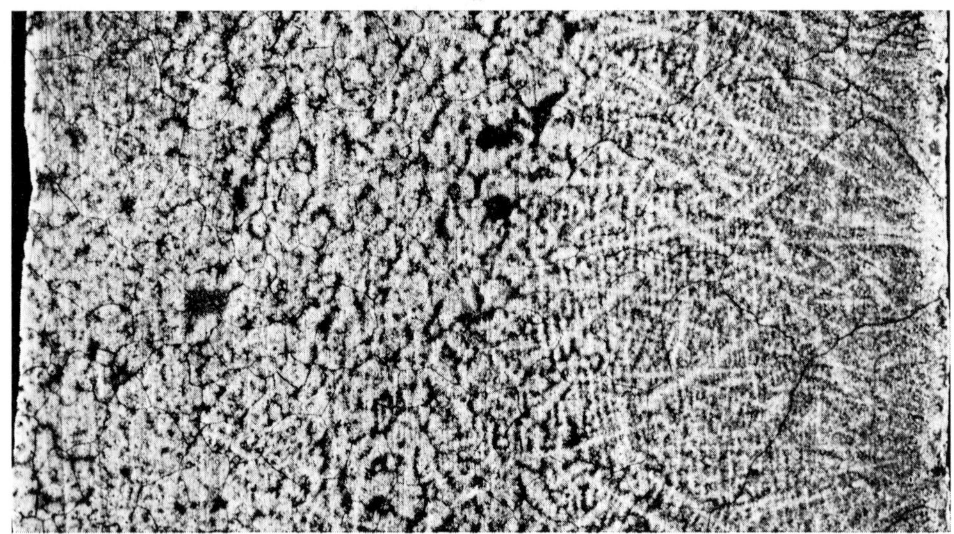

4.55

Fig. 4.53 to 4.59. Armature yoke of cast medium-carbon steel with insufficient strength and elongation

Fig. 4.53. Low deformation fracture, partly "rock candy"-like, of tensile specimen.1 ×

Fig. 4.54. Intergranular fracture of flange disk. 0.7 ×

Fig. 4.55. Section parallel to fracture of flange disk according to Fig. 4.54. Etch: hot 50 % hydrochloric acid. 2.5 ×

Fig. 4.56. Precipitates on twin grain boundaries of former austenite. Etch: Picral. 420 ×

Fig. 4.57. AlN precipitates seen in transmission lacquer replica. 40,000 ×

Fig. 4.58. AlN platelets in carbon film stripped off fracture according to Fig. 4.54. 20,000 ×

Fig. 4.59. Annealing tests with specimens of flange disk. Fracture of specimens. 1 ×

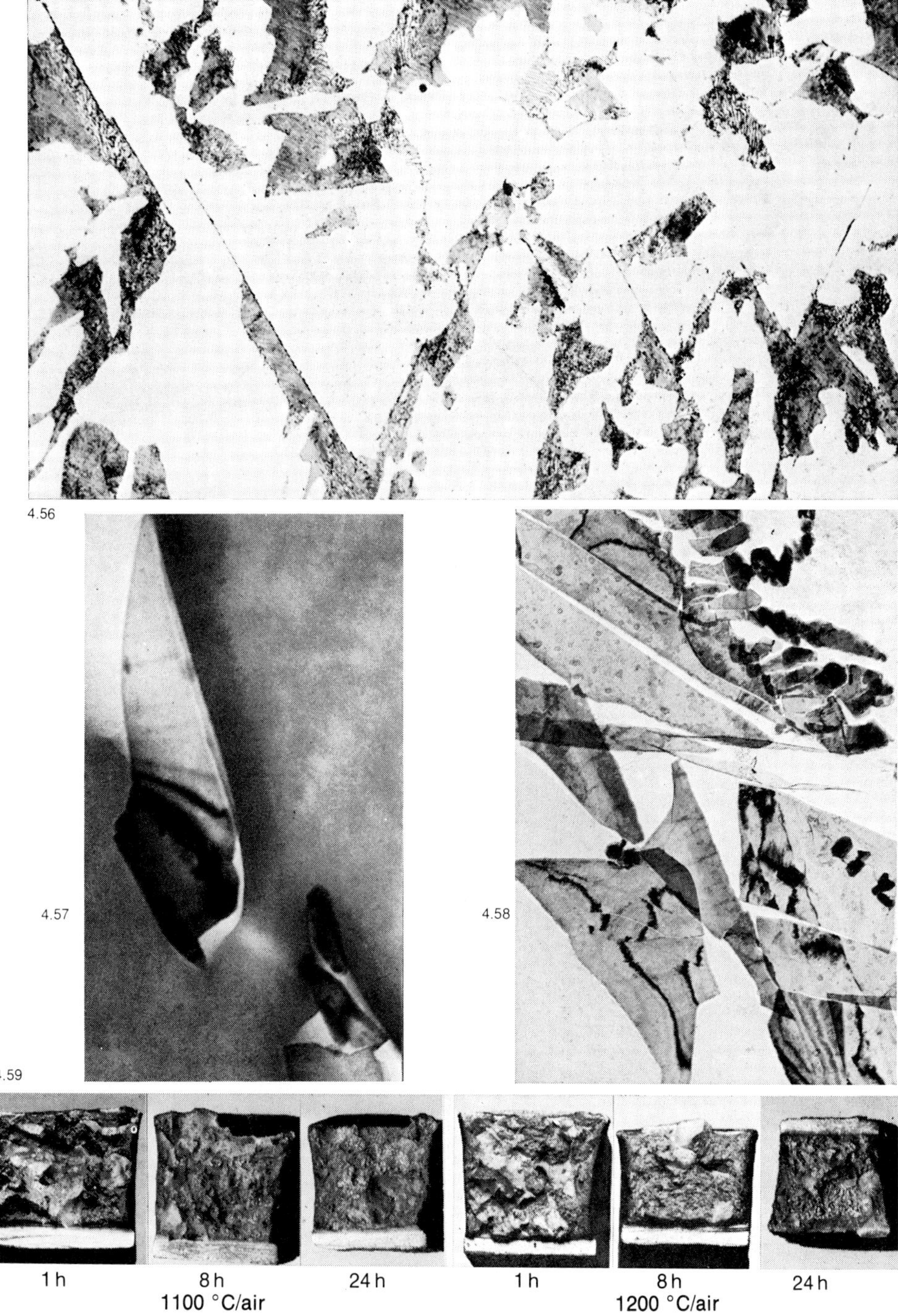
4.56
4.57
4.58
4.59
1 h
8 h
24 h
1100 °C/air
1 h
8 h
24 h
1200 °C/air

sank below the starting value; it was at a minimum of 300 to 400 °C and rose steeply between 400 and 500 °C. Embrittlement also occurred if the plates were cold trimmed and the specimens were subsequently tempered to 200 to 400 °C.

Trimming in the blue brittle range should therefore be avoided. Further laboratory tests showed that the tendency for brittle fracture of the edge zone could be reduced by minor broadening of the cutting gap and a lowering of the rake angle of the knives.

Examples of failures by overloading of parts that were embrittled by strain aging were already discussed in sections 3.2 and 3.3 (Fig. 3.26 to 3.29 and 3.56 to 3.58). In the first example a suspension bar of a concrete structure was discussed that had been made brittle by impact loading. The second example dealt with a bolt whose rupture in a cold rolled thread was facilitated by strain aging. A similar failure occurred on three **screw bolts** from a flange of

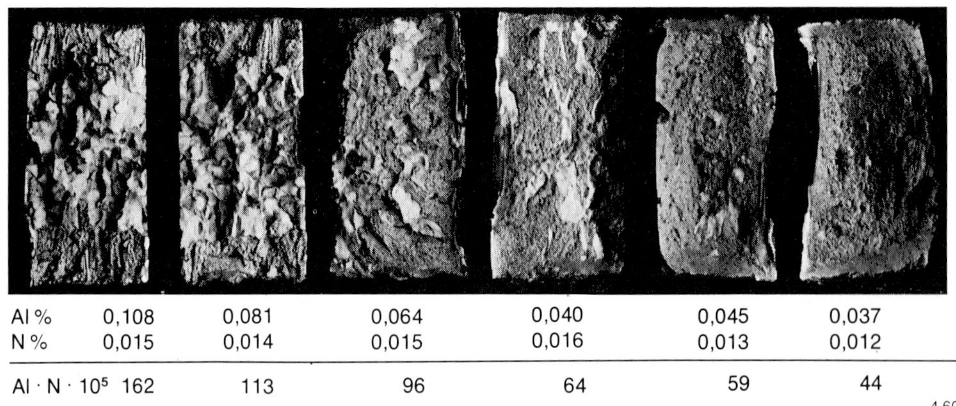

Al %	0,108	0,081	0,064	0,040	0,045	0,037
N %	0,015	0,014	0,015	0,016	0,013	0,012
$Al \cdot N \cdot 10^5$	162	113	96	64	59	44

4.60

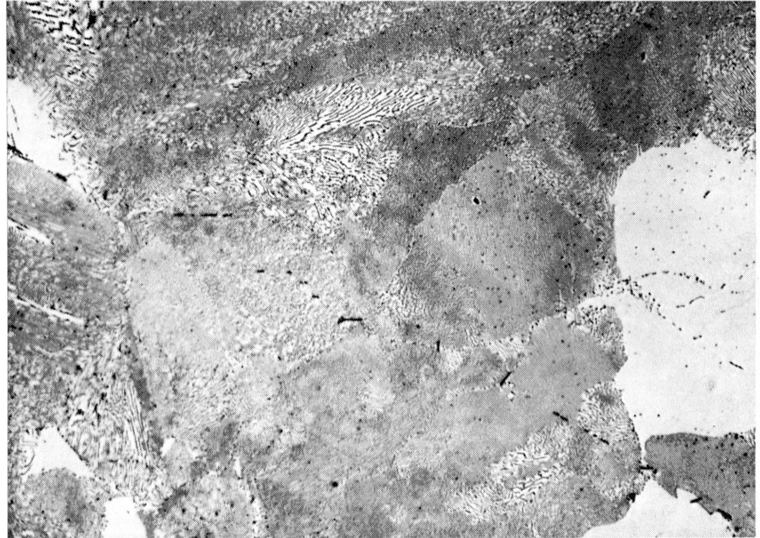

4.61

a rotary kiln furnace in a chemical plant. The deformation-deficient cleavage fractures were located as usual in the first supporting thread. The rimmed steel contained 0.07 % C, 0.01 % Si, 0.46 % Mn, 0.084 % P, 0.001 % Al and 0.030 % N. The phosphorus content was therefore high, nitrogen content still higher, which was unusual for a basic converter steel, even of the old type. A strong tendency for strain aging had to be expected of a steel having this composition, and additionally the somewhat elevated temperature at the furnace flange would have promoted this tendency. That strain aging had indeed occurred was confirmed by longitudinal sections etched according to Fry **(Fig. 4.46).** Force induced flow lines propagated from the root of the thread. This lead to the conclusion that the bolts were overtorqued during tightening.

Mold supports for a concrete structure fractured often when dropped from the scaffolding during construction. They consisted of cap profiles that were cold pressed from a sheet of rimmed basic converter structural steel and were reinforced by a welded-on lift structure of round rods. As can be seen from **Fig. 4.47,** the cracks originated at the welded joints and run mostly along the bent edges of the supports. Analysis of phosphorus and nitrogen in three cracked supports and a profile from a shipment that up to that time had been free of defects, showed the data cited in **Table 3.** They confirmed the suspicion that a phosphorus- and nitrogen-rich steel had been used for the supports. The steel of the defect-free shipment was distinctly purer. Metallographic examination and Baumann replicas indicated that carbon and sulfur content were much lower in the comparison profile than in the cracked supports.

Table 3. Phosphorus and Nitrogen Contents of Cap Profiles

	P %	N %
Fractured Profile 1	0.098	0.018
Fractured Profile 2	0.078	0.015
Fractured Profile 3	0.058	0.011
Profile of good quality shipment	0.036	0.008

In order to test the toughness and aging tendency of the strip material, narrow 30 mm wide sections were cut out of the profiles, and were folded or bent open in a vise by hammer blows **(Fig. 4.48).** In order to reinforce the aging as is the case in the vicinity of the welds, one specimen of each was tempered at 200 and 400 °C, respectively. To neutralize this effect, aged specimens were in part subsequently annealed at 600 °C. As can be seen from Fig. 4.48, the specimens of the fractured profile withstood the 90°-bend folding quite well without cracking in the as-delivered state and after aging at 200 °C (test No. 1), and only became brittle after tempering at 400 °C. But during bending open (tests 2 and 3) they broke prematurely. After annealing at 600 °C all specimens could be bent open or shut 90° without cracking. The specimens of the good quality profile fared better in all respects, even though this steel also tended to strain age. The fractures of the mold supports therefore can be ascribed to aging embrittlement as had been expected.

This test confirmed that information may be gained at times even by simple means, as was mentioned in the introductory chapter.

Fig. 4.60 and 4.61. Pinion gears of cast chromium-molybdenum steel with insufficient elongation

Fig. 4.60. Fractures of heat treated test bars.1 ×

Fig. 4.61. Microstructure of normalized test bar with 0.108 % Al and 0.015 % N. Etch: Picral. 500 ×

4.62

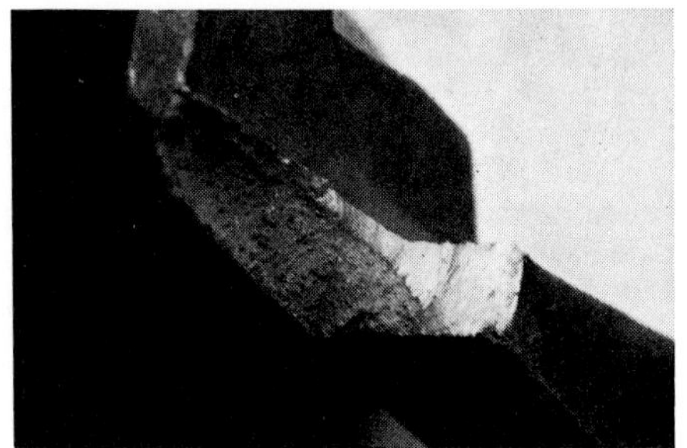
4.63

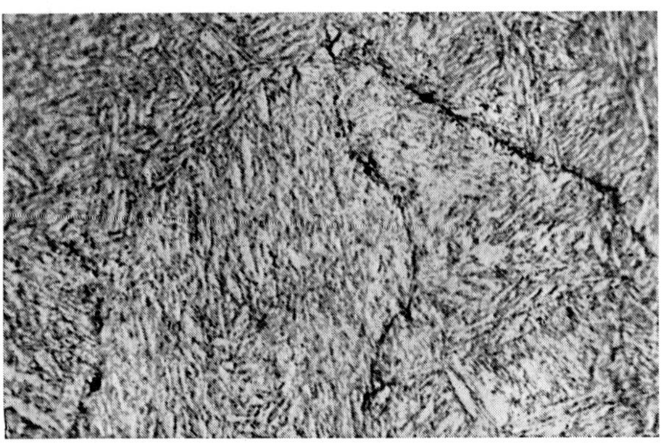
4.65

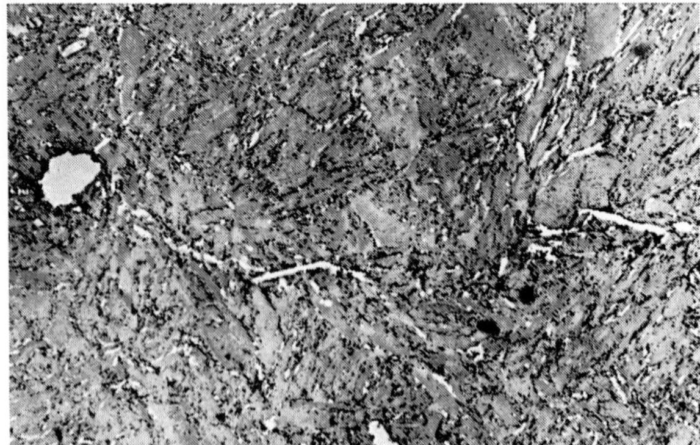
4.66

If cold worked steel parts sensitive to strain aging are later lacquered or hot-galvanized for corrosion protection, there is always the danger of embrittlement due to such strain aging. **Automobile tire rims** that were cold drawn from soft annealed iron bands, bent into rings and brazed, had good deformability but were completely embrittled after lacquering of the inside. This should be done at a temperature between 200 and 300 °C because the outside became stained a straw yellow and cornflower blue color. This embrittlement could be eliminated by annealing for 1 h at 500 °C.

The **cold drawn pipes** shown in **Fig. 4.49** were also made so brittle during lacquering that they split under light hammer blows.

Hot galvanizing pipe of 20 mm O.D. and 2.2 mm wall thickness, made of rimmed basic converter steel with 0.082 % P and 0.014 % N, collapsed during thread cutting **(Fig. 4.50)** or later proved brittle during storage. The fractures had a coarse-grained recrystallized outer zone and weak deformation marks **(Fig. 4.51).** Crack-free pipe sections failed during transverse folding tests after little plastic deformation. In order to answer the question whether embrittlement was due primarily to the coarse grained recrystallization structure or to strain aging, pipe sections were annealed for 2 h at 450 °C. Afterwards they could be completely folded before the cracks started

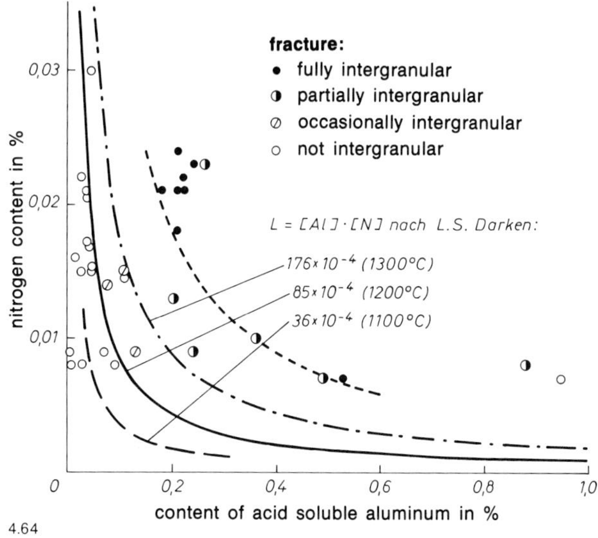

Fig. 4.62 to 4.66. Heat treated precision castings of chromium-molybdenum steel

Fig. 4.62. "Rock candy" fracture of part that broke after a few impact blows. 3 ×

Fig. 4.63. Ductile fracture of a part purposely broken after 10,000 blows. 3 ×

Fig. 4.64. Relation between tendency to intergranular fracture and aluminum and nitrogen contents in chromium-molybdenum precision steel castings

Fig. 4.65 and 4.66 Microstructure of precision casting with intergranular fracture. Etch: Picral

Fig. 4.65. Optical microscope. 1000 ×

Fig. 4.66. Electron microscope. 5000 ×

4.67

4.68

Fig. 4.67 to 4.72. Primary grain boundary precipitate induced ruptures during forging of 13 ton ingot of manganese-molybdenum steel

Fig. 4.67. View

Fig. 4.68. Longitudinal section in 100 mm depth parallel to one plane of octagon. Etch: according to Heyn. 1 ×

Fig. 4.69. Grain boundary under polarized light. 1200 ×

Fig. 4.70. and 4.71 Examination of precipitates by electron microprobe

Fig. 4.70. Electron diffraction pattern

Fig. 4.71. Aluminum K_α-radiation

Fig. 4.72. Precipitates in carbon film stripped off fracture. 5000 ×

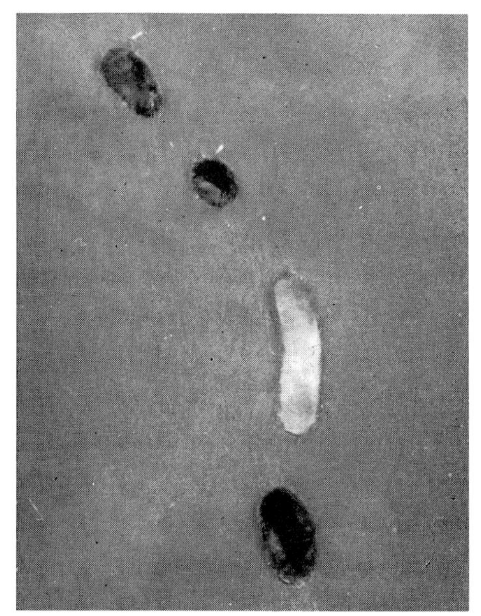

4.69

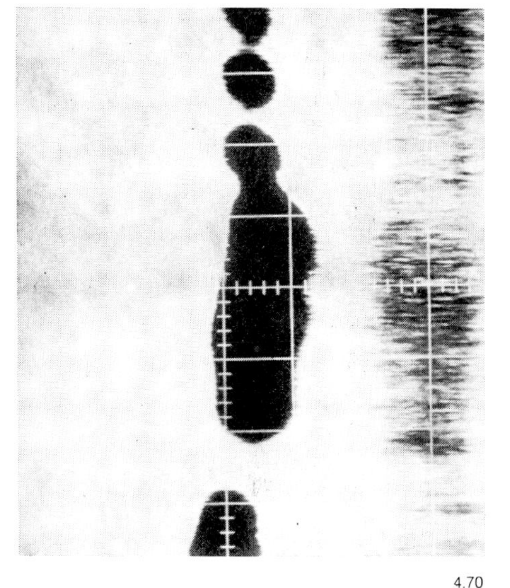

4.70

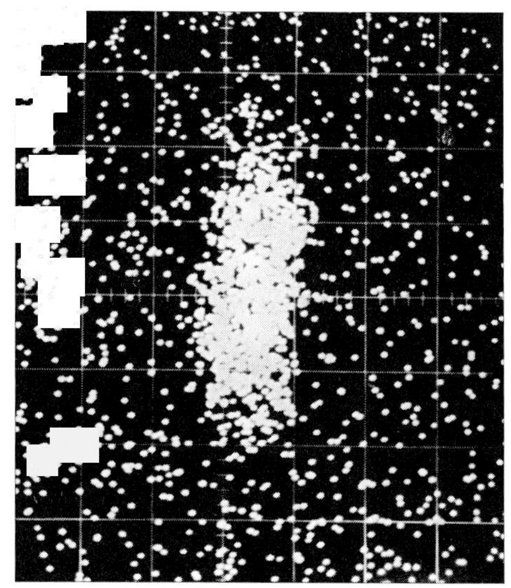

4.71

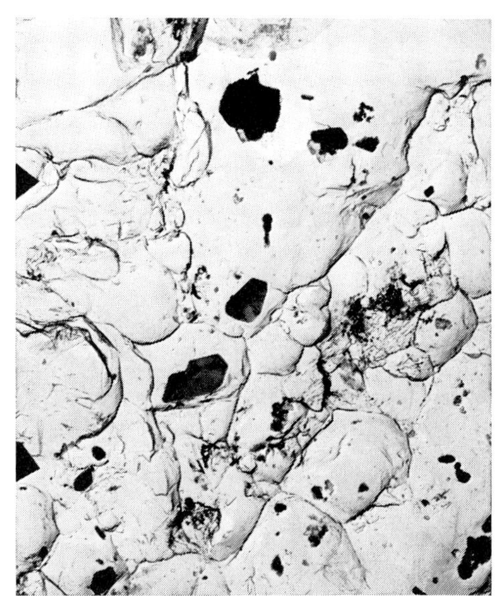

4.72

in the coarse grained outer zone. Embrittlement therefore must have been primarily caused by strain aging. Recrystallization that must have occurred during processing prior to galvanizing merely helped to enhance embrittlement.

As has been shown, strain aging may be eliminated if necessary by heating to 450 to 600 °C without any considerable effect upon strength. However under normal circumstances an age-resistant steel should be used whenever possible where processing methods and plant conditions may lead to embrittlement of the steel through strain aging. Such steels have been available since the early 20's and were first initiated by Fry and developed by Krupp (Izett).

Since that time failures as a result of strain aging have become rarer. If they still exist, as has been shown by the example above, this is due to the fact that many parts date back to that period and are still in use or because an unsuitable steel has been used due to ignorance.

In the year 1931, thirty-five years after it was constructed, the Baaken Bridge in Hamburg was taken down. It is said to have been the first German bridge built of mild steel. **Figure 4.52** shows a section through a riveted connection in an upper top girder joint of one of the principal supports. Angles and joints consisted of basic converter steel with 0,03 % C, trace silicion, 0.25 % Mn, 0.052 % P and 0.062 % S. The rivet showed a typical welded steel structure. Angles and sheets were probably cold worked during straightening and assembly, as was confirmed by Fry etching. The specimen section was heated for 5 h at 200 °C in order to make the region deformed by aging more visible. The angle steel after disassembly had a yield point of approximately 270 MPa (40 ksi), a tensile strength of 380 MPa (55 ksi), an elongation of 28 %, and a reduction of area of 70 % (mean values). This corresponded almost exactly to the acceptance specifications on delivery. An evaluation by the testing personnel considered these properties very satisfactory and stated that the taking down of the bridge had really been unnecessary. No failure was involved. But this analysis may serve as an example for cases in which defects do not lead to failure.

4.3.2 **Intergranular Fracture, Primary Grain Boundary Cracks**

In connection with failures caused by nitrogen, another type should be mentioned, namely the intergranular or primary grain boundary fracture. It is caused by the precipitation of aluminum nitride, as well as of other hard-to-dissolve phases such as oxides, carbides, and sulfides. This type of failure is principally seen in steel castings but also occasionally in cast ingots. It may affect forgeability. After working, a fibrous fracture plane remains. Intergranular fractures cause failures but can also occasionally occur during tensile, bend, and impact tests for quality control. They cause reduction in strength and a considerable loss of toughness. Heats made in the basic electric furnace are particularly susceptible to these types of fractures.

Figure 4.53 shows the tensile specimens of an **armature yoke** made of a medium carbon steel casting with approx. 0.25 % C[18]). They had fractured without noticeable plastic deformation as can be seen from the unchanged fracture cross section. The fracture is in part coarse-grained and intergranular. **Figure 4.54** shows a forcibly ruptured disk from the flange of a yoke. Its fracture is almost entirely intergranular. During pickling of a metallographic section of this disk, a coarse network was etched out; this apparently corresponded to the grain boundaries of the

Fig. 4.73 and 4.74. Metallographic surface of precision castings loaded cathodically with hydrogen, having intergranular fracture. Etched lightly with picral and photographed under alcohol. 100 ×

Fig. 4.73 Evolution of hydrogen from primary grain boundary loaded with AlN precipitates

Fig. 4.74. Primary grain boundaries burst open by hydrogen evolution

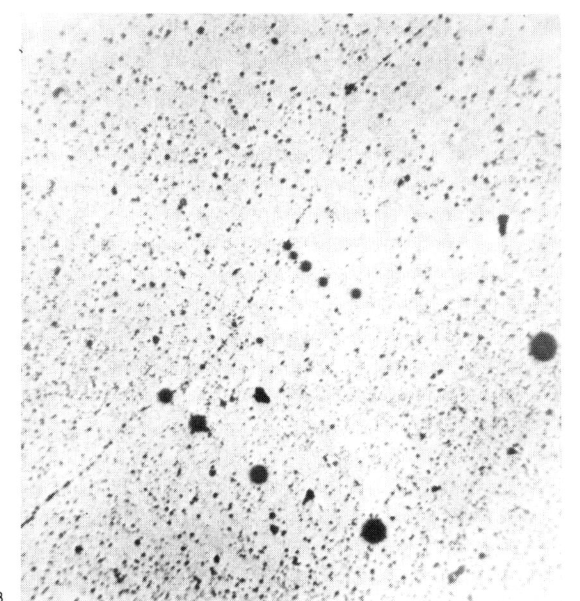

4.73

4.74

4.75

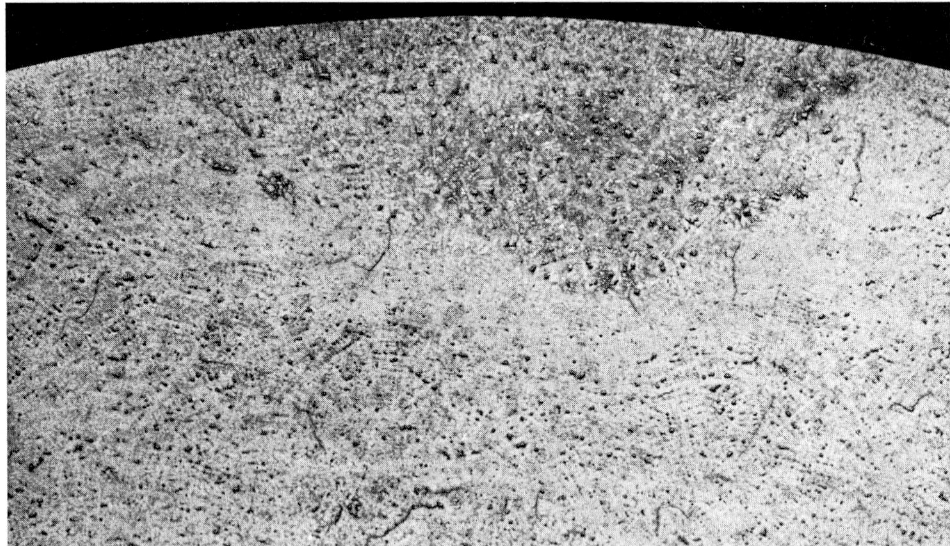

4.76

Fig. 4.75 to 4.78 Broken back-up roll of cast chromium-molybdenum steel

Fig. 4.75. Fracture view. Near surface: Fine-grained fracture with individual brightly shining primary grain boundary cracks. Interior: Deformation fracture with conchoidal areas (dull) and primary grain boundary cracks (shiny)

Fig. 4.76. Transverse section parallel to fracture with primary grain boundary crack and top crust. Etch according to Heyn. 1 ×

Fig. 4.77. Microstructure of rolls. Etch: Picral. 200 ×. Ledeburite and carbide precipitates on primary grain boundaries

Fig. 4.78. Internal crack (stained dark) caused by fast, not penetrating heating of back-up roll. Approx. 0.2 ×

4.77

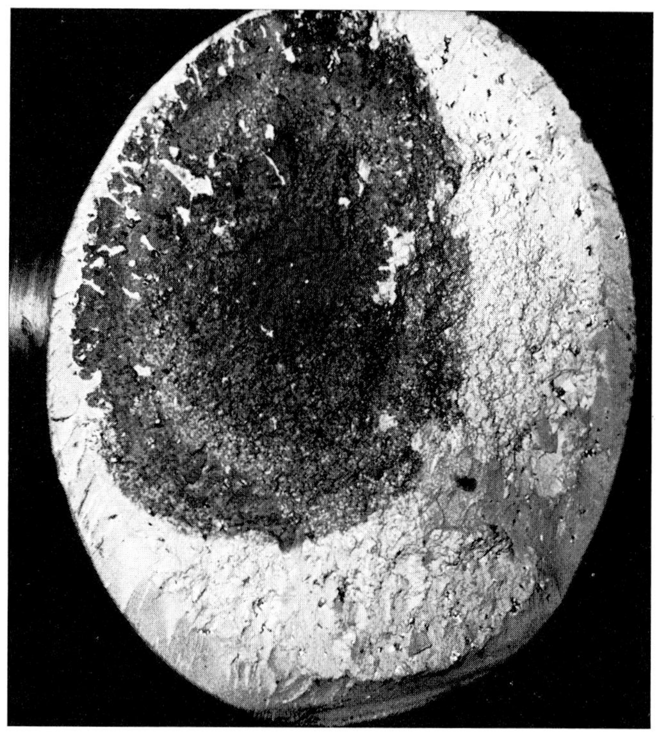
4.78

primary structure, although in the transgranular outer zone no such relationship could be found **(Fig. 4.55).** In the micrograph rod- or platelet-like precipitates could be seen on the primary grain boundaries or on the grain and twin boundaries of the austenite formed after solidification **(Fig. 4.56). Figure 4.57** shows the same in a lacquer replica of a metallographic surface, and **Fig. 4.58** shows a carbon film that had been vapor deposited onto the fracture surface and then stripped from it. In the electron diffraction pattern, the hexagonal aluminum nitride, AlN, could be seen. Chemical analysis of the steel melted in a basic arc furnace showed 0.115 % Al and 0.014 % N.

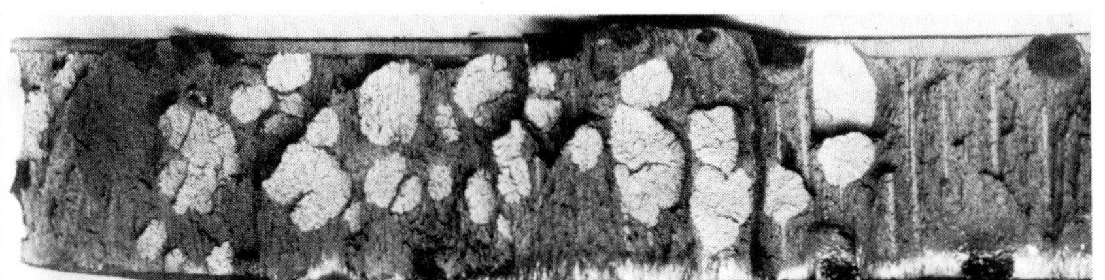

Fig. 4.79. Flakes in fracture of forging of chromium-nickel steel. Approx. 0.5 ×

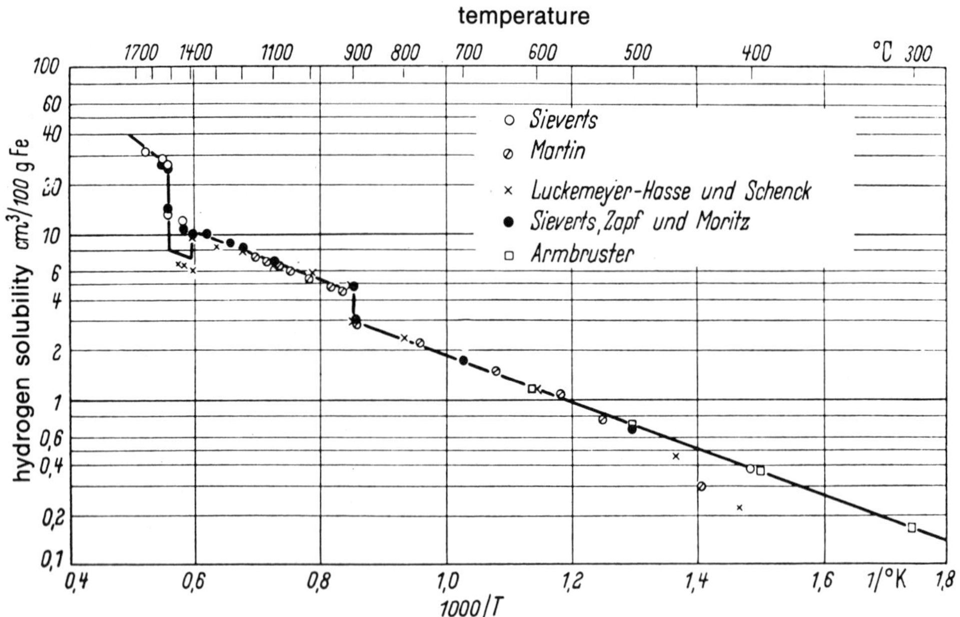

Fig. 4.80. Hydrogen solubility in iron as a function of temperature at atmospheric pressure [according to E. Martin, **Arch. Eisenhüttenw.** Vol. 3, 1929/30, p 407-416; L. Luckemeyer-Hasse and H. Schenck, **Arch. Eisenhüttenw.,** Vol. 6, 1932/33, p 209-214; A. Sieverts, G. Zapf and H. Moritz, **Z. Phys. Chem.,** Vol. 183, 1938/39, p 19-37 with an evaluation of older data of A. Sieverts; M. Armbruster **J. Amer. Chem. Soc.,** Vol. 65, 1943, p 1043, 1054]

Taken from **Die Edelstähle,** F. Rapatz, 5th Ed., Springer Verlag, Berlin, 1962

The melt therefore was heavily deoxidized by aluminum. The high nitrogen content is characteristic for electric arc furnace steel. After reducing the aluminum additions by half, no more rejects occurred due to insufficiently high tensile or bend strength values. An attempt to eliminate the intergranular fracture by solution or diffusion annealing was essentially unsuccessful. It did reduce part of the intergranular fracture, but even after 24 h annealing at 1200 °C, small amounts of it remained, as can be seen from **Fig. 4.59.**

In another foundry some cast steel **pinion gear shafts**[19]) were found to be defective because the specified elongation of 10 % could not always be obtained due to intergranular fractures in the tensile specimens. The shafts were made of cast chromium-molybdenum steel with approximately 1.2 % Cr and 0.3 % Mo. They were melted in a basic electric arc furnace and heat treated to 880 MPa (128 ksi) ultimate tensile strength. In order to test any effect the feed charges had, six normalized specimens were examined from an equal number of melts. In particular, aluminum and nitrogen contents were determined. The data are listed in **Fig. 4.60.** The bar with the highest aluminum content showed the microstructure illustrated in **Fig. 4.61.** It had pronounced precipitates on the primary grain boundaries or on the previous austenite boundaries. Before breaking, all bars were heat treated like the pinion gear shafts. This was advantageous because fracture in the tough state occurs more reliably along the grain boundaries weakened by precipitates than in a steel of inferior deformation capacity. Figure 4.60 shows the fractures. The specimens are arranged in such a way that the portion of intergranular fractures decreases from left to right. A comparison with the analysis data below shows that the tendency for intergranular fracture decreases in the same sense as does the aluminum contents of the steel, or more exactly, as the product Al.N.

When intergranular fracture occurred to a large extent in small **precision steel castings** containing approx. 0.4 % C, 1 % Cr and 0.2 % Mo, the opportunity arose for a systematic analysis to be made of its cause and conditions of formation[17])[20]). The parts were deoxidized with a loosely controlled amount of aluminum in order to obtain a fine grain and were cast in preheated dies (800 °C) and oil quenched to 42 ± 2 HRC. During impact type stressing some parts broke after a few blows, while others withstood more than 10,000 impacts without rupturing. The fracture of the prematurely broken pieces had a "rock candy" appearance **(Fig. 4.62),** while that of the good quality parts, that were purposely broken, was of a ductile nature **(Fig. 4.63).** Nitrogen content was very high throughout with 0.14 % to 0.23 %, and aluminum content varied between 0.031 and 0.27 %. The parts with the intergranular fracture had a high aluminum content, while this was low for those with deformation fractures. The dividing line between the two groups was at about 0.10 % Al.

In order to establish whether this line would shift to higher values with lower nitrogen content, and if so, by how much, a number of castings were made without scrap additions. They contained 0.007 to 0.13 % N. In these cases the aluminum additions were increased to 0.5 and 1 %. **Figure 4.64** shows the results of all tests. It is apparent that aluminum additions may be higher for lower nitrogen contents. In addition, the boundaries between the fields of equal brittle fracture tendencies can be presented by curves of the constant product Al.N. The borderline curve that characterizes the beginning of intergranular fracture coincided under these conditions with the solubility product at 1200 °C, according to L.S. Darken[21]). In this connection the effect of solid solution supersaturation upon precipitation tendency is expressed. The solubility isotherms show that the castings with intergranular fractures have to be solution annealed at a temperature of at least 1200 °C if the intergranular fracture is to disappear. But even this temperature – which is too high for practical application – is insufficient in most cases, as can be seen from the annealing tests given in Fig. 4.59. These results further confirm the old experience gained from nitrided steels, i.e. that the tendency for intergranular fracture disappears again at high aluminum contents. The nucleation of aluminum nitride precipitates probably is accelerated to such an extent that diffusion at grain and twin boundaries becomes superfluous.

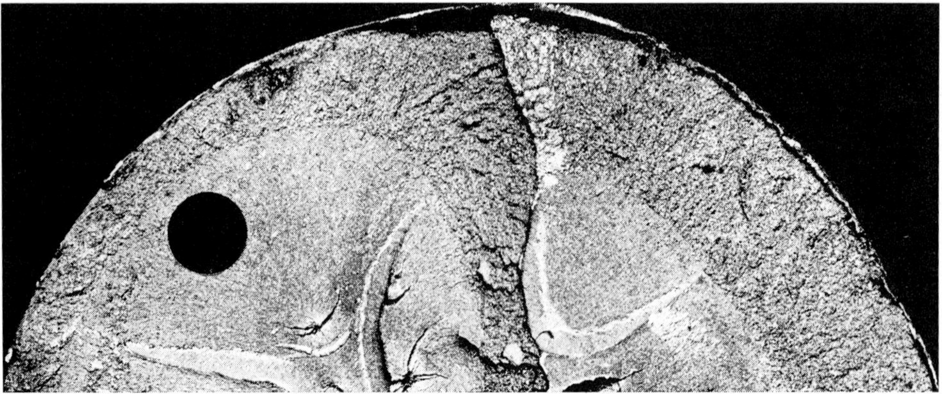

4.81

4.82

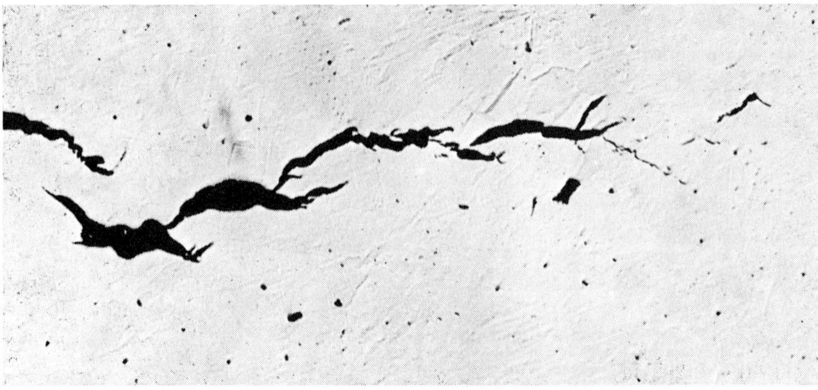

4.83

It should be noted that the limiting values found here for a tendency to intergranular fracture are applicable only to the conditions inherent in that particular investigation. They include small castings and high preheating of the molds and a strong dependence of the cooling rate after solidification because of sluggish precipitation of aluminum nitride.

Figure 4.65 illustrates the very fine precipitates in the heat treated microstructure of a precision casting. They are best made visible by a light etch with picral. **Figure 4.66** reproduces them in an electron micrograph made in from a resin replica. Their structure could not be determined but there is little doubt that the precipitates consisted of aluminum nitride.

But other precipitates that are difficult to dissolve such as carbides, oxides, and sulfides may also cause intergranular fractures[17)] [22)] [23)].

Hot working capacity, too, may be affected by hard-to-dissolve precipitates on the primary grain boundaries. **Figure 4.67** shows, for instance, a 13 metric ton **octagonal ingot** made of a manganese-molybdenum steel, deoxidized by calcium and aluminum[24)], that developed gaping transverse cracks on all eight sides in the forging press during initial pressure application. A longitudinal section taken from a plane 100 mm below and parallel to a flat side had wide cracks at the primary grain boundaries **(Fig. 4.68),** and the fracture was correspondingly intergranular (Fig. 2.20). Metallographic examination of microsections showed a coarse-meshed network of fine precipitates. They were in part transparent and were illuminated four times during turning 360° under polarized light between crossed Nicols. Thus they were anisotropic, i.e. not of cubic structure or glassy **(Fig. 4.69).** Electron microprobe tests showed that the precipitates were not enriched by either iron, silicon, manganese, oxygen or sulfur. In contrast to this, the aluminum radiation showed such high intensity that this metal had to be the principal constituent of the precipitates **(Fig. 4.70 and 4.71).** Platelet-like particles could be discerned on the intergranular fracture planes with the scanning electron microscope (see also Fig. 2.20). They were isolated by vaporizing with carbon and stripping of the film **(Fig. 4.72),** and their lattice structure was examined by electron diffraction. The precipitates proved to be thin and partially transparent platelets of a hexagonal crystal lattice whose parameters resembled those of AlN. It should be noted that this ingot had been cooled in the mold for approximately 11 h before placing into the forging furnace at 1200 °C, while another ingot that was worked after only 5 h remained intact during forging; this, however, had lower nitrogen and aluminum contents. In this ingot the nitrides probably were never precipitated prior to placing into the forging furnace.

4.84

Fig. 4.81 to 4.84. Broken piston rod of drop hammer

Fig. 4.81. Fracture surface. Approx. 0.67 ×

Fig. 4.82. Transverse section parallel to fracture. Etch: Hot 50 % hydrochloric acid. 0.67 ×

Fig. 4.83. Flake crack in unetched transverse section. 100 ×

Fig. 4.84. Fracture of heat treated disk. 2 ×

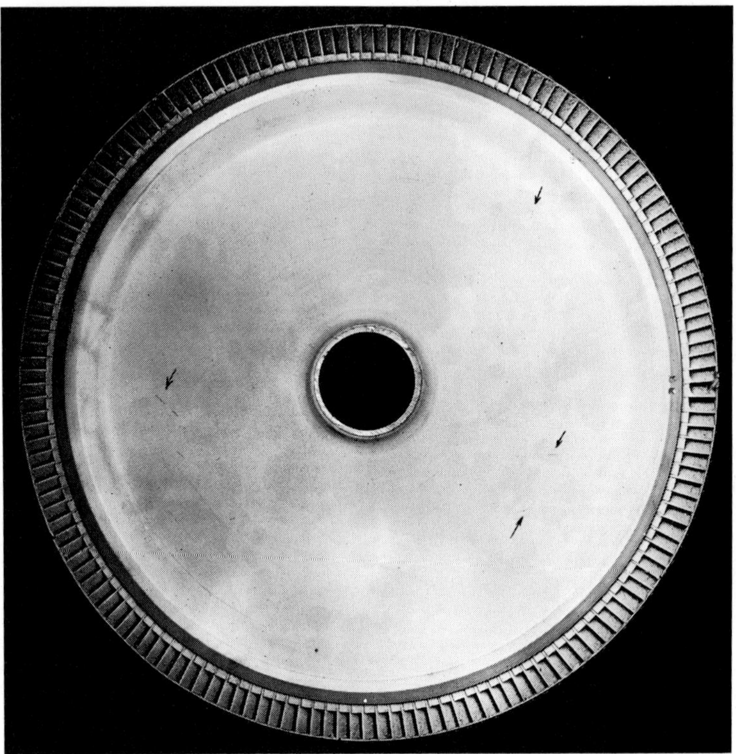

4.85

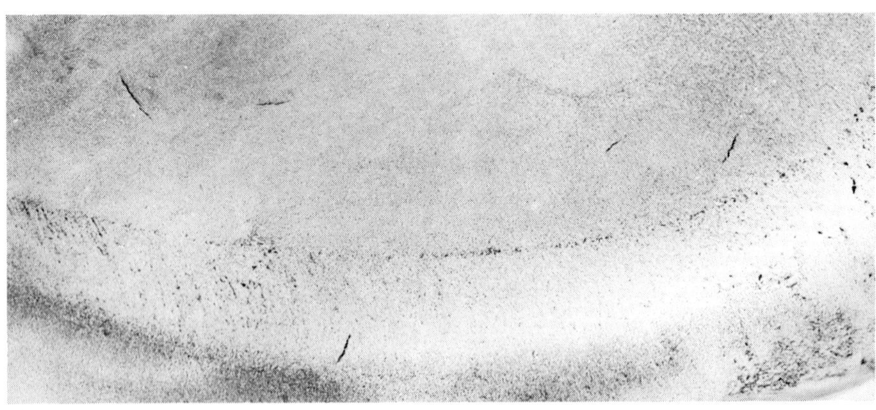

4.86

Fig. 4.85 to 4.88. Turbine rotor of manganese-molybdenum steel with flake cracks

Fig. 4.85. Disk with flake cracks that have been made visible by magnetic particle inspection. Approx. 0.3 ×

Fig. 4.86. Section of Fig. 4.85. 1 ×

Fig. 4.87. Radial section, etched according to Oberhoffer. 1 ×

Fig. 4.88. As in Fig. 4.87. Flake crack in banded structure 12 ×

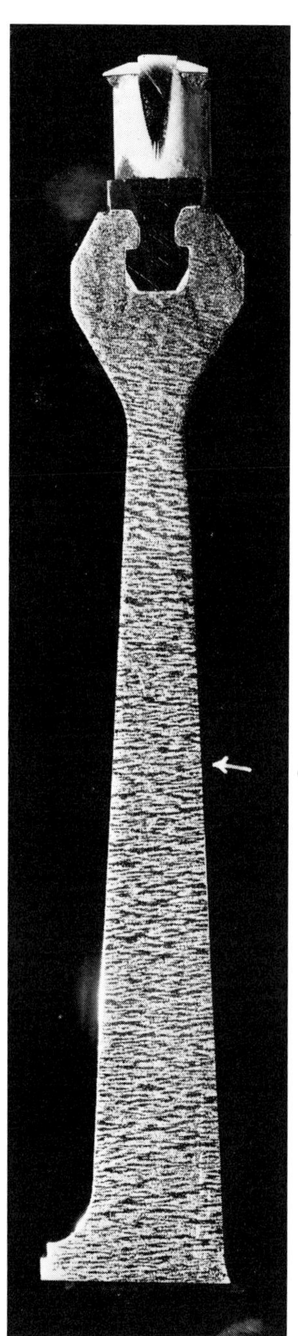

4.87

4.88

4.89

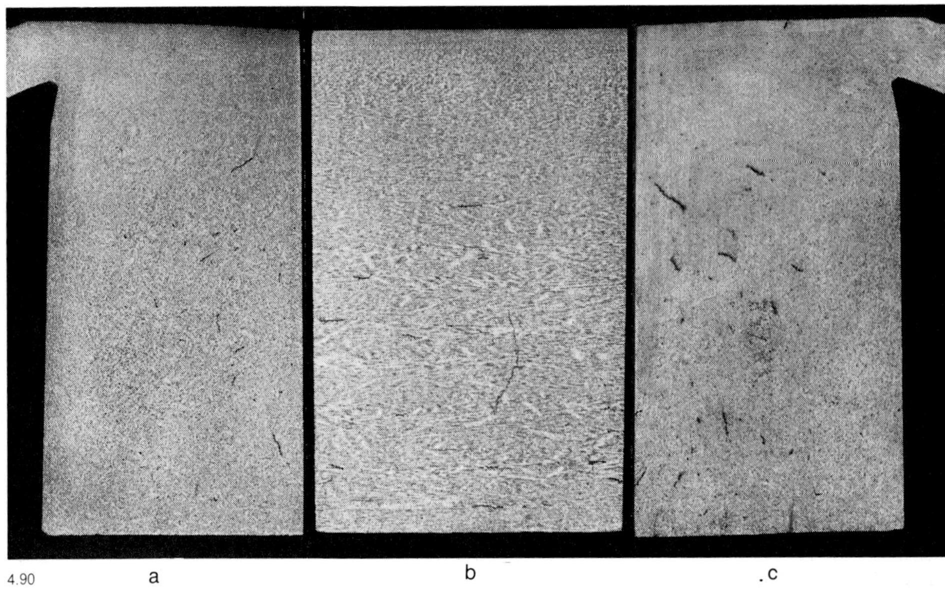

4.90 a b c

Fig. 4.89 and 4.90. Railway wheel rim with flakes

Fig. 4.89. Turned running surface with exposed internal cracks. 0.5 ×

Fig. 4.90. a Transverse section, b Tangential section, c Transverse section. Etch: 50 % hot hydrochloric acid. 1 ×

The precipitation of such phases which result in a low deformation type premature fracture must therefore be the cause for the intergranular fractures. The question, whether hydrogen which according to Z. Eminger[25]) causes intergranular fracturing plays any part, was to be answered by the following test. A metallographic surface of one of the precision castings shown in Fig. 4.62 was cathodically loaded with hydrogen. The polished section was then weakly etched in order to make the grain boundary precipitates visible. It was observed microscopically under alcohol. Immediately after the mount was dipped, a vigorous gas development was observed at serveral places of the coarse-meshed grain boundary network. **Figure 4.73** shows such a grain boundary with two bubbling hydrogen sources. The gas development lasted several hours, but their egress locations changed. Finally it could be seen under the microscope that the grain boundaries had been torn open as shown in **Fig. 4.74.** The precipitates in this case played the part of voids at which hydrogen was precipitated molecularly under high pressure and was able to rupture the steel structure[26]).

Residual stresses of this type may promote the formation of **intergranular fractures** under mechanical stress, but they are not necessarily the cause. They may however be the decisive factor in the formation of similar defects, the primary grain boundary cracks.

Primary grain boundary cracks are short intergranular cracks, that have a brittle cleavage fracture surface, but in contrast to intergranular "rock candy" type fractures they have a bright appearance. The also appear in cast ingots and steel castings and are designated occasionally as casting flakes because of their resemblence to flake cracks which occur in forgings but propagate in a transgranular manner (see also section 4.3.3). **Figure 4.75** shows these flake cracks together with a "rock candy" fracture of a **back-up roll** of cast chromium-molybdenum steel. **Figure 4.76** shows them together with a top crust (see also section 4.4.2) in cross section through the peripheral zone of this particular roll. The brittle fracture in this case was caused by ledeburite in the interdendritic spaces and carbide precipitates on the primary grain boundaries **(Fig. 4.77)** which facilitated the latter's cracking. During heating of the roll – probably at too fast a rate – for annealing or heat treating, an internal crack was formed which propagated from these defects in the interior. It can be seen in **Fig. 4.78** as a dark temper color (see also section 6.1.1). The roll which had a diameter of 1400 mm had operated for a while with this internal crack, but then had fractured during disassembly under flexural stress.

At the end of this section it may be said, that nitrogen should be regarded not only as detrimental to steel, but it may also impart desired properties to steel and therefore may be used as an alloying element. Nitrogen dissolved in gamma-iron increases the stability of the austenite to such an extent that it may be used occasionally as a replacement for nickel in heat treatable stainless steels and in austenitic steels. In carbonitriding, too, the austenite-stabilizing effect of nitrogen is utilized. It facilitates the slower cooling of distortion-sensitive parts subsequent to the customary carburizing without diminishing hardness. Gas nitriding, though, should not be confused with this process. Its hardening effect is based upon the precipitation of special nitrides such as aluminum or chromium nitride in the finest distribution, and that presupposes special alloy steels tailored to the application. Precipitated iron and special nitrides also increase the yield points of ferritic and austenitic steels. The so-called fine-grain size steels are unaffected by overheating due to the presence of several hard-to-dissolve nitrides or carbides (see also section 6.2). This facilitates direct case hardening without intermediate annealing.

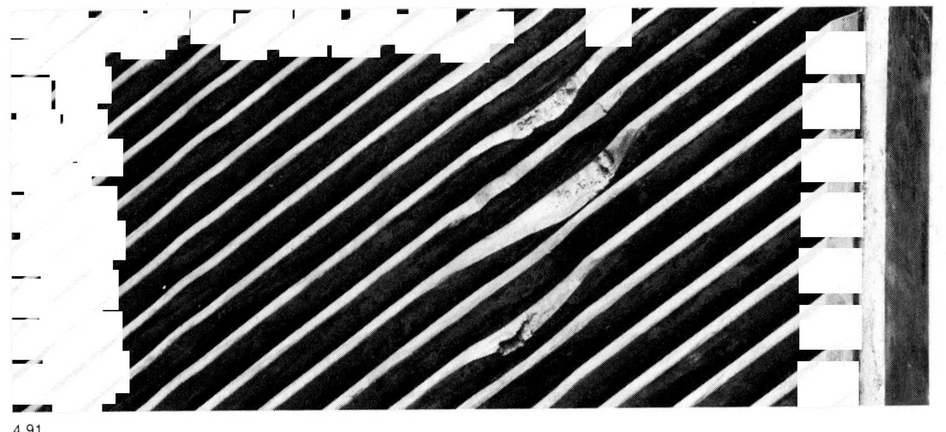

4.91

4.92

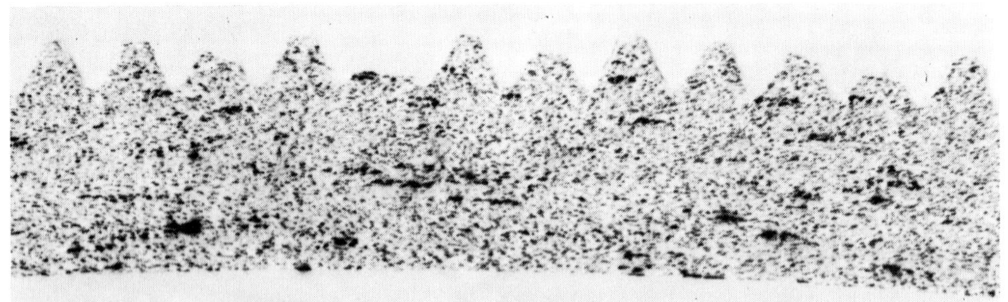

4.93

4.94

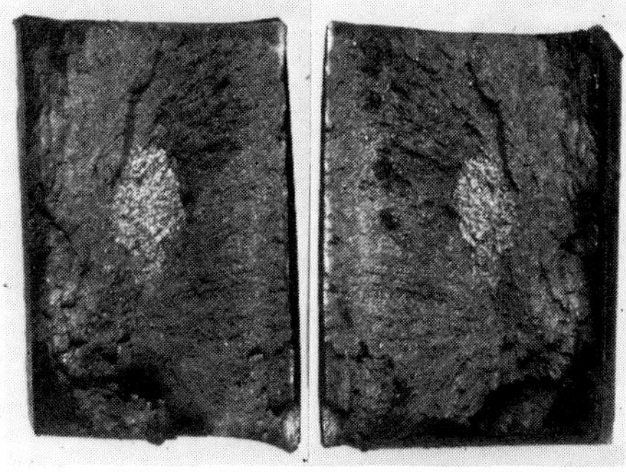

4.95

4.96

Fig. 4.91 to 4.96. Defective gear with obliquely mounted teeth from ship's turbine power transmission

Fig. 4.91. Broken-out teeth. 0.33 ×

Fig. 4.92. Transverse section with flake cracks. Etch according to Heyn. 0.5 ×

Fig. 4.93. Baumann sulfur print of transverse section. 0.5 ×

Fig. 4.94. Flake crack at tooth root. Transverse section. Etch according to Oberhoffer. 3 ×

Fig. 4.95. Flakes in heat treated fracture. 1 ×

Fig. 4.96. Flake cracks in unetched cross section. 75 ×

4.3.3. **Failures Caused by Hydrogen**

While a contributory effect of hydrogen is possible in brittle "rock candy" type fractures and probable in intergranular cracking, the following section will describe a defect that is proven to be caused by hydrogen[27]), i.e. the formation of flake type cracks.

Flakes* are small round or elliptic tension cracks with a transgranular path. In contrast to primary grain boundary cracks, they are formed as a rule not during cooling after solidification, but during cooling following the first forging or rolling operation. In their fracture they are representing the austenitic grain in the forged state. In untreated forgings or rolled blooms they often cannot be distinguished clearly from the surrounding grains. But if the grain has been refined through heat treatment and the fracture has taken on the character of a tough deformation fracture, they stand out as brightly shining spots with cleavage fractures from the surrounding velvety fracture. **Figure 4.79** shows the flakes in the fracture of a heat treated chromium-nickel steel forging. In this illustration the similarity with snowflakes is also apparent. They get their name from this phenomenon.

Electric steels are more strongly flake prone than open hearth steels, and of the latter, basic more so than acid open hearth molten steels. The reason for this is that this type of steel "boils" as a rule up to shortly before tapping, i.e. carbon monoxide is developed which forces hydrogen out, while electric steel is covered with limestone slag for a long period during refining. The use of large amounts of frequently wet limestone is also a reason for the higher flake sensitivity of basic molten steels as compared to acid ones.

Hydrogen is also dissolved in large amounts by solid iron **(Fig. 4.80),** i.e. more so in gamma-iron than in alpha-iron. Solubility decreases with decreasing temperature according to an exponential law, but during the gamma-alpha transformation a jump in solubility occurs again as it does in solidification. But with decreasing temperature the hydrogen has to precipitate in molecular form from the iron lattice. This causes very high gas evolution pressures which may exceed the strength of the steel[28]). These pressures are higher for larger hydrogen concentrations in the steel and for lower precipitation temperatures. E. Houdremont and H. Korschan[29]) established that flakes were formed at temperatures below 200 °C. These technological tests were performed with forged specimens cut from a nickel-chromium steel as well as a ball bearing steel. Alloying elements that decrease the gamma-alpha transformation of the steel at these temperatures may promote flake formation. The enrichment by such elements and of hydrogen by segregation may be reasons why flakes can often be seen preferentially in segregation bands. Thermal stresses may also promote flake formation since they reach a considerable peak during fast cooling of large forgings.

Manganese and silicon-manganese steels are particularly sensitive to flake type cracks, as are nickel-chromium, chromium, and chromium-molybdenum steels, especially if they have an elevated manganese level. Sensitivity also increases with rising carbon content. For instance, ball bearing steels with approximately 1 % C are very sensitive. If carbon is high, a comparatively low manganese content is sufficient to sensitize the steel. It will be shown that flakes were often found in railroad wheel rims of "unalloyed" steel with 0.6 to 0.7 % C and 0.6 to 0.8 % Mn. Track fractures, too, often propagate from flake cracks[30]). On the other hand, 0.2 % C is sufficient to make steel flake sensitive if Mn content is 2 %.

---

* see Appendix I

Fig. 4.97 to 4.99. Crankshaft and raw billet with flakes

Fig. 4.97. Flakes in drop forge die flash of crank arm (arrow) made visible by magnetic particle inspection. 1 ×

Fig. 4.98. Fracture trough arm in plane of die flash

Fig. 4.99. 140 mm square billet with flakes. Transverse section. Etch: Hot 50 % hydrochloric acid. 0.5 ×

4.97

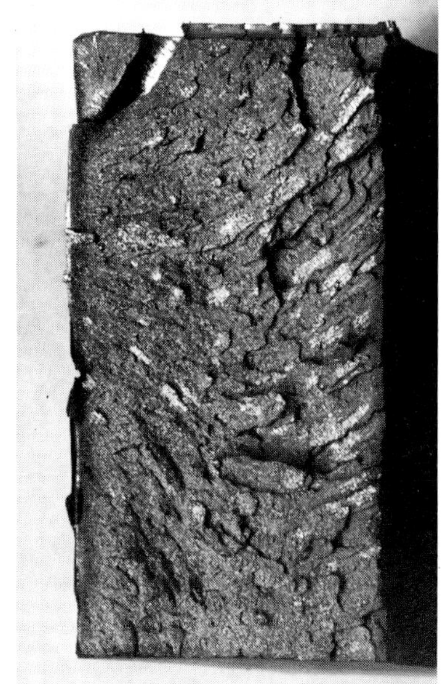

4.98

4.99

Flake cracks as a rule are tightly closed and their determination takes some experience and patience. It is best accomplished by a brief etching which does not penetrate too deeply. A suitable etching solution is hot diluted hydrochloric acid or, even better, a copper ammonium chloride solution prepared according to Heyn which can easily be used for large parts. If the cracks are not immediately visible, it is useful to leave the mount alone for a few hours after rinsing with water and dabbing with blotting paper. If cracks are present, their edges will be corroded by the expelled water, thus making them visible. A fracture test may be used in addition to metallographic examination. For this purpose the specimens may have to be heat treated to achieve a tough and hardened state. Of the non-destructive testing methods, ultrasonic inspection in particular has proved useful in testing for flakes.

Flake formation can be prevented by limiting hydrogen absorption to the maximum extent possible which can be accomplished by having all furnace and ladle additives, and in particular lime, as dry as possible, and preferably red hot. Alternatively, the hydrogen absorbed by the liquid steel is removed by melting and casting in a vacuum. If vacuum installations are not available, it is necessary to let the specimens cool very slowly below 200 °C after the first hot forming. Else extended annealing may be interposed at a temperature at the upper limit of the alpha region in order to keep thermal stresses low and provide a possibility for the outward diffusion of hydrogen. But these procedures do take up much time and furnace space, especially with heavy forgings. It should be mentioned that flake cracks, if they are not in touch with air, may be welded shut by intensive forging.

In the following several **examples** will be cited. **Figure 4.81** shows half the fracture surface of a **piston rod** of a drop forge hammer[31]). Only this part had been sent in for examination and failure analysis. The rod had additionally been damaged by a bore hole that was probably made to obtain chips for analysis. If it was nevertheless possible to establish the cause of failure, it was quite by accident. However, it should be emphasized once more that, in order to insure a successful analysis, it is necessary to protect fracture surfaces from rusting. Furthermore, the failed parts must be put at the disposal of the materials test engineer in their entirety and undamaged.

The piston rod which had a diameter of 180 mm consisted of an unalloyed steel with 0.37 % C and 0.67 % Mn and had an ultimate tensile strength of 550 MPa (80 ksi) at 26 % elongation. The rupture was initiated by several internal fatigue fractures that had combined into two large fractures. The origins of one crack, shown at the left of the picture (with the bore hole), could still be seen on the broken part sent for examination. Apparently there were several flake cracks. Metallographic examination confirmed this. **Figure 4.82** displays the flaky interior of the rod in a transverse section parallel to the fracture plane, while **Fig. 4.83** shows a very small flake crack at large magnification. **Figure 4.84** shows several small flakes in the fracture of the heat treated pickled disk. There were no other defects. The small internal cracks accordingly sufficed to induce fracture during fatigue impact stressing of the piston rod. The stress, however, may have been reinforced by the circumstance that the hammer stood on a very soft base and had sunk somewhat on one side.

Figure 4.85 shows flake cracks that were found during magnetic particle inspection on both sides of a heat treated **turbine rotor** made of steel with 0.25 % C, 1.11 % Mn, and 0.42 % Mo. They had a length of up to 8 mm **(Fig. 4.86).** In metallographic sections taken perpendicular to the surface of the disk it was found by way of the fiber orientation **(Fig. 4.87 and 4.88)** that they

Fig. 4.100 to 4.102. Broken crane track rail with cavity

Fig. 4.100. Longitudinal crack with wood fiber structure. 0,67 ×

Fig. 4.101. Cross section. Etch according to Heyn. 0.5 ×

Fig. 4.102. Peripheral structure at cavity. Transverse section. Etch: Picral. 100 ×

4.100

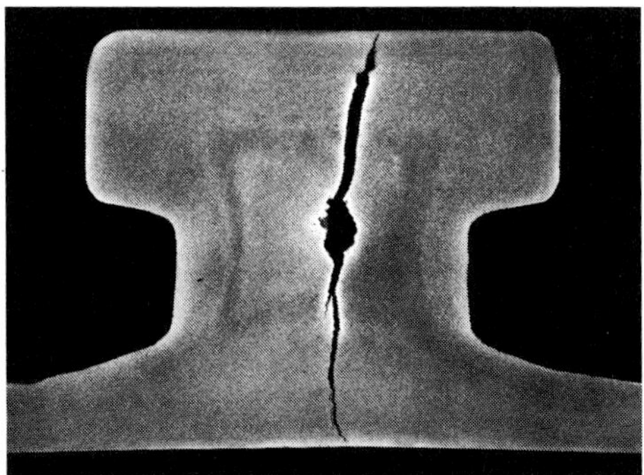

4.101

4.102

were cut from a longitudinally forged piece across the forging direction and subsequently merely formed in the wheel rim for attachment of the blades. The flakes were located in the phosphorus- and certainly manganese-rich zones of the banded structure. Additional forging after the disk was cut might possibly have welded these flakes together. Severe damage could be prevented in this case by magnetic particle inspection prior to assembly.

During turning of the running surface of a **railway wheel rim**[32]) an internal crack was discovered **(Fig. 4.89).** After breaking the crack open, the surface showed a smooth fracture with indications of fatigue striations. At the point of origin an area with a coarse microstructure was visible. Flakes were assumed to be present. The piece remaining after removing the specimen contained two cross sectional surfaces, one at each end; and one tangential section located approximately 20 mm under the running surface. These were then pickled in hot dilute hydrochloric acid **(Fig. 4.90).** Flake cracks became visible on all surfaces. Furthermore, a second fatigue fracture, probably also originating in a flake, was cut off the tangential section.

Several teeth broke off an obliquely toothed **gear rim of a ship's turbine** power transmission. The rims consisted of silicon-manganese steel of 690 to 830 MPa (100–120 ksi) tensile strength. **Figure 4.91** shows the failure sites of the rims. The fractures were located in part at one half the height of the tooth flanks, i.e. not at the place of highest bending stress in the root of the teeth from which fatigue fractures generally propagate in overstressed gears. Also these were not fatigue fractures. In etched longitudinal and transverse sections, the rims were found to contain flake cracks **(Fig. 4.92).** In several cases the cracks were located in segregations that stretched all the way into the teeth **(Fig. 4.93 and 4.94). Figure 4.95** shows them in the fracture of a heat treated specimen. These fractures propagated from the flakes. The interconnection of the flakes with sulfur segregations could be seen in the microstructure **(Fig. 4.96).** Segregations which are not unusual for manganese-alloyed steels apparently had facilitated flake formation, but certainly did not cause it.

During machining of drop forged **two-cylinder crankshafts** made of square billets from three heats, fine cracks appeared in approximately 200 shafts on the pins and crank arms. They were located exclusively in a strip at the surface that corresponded to the flash left by the closed forge dies. The crankshafts were made of a steel with approximately 0.37 % C, 1.3 % Si, and 1.3 % Mn and were heat treated prior to machining. **Figure 4.97** shows the crank arm of one of the shafts. The cracks were a few millimeters long and followed the flash in a string-like pattern. **Figure 4.98** shows them in a fracture specimen that was machined in such a way that the break had to occur in the flash plane. The cracks have the characteristic appearance of flakes. However it is unusual that they were stretched in the direction of forging. That indicates that they were present prior to die forging. This was confirmed by examination of a raw billet. **Figure 4.99** shows an etched transverse section. Thus, during forging of the crankshaft the cracks were not welded shut in the steel volume that could escape into the die flash, where the counterpressure was lower.

These examples confirmed that manganese and silicon-manganese steels have a particular tendency to flake type crack formation and that it is necessary to take precautionary measures during their melting and working.

Fig. 4.103 to 4.106. Forging with cavity burst open during upsetting

Fig. 4.103. Upset end. 3 ×

Fig. 4.104. Flat end. 1 ×

Fig. 4.105. Cross section through flat part. Etch: Picral. 5 ×

Fig. 4.106. As Fig. 4.105. Oxide inclusions in cavity. 75 ×

4.103

4.104

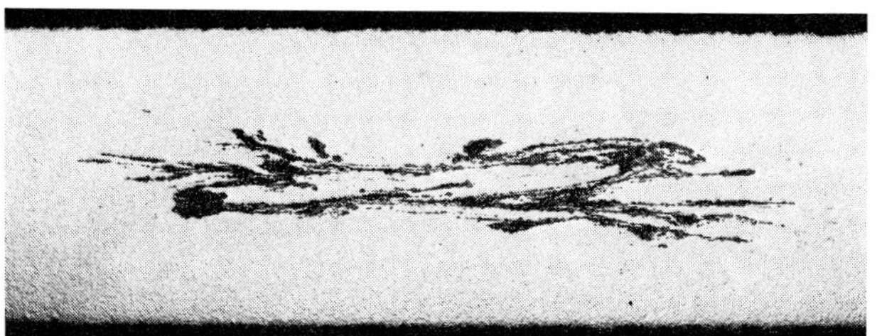

4.105

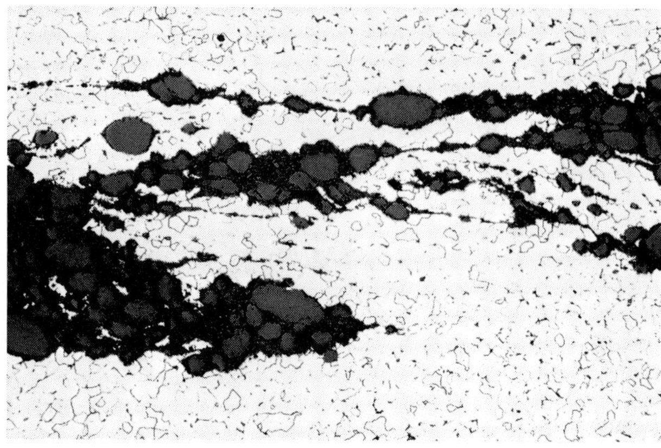

4.106

4.107

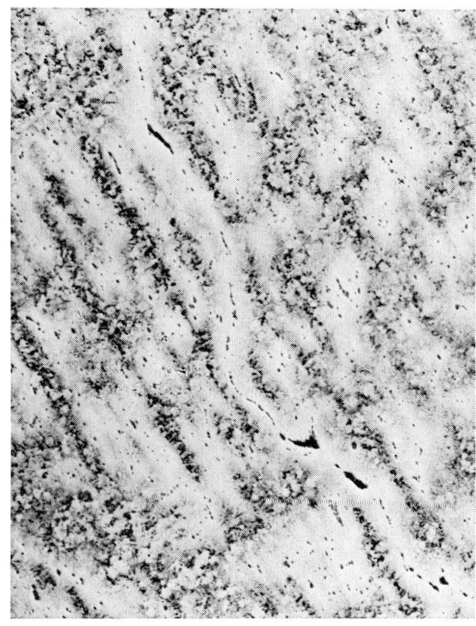

4.108

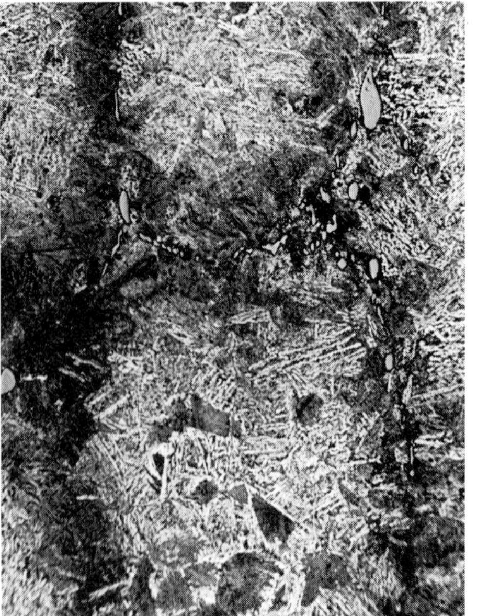

4.109

Fig. 4.107 to 4.109. Broken swivel head with shrinkage cavity

Fig. 4.107. Shrinkage cavity in fracture. 3 ×

Fig. 4.108. Section through cavity zone. Etch according to Oberhoffer. 10 ×

Fig. 4.109. As Fig. 4.108. Etch: Picral. 100 ×

4.4 **Casting Errors**

During casting, errors are frequently made that may lead to failures. The formation of blow holes may be promoted by unsuitable shapes of molds and chill molds. Wet molds cause the absorption of gases. Surface quality of the ingots is affected considerably by steel splash and faulty chill mold painting. Hot tears are promoted by pouring at too high a temperature or too fast, while casting that is too slow or too cold causes the formation of top crusts or boil-over skulls.*

4.4.1 **Blow Holes**

The volume of iron and steel decreases nonuniformly during transition from the liquid to the solid state, as is the case with almost all other metals and alloys. Blow holes (shrinkage cavities) are the result of this shrinkage. If they are limited to the upper part of the casting (head shrinkage cavities), it is possible to remove these blow holes later with the hot top. Concentrating the cavities at the top can be achieved in ingots by the selection of a suitable chill mold shape and heating of the head, or in steel castings by correct construction of the molds and suitable location of the gates and risers. If remnants of open head shrinkage cavities remain in castings or forgings, they can be recognized by the metallographer by oxidation and usually also decarburization of their outer zones. The material is also dirtied by impurities consisting of segregation products around the cavity. This area too must be removed during cropping of the ingots. In section 4.1 a case was described in which this procedure was not followed (Fig. 4.9 to 4.11).

If flow of the liquid metal during pouring is prevented by fast solidification in narrow mold or die cross sections, cavities may also be formed inside the castings. This is easily done in ingots if the mold is either very slender or tapered at the top. Such **"secondary pipes"** can be welded shut during forging or rolling if they have not been exposed to air. They too are characterized generally by segregation products.

Shrinkage cavities which are caused by the inhibited flow of mother liquor after formation of the crystal skeleton, are called "crystal cavities" or "microcavities"*. Such cavities can be recognized metallographically by their location in the dendritic interstices and in the fracture by their dendritic structure.

An example of failure by head shrinkage cavities is shown in **Fig. 4.100,** a longitudinal fracture of a **crane track rail**[33]). The interior of the fracture had a fibrous structure. In the transverse section **(Fig. 4.101)** a cavity could be seen at the crack origin. Its edges were decarburized. **Figure 4.102** shows the edge of the cavity as scaly and the region directly below it permeated with oxides. Accordingly, the cavity showed typical characteristics of a shrinkage cavity. But it should be added that this rail had been highly overstressed, because other rails of the same crane track had been destroyed by fatigue fractures, even though they were not weakened by material or casting defects.

A forge rejected a delivery of **flat steel** because the bars broke open during upsetting on the edge. The steel corresponded according to composition and microstructure to a structural steel with about 410 MPa (60 ksi) ultimate tensile strength. **Figures 4.103 and 4.104** show one of the forgings seen from the upset and the flat end. It showed a hollow space in the upset part and in the center line of the flat part an almost continuous separation. From the transverse section of the flat part it could be observed that the hollow space was filled with oxides and decarburized at the edges **(Fig. 4.105 and 4.106).** Therefore these were shrinkage cavities.

Figure 4.107 shows a dendritic microstructure of a **swivel** head fracture[34]). The section shows the connection of the fracture with the shrinkage cavities in the spaces enriched by phosphorus and sulfur in the residual melt **(Fig. 4.108 and 4.109).**

---

* see Appendix I

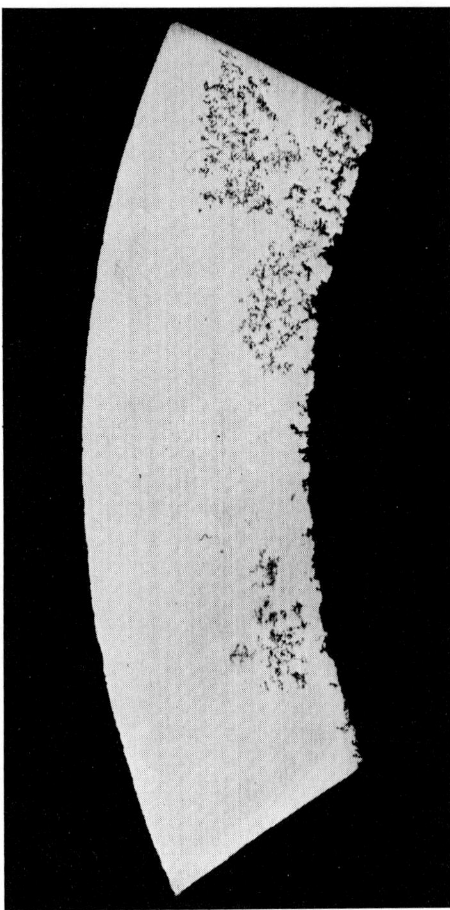

4.110

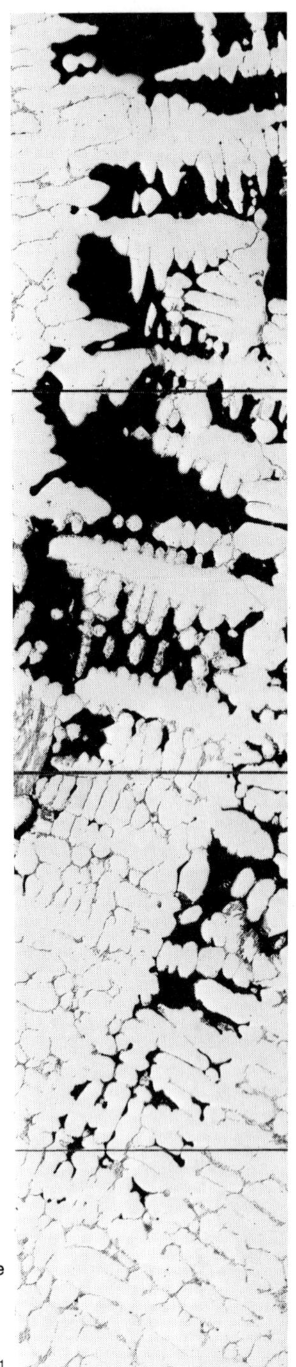

4.111

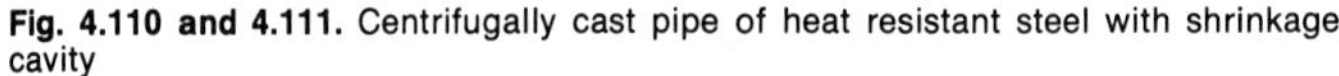

Fig. 4.110 and 4.111. Centrifugally cast pipe of heat resistant steel with shrinkage cavity

Fig. 4.110. Unetched transverse section. 2 ×

Fig. 4.111. Microstructure in the cavity zone. Etch: V2A-etching solution. 50 ×

Figures 4.110 and 4.111 show a characteristic example of shrinkage cavities in centrifugal castings. They represent sections through a cavity containing **centrifugally cast pipe** made of austenitic heat resistant cast steel. The part solidified last is located at the interior of the pipe in this case.

Figure 4.112 presents an excellent example of microcavities. It shows hollow spaces lying 0.05 mm deep, that appeared after one year's service on the running surface of a **Steven pipe** made of a centrifugal steel casting with approx. 0.4 % C, 1.5 % Si, 27 % Cr, and 4.5 % Ni. The microstructure of this alloy consists of ferrite, austenite, and chromium carbide. Electrolytic etching with 10N soda lye that attacks the chromium ferrite more strongly than the nickel-richer austenite[35]), confirmed that the hollow spaces lie in the austenitic regions that solidified last **(Fig. 4.113).**

4.4.2 **Top Crust Pieces**

Pouring that is too hot or too fast may lead to longitudinal cracks especially on large round castings. As a rule the material test engineer need not concern himself with them because such castings usually are scrapped immediately after stripping and are not examined at all. But this is not the case in cast ingots that are poured too cold or too slowly. Under these conditions solidified crusts may form on the rising steel melt top. They are vaulted and shell-shaped so that they swim on the surface. But if they are hit by the pouring stream or stand upright, they sink down and are sometimes found in the lower part of the ingot during testing. Or they may be attached to the chill mold walls and drowned by the rising liquid metal.

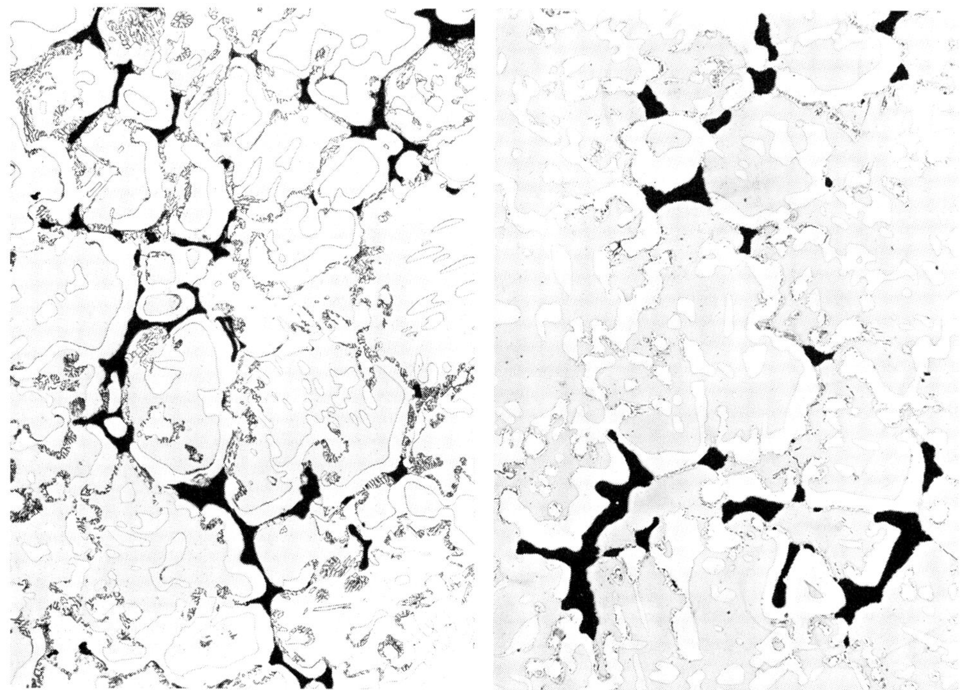

Fig. 4.112 and 4.113. Microstructure of Steven pipe semiferrite heat resistant steel. 80 ×

Fig. 4.112. Etch: V2A-etching solution

Fig. 4.113. Electrolytic etch with 10N caustic soda 1.5 V. Ferrite attacked

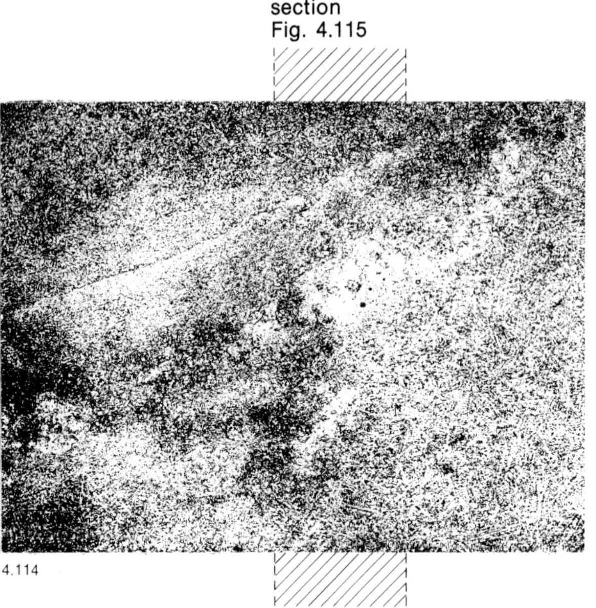

4.114

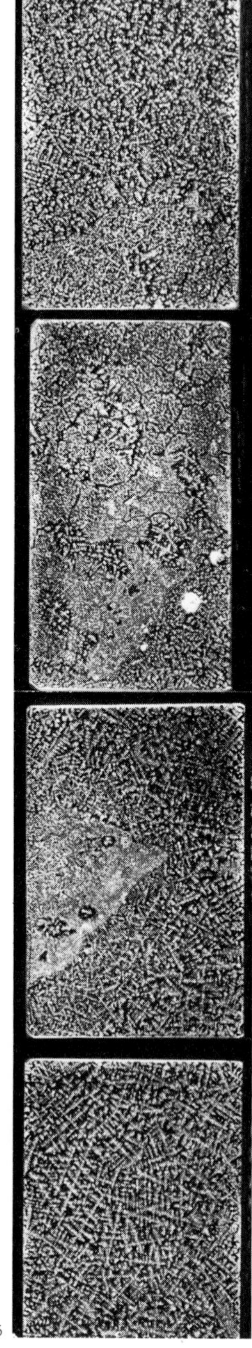

4.115

Fig. 4.114 to 4.119. Trop crust in 6.5 ton cast ingot of medium carbon steel (approx. 0.45 C)

Fig. 4.114. Transverse section. Etch according to Heyn. 0.5 ×

Fig. 4.115. Transverse section according to Fig. 4.114. Etch according to Oberhoffer. 1 ×

Fig. 4.116 and 4.117. Oxide inclusions at fringe of top crust. Unetched polished surface. 100 ×

Fig. 4.118 and 4.119. Microstructure. Etch: Picral. 50 ×

Fig. 4.118. In top crust

Fig. 4.119. In normal region

4.116

4.117

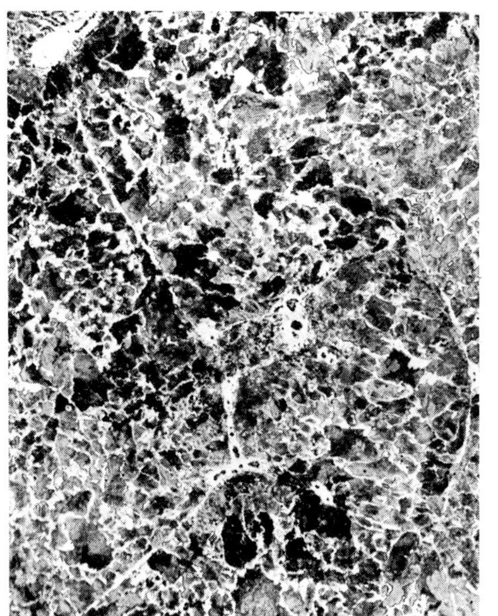

4.118

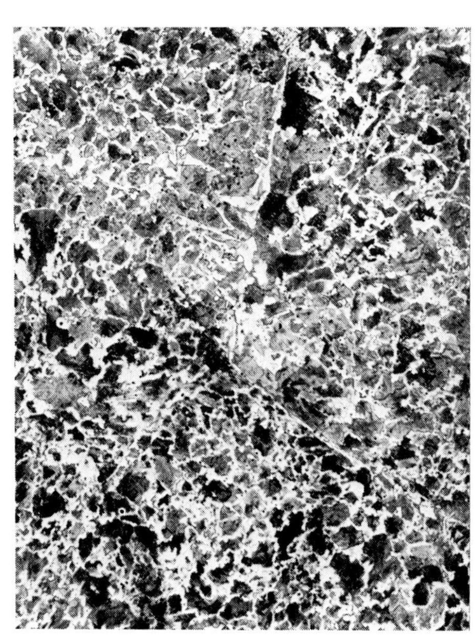

4.119

4.120

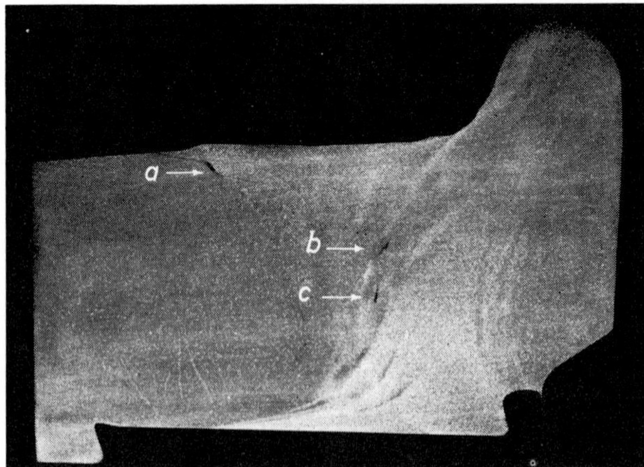

4.121

Fig. 4.120 to 4.122. Railway wheel rim with internal defects

Fig. 4.120. Inernal crack on machined running surface. 0.5 ×

Fig. 4.121. Transverse section A--A according to Fig. 4.120. Etch according to Heyn. 1 ×

Fig. 4.122. Part of transverse section A--A according to Fig. 4.120. Etch according to Oberhoffer. 2 ×. Left: Normal dendritic solidification structure. Right: Top crust with globular primary structure

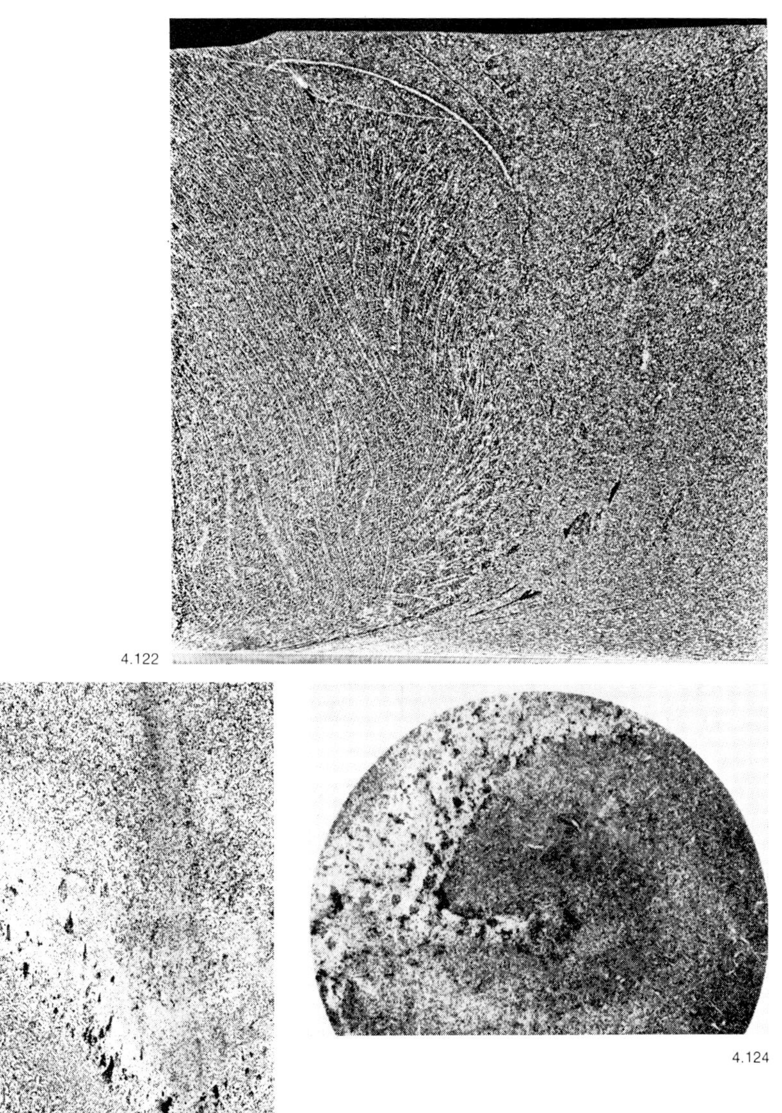

Fig. 4.123 and 4.124. Forging with boil over skulls. Baumann sulfur prints

Fig. 4.123. Surface. Approx. 0.5 ×

Fig. 4.124. Transverse section. Approx. 0.5 ×

Such top crust or boil-over skull pieces can not be differentiated by their composition, nor usually their microstructure from their surroundings, but have a different primary structure corresponding to the differing conditions under which they solidified. Furthermore, they are often permeated and surrounded by oxide slags and segregation products because their top face was in contact with air before sinking, while their bottom side absorbed segregated slags and deoxidation products. A similar picture emerges with runner* sticks, i.e. solidified ladle accretions or other effusions that may drop into the mold if they are not removed in time.

Because the impurities are dragged along, these foreign bodies do not always properly weld all around with the later solidifying material and this may subsequently lead to rejects and failures. On the other hand, the presence of coarse slags and slag stringers also offers the possibility of finding top crusts with the aid of ultrasonic inspection using the pulse-echo method, as will be shown later. A top crust in a cast steel roll was shown already in Fig. 4.76.

Coarse inclusions were exposed in some **crankshafts** made of an unalloyed medium carbon steel (approx. 0.45 % C) during working or testing. Ultrasonic inspection of castings showed a defect trace in the lower part of a 6.5 ton ingot of the same heat of which the crankshafts were made. Metallographic examination of a 15 mm thick transverse disk from this area showed that a top crust with finer primary structure was involved which was situated at an oblique angle in the ingot **(Fig. 4.114).** It also could be clearly distinguished from its surroundings in the Baumann replica. Metallographic sections cut from the region, Fig. 4.114, showed these differences even more distinctly after etching according to Oberhoffer **(Fig. 4.115).** During microscopic examination, large amounts of oxide inclusions were found in the top crust as well as at its boundaries. These appeared as coarse individual units, as clusters **(Fig. 4.116),** and as loose stringers **(Fig. 4.117).** They were probably slagged oxides that sank with the top crust. The pearlitic-ferritic secondary structure of the top crust **(Fig. 4.118)** could not be particularly distinguished from the surrounding steel **(Fig. 4.119).** The coarse-meshed ferritic network that was covered with sulfide inclusions corresponded to the primary grain boundaries and was correspondingly finer in the top crust than in the surroundings.

An internal crack was opened up during machining a **railroad wheel** rim **(Fig. 4.120)**[32]). Examination showed that one of two fatigue fractures had propagated from a slag stinger (point a in **Fig. 4.121).** The slag stringer was interconnected with a top crust whose border zone contained several more coarse imbedded slag inclusions (b and c). The top crust showed a very fine globular grain size in contrast to the coarsely dendritic solidification structure of the base material **(Fig. 4.122).**

Figures **4.123 and 4.124** show boil-over skulls in a **forging**[36]) that were made visible by sulfur prints of the surface and of a cross section. Substantial accumulations of sulfur-rich segregation products at the boundary to the normally solidified material should be noted in addition to the finer solidification structure.

The next example may illustrate how ultrasonic inspection served to find a top crust in a **turbine rotor** made of an AISI type 9840 steel, containing approx. 0.28 % C, 1 % Cr, 1 % Ni and 0.2 % Mo, and to determine its location. During finishing, an 8 mm long crack-like defect was opened on a cylinder surface of a compensating piston of the Pelton wheel. Magnetic particle inspection indicated that the defect had a more substantial dimension than was visible at the outside. Ultrasonic inspection with oblique wave penetration was used to explore the defective region. It is delineated by the dash lines in **Fig. 4.125.** Since the wheel originated in the lower part of the 70 ton ingot and the defect was located within the 5.5 ton forging at the side directed downward in the cast ingot, there were grounds to suspect that the defect was a top crust.

---

*see Appendix I

The defective part was then cut out and the three rings designated A, B and C in Fig. 4.125 were examined in greater detail by ultrasonic inspection with vertical wave penetration at the points 1 to 6, as well as metallographically in the section planes **a** to **l.**

The ultrasonic inspection diagrams showed in part distinct ultrasonic images caused by the defects, and in part greatly extended disturbance zones. The results are shown schematically in Fig. 4.125. **Figures 4.126 and 4.127** are examples of ultrasonic trace diagrams for test sites 6 and 5. The strongest disturbance was found at the upper border of the defective zone which is shaded in Fig. 4.125.

Metallographic examination confirmed the suspicion that the defect was in fact a top crust. The top crust had a triangular cross section **(Fig. 4.128)** in whose upper peak a coarse slag inclusion was located. This had caused the strong image in the ultrasonic diagram. A strongly etched parting line separated the slag inclusion on one side from the base material. A polished section showed that this line or area contained impurities in the form of small spherical inclusions, probably silicates **(Fig. 4.129)**, interrupted by coarser crystalline inclusions, probably spinels **(Fig. 4.130).** The results of the metallographic examination coincided well with those of the ultrasonic inspection. The microstructure of the top crust did not differ from the heat treated structure of the base material. Its chemical composition, too, was the same as that of the wheel, as can be seen from **Table 4.**

Table 4. **Top crust of a turbine rotor.** Chemical composition of top crust and base material

	C %	P %	S %	Cr %	Ni %
Top crust	0.30	0.013	0.006	1.20	1.15
Base material	0.34	0.013	0.006	1.20	1.15

A section of ring A was broken open radially through the defect during heat tinting because then the slag inclusions can be particularly well seen. The fracture propagated exactly through the zone determined by the oblique wave penetration. It had a smooth, somewhat fibrous structure with a coarse "rock candy" type spot under the cone face that was not tinted blue in contrast to the remaining fracture surface.

To answer the question whether such a top crust has any effect upon the strength of the material, tensile specimens of 16 mm diameter were machined in a tangential direction at the points Z1, Z2, and Z3 of Fig. 4.125. They contained the foreign body approximately at the center. The defect became visible during turning of the specimens. Surface etching of the Z3 specimen showed a slag inclusion stringer, while in specimens Z1 and Z2 only individual segregations appeared. For purposes of comparison additional tensile specimens of the same dimensions were taken from a similar location in the defect-free part of the ring (Z4 and Z5). The tensile tests resulted in the data given in **Table 5.** Specimen Z3 failed prematurely along the slag stringer visible on the surface. The fracture had rusted in some places by the penetrating etchant, i.e. in this case there was an open separation prior to the tensile test. The properties of the entire piece would of course not have been diminished to the same extent as the specimens cut from the defective zone. Specimens Z1 and Z2 had almost the same yield point and strength levels as the defect-free specimens Z4 and Z5, but in Z2 the deformation characteristics were adversely affected by defects that were also expressed in the fibrous fracture. Accordingly, in those places where the foreign body was properly welded together with the base material, strength values were not lowered to any great extent by its presence.

Ultrasonic Inspection

Defect region detected by oblique wave penetration

strong ultrasonic image of defect

weak ultrasonic image of defect

Noise

Metallographic Examination

coarse
smaller } inclusions

Stringers

Metallographic sections

Region with gross defects

A B C

Fractured specimen

h i k l b c e g f a d

Z1 Z2 Z3

10 mm

6 5 1 2 3

Ultrasonic inspection sites

4.125

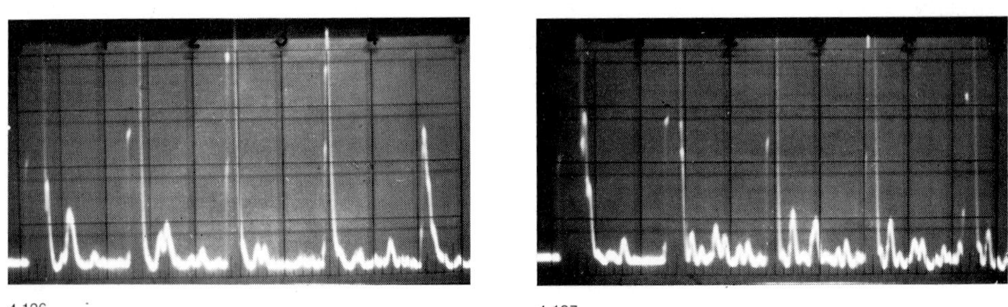

4.126

4.127

Fig. 4.125 to 4.130. Top crust on compensating piston of Pelton wheel

Fig. 4.125. Results of ultrasonic inspection and metallographic examination

Fig. 4.126 and 4.127. Ultrasonic back reflection traces with vertical wave penetration. 0.7 ×

Fig. 4.126 Inspection site 6 according to Fig. 4.125

Fig. 4.127. Inspection site 5 according to Fig. 4.125

Fig. 4.128. Ring B. Etch according to Heyn. 0.7 ×. Top: Section b according to Fig. 4.125. Center: Cylinder surface a according to Fig. 4.125. Bottom: Section c according to Fig. 4.125

Fig. 4.129 and 4.130. Inclusion stringer in section k according to Fig. 4.125. Unetched

Fig. 4.129. Silicates. 100 ×

Fig. 4.130. Oxides. 200 ×

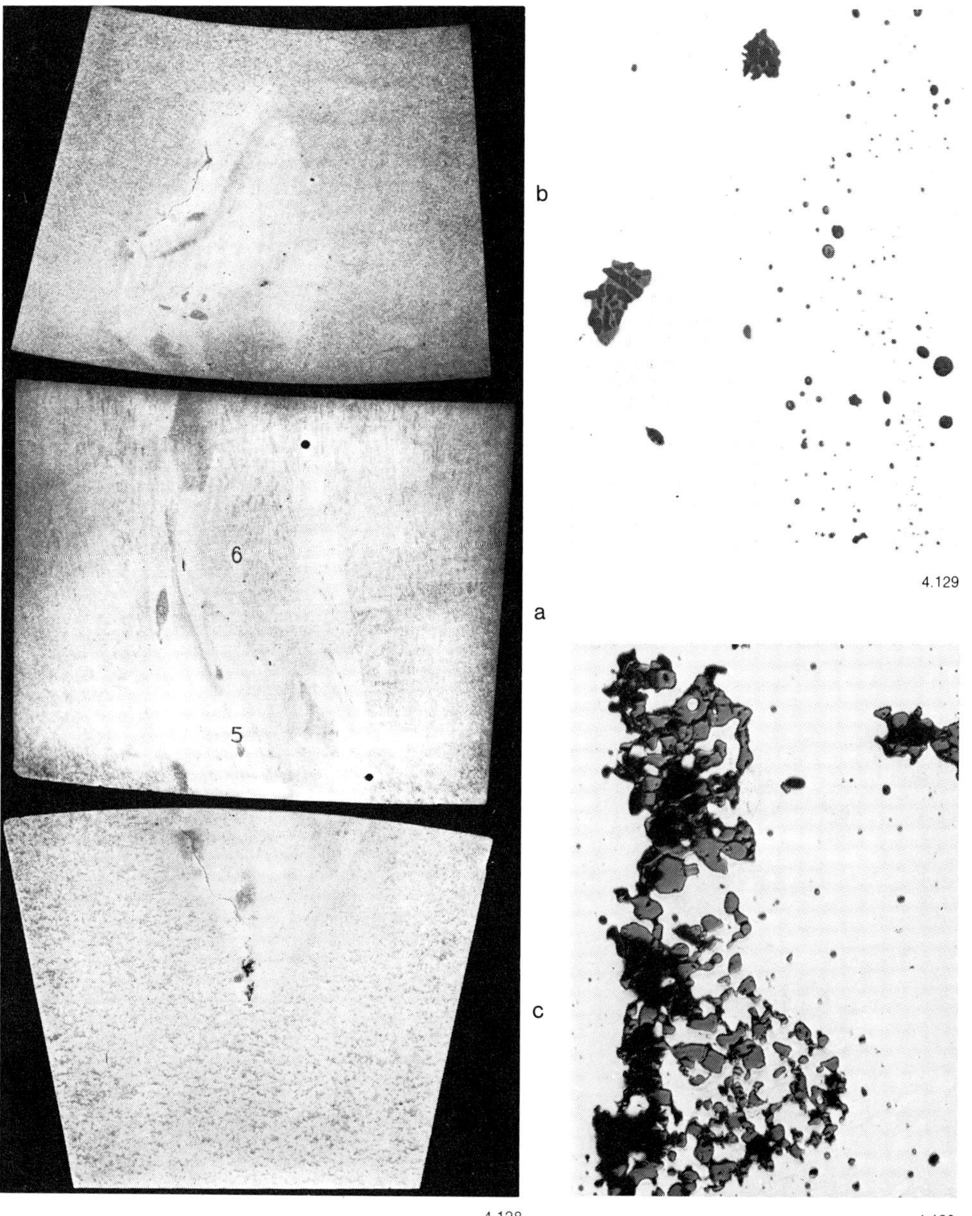

4.128

4.129

4.130

Table 5. Results of tensile tests

Bar	Yield Point MPa (ksi)	Tensile strength MPa (ksi)	Elongation (gauge length = 5 diam.) %	Reduction in area %	Fracture
Z1	440 (63.8)	637 (92.4)	26.8	62	ductile
Z2	430 (62.4)	622 (90.2)	15.2	19	fibrous
Z3	– –	425 (61.6)	4.4	9	defect area
Z4	464 (67.3)	661 (95.8)	30.4	63	ductile
Z5	465 (67.4)	667 (96.7)	29.2	63	ductile

The investigation showed that non-destructive ultrasonic inspection can determine a material defect consisting of or connected with an open separation, both with regard to size and location. It may also shed light on its history.

Sometimes it also was possible to determine analytically the composition of the slags and draw conclusions as to their origin. In connection with the top crust illustrated in **Fig. 4.131** in a case hardened **gear** of a steel with approx. 0.15 % C, 1.2 % Mn and 1.2 % Cr, two different kinds of slags were found; of these one was colored green, and the other black. The structure of the green slag contained coarse mullite crystals **(Fig. 4.132)**, that of the black slag spinel-like segre-

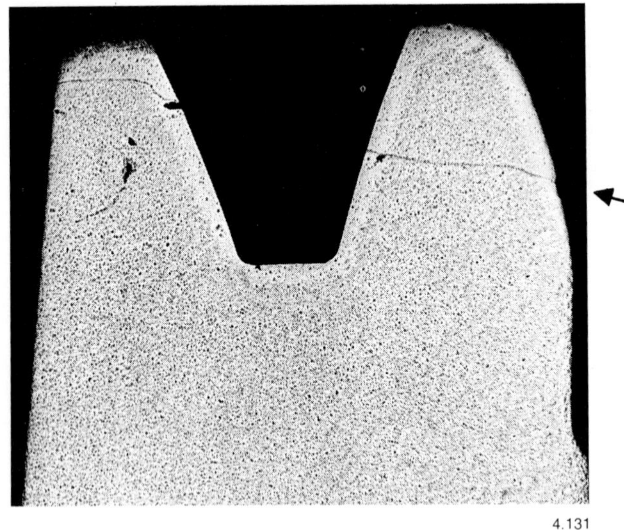

4.131

Fig. 4.131 to 4.134. Top crust in a case hardened gear

Fig. 4.131. Transverse section through teeth. Etch according to Oberhoffer (illuminated vertically). 3 ×

Fig. 4.132 and 4.133. Etch: Nital. 200 ×

Fig. 4.132. Green aluminum oxide-rich inclusion

Fig. 4.133. Black aluminum oxide-deficient inclusion

Fig. 4.134. Inclusion stringer taken from Fig. 4.131 (arrow). Etch: Nital. 100 ×

4.132

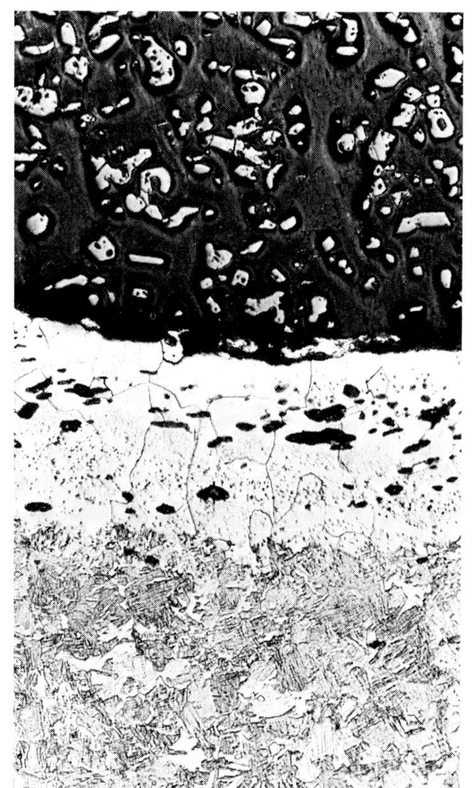

4.133

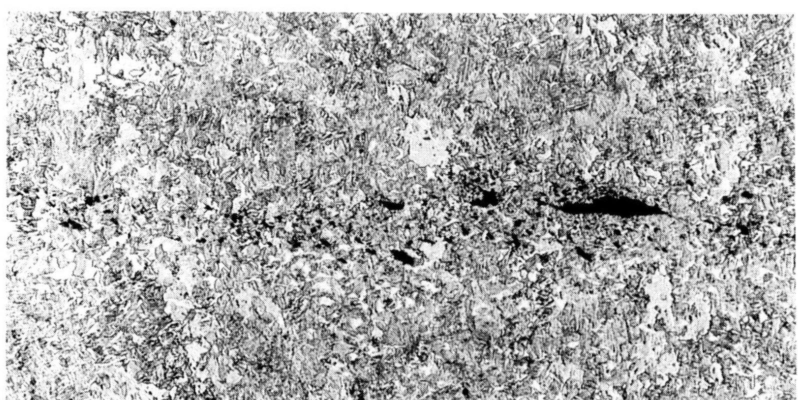

4.134

Table 6. **Top crust in a gear.** Chemical composition of inclusions

	SiO_2 %	FeO %	MnO %	Al_2O_3 %	CaO %	TiO_2 %
Green inclusions	45.0	4.9	18.7	31.7	<1	<1
Black inclusions	24.2	21.7	24.2	<1	<1	<1

gations in a glassy matrix **(Fig. 4.133).** The black slag had reacted with the carbon of the surrounding steel, while the green slag had not. Chemical analysis resulted in the data cited in **Table 6.** The green slag therefore contained substantial amounts of aluminum oxide in addition to silica. This had to be expected considering the high proportion of mullite ($3Al_2O_3 \cdot SiO_2$) in the structure. In contrast, the black slag consisted of about one-fourth each of oxides of silicon, iron, manganese and chromium. Both slags were free of lime, therefore no basic furnace slags were present. Since mullite is a component of the aluminum oxide-rich fire clay, it may be concluded that the green slag came from slagged fire clay fragments of the ladle and channel bricks. However, the still strongly decarburizing oxides of the black slag may have been formed due to the action of oxygen in the air on the top crust metal and may have sunk with it. The finer inclusions in the dividing surfaces between crust and base material consisted of single-phase, glassy, iron-deficient silicates **(Fig. 4.134)** that had no decarburizing effect upon the steel.

Literature Section 4

1) F. Nehl: Die Rotbrüchigkeit kupferhaltiger Stähle und ihre Vermeidung. Stahl u. Eisen 53 (1933) S. 773/78

2) F. Nehl: Die Oberflächenempfindlichkeit von Stählen gegen bestimmte Heizgase. Stahl u. Eisen 58 (1938) S. 779/84

3) H. Buchholtz u. R. Pusch: Ursachen feiner Oberflächenfehler bei der Warmverarbeitung von unlegiertem Stahl. Stahl u. Eisen 73 (1953) S. 204/12

4) K. Born: Die Entstehung von Oberflächenfehlern bei der Warmverarbeitung von Stahl durch Kupfer- und Zinnverunreinigungen. Stahl u. Eisen 73 (1953) S. 1268/80

5) E. Amelung u. H. Schütt: Kupfer- und Zinngehalte im Schrott und ihre Auswirkungen auf den Stahl. Stahl u. Eisen 93 (1973) S. 740/41

6) W. A. Fischer u. H. Bertram: Die Wirkung seltener Erdmetalle auf die Eigenschaften des Baustahles St 52-3. Arch. Eisenhüttenwes. 44 (1973) S. 97/109

7) L. Meyer, H.-E. Bühler, F. Heisterkamp, G. Jackel u. P. L. Ryder: Metallkundliche Untersuchungen zur Wirkungsweise von Titan in unlegierten Baustählen. Arch. Eisenhüttenwes. 43 (1972) S. 823/32

8) F. K. Naumann u. F. Spies: Achsschenkel mit Schlackenzeilen (»Sandstellen«). Prakt. Metallographie 10 (1973) S. 532/38

9) Untersuchung des Einflusses von Oberflächenfehlern auf die Dauerhaltbarkeit von Kurbelwellen. Bericht über eine Gemeinschaftsarbeit des Schmiedeausschusses des Vereins Deutscher Eisenhüttenleute, durchgeführt in der Bundesanstalt für Materialprüfung, Berlin-Dahlem. Erstattet von A. Th. Wuppermann, M. Pfender u. E. Amedick. Verlag Stahleisen m.b.H., Düsseldorf 1958

10) F. K. Naumann, G. Langescheid u. M. Hüser: Die Unterscheidung phosphorarmer und phosphorreicher Bereiche im Eisen durch Ätzen. Arch. Eisenhüttenwes. 38 (1967) S. 145/49

11) E. Plöckinger u. A. Randak: Untersuchungen über das Zeilengefüge in unlegierten und legierten Baustählen. Stahl u. Eisen 78 (1958) S. 1041/58

12) A. Wimmer: Über die Makro- und Mikrostruktur von Gasblasenseigerungen. Stahl u. Eisen 47 (1927) S. 781/86

13) F. K. Naumann u. F. Spies: Blasige Hohlzylinder aus Stahlguß. Prakt. Metallographie 10 (1973) S. 539/41

14) W. Wepner: Der gegenwärtige Stand der Forschung über die Alterung weicher Stähle. Arch. Eisenhüttenwes. 26 (1955) S. 71/98

15) F. K. Naumann: Die Versprödung der Kanten beim Scherenschnitt von Grobblechen aus Thomas- und Siemens-Martin-Stahl. Stahl u. Eisen 81 (1961) S. 1404/09 u. 1464/72

16) A. Kochendörfer u. H. Scholl: Zwillingsbildung und Sprödbruchneigung von Stählen. Arch. Eisenhüttenwes. 28 (1957) S. 483/88

17) F. K. Naumann u. E. Hengler: Muschliger Bruch bei Stahl, seine Ursachen und Bildungsbedingungen. Stahl u. Eisen 82 (1962) S. 612/21

18) F. K. Naumann u. F. Spies: Stahlformguß mit ungenügenden Festigkeitseigenschaften. Prakt. Metallographie 10 (1973) S. 711/16

19) Dieselben: Ritzelwellen aus Stahlformguß mit mangelhafter Dehnung. Prakt. Metallographie 10 (1973) S. 708/10

20) Dieselben: Feingußstücke mit muschligem Bruch. Prakt. Metallographie 10 (1973) S. 704/07

21) L. S. Darken, R. P. Smith u. E. W. Filler: Solubility of gaseous nitrogen in gamma iron and the effect of alloying constituents. Aluminium nitride precipitation. J. Metals 3 (1951) S. 1174/79. Vgl. Stahl u. Eisen 72 (1952) S. 1307

22) F. K. Naumann u. F. Spies: Gebrochene Stützwalzen einer Breitbandstraße. Prakt. Metallographie 9 (1972) S. 658/65

23) F. K. Naumann: Beitrag zur Frage der Bildung von muschligem Bruch in Stahlguß. Arch. Eisenhüttenwes. 35 (1964) S. 1009/10 vgl. Prakt. Metallographie 10 (1973) S. 230/33

24) F. K. Naumann u. F. Spies: Beim Schmieden gerissener Gußblock. Prakt. Metallographie 11 (1974) S. 676/81

25) Zd. Eminger: Muschelbrüche in Nickel-Vanadin-Stählen. Neue Hütte 4 (1959) S. 596/608

26) F. K. Naumann u. W. Carius: Bruchbildung an Stählen bei Einwirkung von Schwefelwasserstoffwasser. Arch. Eisenhüttenwes. 30 (1959) S. 233/38, 283/92, S. 361/70

27) H. Bennek, H. Schenk u. H. Müller: Die Entstehungsursache der Flocken im Stahl. Stahl u. Eisen 55 (1935) S. 321/33. Techn. Mitt. Krupp 3 (1935) S. 74/86

28) L. Luckemeyer-Hasse u. H. Schenck: Löslichkeit von Wasserstoff in einigen Metallen und Legierungen. Arch. Eisenhüttenwes. 6 (1932/33) S. 209/14

29) E. Houdremont u. H. Korschan: Die Entstehungsbedingungen der Flocken im Stahl. Techn. Mitt. Krupp 3 (1935) S. 63/73. Stahl u. Eisen 55 (1935) S. 297/304

30) W. Heller, L. Weber, P. Hammerschmid u. R. Schweitzer: Zur Wirkung von Wasserstoff auf Schienenstahl und Möglichkeiten einer wasserstoffarmen Erschmelzung. Stahl u. Eisen 92 (1972) S. 934/45

31) F. K. Naumann u. F. Spies: Gebrochene Kolbenstangen von Gesenkhämmern. Prakt. Metallographie 11 (1974) S. 357/61

32) Dieselben: Radreifen mit Innenfehlern. Prakt. Metallographie 4 (1967) S. 541/46

33) F. K. Naumann u. F. Spies: Gerissene Kranbahnschienen. Prakt. Metallographie 4 (1967) S. 495/98

34) Dieselben: Gebrochener Gelenkkopf. Prakt. Metallographie 10 (1973) S. 528/31

35) W. Schaarwächter, H. Lüdering u. F. K. Naumann: Die elektrolytische Ätzung mehrphasiger Eisen-Chrom-Nikkel-Legierungen in Natronlauge. Arch. Eisenhüttenwes. 31 (1960) S. 385/91

36) F. K. Naumann u. F. Spies: Pfannenbärte, Deckelstücke, Überwallungen. Prakt. Metallographie 4 (1967) S. 371/72

5. Failures Caused by Processing Errors

Forging and rolling serve not only to give shape to parts, but they may also improve the properties of steel by closing cavities, refining the solidification structure, and equalizing segregations. Cold rolling and drawing serve to improve the surface finish and to increase strength.

5.1 Hot Working Errors

Failures may occur during initial heating of ingots for forging or of castings for annealing. This will be treated in the next chapter (6.1).

Material defects of some significance that are caused by faulty working techniques are **folds, seams, and scabs.** Long stretched folds, so-called laps, for instance, are formed by overfilling of the roll grooves as a consequence of unsuitable shapes and excessive passes. Deep seated folds resembling cracks, the so-called compressive folding cracks, may be formed during passes causing excessive compressive strain. If ingot surface roughness – caused by oxidation during rolling or a defect produced during forging – is smoothed by further working, small shells may be formed. During free form and closed die drop forging, defects may also lead to fold formation.

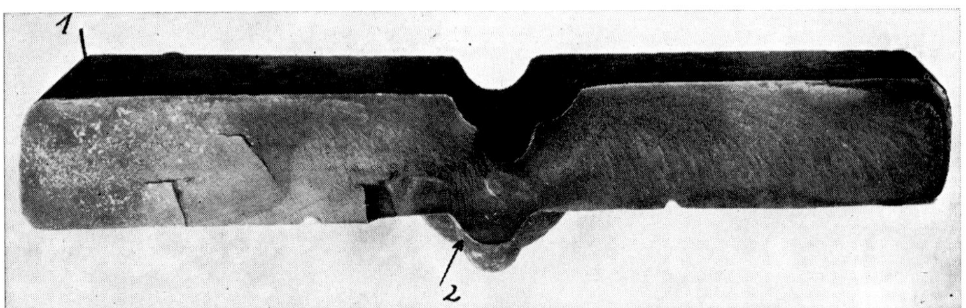

5.1

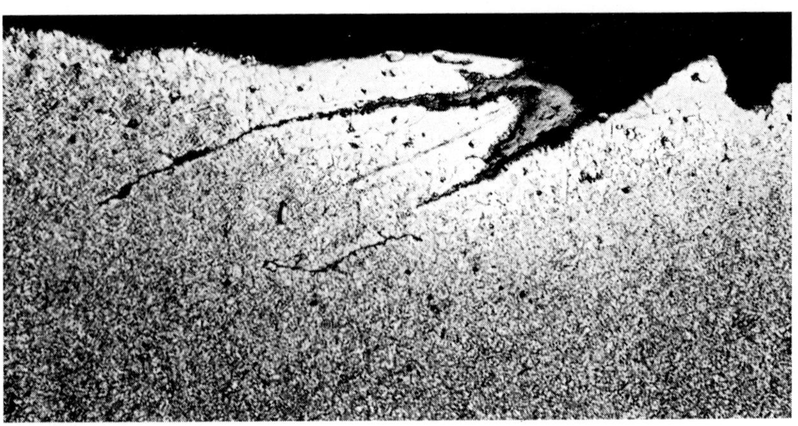

5.2

Fig. 5.1 and 5.2. Fracture of uppermost leaf of railroad car spring

Fig. 5.1. Fracture. 1 x.
1. Origin of primary fracture. 2. Secondary fracture

Fig. 5.2. Lap in fracture origin. Transverse section. Etch: Picral. 100 ×

Folds can be recognized metallographically by the disturbed fiber orientation in the primary structure. They are usually filled with oxide scale and decarburized at the edges. Laps occur frequently at two opposing spots of the cross section. Rolling folds are the harder to interpret and their origins are the more difficult to establish the further the product has been processed. Intermediate pickling, for instance, may serve to dissolve rolled-in oxides. In addition, hydrogen may produce blisters during this operation. It is sometimes difficult to get good starting material, but should be attempted nevertheless.

Folds and scabs form surface defects that may make automobile body sheets unusable[1])[2])[3]). They may also lead to failures, such as fatigue fractures in cyclically loaded parts.

Figure 5.1 shows a fracture of the uppermost leaf of a **railroad car spring** made of medium-carbon steel with approx. 0.45 % C and 1.7 % Si[4]) which was caused by an advanced fatigue fracture at the upper surface of the leaf. Its origin was several millimeters away from the edge shown at the left of the picture. A secondary smaller fatigue fracture on the underside at the cam apparently was formed as a consequence of the first incipient fracture. At the larger fracture initiation site, several longitudinally oriented, line-like defects could be seen at the surface. A transverse section close to the failure showed that these were decarburized laps **(Fig. 5.2)** stretching over the entire length of the leaf; this was also confirmed by additional metallographic sections. The fracture originated in such a rolling fold.

Figure 5.3 shows the torsion fracture in the third and tenth turns of a **helical compression spring**[5]) made from 1.8 mm diameter wire. The failure occurred during the application of a compressive stress to a valve rocker arm. This failure too, was initiated by a fatigue fracture that propagated from the most highly stressed internal fiber. In the fracture origin there was a longitudinal fold **(Fig. 5.4)** which could still be observed at the fourth and fifth turns where a further incipient fracture originated **(Fig. 5.5).** A transverse section adjacent to the fracture also showed that this fold was a decarburized lap **(Fig. 5.6).** This fault helped to promote the failure of this highly stressed spring.

The following case may serve as an example of **compression folds.** A bright cold drawn basic converter steel **wire** having a 3.1 mm diameter suffered a large longitudinal crack during twist tests; these tests were designed and routinely used for the determination of longitudinal defects. A transverse section through the untwisted grip section showed longitudinal cracks at two opposing spots. In contrast to the laps of the type previously described, these cracks did not propagate obliquely and arc-shaped, but almost in a straight line and radially toward the center of the cross section **(Fig. 5.7).** It was shown after Oberhoffer etching that the fibers in this area were compressed and oriented almost perpendicular to the surface. The pattern of the core segregation indicated that these were the side faces of the original billet profile, probably those sides at which the roll grooves were open. Experience has shown that during rolling compression fold cracks are easily formed at these sites[6]).

Bolts of 2.6 and 3.4 mm diameter low-carbon steel wire tore open during upsetting of the heads. The bolts showed longitudinal defects in the extension of the cracks that could be seen better after light pickling with diluted hydrochloric acid **(Fig. 5.8).** Wire specimens, from which the bolts were produced, also showed the same defects **(Fig. 5.9).** Metallographic examination through transverse sections of the bolt shafts and wires showed that no material defects per se were present at the surface **(Fig. 5.10).** The failure in this instance was caused by the opening of rolling folds **(Fig. 5.11).** The difference between this and similar phenomena due to surface defects can be seen from a comparison with Fig. 4.31 to 4.34.

In the following some examples are still cited for rolling defects on **strips and sheets.** A **cold rolled strip** of 50x1 mm cross section was rejected because of scale-like surface defects; these are shown in **Fig. 5.13.** The 3 mm thick **hot rolled starting strip** already had similar and even stronger imperfections which were arranged axially in the rolling direction **(Fig. 5.12). Figure 5.14** shows these in transverse section. They consisted of rolling folds filled with scale. The

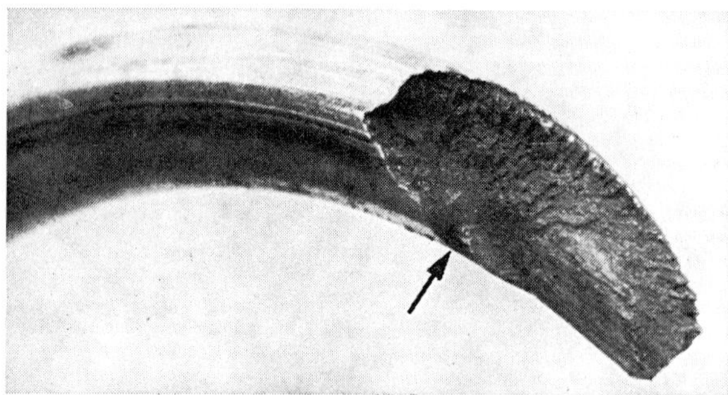

5.3

5.4

5.5

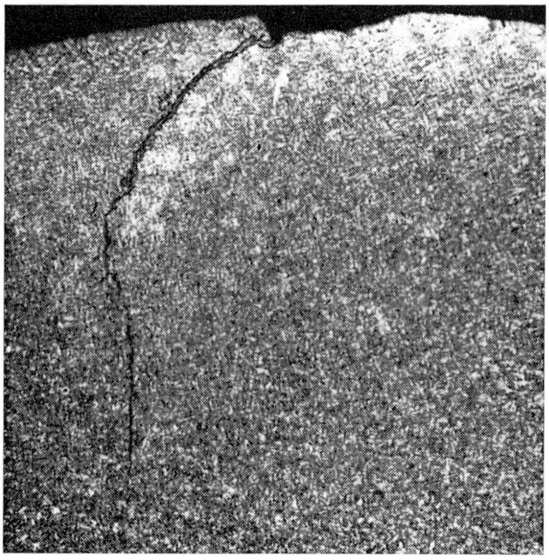

Fig. 5.3 to 5.6. Fracture of helical compression spring

Fig. 5.3. Torsion fracture. 10 x. Fracture origin shown by arrow

Fig. 5.4. Inner view of third turn with fracture path (arrow). 10 ×

Fig. 5.5. Incipient fracture in fourth turn. 10 ×

Fig. 5.6. Transverse section with decarburized rolling fold next to fracture. Etch: Picral. 200 ×

5.6

defects in the cold rolled strip were of the same type, but on a smaller scale **(Fig. 5.15).** They are a direct consequence of the imperfections caused by hot rolling of the strip.

Surface defects in the form of small flat metal scales oriented in the rolling direction **(Fig. 5.16)** occurred in **outer car body sheets** made of aluminum-killed steel. The rolling mill operator believed that they were caused by aluminum oxide inclusions. Metallographic examination of longitudinal sections showed that they were rolling folds **(Fig. 5.17).** No aluminum oxide inclusions could be found in any case connected with these failures.

Sometimes **impressions of inclusions** such as scale can be found in the surface of sheets and strips. **Figure 5.18** shows such an impression in a hot rolled thin gage sheet. The longitudinal section **(Fig. 5.19)** showed that pressed-in scale was adhering to its base. In cold rolled sheets or strips, the scale as a rule is removed by pickling. This leaves empty dimples whose fringes sometimes display grains elongated by cold deformation. **Figure 5.20** shows such an impression of a bright normalized strip steel of 170 x 5 mm cross section.

Unsuitable **forging techniques** also may cause failures as is shown in the following case. During forging of hexagonal ends of 30 mm width on 35 mm round bars made of a heat resistant steel with approx. 0.15 % C, 2 % Si, 25 % Cr and 20 % Ni (type 314), the bars tore open from the inside out **(Fig. 5.21).** The steel manufacturer believed that thermal stress cracks from the cutoff of the bars caused the defective forgings.

Iridium 192 radiation of a number of bars disclosed that the internal forging faults extended in part to three-quarters of the hexagon length **(Fig. 5.22).** Metallographic examination showed no material defects in connection with the forging cracks. The joint faces were cold worked and in part recrystallized, but not cracked.

Forging tests at starting temperatures of 1200, 1050 and 900°C resulted in the tears always occurring when the edges – that had formed during prior blows – were hit. If this was avoided by careful and exact turning of 60° between the blows, a crack-free hexagon could be forged. During forging of square shapes this was always possible without any special precautions.

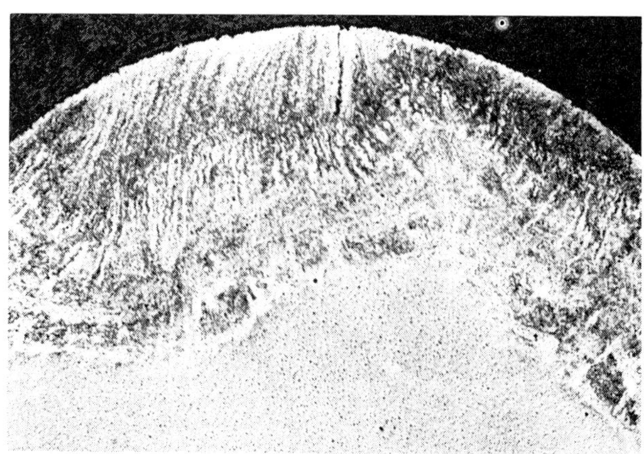

Fig. 5.7. Compression fold crack in basic converter steel wire of 3.1 mm diameter. Transverse section. Etch according to Oberhoffer. 40 ×

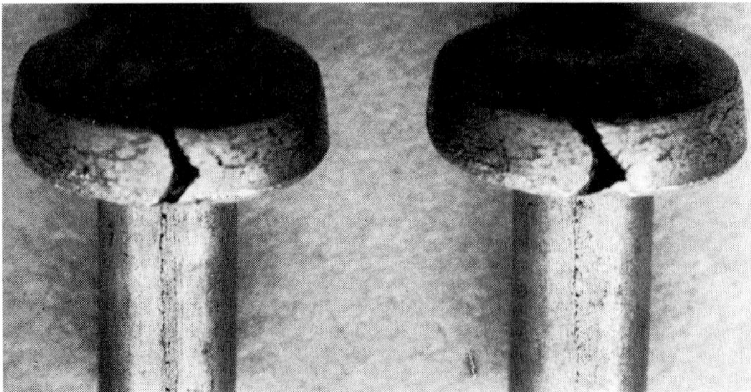

5.8

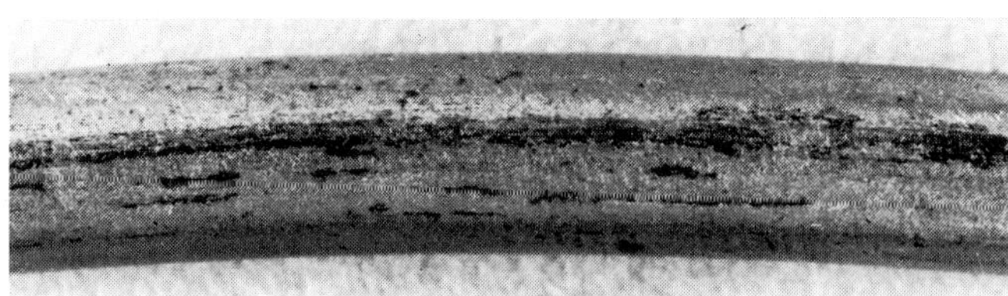

5.9

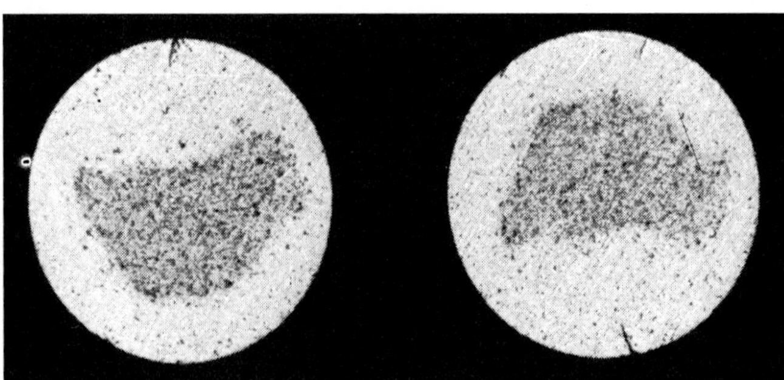

5.10

Fig. 5.8 to 5.11. Bolts of low carbon steel torn open during upsetting of the heads

Fig. 5.8. Side view. Surface pickled with diluted hydrochloric acid. 5 ×
Fig. 5.9. Wire from which bolts were produced. Surface pickled with diluted hydrochloric acid. 5 ×
Fig. 5.10. Transverse section through wire. Etch according to Heyn. 10 ×

Fig. 5.11. Rolling fold in transverse section through wire. Etch: Nital. 100 ×

Crack formation should also be avoidable by forging the hexagon in a closed die drop forge.

The amount of plastic deformation produced during the forging and rolling reduction is also of importance for the properties of the steel. For instance, too little deformation in carbon-rich alloyed tool steels and high speed steels cannot destroy the coarse-meshed network of eutectic (ledeburitic) carbides of the cast state which reduces the toughness of the tools. Therefore thick tools, such as large mills, should not be cut from bars that would require heavy castings with correspondingly strong segregations in their manufacture, but should be forged individually on all sides.

A **hob mill cutter,** which failed axially during service, was made of 3-2.6-2.4 W-Mo-V high speed steel. The fracture at one end face **(Fig. 5.23)** evidently originated from the right angle of the keyway root; it then propagated along the keyway. On the base of the latter, machining grooves could be seen that contributed further to the already high notch effect caused by the shape of the keyway. The fracture in part indicated a coarse grain or network.

Metallographic examination showed a coarse-meshed carbide network that indicated no visible elongation in the longitudinal section **(Fig. 5.24 and 5.25).** The forging therefore must have been less complete than is customary and appropriate for such highly stressed mills. The brittleness caused by this carbide network undoubtedly facilitated the failure of the milling cutter – most likely due to impact – as a result of the notch stress concentration effect of the keyway.

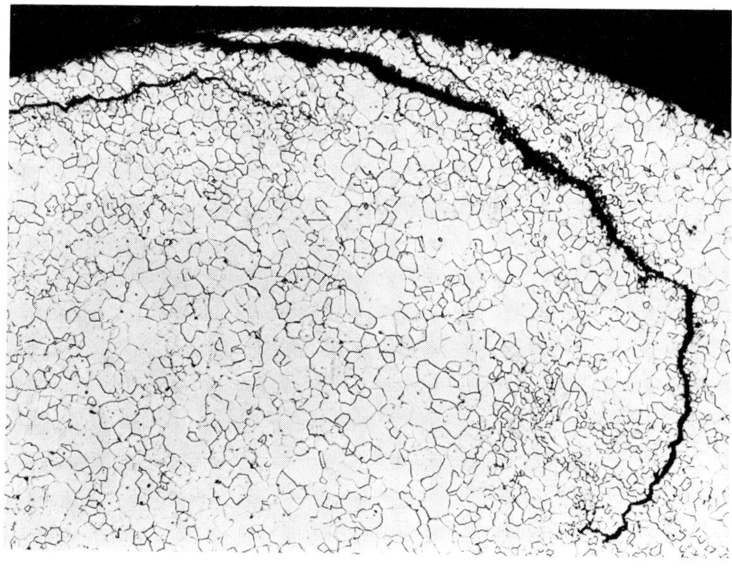

5.11

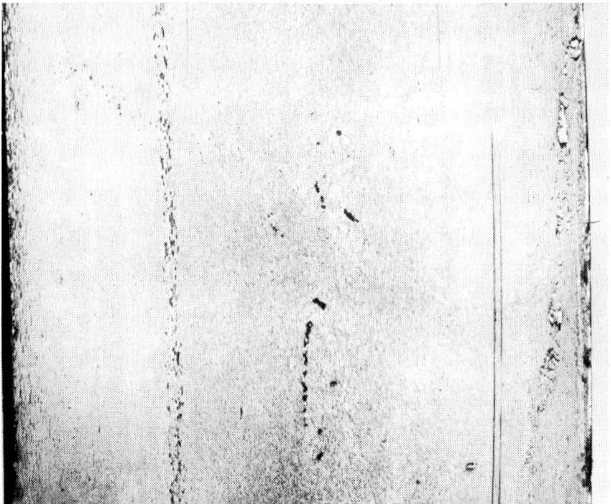
5.12

5.13

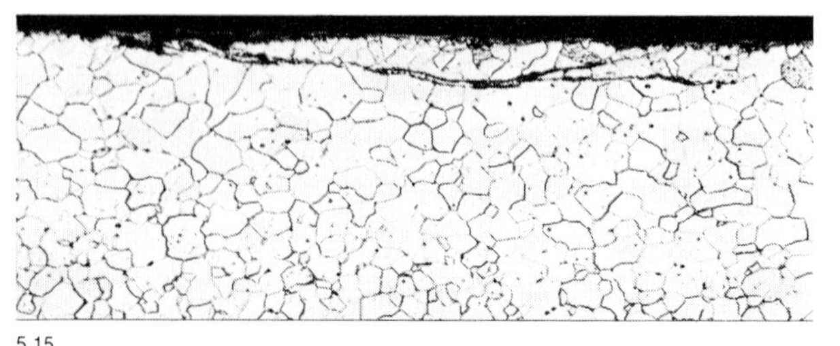
5.15

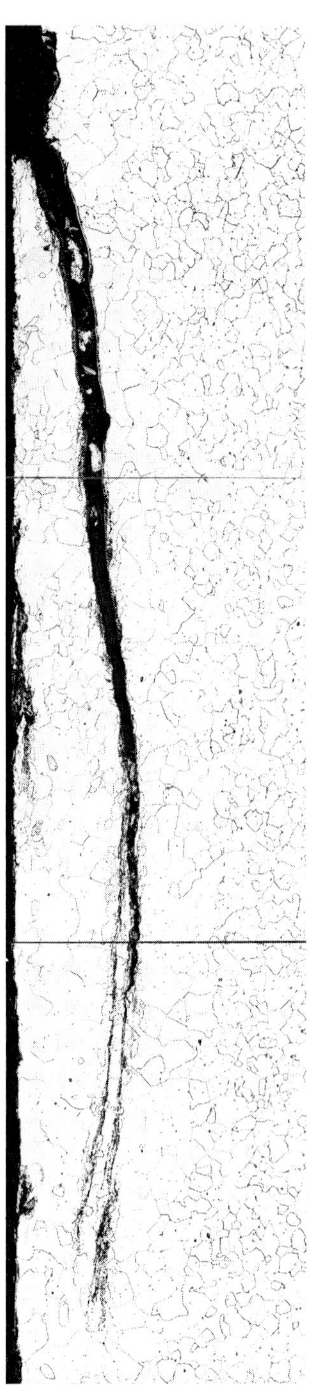
5.14

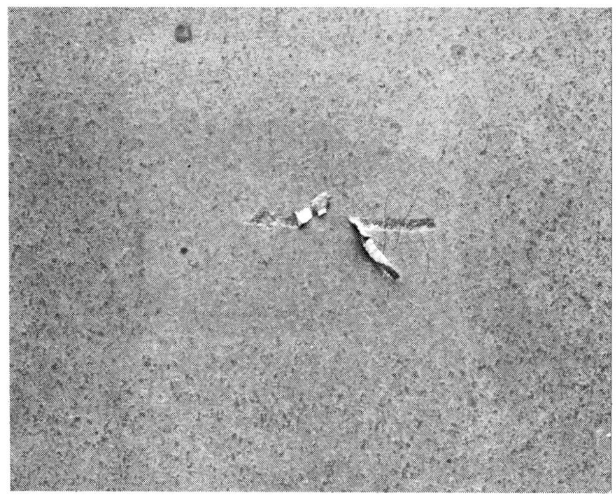

5.16

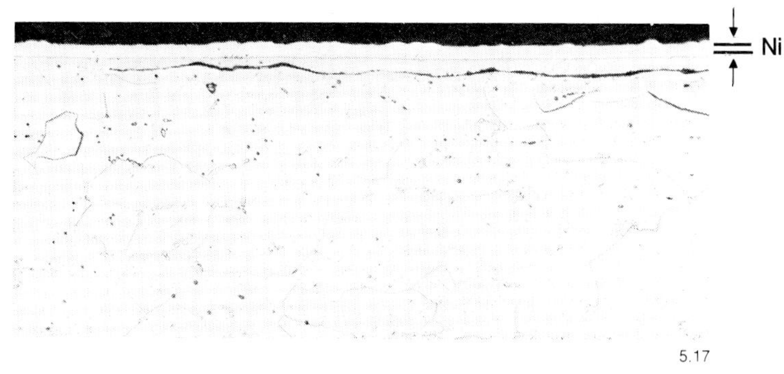

5.17

Fig. 5.16 and 5.17. Surface defects in outer automobile body sheet of 2 mm thickness

Fig. 5.16. Surface. 1 ×

Fig. 5.17. Longitudinal section. Etch: Nital. 200 ×. Surface plated with nickel before preparation of metallographic section

Fig. 5.12 to 5.15. Surface defects on hot rolled strip and subsequently produced cold rolled strip

Fig. 5.12. Surface of hot rolled strip 78 × 3 mm. 1 ×

Fig. 5.13. Surface of cold rolled strip of 50 × 1 mm. 3 ×

Fig. 5.15. Transverse section through cold rolled strip. Etch: Nital. 100 ×

Fig. 5.14. Transverse section tnrough hot rolled strip. Etch: Nital. 60 ×

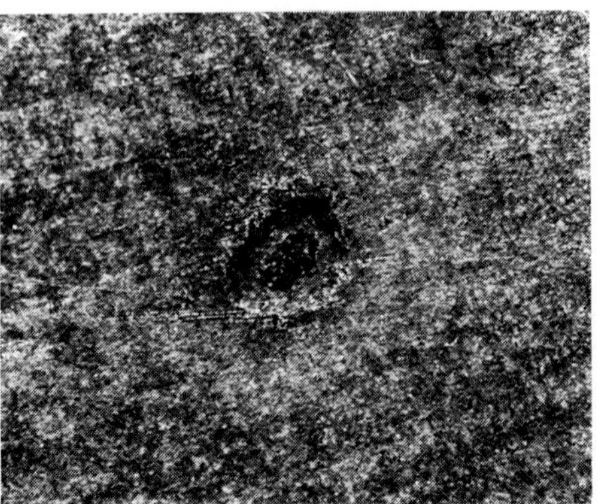

Fig. 5.18. Impression on surface of hot rolled thin gage sheet. 3 ×

Fig. 5.19. Scale rolled at base of impression. Unetched longitudinal section. 200 ×

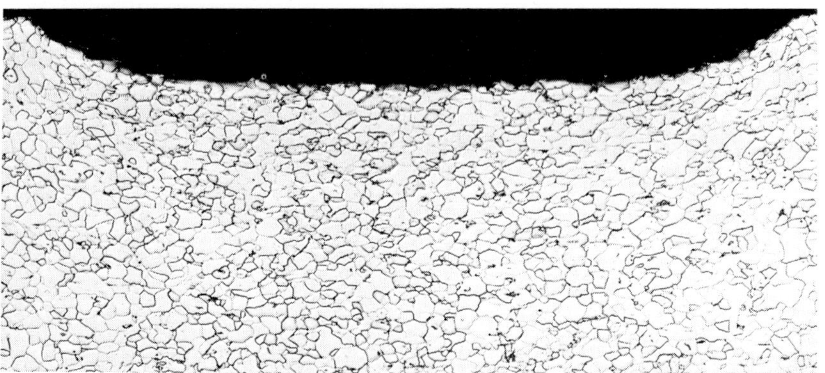

Fig. 5.20. Transverse section through impression at surface of cold rolled strip steel. Etch: Nital. 50 ×

5.21

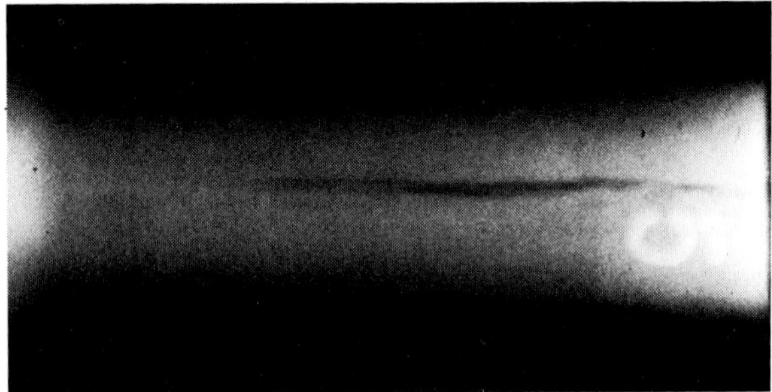

5.22

Fig. 5.21 and 5.22. Faulty forging of hexagonal bar of heat resistant steel. 1 ×

Fig. 5.21. End face

Fig. 5.22. Defect indication by Ir 192 radiation

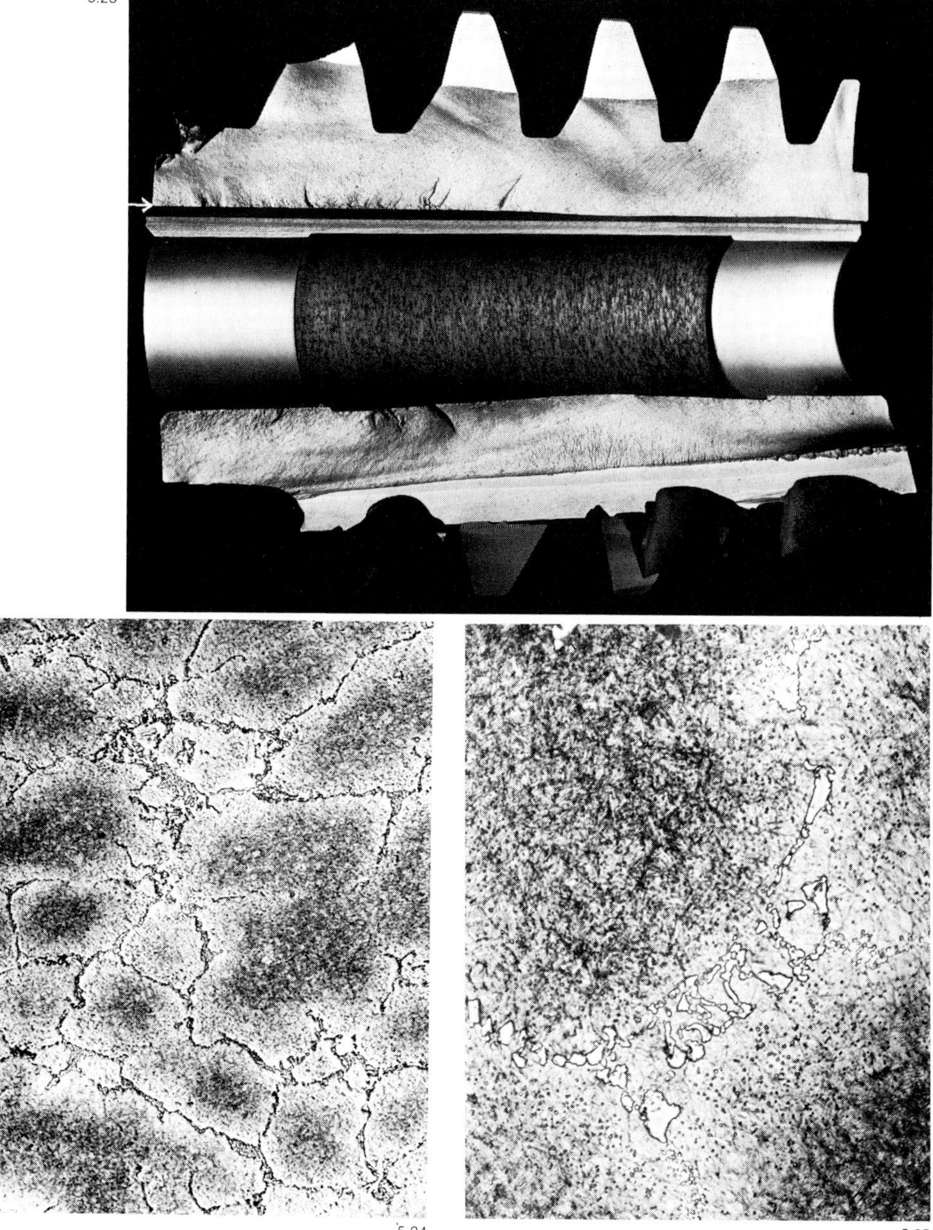

Fig. 5.23 to 5.25. Broken hob type milling cutter of 3-2.6-2.4 W-Mo-V high speed steel
Fig. 5.23. Fracture. 0.5 ×

Fig. 5.24 and 5.25. Coarse mesh carbide network not destroyed by forging. Longitudinal section. Etch: Picral
Fig. 5.24. 100 ×
Fig. 5.25. 500 ×

Errors in thermomechanical treatment may also be due to the application of **inappropriate temperatures.** Excessively high forging or rolling temperatures lead to the formation of coarse grains that cannot be removed by subsequent heat treatment in non-transformation steels such as low-carbon ferritic stainless steels and heat-resistant chromium steels. Parts made of such steels have a tendency toward brittle fracture. Other steels may be regenerated by normalizing if overheating was not too severe. But this is not always true as can be seen from the following failure case.

The pin of an **embossing roll** of cellulose cotton, that was said to be made of a carbon steel with 500 to 600 MPa (73 to 87 ksi) tensile strength, broke without visible deformation. The fracture occurred in the sharp-edged transition from the pin to the barrel of the roll. Metallographic examination showed that the steel had a high sulfide inclusion content **(Fig. 5.26)** and an extraordinarily coarse grain size **(Fig. 5.27).** Therefore, the roll had been severely overheated during forging. Tensile and notch impact tests on longitudinal specimens showed a 720 and 730 MPa (104 and 106 ksi) tensile strength, 4 and 10 % elongation, 10 and 12 % reduction in area, and 3 J notch toughness. All deformation characteristics were therefore too low for the higher strength steel actually used. The specimen fractures were partly of a "rock candy" nature. Such an unsuitably shaped roll, made coarse-grained and brittle as a result of overheating during forging, had to break without plastic deformation under impact-type loading.

5.26

5.27

Fig. 5.26 and 5.27. Microstructure of overheated forged embossing roll. 100 ×

Fig. 5.26. Unetched transverse section

Fig. 5.27. Longitudinal section. Etch: Nital

A low working temperature may also produce an adverse effect; this leads to grain elongation with strengthening or recrystallization[7]). During low deformation forging or rolling, such as in finishing passes (so-called polishing), considerable grain growth, and as a consequence a lowering of the yield point and toughness, may result. Tool steels may become sensitive to overheating if forged or rolled at temperatures that are too low. The consequences of deformation in the blue brittle temperature range were described already in section 4.3.1.

In this section we should mention a special group of **failures in closed die drop forgings.** These are connected with the flash squeezed out of the gap between the upper and lower dies during forging. This flash zone is, per se, a weakened area because segregation products and material defects from the core of the forging may reach the surface at this point in the absence of an opposing pressure. The failure of a crankshaft due to flake cracks in the flash seam which became apparent during machining was already described in section 4.3.3.

Mechanical and thermal stresses between the thinner and colder flash and the core material may lead to the formation of transverse cracks in the seam zone during die forging. **Figure 5.28** shows a drop forged **ball housing** made of AISI 1035 carbon steel, that broke from such cracks after a short period of service. The cracks and a large part of the fracture **(Fig. 5.29)** were heavily scaled and decarburized and therefore were probably generated while hot **(Fig. 5.30).**

5.28 5.29 5.30

Fig. 5.28 to 5.30. Drop forged ball housing of 1035 steel that failed due to cracks in closed die seam

Fig. 5.28. Fracture with cracks. 3 ×
Fig. 5.29. Fracture with partly "rock candy"-like structure. 3 ×
Fig. 5.30. Unetched longitudinal section through cracks according to Fig. 5.28. 20 ×

5.31

5.32

5.33

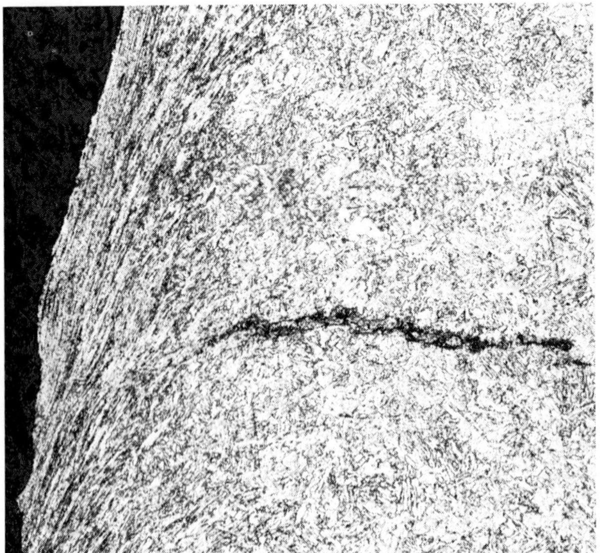

5.34

Fig. 5.31 to 5.34. Drop forging with longitudinally torn open deburring seam

Fig. 5.31. Side view. 0.5 ×
Fig. 5.32. Site of crack from Fig. 5.31. 2 ×
Fig. 5.34. As Fig. 5.33. Etch: Nital. 100 ×

Fig. 5.33. Unetched section transversely through deburred seam. 3 ×

Another type of failure may be caused by deburring, especially if this is done cold. This failure type is generally indicated by the formation of longitudinal cracks in the seam or next to it. **Figure 5.31** shows a drop forging of AISI type 1023 carbon steel with longitudinal cracks in the deburred seam **(Fig. 5.32).** The cracks follow the fiber that was strongly deflected by cold deformation according to **Fig. 5.33 and 5.34,** and probably were caused by shear stresses.

Parts made from a high hardenability steel should be cooled slowly, e.g. in a bed of sand, after forging or rolling so that no high residual stresses or cracks are formed. But failures due to cooling that is too fast after hot deformation are rarely seen by the metallographer because such cracked parts are usually discarded beforehand.

5.35

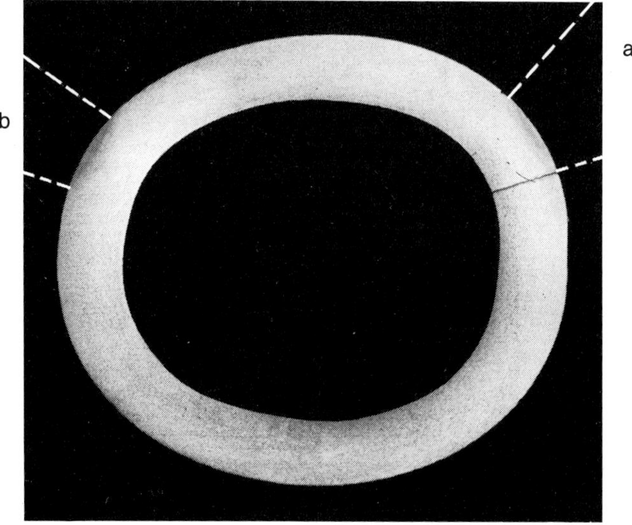

5.36

Fig. 5.35 and 5.36. Cracked high pressure pipe of oil hydraulic shear installation

Fig. 5.35. Crack on outside surface. 4 ×

Fig. 5.36. Transverse section. Etch: Nital. 1 ×. a and b are zones of fine-grained microstructure

5.37

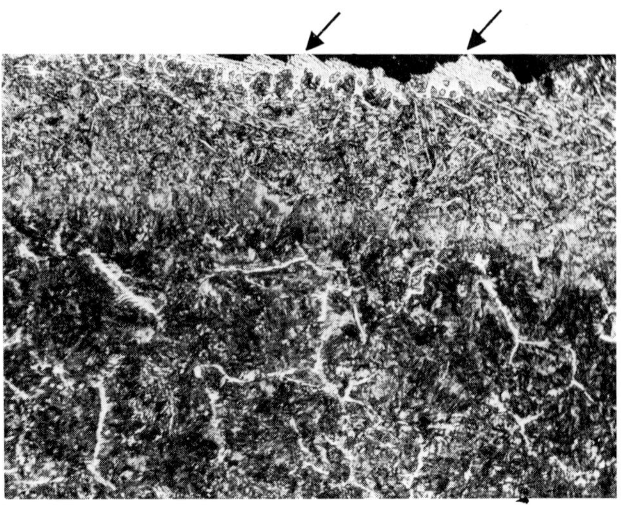

5.38

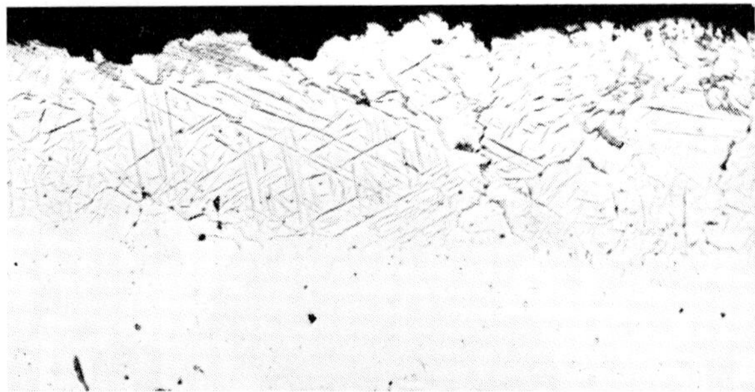

5.39

5.40

Fig. 5.41. Drawing tears in hoisting cable wire. Longitudinal section. Etch: Nital. 10 ×

5.42

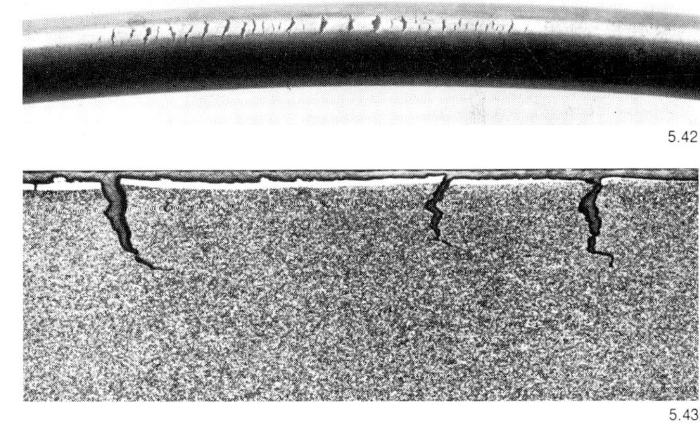

5.43

5.44

Fig. 5.42 to 5.44. Cold drawn 6 mm square wire with edge cracks

Fig. 5.42. Edge with cracks. 2 ×

Fig. 5.43. Longitudinal section through edge with cracks. Etch: Picral. 8 ×

Fig. 5.44. Martensitic peripheral zone in longitudinal section. Etch: Picral. 500 ×

Fig. 5.37 to 5.40. Sprocket gear broken from carburized flame cut edge

Fig. 5.37. Fracture. 1 ×. Flame cut edge and fracture origin at top

Fig. 5.38 and 5.39. Peripheral structure under flame cut edge. 200 ×

Fig. 5.38. Etch: Nital. Arrows indicate ledeburite

Fig. 5.39. Etch: Alkaline sodium picrate solution

Fig. 5.40. Stress crack at flame cut edge. Unetched. 200 ×

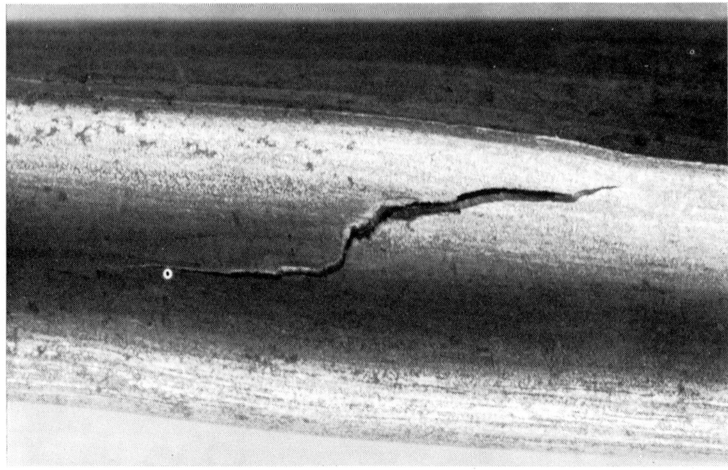

5.45

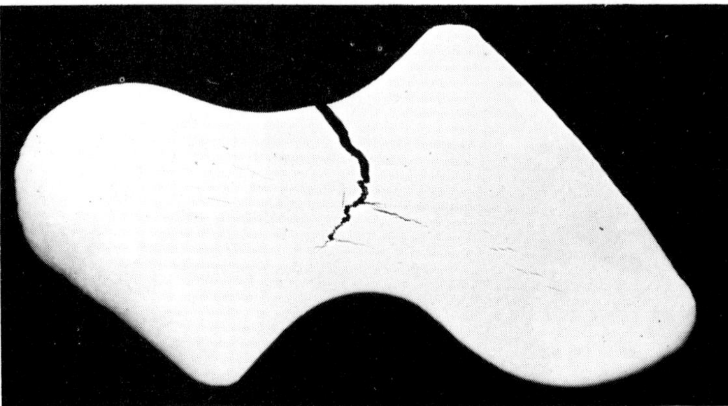

5.46

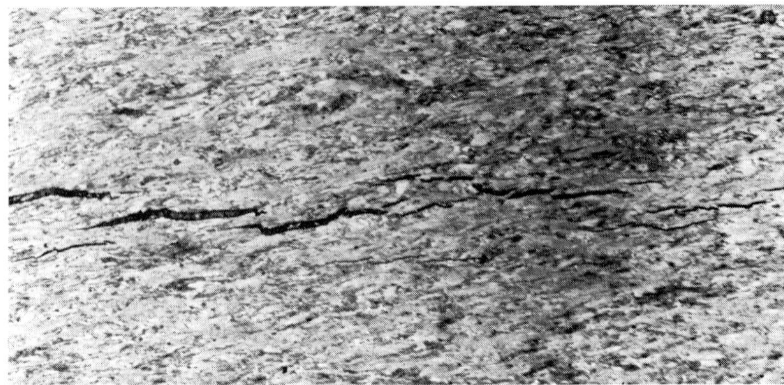

5.47

In special cases, failures have occurred due to slow cooling from the forging temperature. An example was a **drum cover** of a centrifuge that cracked radially at several places. It consisted of the semi-ferritic heat resistant steel containing approx. 0.2 % C, 1 % Si, 26 % Cr, and 4.5 % Ni, that has a strong tendency toward σ-phase formation[8]) during slow cooling or annealing in the temperature range under 850 °C. The cracks were found only after heat treatment. Metallographic examination did not give any indication of the origin of this failure. It was shown only later that the cover had been cooled so slowly after forging that sigma phase could form; this led to stress and crack formation through the contraction during precipitation. The phase itself had been dissolved again during heat treatment and was transformed into chromium ferrite.

The following failures are also due to hot working errors in a wider sense. A **pipe bend** (70 mm O.D. and 10 mm wall thickness) made from a steel which had a tensile strength of about 350 MPa (50 ksi) was used in an oil hydraulic shear installation at a working pressure of 315 atm. This bend broke open longitudinally after a short period of service without visible deformation **(Fig. 5.35)**[9]). The crack was located sideways inside the bend. In the etched transverse section of the completely defect-free material, deep zones of 10–15 mm width with altered microstructure **(Fig. 5.36** points a and b) stood out at both sides of the bend plane. The crack was located at the periphery of one of these. In these zones the ferritic-pearlitic microstructure was much more finegrained than the base structure and also showed other signs of more rapid cooling. The pipe probably had been placed upon a cold support during bending or came in touch with a cold tool. The pipe had cracked in this transition zone that had been cooled earlier and faster by way of thermal and transformation cracks. A similar case was reported in section 1 (Fig. 1.23 to 1.25).

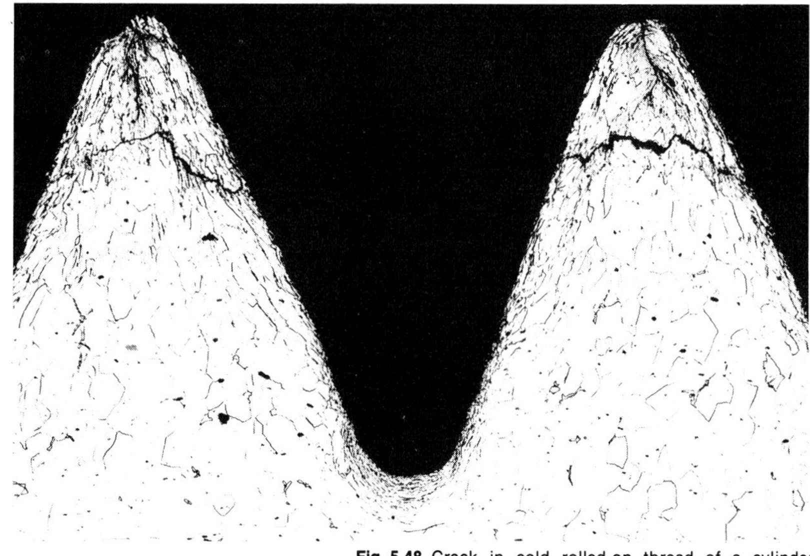

Fig. 5.48. Crack in cold rolled-on thread of a cylinder bolt. Etch: Picral. 100 ×

Fig. 5.45 to 5.47. Crack in a Z-profile wire produced during rolling of the profile

Fig. 5.45. Crack site. 7 ×
Fig. 5.46. Unetched transverse section. 10 ×
Fig. 5.47. Internal crack according to Fig. 5.46. Transverse section. Etch: Picral. 200 ×

5.49

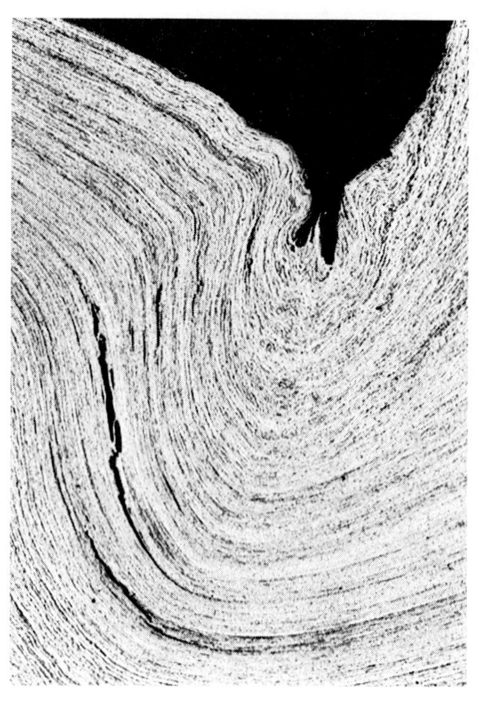

5.50

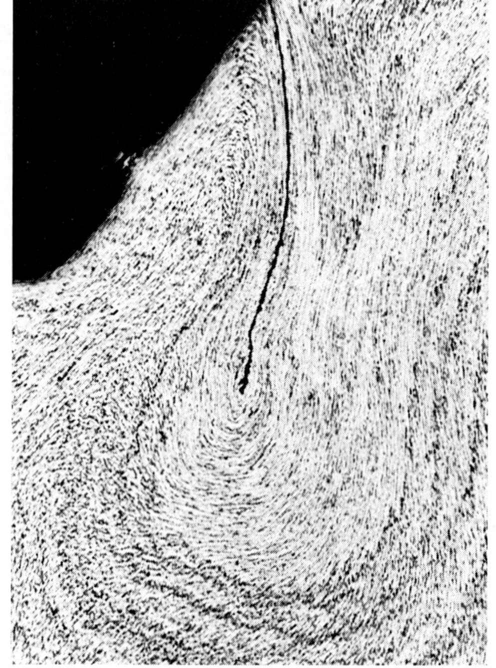

5.51

Some **sprocket gears** that were produced by flame cutting the teeth from heat treated preforms of carbon steel type 1055 cracked when being dropped during handling. Both fracture halves were noticeably offset against each other which indicated the presence of substantial residual stresses. **Figure 5.37** shows one of the fracture halves. The failure occurred at the deepest place of the tooth root without any recognizable deformation and, according to the orientation of the fibers, propagated from the flame cutting surface. The fracture was grainy. Bend tests with sections that were bent with the tooth root on the tension side, showed similar grainy and deformation-deficient fractures. When the gas heat-affected tooth root zone was machined off, the steel could be deformed. A ring that had no gear teeth also broke in a typical deformation fracture during bending over the side plane, but only after extensive deformation.

During metallographic examination of one of the sprockets the flame-cut edge showed the microstructure of a hypereuteotoid steel with secondary cementite in the shape of platelets and grain boundary networks, and in part also the ledeburitic eutectic of white cast iron **(Fig. 5.38 and 5.39).** The strong carburization illustrated here proves that the flame cutting took place either under wrong burner guidance or with significant gas excess. The strong carburization resulted in a brittle steel causing partial cracking **(Fig. 5.40).** The rupture of the gears under impact-type stress therefore was facilitated by strong carburization at the flame-cut edge and the formation of fine quench cracks.

5.2 **Cold Working Errors**

During cold working, failures may also occur through errors in **deformation technique.** During the drawing of wires with excessive reduction in cross section, too sharp an entrance angle of the draw plate bore or insufficient lubrication may result in the formation of internal cracks. This can be seen, for example, from **Fig. 5.41** which is a longitudinal section of a patented **wire** drawn to 3.1 mm diameter that broke during stranding into a hoisting cable.

If lubricating film breaks, thin martensitic layers may be formed at the surface during **drawing of steel wire**[10]). This may also happen in guides, reverse rollers and similar parts due to friction. Even in razor blade steel such grinding checks were found. **Figure 5.42** shows a host of edge cracks in a 6 mm **square wire** that was drawn from an 8 mm square rolled wire. As can be seen from **Fig. 5.43 and 5.44** these cracks propagated from a thin martensitic surface zone that was caused by thermal friction during drawing.

The crack shown in **Fig. 5.45** in a **Z-profile wire** evidently occurred during rolling of the profile. As can be seen from the transverse section in **Fig. 5.46,** a group of short internal cracks are visible in addition to one large crack, all of which run parallel to one another. The pronounced grain elongation in this zone **(Fig. 5.47)** proves that this was the direction of strongest deformation. The previously cold drawn wire had been stressed beyond its deformation capacity during profile rolling.

The failures shown in **Fig. 5.48** occurred during **cold rolling of a thread.** They led to a tear-out of the thread of a bright galvanized slotted **head screw.***

---

* AM 4 x 22 mm according to DIN 84.

Fig. 5.49 to 5.51. Folds in flow formed part

Fig. 5.49. Longitudinal section. Etch according to Heyn. 1 ×

Fig. 5.50 and 5.51. Longitudinal sections. Etch: Nital. 50 ×

Fig. 5.50. Crack at base of hollow
Fig. 5.51. Crack at open end of hollow

The laps and tears at the edge of the inner space of a **flow formed part** shown in **Fig. 5.49 to 5.51** may also have been caused by overloading during pressing. No material defects or treatment errors were found in any of these parts.

Shot peening of the surface of cyclically stressed parts is an established method to increase fatigue strength because it strengthens the outer zone by placing it under compressive stress. But if this operation is conducted under unsuitable conditions, it may also have the opposite effect and cause failures[11]). **Figure 5.52** shows the shot peened surface of a **valve spring** that developed transverse incipient cracks in several places; starting from such an incipient crack the spring then broke prematurely at a point that was not the most highly stressed[12]). From the longitudinal section of **Fig. 5.53** it can be seen that the cracks were caused by cold deformation during shot peening under the additional influence of shear stresses.

Figure 5.54 shows fatigue fractures in a cold worked low-carbon unalloyed steel **tire rim** that was strengthened in the fillets by shot peening with wire shot. The metallographic section showed the surface to be very rough at the shot peened places with incipient cracks and folds **(Fig. 5.55);** this can be seen particularly clearly in a comparison with the smooth surface of the non-strengthened part **(Fig. 5.56).**

According to A. Pomp and M. Hempel[11]) the most uniform results with regard to an improvement in fatigue strength can be obtained during blasting with fine grained steel shot at low blasting pressure but extended blasting time.

Accidental cold deformation and its consequential defects may occur during processing, trimming, and punching with dull tools. **Figure 5.57** shows the fracture of an **angle bracket** of rimmed cast basic converter steel with 370 to 450 MPa (53 to 63 ksi) tensile strength. The fracture propagated from two opposing points at the edge of a rivet hole. In one of the other rivet holes another incipient crack was found to have dark discoloration on the fracture surface. This suggests that the crack was present some time prior to the rupture of the angle **(Fig. 5.58).** Sections through the rivet holes showed strong deformation of the peripheral structure, and in the deformed zone incipient cracks could already be detected **(Fig. 5.59).** Very probably the fracture propagated from such an incipient crack. Etching according to Fry showed that the angle bracket was strain aged – probably through straightening deformation **(Fig. 5.60).** Aging embrittlement made crack tip deformation impossible, and as a result could not prevent crack growth.

If the cold worked sections are softened by annealing, a coarse grain size may be produced by **recrystallization** in regions that have slight plastic deformation as long as the deformation exceeds a certain threshold value[7]). In low-carbon unalloyed steel the critical level of strain, i.e. the degree of deformation that leads to the coarsest grain size, is around 10 %. The grain size grows with increasing temperature for the same degree of deformation. In steels with α-γ-transformation, the upper temperature region of recrystallization is limited by the transformation temperature. Therefore, any deliberate or nondeliberate slight cold work should be avoided prior to soft annealing and stress relieving.

The fracture and metallographic section in **Fig. 5.61** illustrates the coarse grain structure of a stress relieved steel **wire** that had been cold drawn from 11 to 10 mm diameter. The wire could still be bent slowly and with modest effort, but broke upon impact-induced bending.

Coarse grains lead not only to embrittlement but also to the formation of a crinkled surface during plastic deformation as can be seen in **Fig. 5.62** on a **bent tubing** for a bicycle handlebar. This should be taken into consideration during rolling of thin gage sheet, especially of deep drawn and automobile body sheet, where a smooth surface is of particular importance.

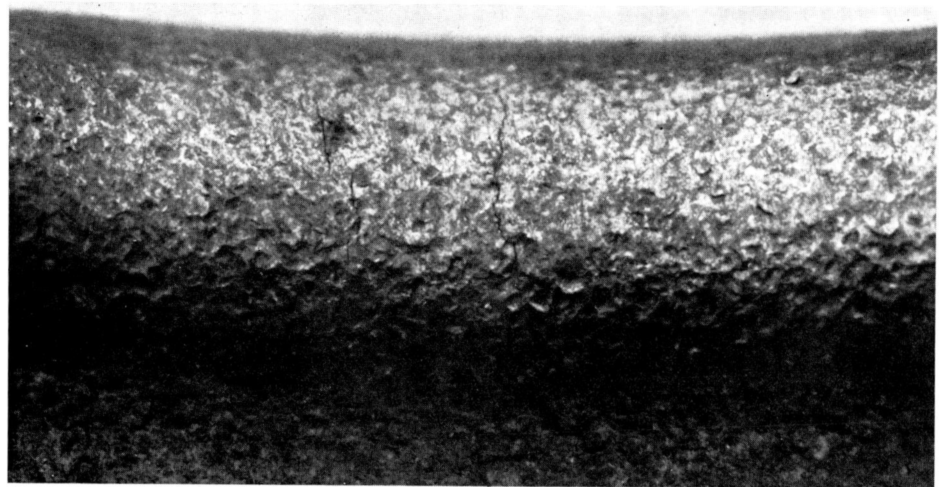

5.52

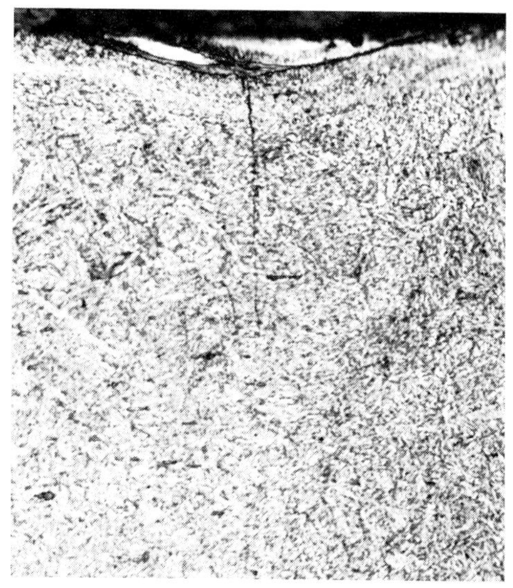

5.53

Fig. 5.52 and 5.53. Valve spring with incipient crack and failure originating at shot peened surface

Fig. 5.52. Fatigue fractures at surface. 20 ×
Fig. 5.53. Longitudinal section through incipient crack.
Etch: Picral. 500 ×

5.54

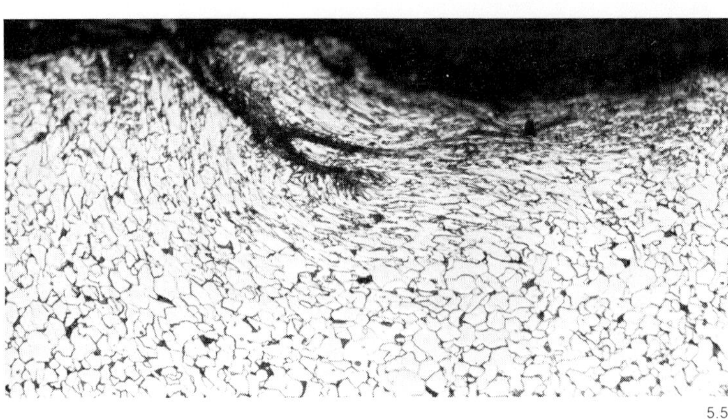

5.55

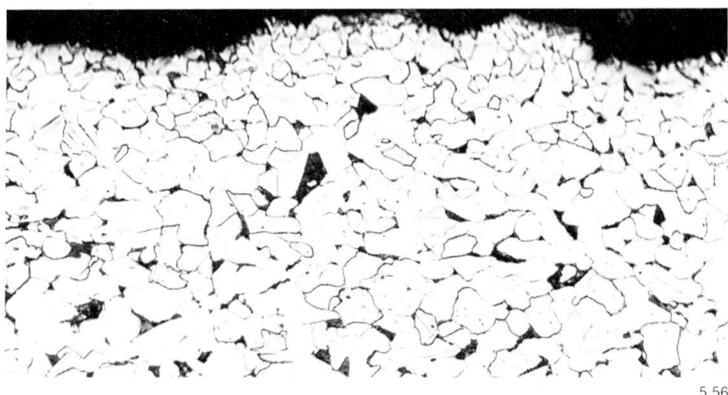

5.56

Fig. 5.54 to 5.56. Prematurely broken shot peened tire rim

Fig. 5.54. Failure that was induced by several fatigue fractures

Fig. 5.55 and 5.56. Transverse section. Etch: Nital. 200 ×

Fig. 5.55. Microstructure below the shot peened face
Fig. 5.56. Microstructure below opposite, non-strengthened surface

Fig. 5.57 to 5.60. Angle bracket of rimmed cast basic converter steel that broke from a cold punched rivet hole

Fig. 5.57. Fracture with origin at rivet hole. 0.8 ×
Fig. 5.58. Incipient fracture in another rivet hole. 1 ×
Fig. 5.59. Metallographic section through a rivet hole that displayed incipient cracks in the cold deformed punched edge. Etch: Nital. 100 ×
Fig. 5.60. Cross section through angle bracket. Etch according to Fry. 0.8 ×

5.57

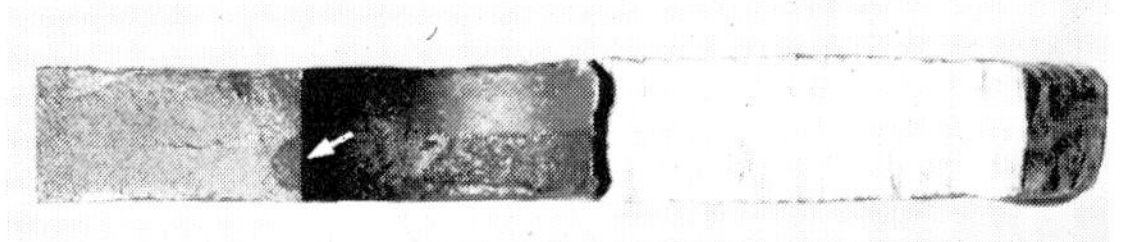

5.58

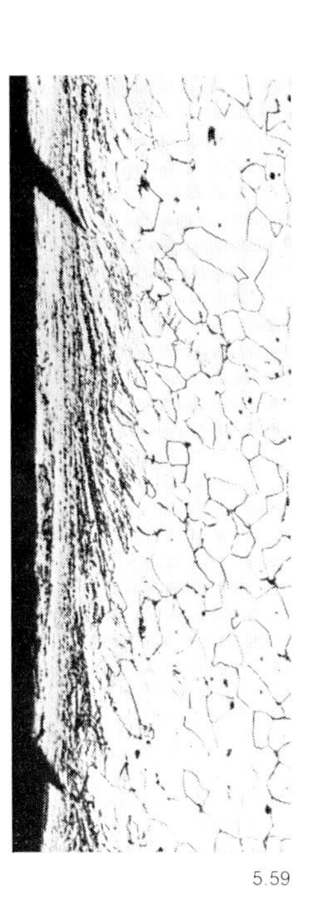

5.59

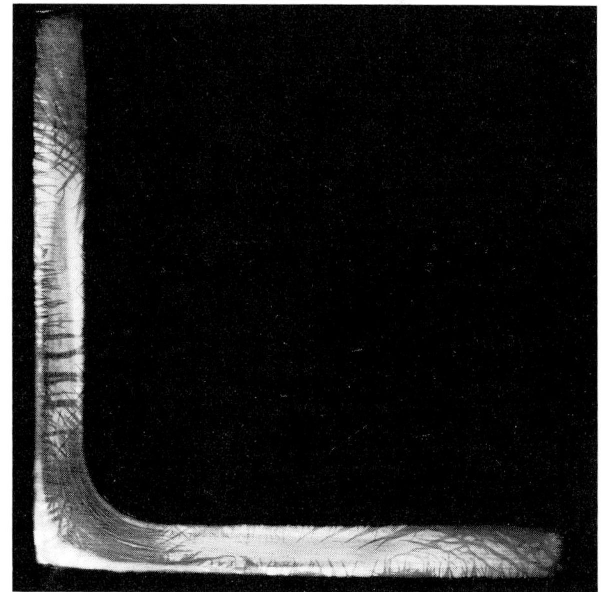

5.60

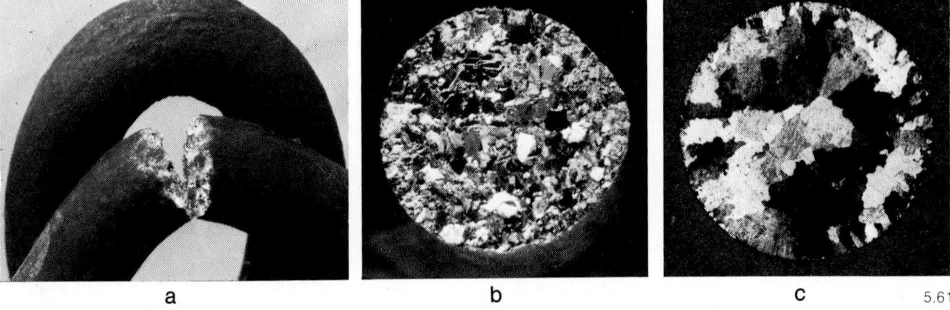

Fig. 5.61a to c. Stress relieved steel rod drawn from 11 to 10 mm

Fig. 5.61a. Bend specimens. 1 ×. Top: Bent slowly. Bottom: Bent by impact loading

Fig. 5.61b. Fracture. 2 ×

Fig. 5.61c. Transverse section. Etch according to Heyn. 2 ×

Fig. 5.62. Bicycle handlebar bent from a coarse-grained recrystallized tubing. 1 ×

Literature Section 5

1) Oberflächenfehler an kaltgewalztem Band und Blech. Herausgegeben vom Verein Deutscher Eisenhüttenleute. Verlag Stahleisen m. b. H., Düsseldorf 1967

2) Walzdrahtfehler. Herausgegeben vom Verein Deutscher Eisenhüttenleute. Verlag Stahleisen m. b. H., Düsseldorf 1973

3) G. Oehler: Oberflächenfehler an kaltgewalzten Tiefziehstahlbändern für den Karosseriebau. Mitt. Forsch.-Ges. Blechverarbeitung 1953, S. 169/74

4) F. K. Naumann u. F. Spies: Gebrochenes Federblatt. Prakt. Metallographie 11 (1974) S. 227/30

5) F. K. Naumann u. F. Spies: Gebrochene Schraubendruckfeder. Prakt. Metallographie 10 (1973) S. 421/22

6) H. Cramer: Einfluß einiger Kalibrierungsarten auf das Entstehen von Druckfaltungsrissen in Walzstäben. Stahl u. Eisen 55 (1935) S. 797/805

7) F. K. Naumann u. F. Spies: Grobkornbildung durch Rekristallisation. Prakt. Metallographie 4 (1967) S. 423/25

8) F. K. Naumann: Beitrag zum Nachweis der σ – Phase und zur Kinetik ihrer Bildung und Auflösung in Eisen– Chrom- und Eisen–Chrom–Nickel–Legierungen. Arch. Eisenhüttenwes. 34 (1963) S. 187/94

9) F. K. Naumann u. F. Spies: Gerissener Rohrbogen einer hydraulischen Anlage. Prakt. Metallographie 10 (1973) S. 648/50

10) A. Pomp: Örtliche Martensitbildung bei Stahldraht. Mitt. Kais.-Wilh.-Inst. Eisenforsch., Düsseldorf, 16 (1934) S. 15/19. Vgl. Stahl u. Eisen 54 (1934) S. 297

11) A. Pomp u. M. Hempel: Dauerfestigkeit von Schraubenfedern mit unterschiedlicher Fertigungsart. Arch. Eisenhüttenwes. 21 (1950) S. 243/62

12) F. K. Naumann u. F. Spies: Vorzeitig gebrochene Ventilfeder. Prakt. Metallographie 7 (1970) S. 115/17

6. Failures Through Errors During Heating and Thermal Treatment

6.1 Heating Errors

The following errors occur most frequently during heating of parts for forging or heat treatment:

1. They are heated too fast causing stresses in the outer zones, especially in thick specimens, which is the result of a large temperature gradient and a large difference in thermal expansion.

2. They are heated either non-uniformly, or on one side, or locally overheated.

3. They are not thoroughly heated, especially if made of steels with low thermal conductivity which results in the tearing open of the core during working.

4. They are heated at too high a temperature or for too long a time, so that they scale heavily, decarburize, or burn.

6.1.1. Errors During Heating

A typical failure case due to heating that was too fast and not thorough enough was the fracture of a **pinion shaft** of a spur gearing **(Fig. 6.1).** The shaft consisted of a 4150 type chromium-molybdenum steel, and broke after only 3 months' service even though the maximum bend stress was said to have been only 33.4 MPa (4840 psi). During an attempt to repair the shaft temporarily, at which time it was provided with a central bore, a second internal crack was detected about 520 mm distant from the previous fracture. The latter originated at point a, designated by an arrow in Fig. 6.1, and propagated over most of the cross section. Service stresses caused two fatique fractures which propagated from the internal crack (left and right in Fig. 6.1), that led to failure in a short time. A longitudinal section through the fracture origin showed that a segregation streak was present at this point **(Fig. 6.2).** It was of a dimension considered normal for this size of forging and could not have been the cause of failure; however this streak was the point of lowest resistance, from which each fracture originated during overloading.

This was not the case in the **back-up roll** discussed in section 4.3.2 (Fig. 4.75 to 4.78) in which the formation of the cracks was decisively facilitated by internal defects of different types.

A number of thick-walled **housings** of 4140 steel showed longitudinal cracks after heat treatment in the interior of the U-shaped part. They originated in the fillets **(Fig. 6.3)** and were located at both sides of a transverse bore hole. During finishing and rectangular machining of the fillet they were removed except for minor traces. Two housings and one raw forging, from which they had been drop forged, were examined. The raw forging as well as the housings were found to be clean and free of defects during metallographic examination, as was proven by Baumann sulfur prints of the transverse section of one of the housings **(Fig. 6.4).** The cracks were strongly oxidized and decarburized **(Fig. 6.5 and 6.6).** Therefore they must have been present before hardening. Since the raw billets were free of cracks it may be concluded that these occurred during heating for hardening. Cold parts were probably put into the furnace that was held at hardening temperature. The bores restricted elongation and facilitated crack formation. Perhaps failure could have been prevented if the hole had been drilled after heat treatment.

6.1

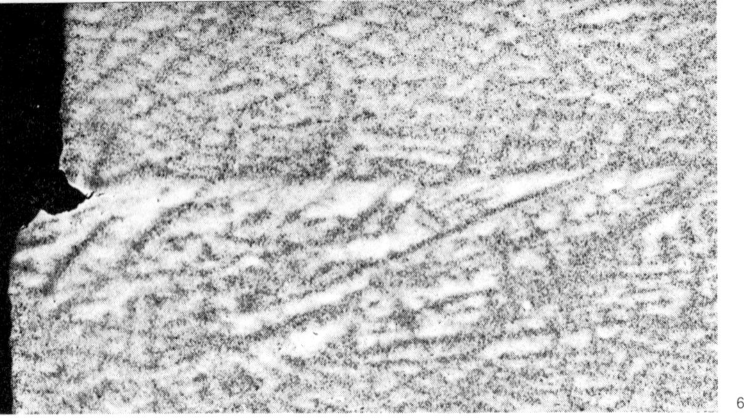

6.2

6.3

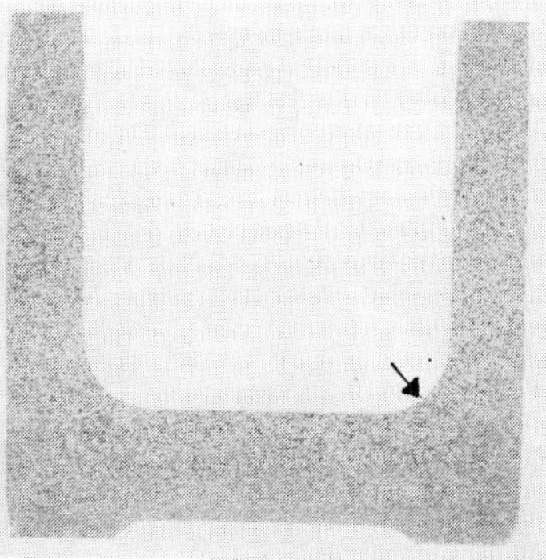

6.4

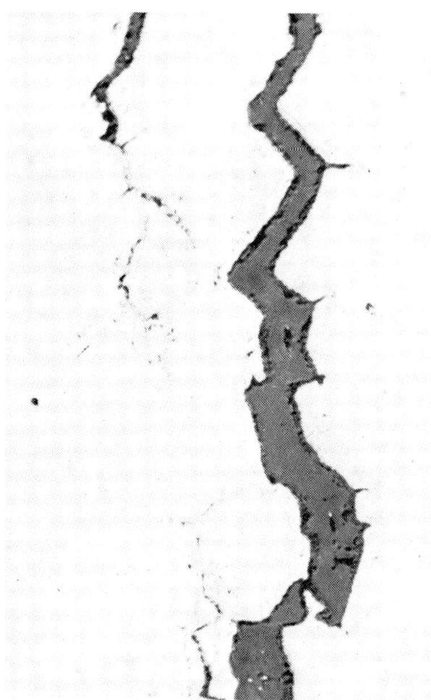

6.5

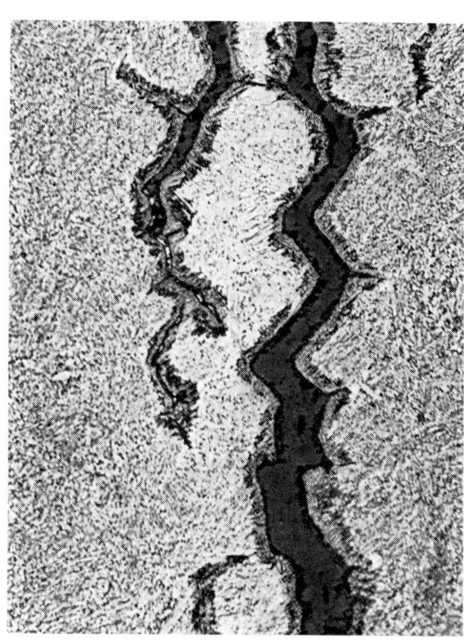

6.6

Fig. 6.3 to 6.6. Heat treated housing of 4140 type steel with longitudinal cracks in fillets

Fig. 6.3. Internal view. 0.5 ×

Fig. 6.4. Baumann sulfur print of transverse section through housing. Arrow points to crack site. 0.5 ×

Fig. 6.5 and 6.6. Crack propagation in transverse section through housing. 500 ×

Fig. 6.5. Unetched

Fig. 6.6. Etch: Nital

Fig. 6.1 and 6.2. Pinion gear shaft of 4150 type steel broken after three months of service

Fig. 6.1. Fatigue fractures propagating from an internal heat-up crack. 0.2 ×. Fracture origin designated by arrow

Fig. 6.2. Longitudinal section through crack origin. Etch according to Oberhoffer. 3 ×

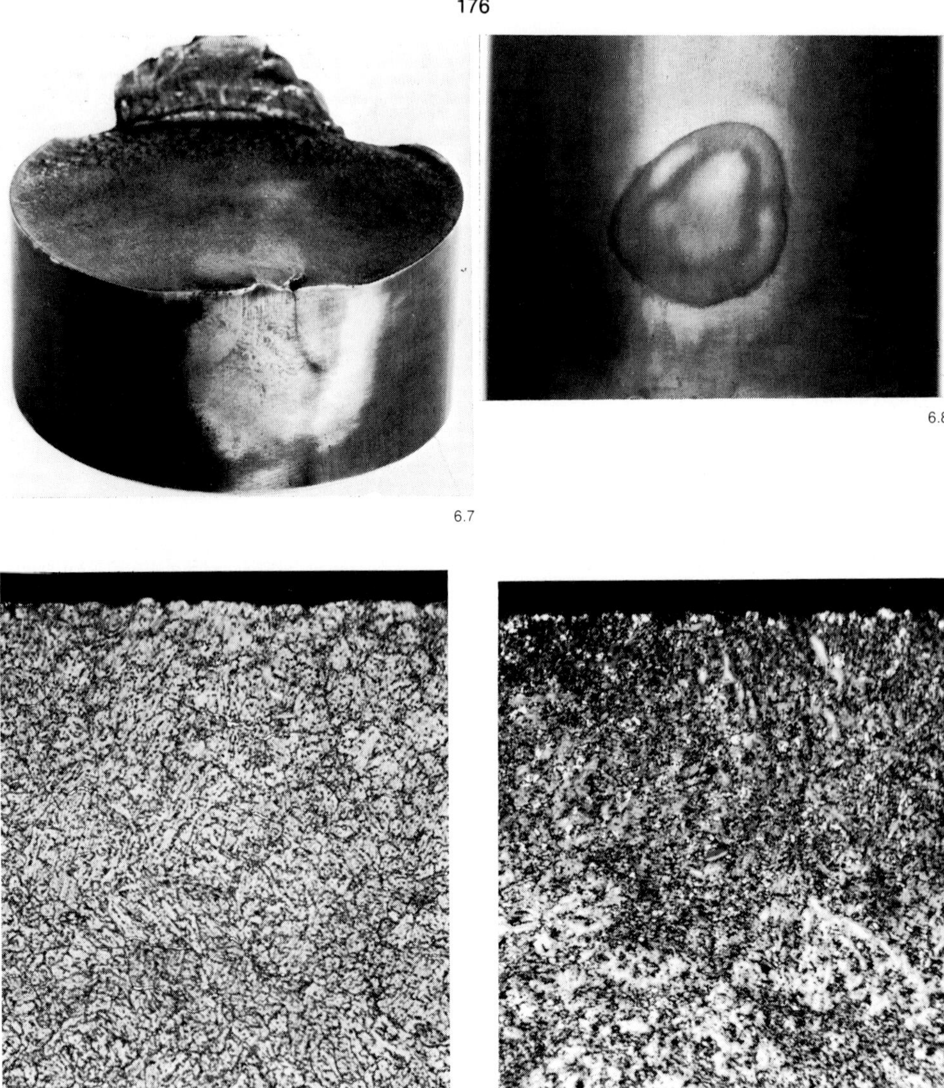

Fig. 6.7 to 6.10. Nitrided piston rod that was heated locally by a flame for straightening and broke at this spot

Fig. 6.7. Fatigue fracture originating in a burned area. Surface etched with Nital. 0.8 ×

Fig. 6.8. One of several other burned areas. Surface etched with Nital. 0.8 ×

Fig. 6.9 and 6.10. Edge structure of piston rod. Etch: Picral. 500 ×

Fig. 6.9. Unheated area

Fig. 6.10. Area heated by burner

An example for failures that occurred due to local heating had previously been shown in section 3.3 (Fig. 3.78 and 3.79) featuring a cracked **extruder worm screw** of stainless steel that apparently was distorted during heat treatment and was supposed to be straightened by local heating at various places[1]).

The same mistake was made during straightening of a **nitrided piston rod** of a steel containing approx. 0.3 % C, 2.5 % Cr, 0.2 % Mo, and 0.15 % V. It failed as a result of a large bend fatigue fracture that originated in a longitudinal crack **(Fig. 6.7).** After surface etching it could be seen that the rod had been spot heated at this place as well as others **(Fig. 6.8).** The heat treated microstructure was changed by precipitation of ferrite **(Fig. 6.9 and 6.10,** respectively); accordingly the temperature exceeded 750 °C. The surface hardness had decreased from 750 to 450 HV 10 as a result of an agglomeration of the nitrides (normally finely dispersed).

If straightening of the piston rod was indeed necessary, an unsuitable method was certainly used in this case with the point-like rapid heating. Nitriding layers are temper resistant up to 500 °C. A uniform heating to lower temperatures would have been permissable. Under normal circumstances straightening of nitrided parts is not necessary because distortions can be avoided by slow cooling after hardening that keeps residual stesses low, especially since the nitriding temperature is relatively moderate. But this assumes that the parts are free of stress at insertion into the nitriding furnace and that they are placed or hung in it in such a way that they do not bend under the load of their own weigth.

Especially **high speed steels** must be heated slowly and thoroughly for forging and hardening due to their poor heat conductivity.

An example of failure due to insufficient through-heating is shown in **Fig. 6.11.** It is the transverse section of a **high speed steel billet.** In the interior it had a cross-shaped cleavage that is characteristic of faulty forging. The microstructure was coarsely dendritic in the interior and had ledeburitic carbides **(Fig. 6.12),** while it was uniformly fine at the exterior **(Fig. 6.13).** There were no indications of a cavity.

6.1.2 **Temperature Errors: Scaling, Burning**

Scaling during heating is a normal occurrence and if it is kept within limits, is not considered a defect within the meaning of these discussions. However, considerable amounts of metals are lost during this process[2]). The amount or depth of the oxide scale increases with time according to a parabolic law. Excess oxygen as well as carbon dioxide and steam in heating gases have an oxidizing effect. Oxidation attack is increased by sulfur-containing gases, especially by hydrogen sulfide.

As a rule the oxide scale is removed later by pickling or mechanical treatment. Under special circumstances, in particular in copper-containing steels, descaling by pickling is difficult and surface defects may be formed, that may make sheets unusable for automobile body making. This had been described already in section 4.1.

In steam-rich but oxygen-deficient heating gases which can be generated during the burning of illuminating gas without excess air, an oxide layer may be formed at high temperatures that is firmly anchored in the metal by penetration of the oxygen into the austenitic grain[3]). This not only causes surface defects that are designated as "orange peels" or "alligator skins" because of their characteristic appearance, but also incipient cracks during hot or cold deformation[4]). These, however, are only of shallow depth. These failures, toc, are reinforced by the presence of copper in the steel[5]). In addition to copper, other elements may also accumulate below the oxide layer whose oxides have a lower heat of formation than iron oxide. Examples are nickel, tin, phosphorus, sulfur, arsenic, antimony and even carbon[6]). To prevent this, it is recommended

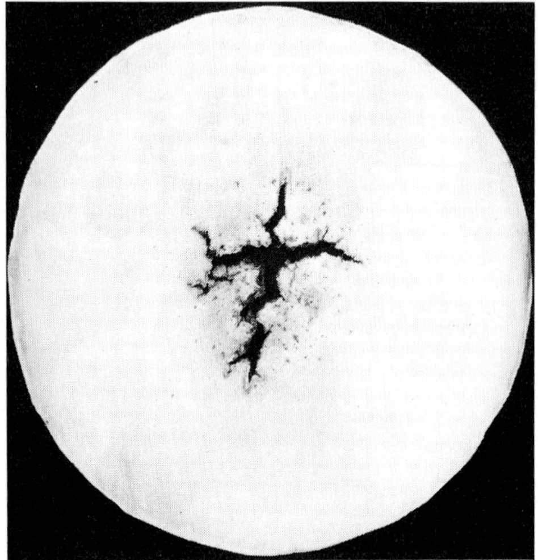

6.11

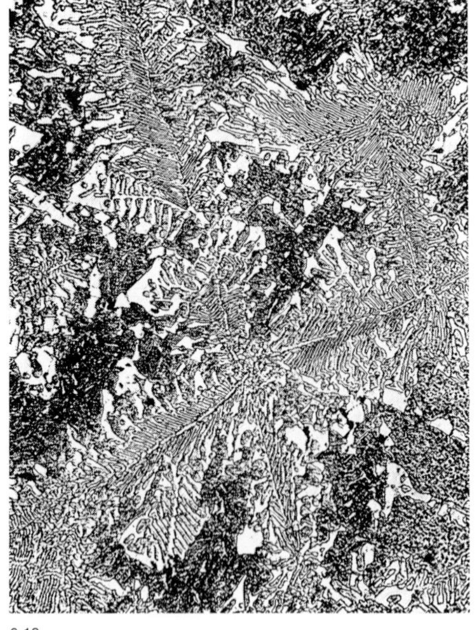

6.12

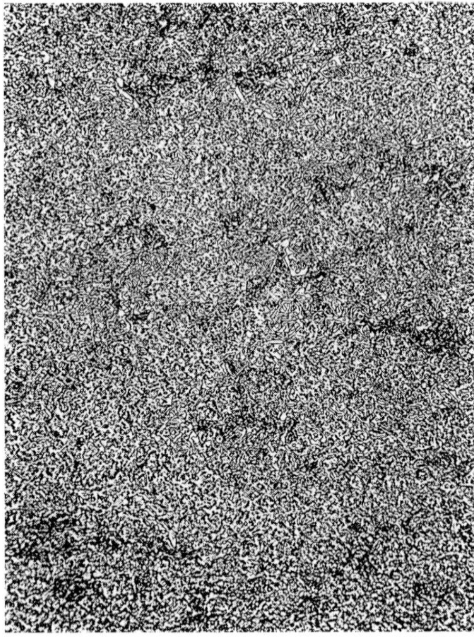

6.13

Fig. 6.11 to 6.13. Billet of high speed steel with internal forging fault

Fig. 6.11. Transverse section etched lightly with diluted hydrochloric acid. 0.5 ×

Fig. 6.12 and 6.13. Microstructure of billet. Transverse section. Etch: Nital. 200 ×

Fig. 6.12. Interior

Fig. 6.13. Below surface

that overheating and long times be avoided; in addition, combustion should be adjusted so that oxygen in the exhaust gas be limited to 1 to 2 %. Also steel should be used that contains no more than 0.15 % Cu or Ni[7]).

During very strong overheating, fine oxidic or silicate precipitates may be deposited on the boundaries of the overheated grains[8]) in the interior of the steel component. They are torn open during forging. This phenomenon also may be designated as burning. It may be removed by hot working but not by heat treatment as in the case of overheating.

An example for grain boundary oxidation in an early stage is given in **Fig. 6.14.** The peripheral structure of a patented open hearth **steel wire** with 5.8 mm diameter tore open during drawing. The steel contained 0.16 % Cu. The broadened austenitic grain boundaries of the outer grains appeared brownish in the optical microscope, but no metallic or nonmetallic inclusions could be seen.

A stronger surface oxidation was shown in an **automotive steering lever** made of AISI type 5140 steel. This drop forged part had cracked superficially during forging in a network-like pattern **(Fig. 6.15).** In the metallographic mount a 0.1 to 0.15 mm zone could be seen in which the austenitic boundaries were decarburized and covered with oxide precipitates **(Fig. 6.16).** Therefore this was a burning of the peripheral zone which apparently was caused by a torch flame during heating.

A more penetrating burn which was accompanied by precipitation of copper occurred in a seamless hot drawn **boiler pipe** of 300 mm NB with 9 mm wall thickness. This pipe of heat resistant low-carbon steel, containing approx. 0.15 % C and 0.3 % Mo, cracked repeatedly in the tension fiber during hot bending while it was filled with sand **(Fig. 6.17)**[9]). The steel contained 0.13 % C, 0.17 % Si, 0.53 % Mn, 0.032 % P, 0.022 % S, 0.26 % Mo and also a non-deliberate addition of 0.26 % Cu. The pipe bend had a very coarse bainite structure with little ferrite, whereas the straight part had a very fine grained microstructure of ferrite and pearlite. This suggests that the

Fig. 6.14. Edge structure of a patented steel wire torn during drawing. Transverse section. Etch: Picral. 1000 ×

6.15

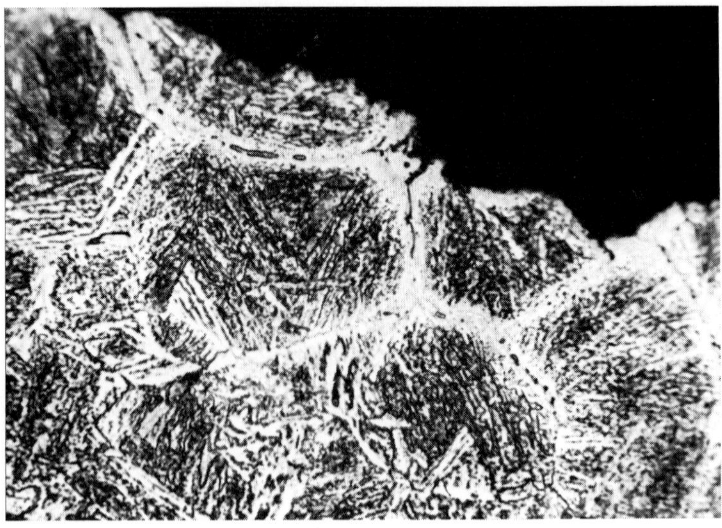

6.16

Fig. 6.15 and 6.16. Automobile steering lever of 5140 type steel that was burned at the surface and developed incipient cracks during drop forging

Fig. 6.15. View. 4 ×
Fig. 6.16. Structure at edge with oxide precipitates on the austenitic grain boundaries. Etch: Nital. 1000 ×

pipe had been overheated during bending. The surface was strongly scaled **(Fig. 6.18).** Under the oxide layer metallic copper had precipitated in the iron **(Fig. 6.19).** Here, dispersions of fine oxide and copper precipitates were found. Along the austenitic grain boundaries oxidation had penetrated up to 3 mm depth **(Fig. 6.20).** This failure therefore may be designated as red or hot shortness.

Figure 6.21 shows a typical case of **hot shortness** or braze-induced fracture sensitivity. It is the microstructure in the outer zone of a **forging** of unalloyed steel that contained 0.29 % Cu and that cracked in numerous places on the surface during forging. The surface-precipitated copper had deeply penetrated the austenitic grain boundaries, and the cracks followed along these grain boundaries.

A **round structural steel rod** of 370–450 MPa (53–65 ksi) ultimate tensile strength showed strong signs of burning. A cutting edge had been forged at an apparently uncontrolled forging temperature which resulted in a failure **(Fig. 6.22).** As can be seen from a longitudinal section through the rod's cutting edge **(Fig. 6.23),** the steel had become very coarse-grained. It had torn open at the austenitic grain boundaries that were covered with round inclusions **(Fig. 6.24 and 6.25).** Judging from the microstructure, the edge was quenched in water after forging and became purely martensitic as a consequence of the coarse grain and in spite of the rather low carbon content of 0.17 %.

A **ring** of 1015 mm O.D., 830 mm I. D. and 335 mm thickness made of heat treated AISI grade 6150 steel showed flaws in the inner part of the wall. It was produced from a raw billet of 355–385 mm diameter by upsetting, punching under a forging press, and rolling in a ring rolling mill. A section that the sender had already machined 15 mm from both the inside and outside is shown in **Fig. 6.26.** Judging by the location of the voids, the flaws could not be any shrinkage cavities. In radial sections (A--A and B--B), too, no signs of segregation or decarburization were found, such as occur in conjunction with cavities. A fracture in the cross section C--C **(Fig. 6.27)** was partly intergranular **(Fig. 6.28).** Trenches in the sections were eroded by etching with dilute hydrochloric acid between the voids **(Fig. 6.29).** In the micrographs these appeared as austenitic grain boundaries that were covered with round inclusions consisting of either oxides or silicates **(Fig. 6.30 and 6.31).** These failures accordingly should be interpreted as faulty forgings as a consequence of prior burning.

6.1.3. **Decarburizing, Carburizing**

Decarburization often occurs in conjunction with oxidation[10]). It may be caused by oxygen or steam just as in scaling, and in some special cases also by hydrogen. These will be described later in section 15.2. Decarburization reaction by oxidizing gases takes place at the surface of the steel. The removed carbon must be replaced by diffusion. If the removal of the reaction product, carbon monoxide, is not impeded by scale formation – as in decarburizing by weakly oxidizing gases such as moist hydrogen –, then this is the rate-determining process. Therefore the amount of carbon lost is proportional to the square root of time **(Fig. 6.32)**[11]). No simple relation exists for the temperature dependence of carbon diffusion because the diffusion coefficient of carbon in γ-iron is lower at the same temperature than in α-iron. The amount of carbon removed decreases after passing through a maximum around 800 °C, where the product of diffusion coefficient and concentration gradient attains its highest value in the decarburized layer. An increase of this product occurs again with rising temperature after entry into the γ-region **(Fig. 6.33)**[11]).

During annealing in more strongly oxidizing gases or gas mixtures such as oxygen or air, the time dependence is different and the inflection point in the temperature dependence does not occur, because decarburization is noticeable only in the γ-region due to the effect of the protective scale.

In addition to the depth of the decarburized layer, its microstructure is of importance. It is a function of temperature at which decarburization takes place. During annealing in the temperature range of the α or $(\alpha + \gamma)$ region (conditions a and b in **Fig. 6.34**) a zone of α-crystallites is formed at the outer edge of the component. The crystallites are often columnar **(Fig. 6.35a).** This state is designated as "full decarburization". If decarburization commences in the γ-region and traverses the $\alpha + \gamma$ region (condition c in Fig. 6.34), then a zone of decreasing ferrite is attached to the completely decarburized layer **(Fig. 6.35b).** If the entire decarburization takes place in the γ-region (condition d in Fig. 6.34), a ferritic peripheral zone is not necessarily formed, but the carbon content decreases steadily from inside to out **(Fig. 6.35c).** This incomplete decarburization is designated as "partial decarburization".

The purely ferritic case in columnar form is rarely seen except in special situations because the commonly present strongly deoxidizing gases do not decarburize at the low temperatures of the α-region due to the presence of scale. It can occasionally be found on the surfaces of alloys with narrowed or constricted γ-region at which the α-iron is constant up to higher temperatures, e.g. in silicon spring steels or aluminum-alloyed nitrided steels; or in densely closed cracks in which a mildly oxidizing atmosphere may be present that prevents scaling or makes it more difficult.

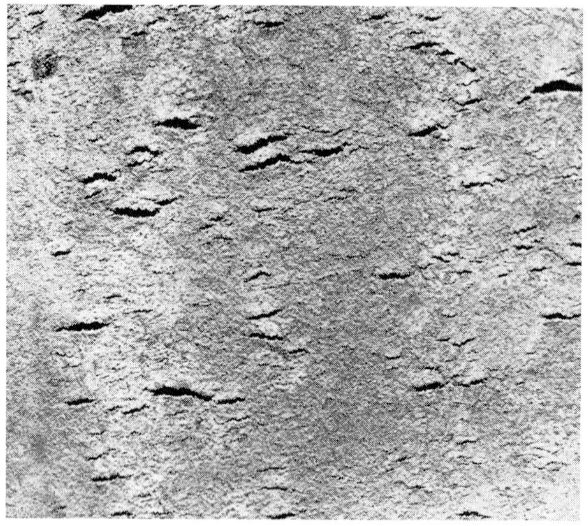

6.17

Fig. 6.17 to 6.20. Boiler pipe of 0.3 % Mo containing low-carbon steel that developed incipient cracks during hot bending

Fig. 6.17. Torn open surface. 1 ×

Fig. 6.18 to 6.20. Structure at edge in the tension fiber of the pipe bend. Longitudinal section. Etch: Nital

Fig. 6.18. Scaling and internal oxidation. 500 ×
Fig. 6.19. Precipitates of metallic copper. 1000 ×
Fig. 6.20. Oxide precipitates on the grain boundaries of coarse austenitic grain. 200 ×

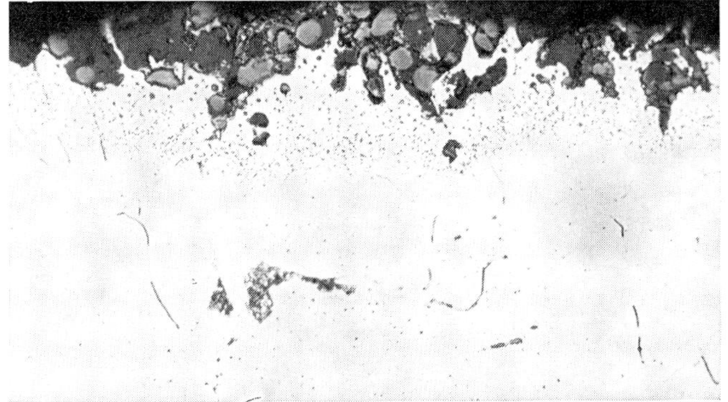

6.18

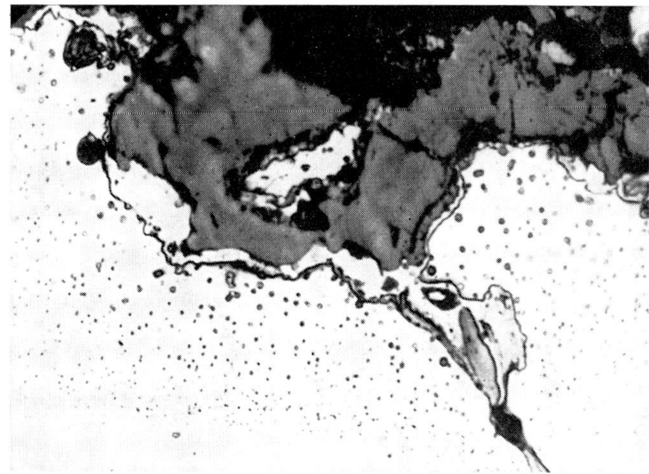

6.19

6.20

From the above discussion it is evident that annealing at low temperatures in lightly oxidizing gases may prevent scaling, but may have undesired consequences regarding decarburization. During soft annealing in boxes or in protective atmospheres these conditions are often present. If decarburization has to be avoided, e.g. in spring steels, it would be better to anneal under strongly oxidizing conditions, i.e. in air, and to tolerate a certain amount of scale. In this case it may even result in an accumulation of the carbon from the burned iron in the steel and thus lead to apparent recarburization as has been shown by W. Oelsen[6]).

For an evaluation of crack failures it is important to know when or at which processing step the crack has formed. An indication of this may be given by the microstructure at the crack peripheries. If the crack has been strongly decarburized along its entire length **(Fig. 6.36)** it may be concluded that it did not occur during quench hardening, but was present prior to it and was formed at the latest during heating for final treatment. But if a hardened or heat treated part has a wide crack which is strongly decarburized on the outside and ends in a tightly closed non-decarburized inner fissure **(Fig. 6.37),** then it probably is a quench crack that originated in a surface defect, such as for instance, a rolling fold. But the reverse case also occurs, i.e. that a crack is more strongly decarburized in its propagation than directly below the surface. If decarburization then has taken place in the completed form characteristic for low temperatures, it may be concluded that it is a quench crack whose borders were decarburized during tempering (Fig. 6.103 and 6.104).

Superficial decarburization is damaging for several reasons. It decreases not only hardness, hardenability and wear resistance of the surface zone, but also, to a large extent, fatigue strength, especially for applied bend and torsion stress (see Fig. 2.26, curve e). During hardening, stresses may form by the advanced γ–α transformation in the decarburized case layer that lead to distortion or cracking. In surface decarburized tools or structural parts of austenitic manganese steel with approx. 1 % C and 12 % Mn, crack and fracture failures may occur by local martensite formation as a result of stresses in the microstructure or embrittlement[12]). Examples of failures by decarburization will be cited below.

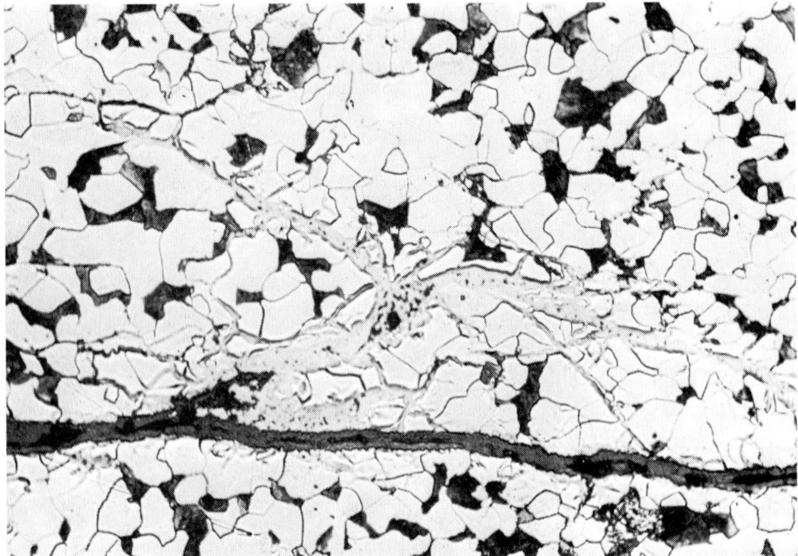

Fig. 6.21. Microstructure in the outer zone of a forged billet with 0.29 % Cu that was torn open due to braze-induced fracture sensitivity. Etch: Nital. 200 ×

The following two failures may serve as examples for fatigue fractures as a result of decarburization. **Figure 6.38** shows the rupture of a **leaf for an automotive spring**[13]) that was heat treated to a strength of 1270 MPa (185 ksi). The rupture consisted of numerous fatigue fractures that had primarily propagated in the main probably from the upper side of the leaf; also there was a small residual fracture displaced to one side. The core consisted of a perfectly fine-acicular structure which was free of ferrite, but the case was clearly decarburized **(Fig. 6.39).** This decarburization certainly must be regarded as the cause of failure since no other defects were found and stress was not overly high considering the small residual fracture.

Prongs of a hay turner[13]) from a shipment of patented spring steel wire drawn to 9 mm diameter broke on the test stand after 15,000 load cycles whereas they usually withstood over 40,000 cycles under similar severe test conditions. All ruptures were located in the second turn. **Figure 6.40** shows one of the ruptures. It was composed of a smooth fatigue fracture which had started from the most heavily loaded inner fiber and a strongly deformed ductile final fracture. After removal of the lacquer coating, several cracks could be seen next to the fracture and also in the interior of the first turn **(Fig. 6.41).** Upon breaking these open, the cracks were likewise found to be fatigue fractures. Metallographic examination of the prongs, as well as of wires from both the good and rejected shipments, showed that the broken prongs and the wires from which they were produced were surface decarbürized, whereas the good wires were not or only to a small extent. **Figure 6.42** shows the surface microstructure of the zone full of cracks. The connection of the fatigue fractures with the surface decarburization is here most apparent.

Drawing capacity of wires is a function of their microstructure. Lamellar pearlite is harder to deform than a microstructure of spheroidal cementite in a ferrite matrix. During the customary soft annealing of hypereutectoid steels by a cyclic heating around the A_1 temperature, a fast spheroidizing of cementite is achieved, while during the respective cooling from the (γ+cementite) region, carbon precipitated onto the undissolved carbide nuclei. If such a steel is decarburized superficially this process cannot take place for lack of nuclei[14]). Such was the cause of fracture of AISI type 50100 **ball bearing steel wires** during drawing[15]). **Figure 6.43** shows a longitudinal section through one of the fractures. The surface structure is clearly changed as compared to the core structure. In addition, several other incipient cracks are visible in this zone. Microscopic observation showed that the pearlite in this zone was lamellar while it was grainy in the core structure **(Fig. 6.44 and 6.45).** In the outer zone, with the exception of a minor residue of a coarse-meshed network, the hypereutectoid carbides expected in this steel were absent. The steel therefore had been decarburized superficially, which was the cause of its poor drawing capability.

In the production of thread-rolled **heat treated bolts** of type 5140 steel[16]), some parts of each shipment had to be rejected due to quench cracks in the thread. To determine the cause, 30 wire specimens in each of two "bad" (S1 and S2) wire shipments and 27 specimens of a "good" (G) shipment were examined. The composition of all wires corresponded to specifications. But a difference existed inasmuch as the steel of the good shipment, based on its higher aluminum and nitrogen contents, was considered a fine-grained steel in contrast to the others (see also section 6.2). Metallographic examination of the wires in the soft annealed state showed that those of the crack-sensitive shipment were surface decarburized much deeper than those of the good shipment as is shown in the following Table of mean values:

Shipment	S1	S2	G
Depth of fully decarburized zone (mm)	0.05	0.05	0
Total decarburized depth (mm)	0.14	0.12	0.05

6.22

6.23

Fig. 6.22 to 6.25. Structural steel rod burned during heat-up for forging of a cutting edge and torn open during forging

Fig. 6.22. View of cutting edge. 2 ×
Fig. 6.23. Longitudinal section through cutting edge with coarse grain structure and tear-outs at the surface. Etch: Nital. 10 ×

Fig. 6.24 and 6.25. Microstructure in the crack region. Etch: Picral
Fig. 6.24. 100 ×
Fig. 6.25. 500 ×

In particular, no completely decarburized zone was present in the wires of the good shipment. The photomicrographs **6.46a and b** characterize this state. The frequency curves for total decarburization depth **(Fig. 6.47)** had a maximum of about 0.05 mm for the good shipment and more than double that (about 0.11 and 0.18 mm) for the bad shipments.

The higher quench crack sensitivity of the bad wires may have been partially due to the fact that they consisted of a more readily hardened coarse-grained steel (see section 6.2). Hardness tests with notched wire specimens in which the notch was machined in one, while in another it was beaten in by a chisel, showed that quench cracks occurred only in the latter case thus confirming the participation of surface decarburization in the quench crack formation.

Difficulties and failures may also occur due to **surface carburization** or the previously mentioned **carbon accumulation** under the surface. Failures which can occur during wire drawing may be caused by carburization from unremoved stearate. These have already been discussed in section 1.2 (Fig. 1.12). During working of type 1095 tool steel **wires,** an unusually high wear was found in the drawing tools of a wire drawing plant. The same wire showed a bright surface after pickling because hydrochloric acid did not attack it. The difficulties were confined to the wire shipped by one of two suppliers. Metallographic examination of one soft annealed finished wire of 1.43 mm diameter showed that the surface was covered with a closed coating of cementite **(Fig. 6.48).** The time when this carburization may have taken place could not be established from the information provided by the wire drawing plant. Since it occurred only in the wire of one supplier, indications were that it was already present in the rolled product. This was confirmed by an examination of the 5 mm thick rolled wire. In the transverse section a network of cementite was found under a dense scale **(Fig. 6.49).** This is a typical picture of a carbon accumulation. Subsequent chemical analysis showed that carbon contents was 1.00 % in the core of the wire but had risen to 1.49 % in the outermost surface layer.

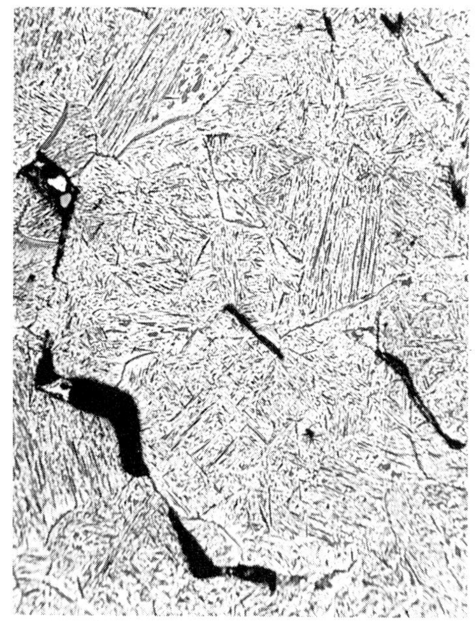

6.24

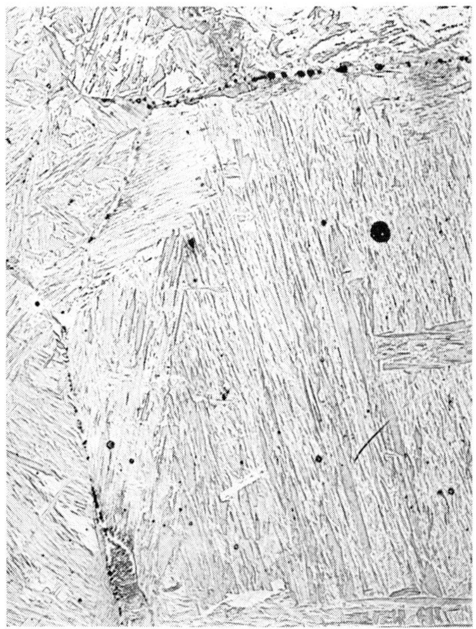

6.25

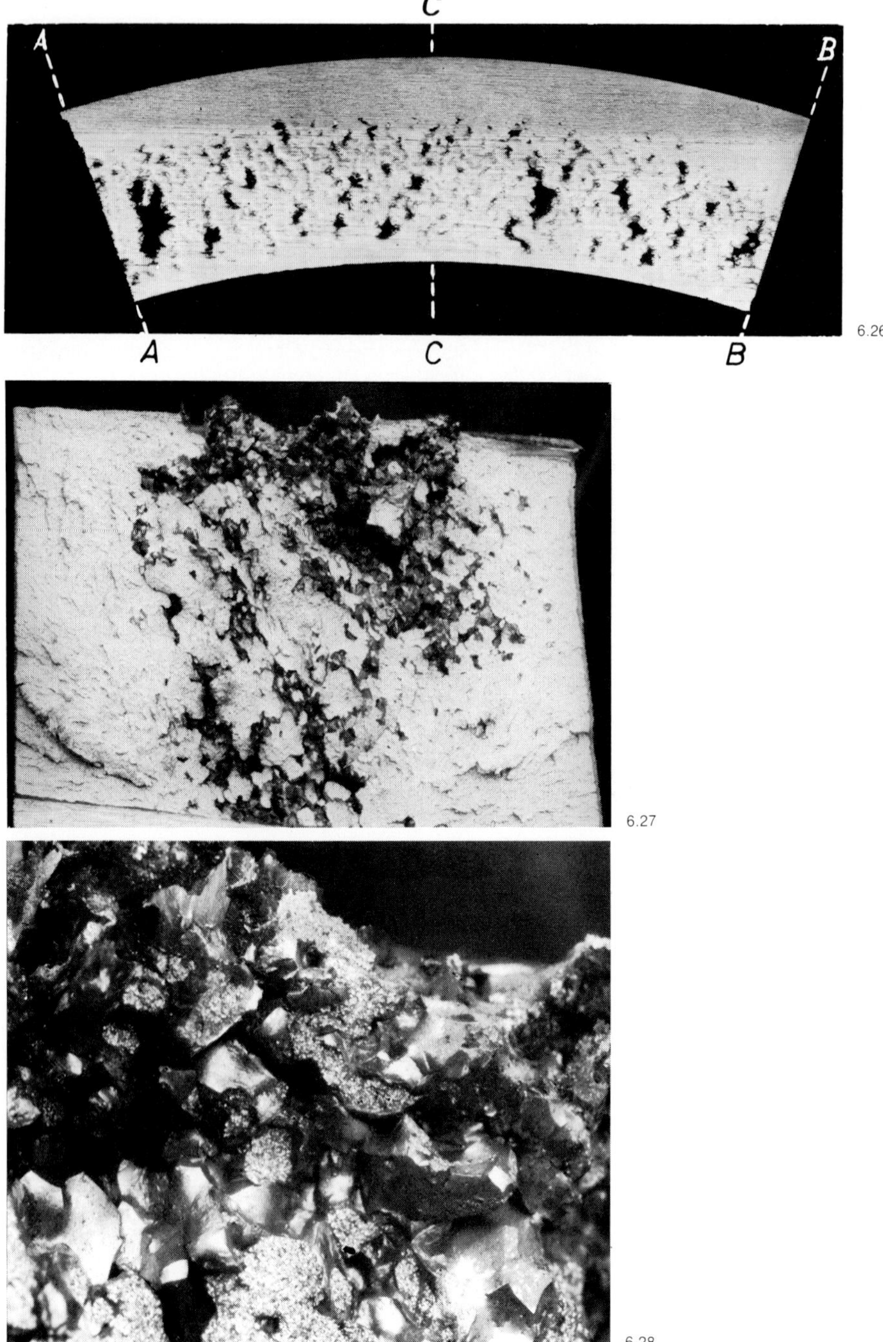

6.26

6.27

6.28

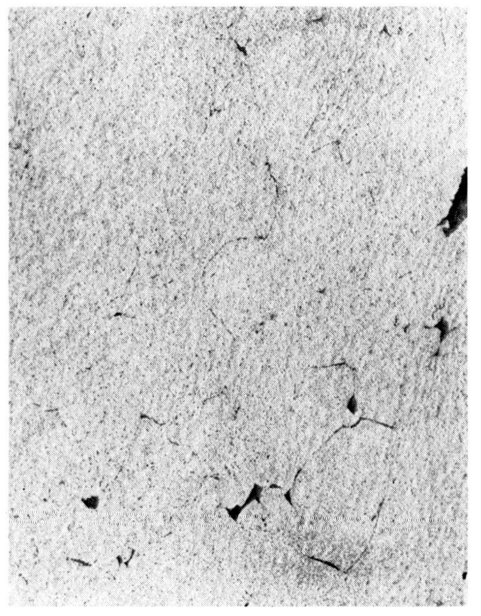

6.29

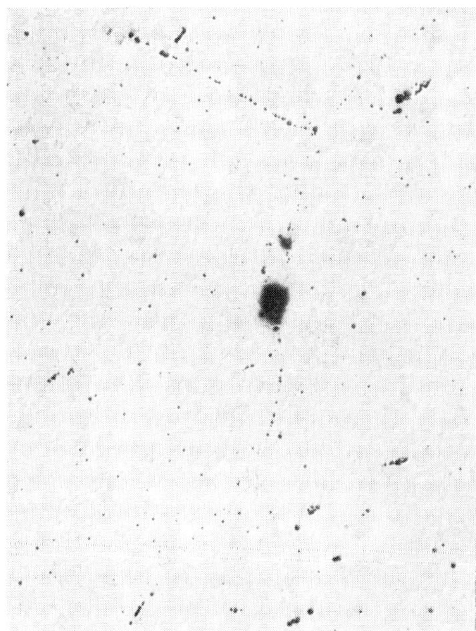

6.30

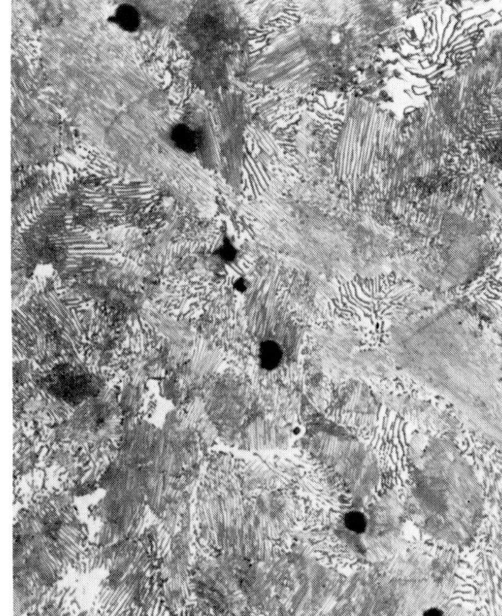

6.31

Fig. 6.26 to 6.31. Burnt and malforged ring of 6150 type steel

Fig. 6.26. Transverse section. Unetched. 0.33 ×
Fig. 6.27. Voids exposed by breaking in plane C--C. 1 ×
Fig. 6.28. Cut-out of Fig. 6.27. 5 ×

Fig. 6.29 to 6.31. Section A--A according to Fig. 6.26.

Fig. 6.29. Austenitic grain boundaries made better visible by lightly etching with hot diluted hydrochloric acid. 5 ×
Fig. 6.30. Unetched section. 50 ×
Fig. 6.31. Etched with Nital. 1000 ×

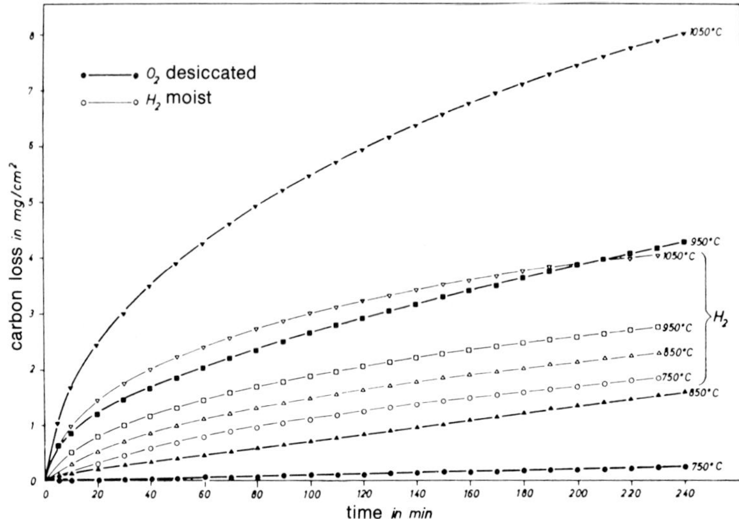

Fig. 6.32. Time dependence of decarburization of an alloy with 0.82 % C subjected to desiccated oxygen and moist hydrogen (according to W. Oelsen, K.-H. Sauer, and G. Naumann[11])

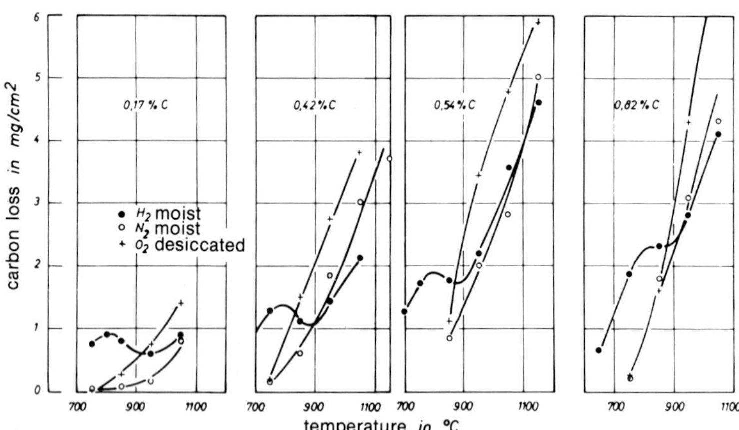

Fig. 6.33. Temperature dependence of decarburization during annealing in various gases. Heating time 4 h (according to W. Oelsen, K.-H. Sauer and G. Naumann[11])

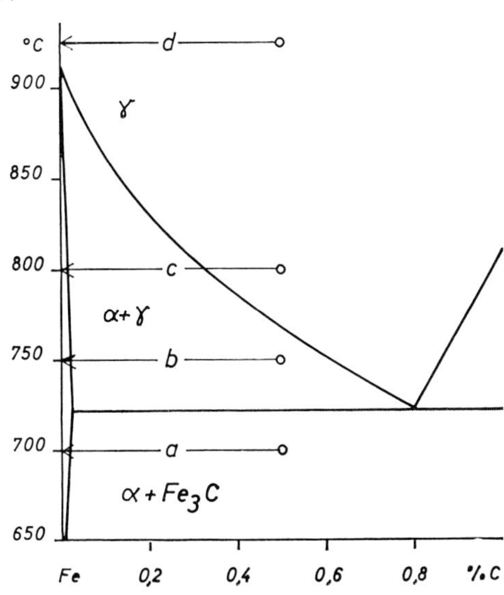

Fig. 6.34. Schematic presentation of potential cases of decarburization in iron-carbon phase diagram

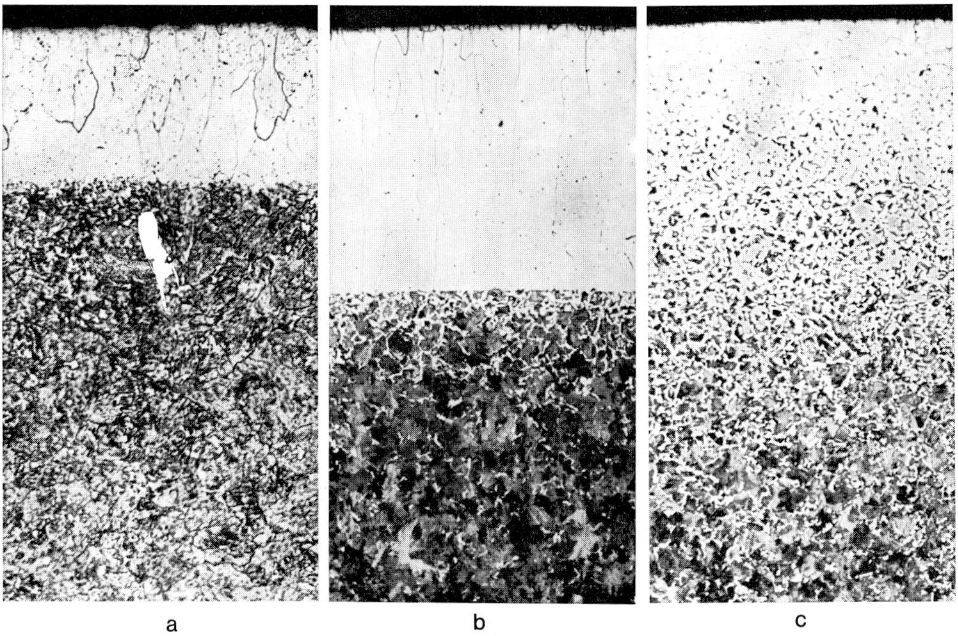

a b c

Fig. 6.35a to c. Outer edge microstructure of steel with 0.73 % C after 4 h annealing in moist hydrogen. Etch: Picral. 100 × a. 700 °C (condition a), b. 800 °C (condition c), c. 900 °C (condition d)

6.2. **Heat Treating Errors**

Errors in heat treatment may occur during heat-up, during selection and control of temperature and time, and during cooling. Errors in heating were already discussed in the last section.

The **temperature** of all heat treatments must be selected according to the particular type of steel. They are given in steel specifications and materials test sheets. Normalizing for grain transformation and hardening requires heating to the temperature of the austenite or austenite + carbide region. Unalloyed steels that do not contain any hard-to-dissolve carbides or nitrides are sensitive to overheating and therefore should not be heated too high because a coarse grain size makes steel sensitive to brittle fracture during impact-type loading. By addition of special nitride or carbide formers, such as aluminum or vanadium, overheating sensitivity, i.e. the tendency to coarse grain formation, may be substantially reduced. Undissolved nitrides or carbides act in this case as a barrier to the growth of the austenitic grain. But with austenite grain size, hardenability characterized by depth of hardness decreases concurrently. This is so because transformation tendency increases with the total length of the grain boundaries that act as transformation nuclei, thus making the delay of the γ–α transformation into martensite or bainite, i.e. hardening, much more difficult.

Steels insensitive to overheating typically are **fine grained steels.** The formation of a fine grain is linked to a very fine distribution of grain growth inhibiting precipitates. Final forging or rolling at too cold a temperature may lead to undesirable agglomeration of the precipitated particles and either influence the insensitivity or nullify it, according to tests conducted by the author.

Whether and to what extent a steel or heat is sensitive to overheating can be determined most quickly und surely by hardening of a number of specimens from incremental temperature increases. Incipient coarse grain formation can be reliably observed in hardness fractures by a bright reflection of light at grain cleavage planes. The hardening temperature at which the first

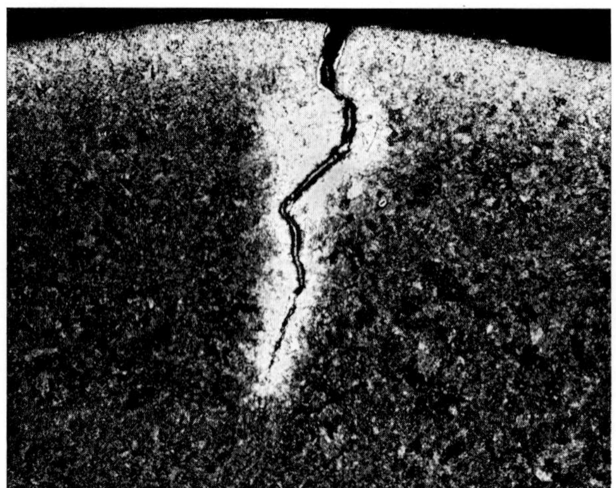

Fig. 6.36. Longitudinal crack in rod of heat treated nickel-chromium steel. Transverse section. Etch: Nital. 10 ×

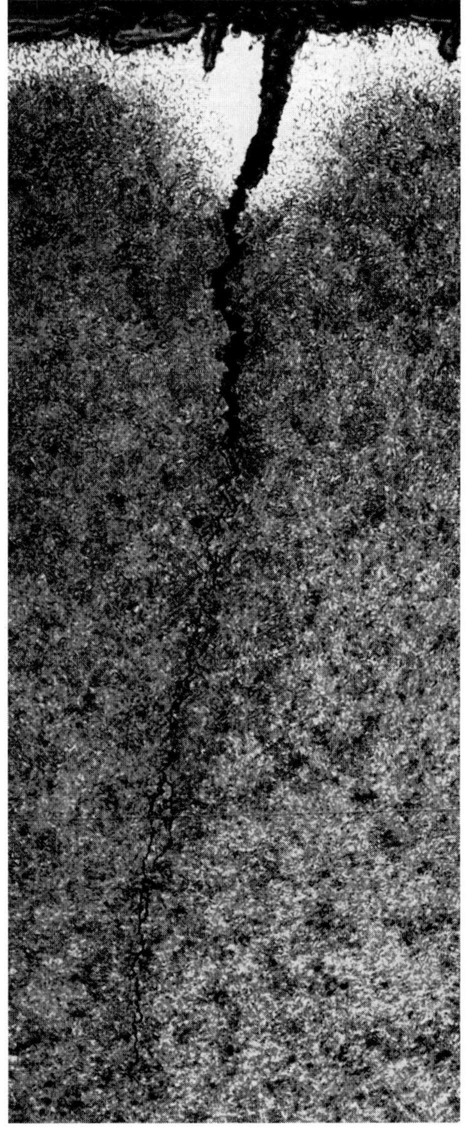

Fig. 6.37. Quench crack in a heat treated lock washer of silicon steel, propagating from a decarburized surface defect. Etch: Picral. 100 ×

coarse grain formation occurs, or more exactly, its distance from the transformation temperature, gives an indication of the sensitivity to overheating of the steel or heat. This process is preferable to the well known McQuaid-Ehn test which only indicates whether a steel shows a more or less coarse grain after cementation at 925 °C.

On the other hand special carbide forming alloying elements are added to the steel in order to improve its hardenability or temper stability. In this case the hardening temperature naturally must be raised sufficiently for the special carbides to go into solution. High speed steels in which temper stability is of decisive importance therefore are hardened at temperatures just below the solidus plane.

Solutioning as well as spheroidizing processes take **time,** the more so, with lower temperatures. The heating period therefore must not be too short. On the other hand, undesirable situations such as overheating and burning are intensified with increasing heating time. For instance, an extended holding time at temperature during hardening of high speed steels easily leads to incipient melting of the segregation zones.

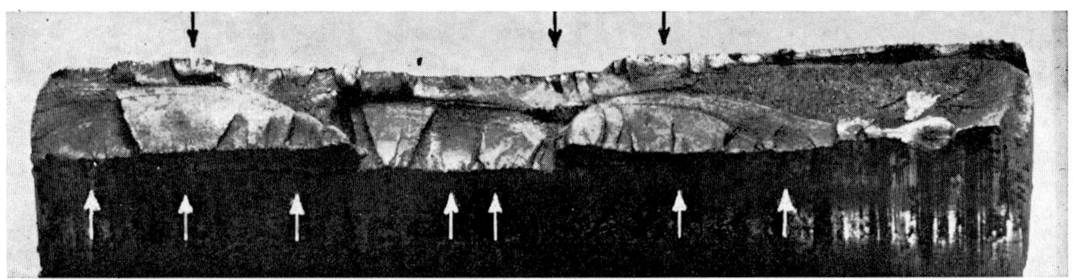

6.38

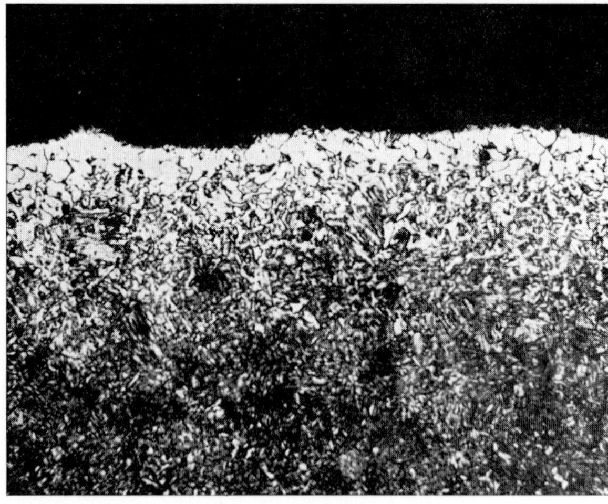

6.39

Fig. 6.38 and 6.39. Fracture of an automobile leaf spring with decarburized edge

Fig. 6.38. Fracture. 2 ×. Arrows indicate fracture origins of fatigue failures
Fig. 6.39. Structure at edge. Transverse section. Etch: Nital. 200 ×

6.40

6.41

6.42

Fig. 6.40 to 6.42. Fatigue fractures in a hay turner prong

Fig. 6.40. Fracture. 5 ×
Fig. 6.41. Incipient fatigue failure cracks in vicinity of fracture. 5 ×
Fig. 6.42. Decarburized surface structure with incipient fatigue failure cracks. Longitudinal section. Etch: Picral. 100 ×

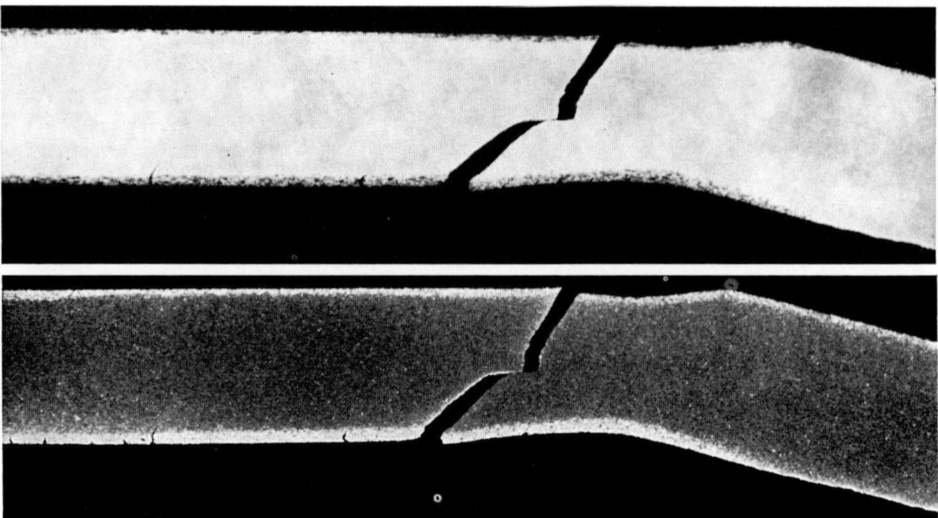

6.43

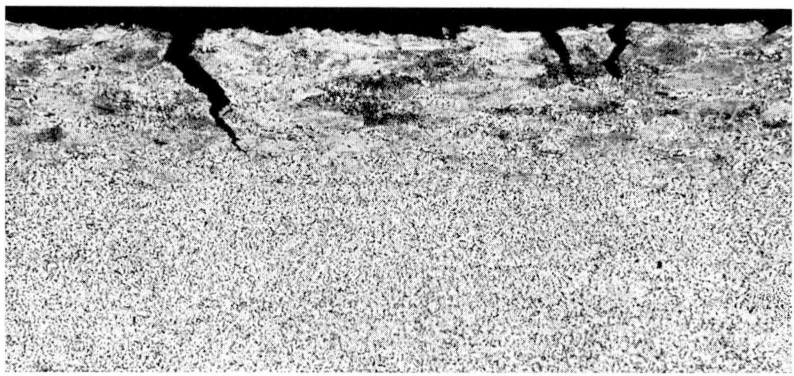

6.44

Fig. 6.43 to 6.45. Type 50100 ball bearing steel wire torn during drawing

Fig. 6.43. Longitudinal section through a fracture region. Etch; Picral. Top: Illuminated vertically. Bottom: Illuminated obliquely. 4 ×

Fig. 6.44. Microstructure of the outer surface with cracks. Longitudinal section. Etch: Picral. 100 ×

Fig. 6.45. As Fig. 6.44. 500 ×

The **cooling** rate from the treating temperature is dependent upon the purpose of the treatment, the steel alloy, and the thickness and shape of the parts. Fast cooling promotes the formation of stresses and thus distortion and cracking. Therefore the rate of cooling should not be higher than necessary for the particular purpose of the treatment. Steels with high hardenability must be cooled slower than steels with low hardenability. Parts with sharply angular cross-sectional transitions as well as those with large differences in cross sections and complicated design are more sensitive than simple shapes. Long slender parts should preferably be submerged in the quench bath vertically rather than flat for the sake of dimensional stability. All parts should be moved against the bath – or vice versa – in order to avoid fixation of steam bubbles that leads to the formation of soft spots, particularly in low hardenability fine-grained steels. For the same reason scale adhesions should be removed prior to quenching.

Surface hardening poses some special problems. During **case hardening,** steel, carburizing agent, and temperature should be selected in such a way that no interconnected network of hypereutectoid carbides are formed because this makes the case crack-sensitive and brittle. High-stressed parts may be quench hardened directly from the high case hardening temperature only if they are produced from a fine-grained steel that is insensitive to overheating. Otherwise the carbon-rich steel becomes coarse-grained and impact-sensitive in the case, and also does not attain full martensitic hardness because of strong residual austenite. The case hardening temperature and the rate of cooling can be lower if the carburizing agent is mixed with a nitrogen-evolving gas. This is due to the nitrogen absorption displacing the α–γ transformation to lower temperatures and stabilizing the austenite. In particular, this lowers the danger of distortion, so that **"carbonitriding"** is especially suitable for surface hardening of complex-shaped parts.

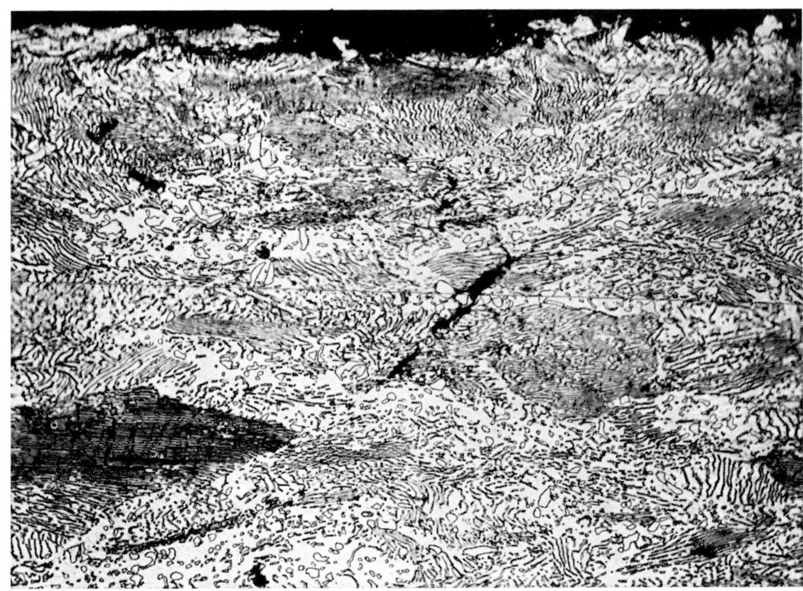

6.45

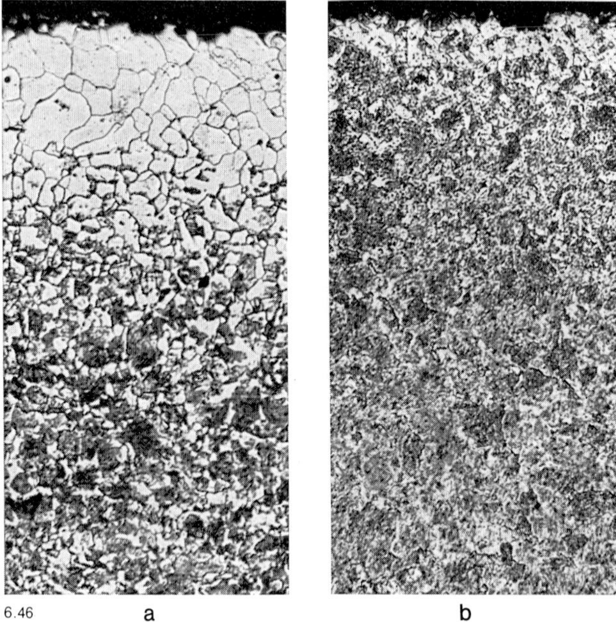

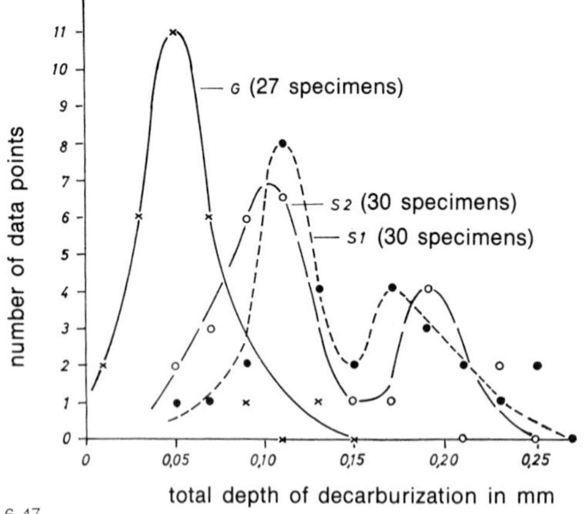

Fig. 6.46 and 6.47. Bolts of 5140 type steel that were in part quench crack sensitive (S1 and S2) and in part not quench crack sensitive (G)

Fig. 6.46a and b. Structure at surface of wires of bad (S2) and good (G) shipment. Transverse section. Etch: Picral. 200 ×

Fig. 6.46a. Wire S2

Fig. 6.46b. Wire G

Fig. 6.47. Frequency distribution of decarburization depth

The lowest distortions for the highest surface hardness and wear resistance are possible by nitriding. The temperature should be around 500 °C and at most 550 °C. This process requires the use of special steels alloyed with aluminum, chromium or vanadium. Due to the low temperature, the time input is high and core strength is limited because of the long-time heating of the previously heat treated parts inherent in this method. Therefore nitrided parts do not stand up well to high face pressure. If this fact is disregarded, failures occur in the base material through breakthroughs in the nitrided layer.

The opposite, namely the formation of a hardened case that is too deep in relation to the total cross section or even a complete through hardening of the parts occurs often during case, flame, or induction hardening. In that event, the prime purpose of case hardening is missed, namely to produce a wear resistant surface on a tough core that is capable of accomodating jolting and impact-type stresses by plastic deformation. As a consequence, not only the case may crack during such overloads, but the entire part may fracture without deformation. In surface hardened gears, the root of the teeth is sometimes not hardened. This may be justified as a precautionary measure in switch gears subject to mechanical shock. But in other gears with smooth operation this should be considered a mistake because it eliminates protection against fatigue cracks in the root of the teeth that are particularly highly stressed in bending.

6.48

6.49

Fig. 6.48 and 6.49. Wire of 1095 steel that caused an unusually high wear of the drawing tools. Surface structure. Etch: Alkaline sodium picrate solution

Fig. 6.48. Cementite layer at surface of finished drawn wire of 1.43 mm diameter. 1000 ×

Fig. 6.49. Cementite network under scaled surface of 5 mm rolled wire. 500 ×

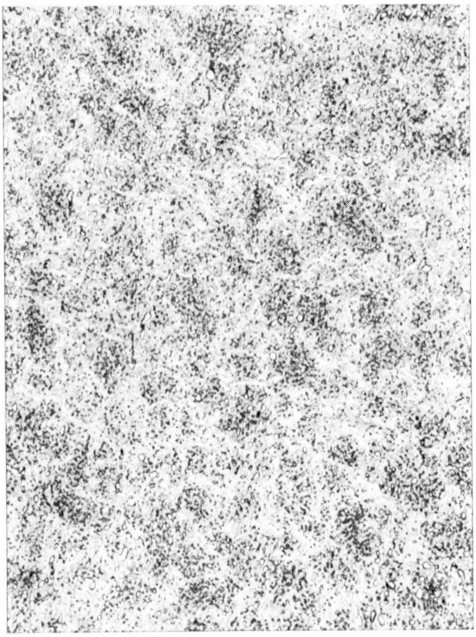

6.50

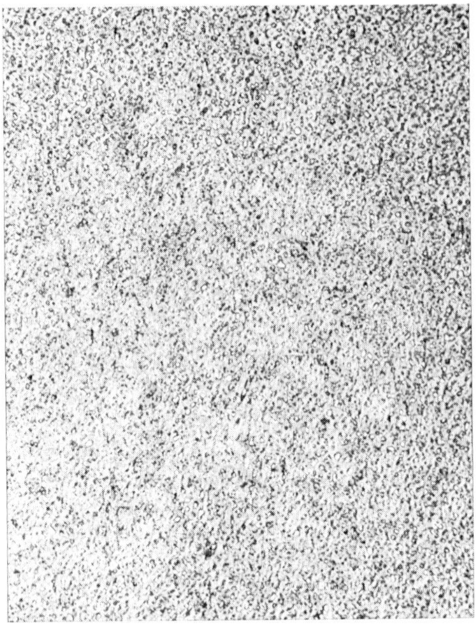

6.51

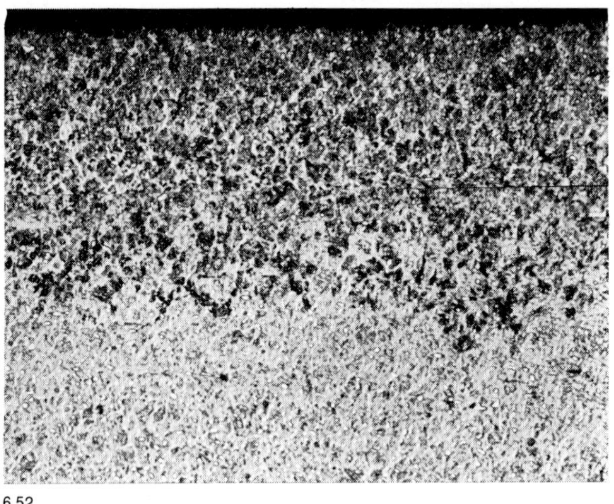

6.52

Fig. 6.50 to 6.52. Needles of needle bearings made of type 52100 ball bearing steel. Longitudinal sections. Etch: Picral

Fig. 6.50 and 6.51. Core structure of a rejected and a good needle. 500 ×

Fig. 6.50. Too soft

Fig. 6.51. Normal

Fig. 6.52. Soft spot in surface structure of rejected needle. 500 ×

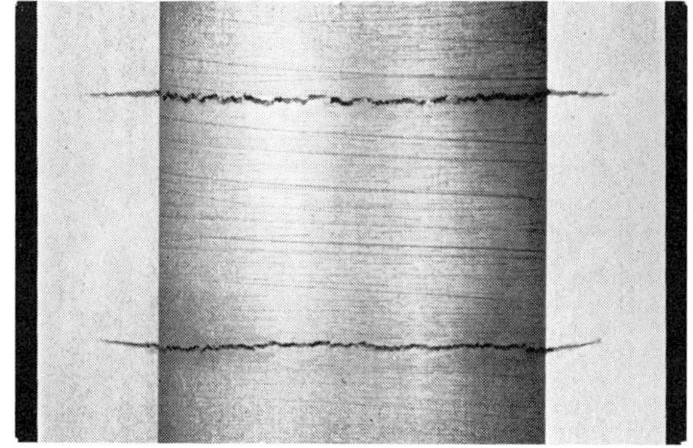

6.53

6.54

6.55

Fig. 6.53 to 6.55. Unalloyed steel rods with drawing tears

Fig. 6.53. Rod that showed drawing tears during drilling

Fig. 6.54. Rod that broke apart during machining. 1 ×

Fig. 6.55. Microstructure of torn rod. Longitudinal section. Etch: Nital. 100 ×

6.56

a. overheated during hardening

b. post-treated for 10 min at 780 °C/water

c. post-treated for 10 min at 780 °C/air + 10 min at 780°C/water

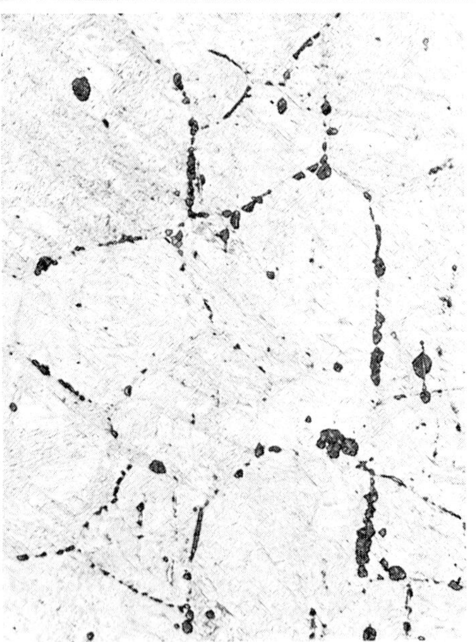

6.57

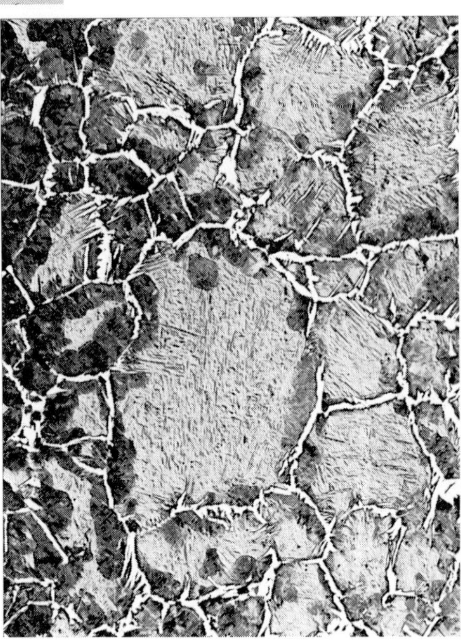

6.58

Fig. 6.56 to 6.60. Triple-ply plow board laminate which was overheated during hardening and broke during shipping

Fig. 6.56a to c. Fractures in the as-received condition and after correct hardening

Fig. 6.57 and 6.58. Microstructure of the broken board. Etch: Picral. 100 ×

Fig. 6.57. Outer ply

Fig. 6.58. Inner ply

Fig. 6.59 and 6.60. Microstructure of the board post-treated for 10 min at 780 °C/air + 780 °C/water. Etch: Picral. 500 ×

Fig. 6.59. Outer ply

Fig. 6.60. Inner ply

6.2.1. **Temperature Errors**

A few examples of temperature errors are cited below. The first concerns a failure caused by quench hardening from **too low a temperature.** The needles of **needle bearings** of an AISI 52100 type ball bearing steel were found too soft during hardness testing. Four of these were examined and compared with four harder ones. The hardness of the rejected needles was found to be between 681 and 813 with a mean of 740 HV 10, while the comparison needles had a hardness of 882 to 946 with a mean of 913 HV 10. Metallographic examination showed that the soft needles had a substantially higher amount of fine undissolved carbide in the microstructure **(Fig. 6.50)** than the harder ones **(Fig. 6.51).** They also showed spots in the surface structure that consisted of very fine pearlite **(Fig. 6.52)** which further served to reduce the average hardness value.

The following failure was due to annealing and rolling at **too high a temperature.** A drawn 65 mm diameter **round rod** of unalloyed steel with 700–850 MPa (100–124 ksi) tensile strength, showed internal cracks transversely to the axis **(Fig. 6.53)** after a central longitudinal bore of 40 mm diameter was drilled. Another section broke transversely during an attempt to machine the rod down to 48 mm diameter **(Fig. 6.54).** The cracks and fractures had the characteristic in the axial direction vaulted shape of **drawing tears.** The microstructure of the rod consisted of pearlite with a very coarse-meshed ferrite network **(Fig. 6.55).** Normalizing at a low temperature or, even better, soft annealing to grainy pearlite would have improved drawing capacity.

A **board of a plow** broke during shipping. These tools, subject to stress by wear and impact, were made from laminated plates that had two plies of hardenable tool steel on the outside to withstand the wear stresses and inside a ply of low carbon steel to absorb shocks. **Figure 6.56a** shows the fracture being grainy throughout, an indication of pronounced overheating. The microstructure was accordingly coarse-grained in the core sheet as well as in the outer sheets **(Fig. 6.57 and 6.58).** The predominantly martensitic microstructure of the inner ply proved to be

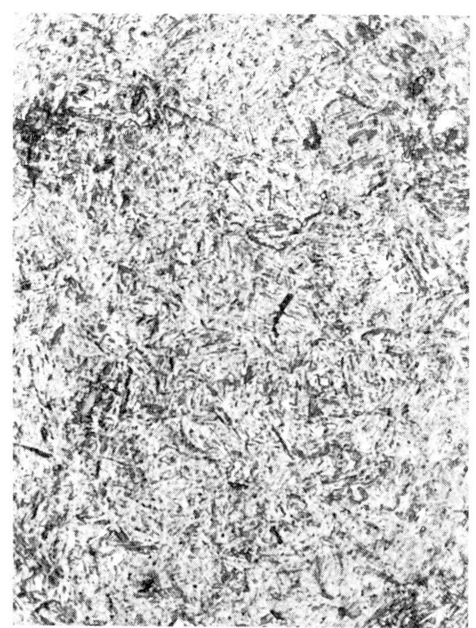

6.59

6.60

6.61

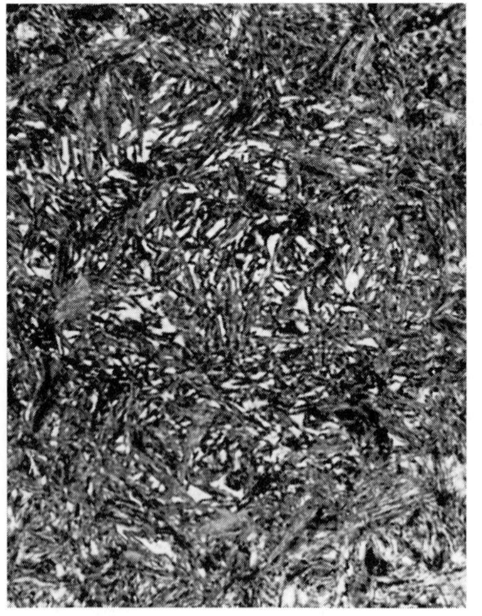

6.62

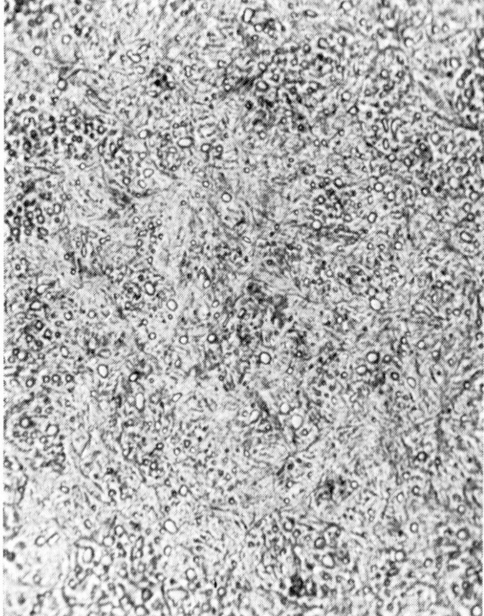

6.63

hardened too. This was probably not only the consequence of the coarse grain but also was caused by carbon absorption by diffusion from the outer sheets. The reason why the plow board broke even before service is thus explained. In order to demonstrate how fracture and microstructure should look after the correct treatment, we have water quenched from 780 °C specimens taken from the plates with and without preceding normalizing. This resulted in the fractures b and c of Fig. 6.56 and the microstructures of **Fig. 6.59 and 6.60** (please note the fivefold magnification increase as compared to Fig. 6.57 and 6.58). The fracture grain of the outer plies is now velvety fine and the inner ply, too, is more fine-grained and not hardened. Subsequent treatment substantially increased ultimate strength and flexing capacity as established by bend tests.

Even the formation of quench cracks is facilitated by quenching temperatures that are too high. **Quench cracks** often cannot be distinguished from other types of stress cracks. But if they are connected so clearly with overheating as in the following case, then they are fairly certain to be quench cracks. Two hardened **segmented knives** used for slitting corrugated cardboard broke during mounting. They consisted of an unalloyed steel with approximately 0.8 % C and were tempered to a hardness of 56 to 58 HRC. They were examined together with a knife of chromium steel that had proved satisfactory. The fractures started from the milled slot needed for gripping at that point where the transition from straight to circular profile occurs **(Fig. 6.61,** bottom). They consisted of a quench crack that propagated through a break in its path into the residual fracture that originated during gripping. One of the two broken knives showed an incipient crack in addition to the fracture **(Fig. 6.61** top) originating at a corresponding point. The fracture of the broken knife was grainy, while that of the good quality comparison knife was velvety. The microstructure of the broken knives consisted of coarsely acicular, low or untempered martensite with a substantial amount of residual austenite **(Fig. 6.62),** whereas that of the good quality knife consisted of fine acicular martensite with undissolved carbides **(Fig. 6.63).** According to spectral analysis, this knife consisted of a chromium alloy steel.

As already stated, tools of high speed steel are heated for hardening close to the melting point and occasionally melt at the segregation zone.

Figure 6.64 shows quench cracks on the face of a **gear** of 25 mm diameter made of high speed steel grade M 2 that was machined out of a rod of 27 mm diameter. After a three-step preheating, the gear was quenched from 1200 °C in a bath of used-up nitriding salt heated to 550 °C and tempered three times in the same bath at 580 °C. The end face was ground down by 0.2 to 0.25 mm. The cracks follow along the carbide streaks that were permeated by shrinkage voids **(Fig. 6.65).** In the outer part their peripheries were surrounded by austenite that was stabilized by absorption of nitrogen from the nitriding bath **(Fig. 6.66 and 6.67).** This proves that these cracks were not grinding checks which they resemble. The formation of the quench cracks was caused or facilitated by the excessive hardening temperature that led to incipient fusion in the grain boundary segregation regions.

The solidified molten eutectic and its location in the segregation zones that solidified last are still more distinct in the microstructure of three **reamers** of 65 mm diameter made of a high speed steel containing 12 % W, 2.5 % V and 0.8 % Mo. These had broken longitudinally in the roots of the teeth **(Fig. 6.68 and 6.69).**

Fig. 6.61 to 6.63. Segmented knife of steel with 0.80 % C for the slitting of corrugated cardboard

Fig. 6.61. Cracked and fractured segmented knife. 0.67 ×

Fig. 6.62 and 6.63. Transverse sections. Etch: Picral. 1000 ×

Fig. 6.62. Overheated microstructure of broken knife

Fig. 6.63. Normal microstructure of good knife made of chromium steel

6.64

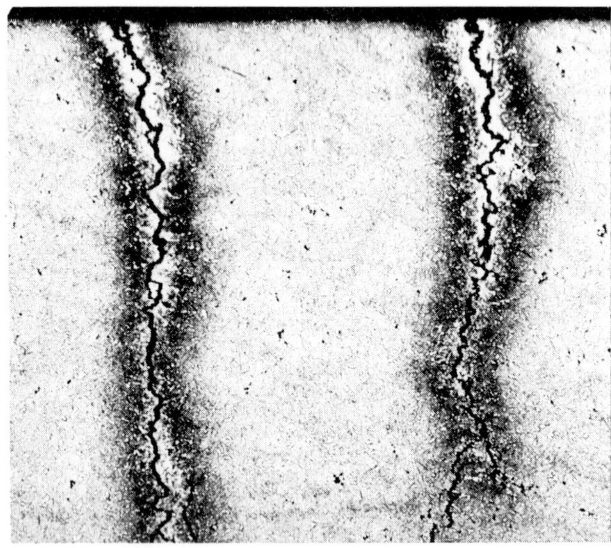

6.66

Fig. 6.64 to 6.67. Gear of high speed steel grade M 2 overheated during hardening, causing incipient melting and cracks

Fig. 6.64. End face with cracks. 2 ×

Fig. 6.65. Incipient melt in a carbide streak. Longitudinal section. Etch: Nital. 500 ×

Fig. 6.66. Longitudinal section through end face (top). Etch: Nital. 100 ×

Fig. 6.67. Nitrided crack origin according to Fig. 6.66. 500 ×

Prior to the use of fine grained steels, overheating during direct hardening after carburization occurred quite frequently. A case hardened **automotive crankshaft** of AISI type 5115 became unusable after only 3000 kilometers road service due to breakouts and heavy wear of a pinion. It was examined metallographically together with a comparison crankshaft that had been removed from service after 20,000 km without apparent damage except for normal wear. **Figure 6.70a** illustrates the damaged crankshaft and **Fig. 6.70b** the one slightly worn. The microstructure in the case of the teeth of the small pinions consisted of coarse acicular martensite with a high fraction of residual austenite in the gear with the strong fragmentations **(Fig. 6.71).** In the undamaged pinion, on the other hand, it consisted of a considerably finer, but not an ideal quench structure **(Fig. 6.72).** The coarse grained overheated microstructure had to have a particularly damaging effect in such gears that are subject to shock.

It had previously been mentioned that a steel with coarse austenitic grain case hardens deeper than a fine grained one. This is exemplified in the following failure. A machine factory sent two case hardened **bevel gears,** that had broken during service, for examination. The fractures were uniformly coarse grained from the periphery to the core **(Fig. 6.73).** Metallographic investigation showed that the gears were case hardened to a depth of 1.5 to 2 mm. The structure in the case consisted of a very coarse grained martensite with a high amount of residual austenite **Fig. 6.74)** in the core of acicular bainite **(Fig. 6.75).** The through hardening led to a deformationless breakout of the teeth, thus missing the purpose of case hardening. The case hardened layer was very deep in relation to the tooth thickness, but without the strong overheating this would probably not have led to the brittle fracture in the core region.

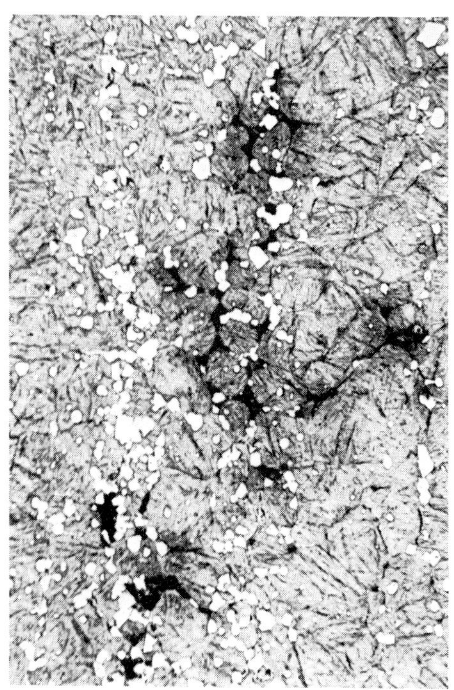

6.65

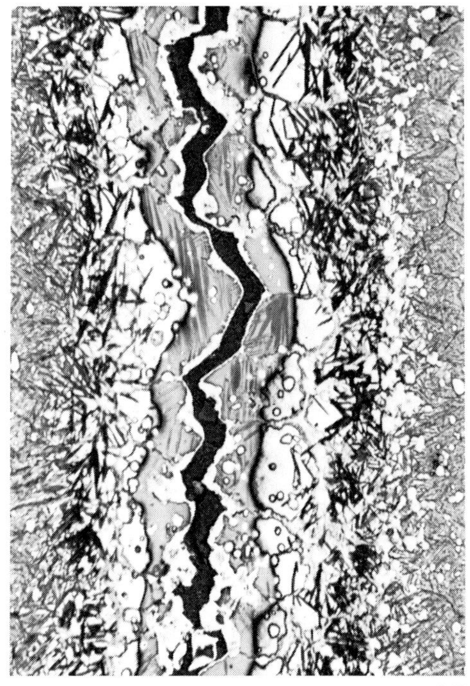

6.67

As an example of the effect that hard-to-dissolve carbides have relative to grain refining and hardenability reducing, the following case will be discussed. After a change of steel supplier, endless **conveyor chain links** broke several times, even at moderate stress, with a coarse grained fracture. Three of these were examined together with three links of a conveyor furnished by the old supplier. Spectrographic analysis showed that the broken links consisted of an unalloyed steel. The links of the satisfactory shipment contained about 0.1 % V in addition to the otherwise comparable composition. Metallographic examination showed that the broken links were case hardened to 0.7 to 1.3 mm at the points where they intertwine. The coarse acicular, austenite-rich microstructure of the surface was an indication that hardening took place from the high carburizing temperature withouth intermediary cooling and reheating to a lower hardening temperature appropriate for the carbon rich case **(Fig. 6.76).** Even the core structure was still predominantly martensitic **(Fig. 6.77).** Core hardness of 350 to 380 HV 10 was accordingly high. In contrast the surface structure of the chain links from the good quality shipment was finely acicular and low in austenite **(Fig. 6.78),** and the core was overwhelmingly transformed to bainite **(Fig. 6.79).** Core hardness was only 260 to 290 HV 10. The fractures of the chain links therefore were caused by overheating before quenching; the better behavior of the links of the old shipment was probably due to the use of a vanadium-containing case hardened steel. A less expensive fine grained steel deoxidized with aluminum would have behaved similarly.

Since fine-grained steels have low hardenability, they require intensive and uniform quenching for hardening. Otherwise they would form soft spots. During very slow cooling of hypereutectoid steels the precipitated carbon combines with the pre-eutectoidal cementite precipitate so that the latter seems to be surrounded by a ferrite border, and the cementite lamellae of the pearlite show a tendency to agglomerate. This microstructure has been designated as anomalous even though it comes closer to equilibrium than the conventional one. Whether a steel has a tendency to coarse grain formation or to develop an abnormal microstructure can be established by cementation tests according to McQuaid-Ehn.

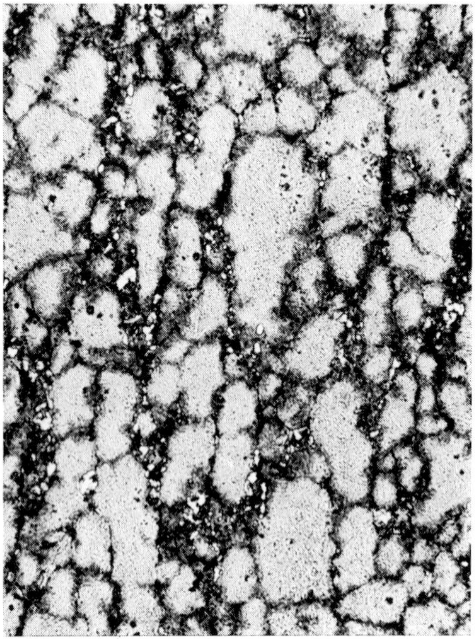

6.68

6.69

Fig. 6.68 and 6.69. Microstructure of reamers of 12-2.5-0.8 W-V-Mo high speed steel that cracked during hardening. Longitudinal section. Etch: Nital

Fig. 6.68. 100 ×

Fig. 6.69. 500 ×

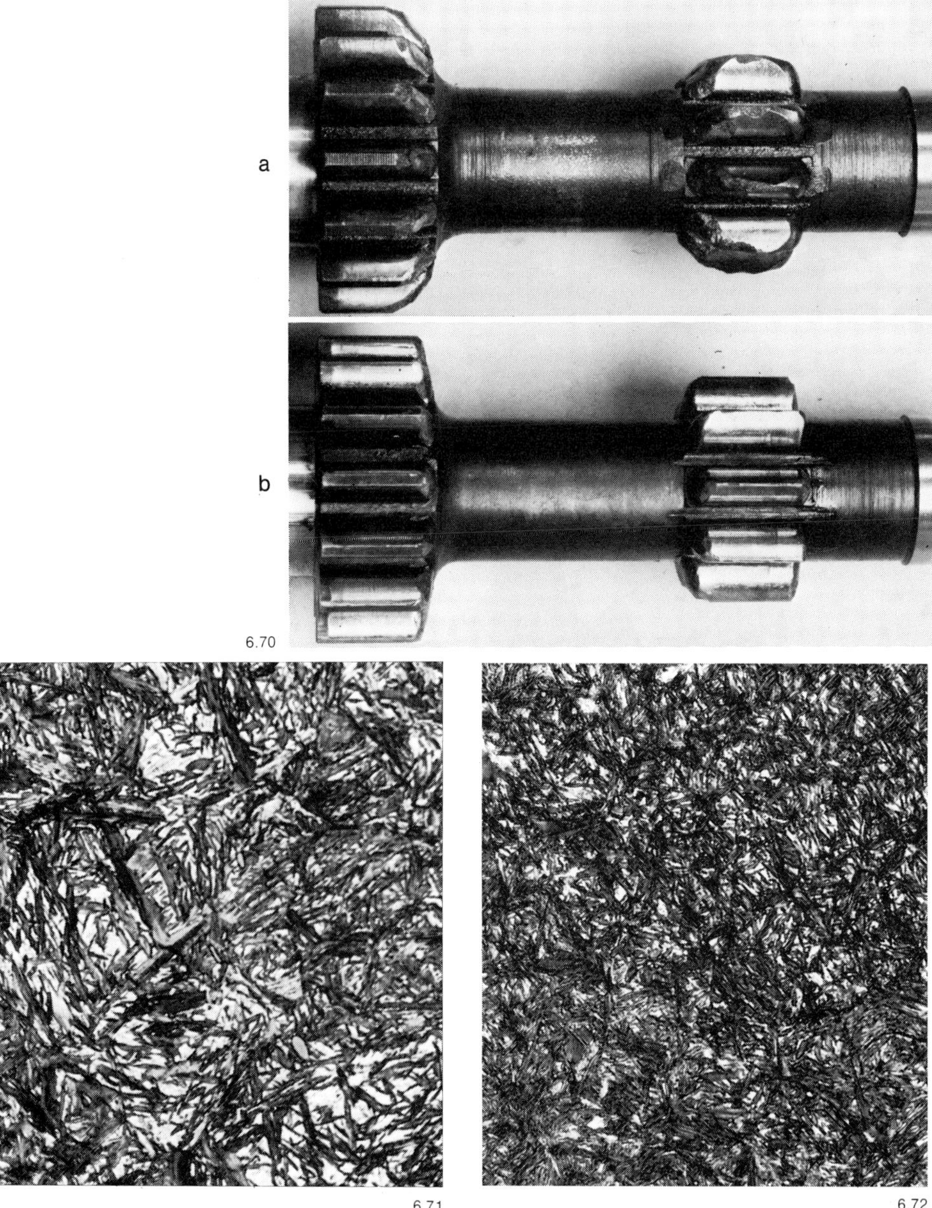

Fig. 6.70 to 6.72. Automotive engine crankshaft carburized and quenched from too high a temperature

Fig. 6.70. Case hardened crankshafts. 1 ×
a. Breakouts after 3000 km driving, b. Normal wear after 20,000 km driving

Fig. 6.71 and 6.72. Structure at surface of teeth of small pinion. Transverse section. Etch: Picral. 500 ×

Fig. 6.71. Heavily worn pinion according to Fig. 6.70a

Fig. 6.72. Normally worn pinion according to Fig. 6.70b.

6.73

6.74

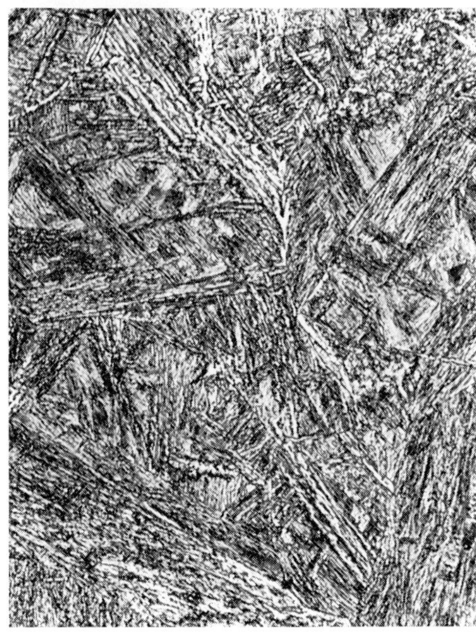

6.75

Fig. 6.73 to 6.75. Bevel gears overheated during case hardening with broken-out teeth

Fig. 6.73. View. 1 ×

Fig. 6.74 and 6.75. Microstructure inside the teeth of damaged bevel gears. Transverse section. Etch: Picral. 500 ×

Fig. 6.74. Case structure

Fig. 6.75. Core structure

Fig. 6.76 to 6.79. Chain links of endless conveyor belt. Longitudinal sections. Etch: Picral. 500 ×

Fig. 6.76 and 6.77. Broken chain links

Fig. 6.76. Case structure

Fig. 6.77. Core structure

Fig. 6.78 and 6.79. Chain links of good quality shipment (steel contained approximately 0.1 % V)

Fig. 6.78. Case structure

Fig. 6.79. Core structure

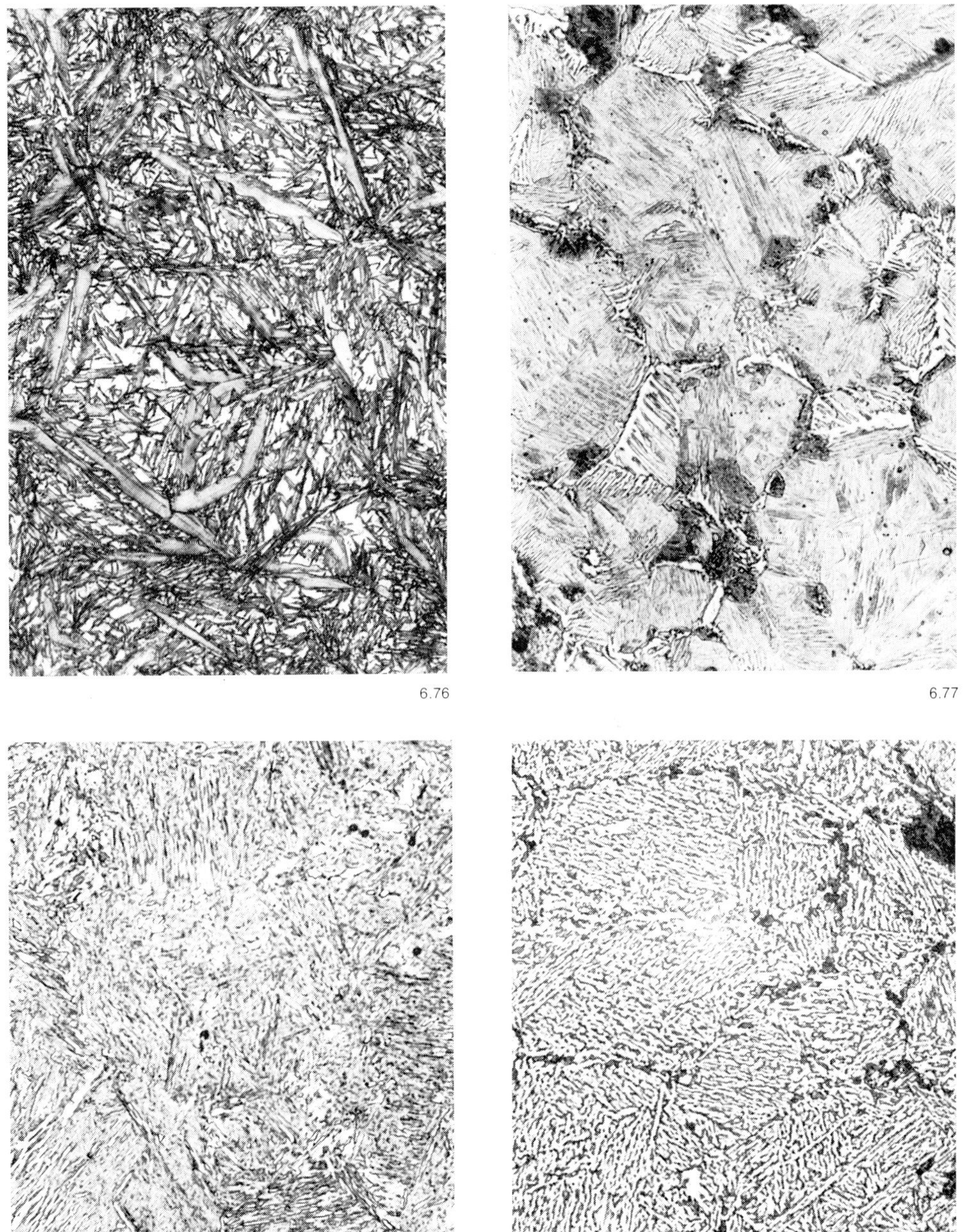

6.76

6.77

6.78

6.79

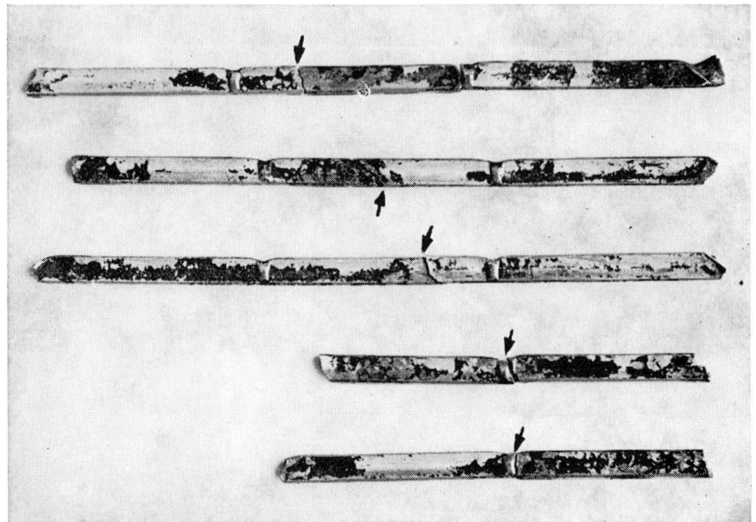

6.80

6.81

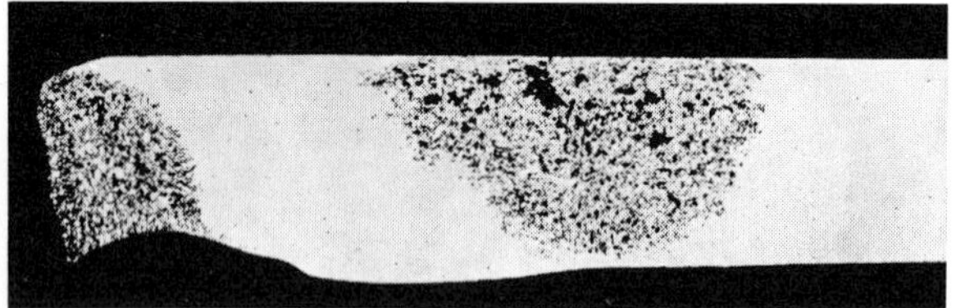

6.82

Fig. 6.80 to 6.86. Granular disintegration in screen bars of austenitic steel that were treated wrong. Bars were exposed to brackish water of river mouth

Fig. 6.80. Broken screen bars. Fractures designated by arrows. 1 ×

Fig. 6.81. Fracture. 15 ×

Fig. 6.82. Longitudinal section through fracture with corrosion spots. Unetched. 5 ×

Fig. 6.83. Microstructure of defective region. Electrolytically etched with 50 % diluted nitric acid. 100 ×

Fig. 6.84 and 6.85. Microstructure of unattacked part

Fig. 6.84. Etch: V2A-etching solution. 200 ×

Fig. 6.85. Electrolytically etched with aqueous solution of ammonia (carbide etch). 500 ×

Fig. 6.86. Microstructure in the unattacked part after heat treatment for ½ h 1050°C/water. Etch: V2A-etching solution. 200 ×

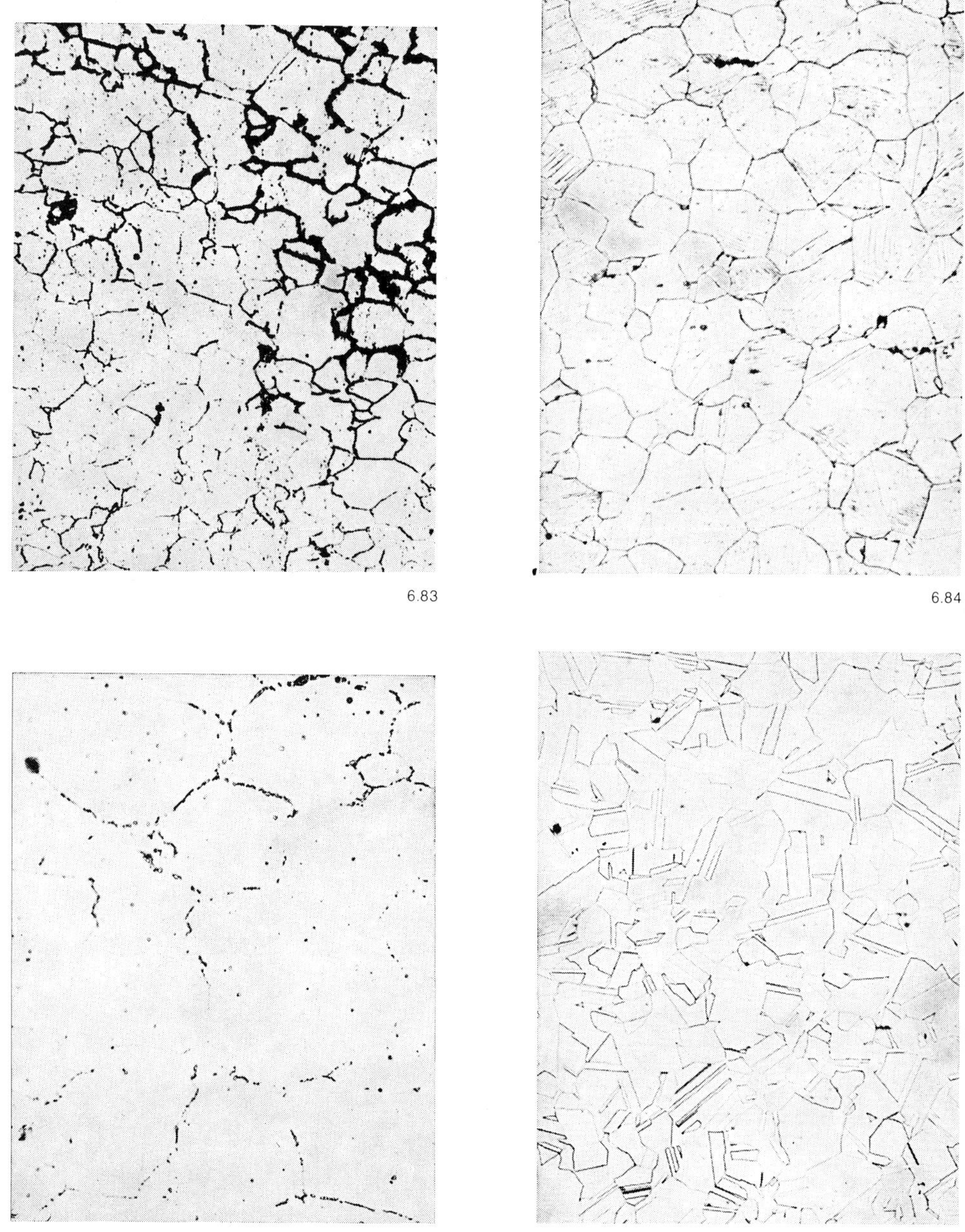

6.83

6.84

6.85

6.86

Examples of soft spots and microstructural anomalies will be cited in the discussion on cooling. But first some more examples of failures through temperature errors will be shown.

Ferritic steels must be kept for some time below the A_{C1} point for soft annealing after cold deformation. Austenitic steels attain their highest softness and best deformability, however, if they are cooled fast from the γ-region. In titanium-free and niobium-free (not stabilized) stainless chromium-nickel and chromium-nickel-molybdenum steels, heating to 500 to 800 °C must be avoided because in this temperature range chromium carbides are precipitated onto the austenitic grain boundaries which makes these steels susceptible to intergranular corrosion (see section 15.3.3.3). The following failure is due to the fact that this rule was violated.

Screen bars of stainless steel type 316 **(Fig. 6.80)** broke with a low deformation-type fracture **(Fig. 6.81)**[17]) in the brackish water of a river mouth that contained 10,000 mg Cl' and 1440 to 1520 mg SO_4'' per liter and had a pH value of 6.6 to 6.7. In longitudinal and transverse sections they showed the non-cohesive microstructure characteristic for grain boundary disintegration. The attack originated at individual points at which the passive layer of the surface was pierced and had penetrated deeply into the wires. The wires had broken at these sites **(Fig. 6.82).** Here, the microstructure had disintegrated due to intergranular corrosion **(Fig. 6.83).** In the regions not attacked, the wires showed an austenitic structure with precipitates of chromium carbide on the grain boundaries **(Fig. 6.84 and 6.85).** This microstructure is characteristic of the susceptibility to intergranular corrosion or grain boundary disintegration. During the precipitation of the chromium-rich carbide $Cr_{23}C_6$ so much chromium is taken from the zones near the grain boundaries that the resistance limit is exceeded. The carbides could be brought into solution by the specified treatment for this steel at 1050 °C / ½ h, quench in water, and the normal microstructure could be reconstituted in those places that were not attacked **(Fig. 6.86).** As is known from a prior investigation and judged cause for rejection at that time, these screen bars were originally made from cold drawn steel whose corrosion resistance had been reduced by cold deformation (see also section 3.1, Fig. 3.15 to 3.19). The manufacturer of the screens apparently believed that he needed to eliminate the deformation stresses in the manner customary for steels of common type by soft annealing at about 700 °C and he thus came from the frying pan into the fire.

If heat treated steels with high chromium content are annealed at temperatures from 600° to 800 °C, failures may occur through the formation of σ-phase[18]). An example for this has been cited already in section 3.2 (Fig. 3.33 to 3.35)[19]).

The following failure also is due to annealing at an unsuitable temperature. **Generator sheets** with approximately 2.5 %Si were annealed by the user for a long period at 800 °C under dissociated ammonia – probably to lessen eddy current losses – after which they emerged heavily embrittled and fragile. As could be seen by metallographic examination, the fractures followed along the boundaries of the coarse grain **(Fig. 6.87).** Oxide precipitates had formed on these boundaries **(Fig. 6.88).** They consisted of crystals **(Fig. 6.89)** that showed a distinct spot pattern during electron diffraction. It was probably the iron silicate Fayalite ($2FeO \cdot SiO_2$). This constituted a typical example of internal oxidation. It occurs in cases where the annealing atmosphere is constituted in such a way that only alloying components that have a higher oxygen affinity than the base metal are oxidized. This type of grain boundary oxidation observed here and responsible for the failure is based upon the fact that the low annealing temperature favored grain boundary diffusion over lattice diffusion. Sheets of the same shipment that were annealed at approximately 900 °C in a reducing hydrogen-containing atmosphere could be bent 180° without cracking.

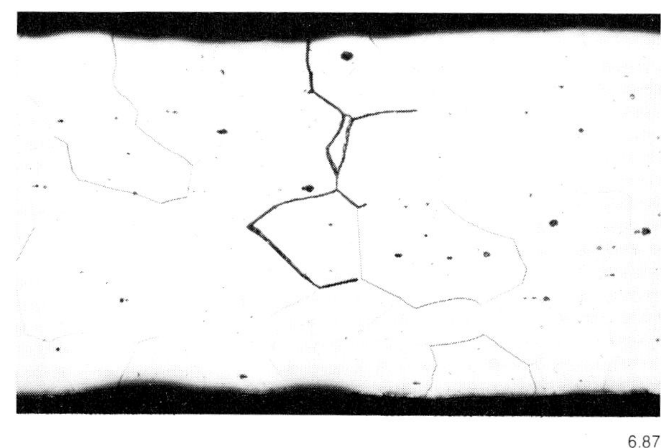

6.87

6.88

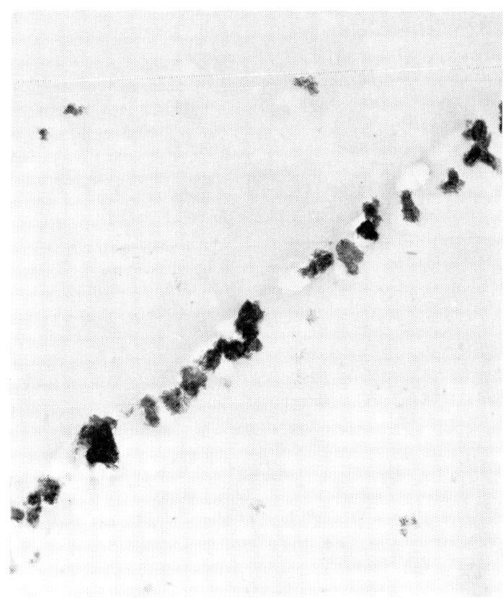

6.89

Fig. 6.87 to 6.89. Embrittled generator sheet after annealing at 800°C in dissociated ammonia

Fig. 6.87. Fracture in the ferritic grain boundaries. Unetched section. 200 ×

Fig. 6.88. Oxide precipitates on ferritic grain boundaries. Unetched section. 1500 ×

Fig. 6.89. Grain boundary precipitates in extraction replica. 15,000 ×

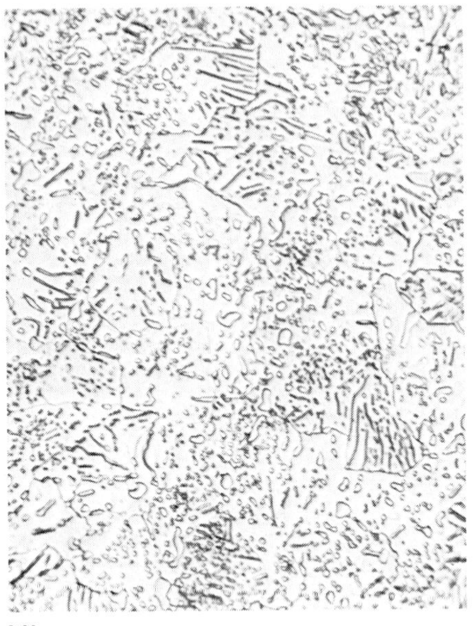

6.90

6.91

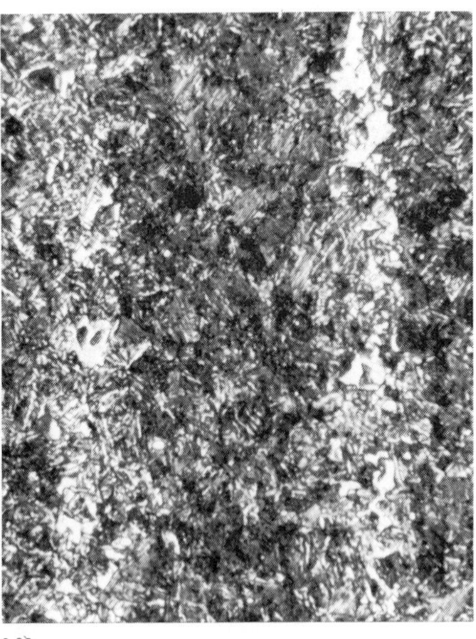

6.92

Fig. 6.90 to 6.92. Circular saw blades of a 0.8 % C, 0.5 % Cr and 0.2 % V containing steel with soft spots. Transverse sections. Etch: Picral. 1000 ×

Fig. 6.90 and 6.91. Intitial microstructure of blades

Fig. 6.90. Blade with soft spots

Fig. 6.91. Blade that could be hardened well

Fig. 6.92. Hardening structure of soft-spotted blade

6.2.2 **Timing Errors**

By holding at temperature for too long or too short a period, errors in timing may cause failures. If the time at temperature is too short, the effect will be similar to that of too low a temperature and lead to an insufficient solutioning of carbides. This causes inadequate strength or temper resistance. An example was cited in the preceding section.

This is particularly true if the carbides in the initial state are agglomerated into thick particles. A heat treating plant rejected a shipment of 4 mm thick plates of a steel containing approx. 0.8 % C, 0.5 % Cr and 0.2 % V that had been selected for **circular saw blades,** because the blades developed soft spots during normal oil quench hardening from 840 °C. Examination of a soft-spotted blade and one specimen each of well and poorly heat treatable plates showed that the two plate specimens had a very different initial microstructure. The rejected plate consisted of grainy pearlite, while the good displayed a fine lamellar pearlite **(Fig. 6.90 and 6.91).** The one was therefore soft annealed, the other normalized. During hardening of the soft annealed plate it was not possible to bring all the carbides into solution in order to undercool during martensitic transformation, as the microstructure of the soft annealed plate indicates **(Fig. 6.92).** Hardness tests confirmed that the holding time of 15 min was insufficient. By increasing the time, raising the temperature, and using a faster quenching oil, it was possible to harden the soft annealed plate without a problem.

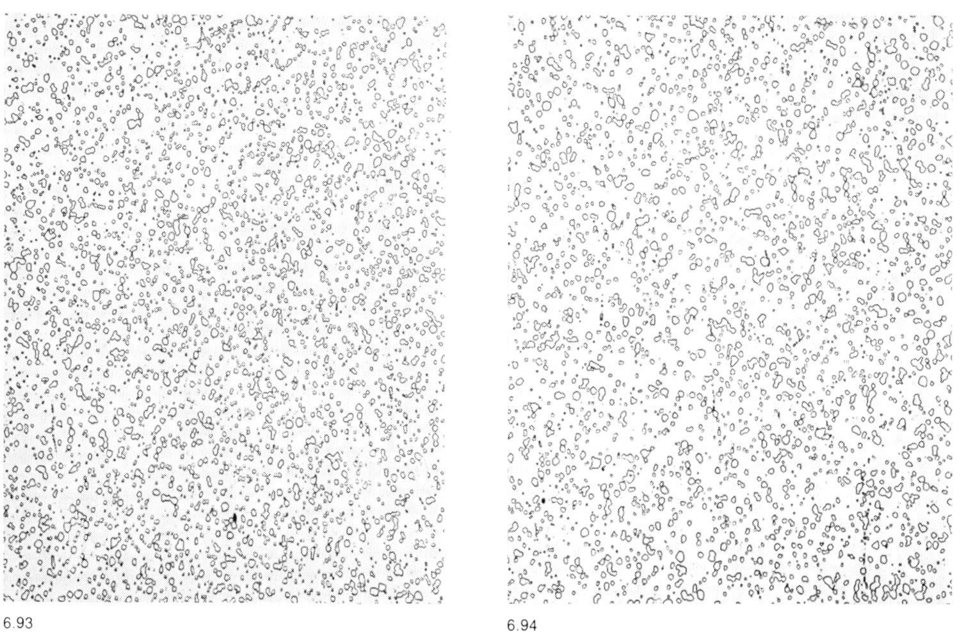

6.93

6.94

Fig. 6.93 and 6.94. Microstructure of steel strips with 0.85 % C of varying hardness and cold workability. Transverse sections. Etch: Picral. 500 ×

Fig. 6.93. Poorly rollable strip whose cold deformation caused heavy wear of the rolls

Fig. 6.94. Easily rollable strip

If the time is too short for soft annealing, the carbides are not adequately spheroidized, just as is the case for annealing at too low a temperature. This prevents the steel from attaining the necessary deformation capacity for cold working. During the production of sheet bars from cold rolled steel strip with 0.85 % C, exceptional wear occurred on the rolls. Examination of two hard-to-roll specimens and one of good quality showed that the former were definitely harder with a mean of 182 to 186 HV 1, as compared to the well deformable specimen that had a hardness of only 174 HV 1. This was expressed in the microstructure by way of a higher proportion of fine, not agglomerated, carbides **(Fig. 6.93 and 6.94).** Measurements with the quantitative television microscope Quantiment showed for the rejected steel a mean particle diameter of the carbides of 0.882 and 0.815 μm as against 1.015 μm for the well rollable material.

At high temperatures a holding time that is too long may lead to burning and in high speed steels may also result in melting. Examples for this have also been cited previously. In case hardening, an excessive holding period may lead to the absorption of too much carbon into the case or too great a case depth. The first is always undesirable, because the case becomes shock and impact sensitive, especially if the hypereutectoidal carbides are in the form of a closed network at the austenitic grain boundaries. The case depth must have a reasonable relation to the cross sectional thickness. It must not be too thin, because it could crack under high specific pressure, nor can it be too heavy so that a sufficiently thick tough core remains to absorb impact-type overloads.

Fig. 6.95. Inductively surface hardened cam shaft with soft spots. Surface lightly etched with nital. 0.75 ×

Fig. 6.96 to 6.99. Case hardened box of hammer mill with soft spots

Fig. 6.96. Longitudinal section. Etch: Nital. 5 ×

Fig. 6.97 and 6.98. Microstructure of surface layer. Etch: Nital. 500 ×

Fig. 6.97. Normal area

Fig. 6.98. Soft spot

Fig. 6.99. Microstructure of surface layer of cementation specimen after 8 h/925°C in charcoal/barium-carbonate and cooling in furnace (according to McQuaid-Ehn). Etch: Picral. 500 ×

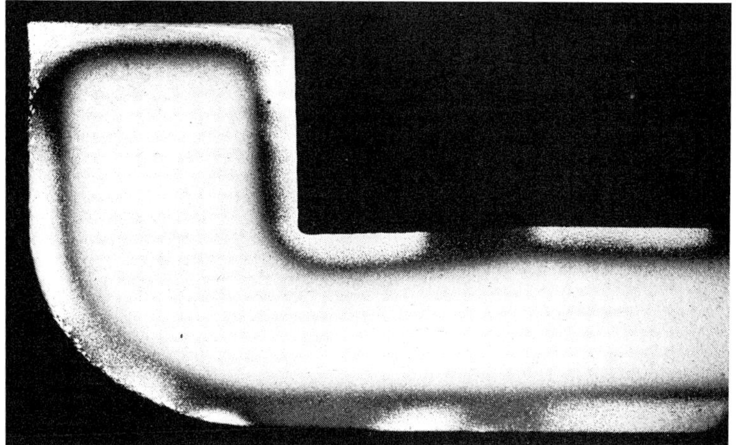

6.96

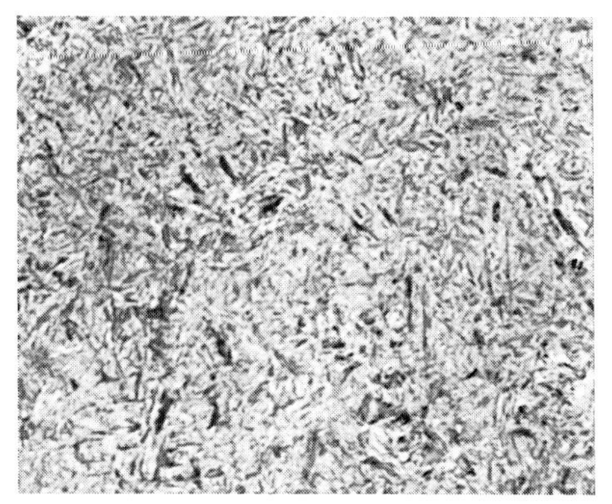

6.97

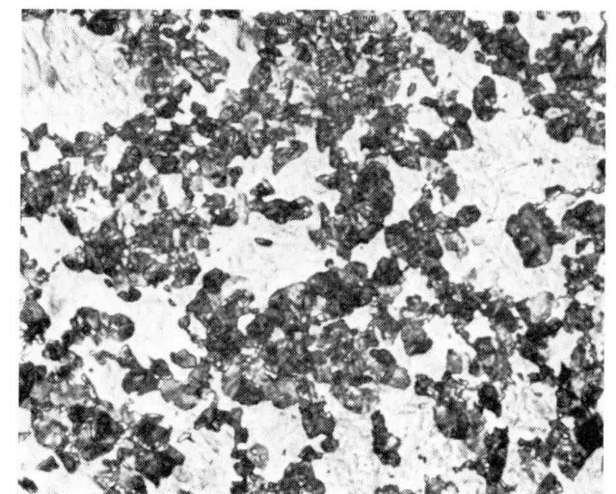

6.98

6.99

6.2.3 **Cooling Errors**

The rate of cooling must be appropriate for the steel as well as the shape and thickness of the workpiece, as has been mentioned previously. Steels with a high transformation tendency require a faster quench for hardening than those having greater hardenability. If the access of the cooling medium is locally prevented, i.e. by the jaws of the hardening tongs, by oxide adhesions, or by the formation of vapor bubbles among others, these low hardenability steels have a higher tendency to form soft spots than other types.

Soft spots may be proved readily by surface etching of the work piece. For instance, **Fig. 6.95** illustrates the etched surface of an inductively surface hardened **camshaft** of a steel similar to AISI type 1055 with soft spots. The surface hardness shows a frequency maximum of 800 HV 0.5, but in the dark spots it has decreased to values of 300 to 400 HV 0.5.

Case hardened surface layers are particularly susceptible to soft spots. **Figure 6.96** shows an etched longitudinal section trough a **box of a hammer mill** of unalloyed case hardening type 1010 steel. While the surface structure of the case is purely martensitic in the light zones **(Fig. 6.97),** it consisted entirely or in part of fine-lamellar pearlite in the dark appearing soft spots **(Fig. 6.98).** A cementation test according to McQuaid-Ehn showed a very fine-grained anomalous microstructure of cementite and degenerated pearlite **(Fig. 6.99).** The steel therefore was certainly a fine-grained type.

Cooling after hot deformation or annealing occurs faster in the outer zone than in the core, especially in heavy parts, and in parts with different thicknesses it occurs faster in the thin cross section than in the thick ones. This leads to thermal stresses and additionally to transformation stresses caused by the differing undercooling of the γ-α transformation. Consequently distor-

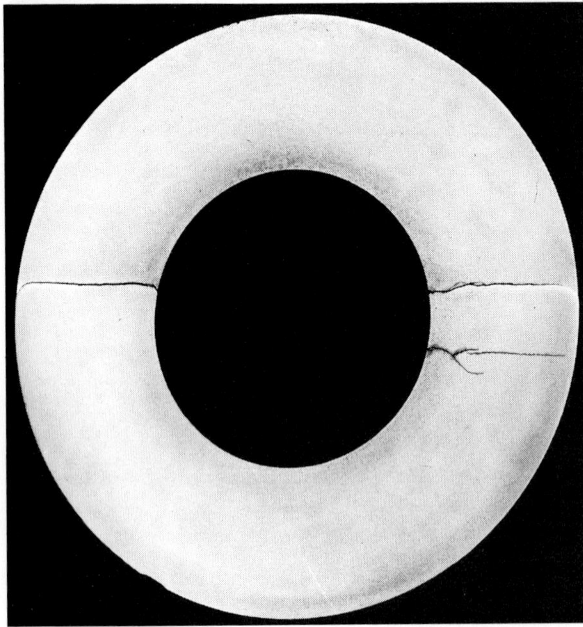

Fig. 6.100. Quench cracks in transverse section of heavy rod of steel grade 4140. Etch according to Heyn. 0.5 ×

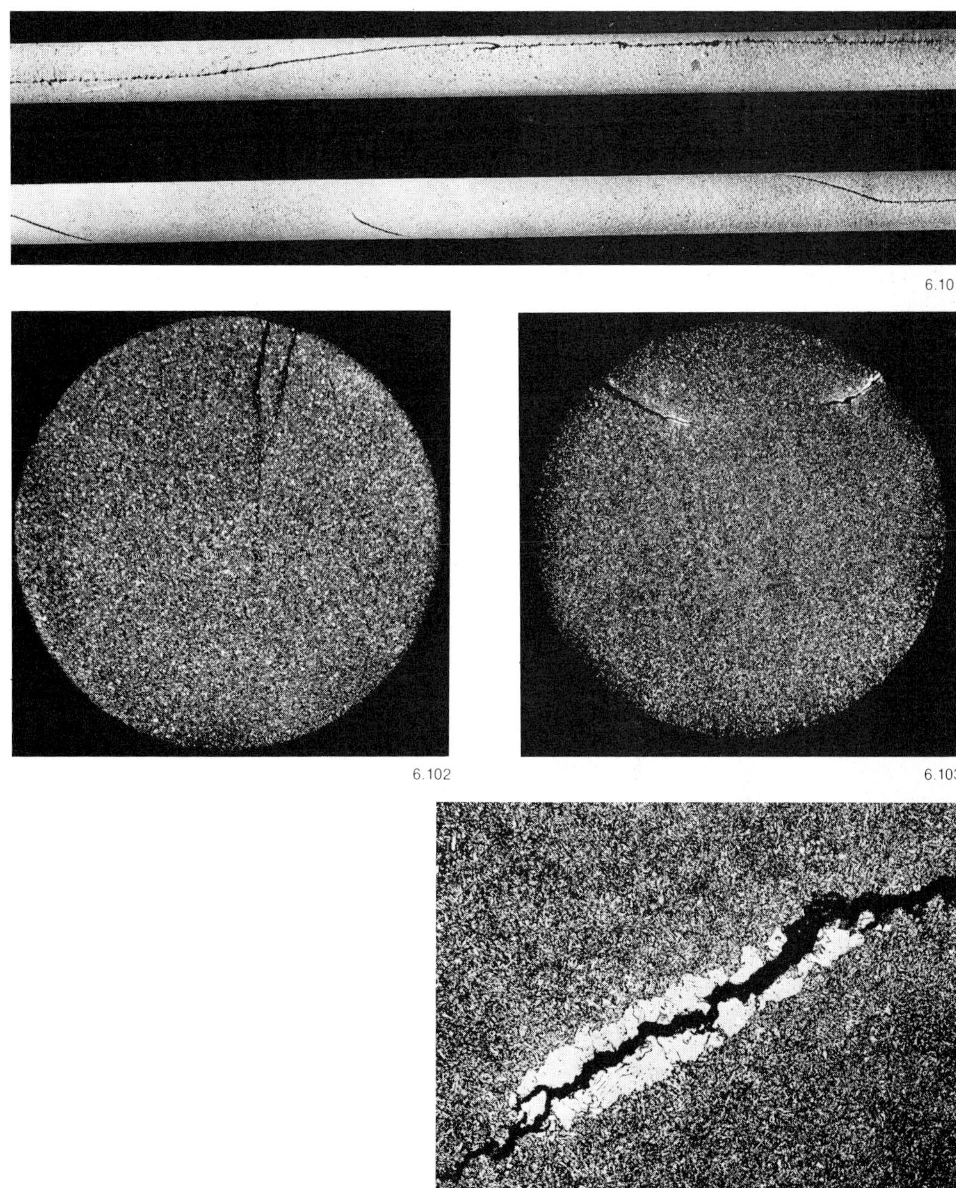

6.101

6.102

6.103

6.104

Fig. 6.101 to 6.104. Water quenched heat treated bars with quench cracks

Fig. 6.101. Surface. Lacquered white and magnetic particle inspected to show cracks better. 1 ×

Fig. 6.102 and 6.103. Transverse sections. Etch: Nital. 7 ×

Fig. 6.104. Decarburized crack branch according to Fig. 6.103. 100 ×

6.105

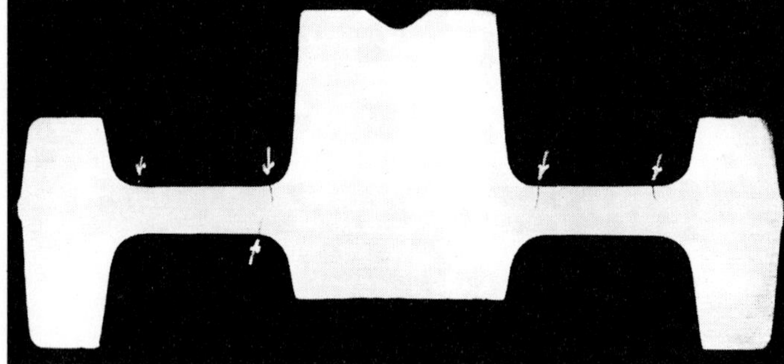

6.106

Fig. 6.105 and 6.106. Driveshaft blank of 1045 steel broken during quenching

Fig. 6.105. View. Approx. 0.75 ×. Cracks made visible by magnetic particle inspection

Fig. 6.106. Transverse section, lightly etched with diluted hydrochloric acid. 0.75 ×

tions or cracks may be formed. The stresses are greater the higher the temperature was, and the faster cooling takes place. Especially sensitive are parts with rectangular and acutely angular edges in which the heat is withdrawn from two sides. Thermal stress cracks caused by cooling as a rule propagate deeply and smoothly.

If no further heat treatment is applied after quenching, these cracks are not decarburized in contrast to incipient cracks produced by heating. If the parts are tempered after hardening, the crack edges may display decarburization in the form characteristic for low decarburization temperatures (see also section 6.1).

Figure 6.100 shows typical hardness cracks in **heavy rods** of 160/79 mm diameter of chromium-molybdenum heat treated steel grade 4140 with 0.42 % C, 0.34 % Si, 0.95 % Mn, 0.98 % Cr and 0.18 % Mo. Previously, the rods were shipped by the sender to be quenched in oil for heat treating as was customary for carbon- and maganese-rich steel. Not one of the rods developed cracks during this treatment. Now, when attempting to harden 21 rods in water, 18 developed longitudinal cracks of the type shown. Upon examination no material defects or inhomogeneities were found. The edges of the cracks were not decarburized. The cause of crack formation therefore was too rapid quenching.

Figure 6.101 illustrates **drawn bars** of a steel with approx. 0.35 % C and 0.8 % Mn that had longitudinal and spiral cracks after the water quench heat treatment. Transverse sections **(Fig. 6.102 and 6.103)** indicated that the cracks propagated in part radially and in part in flat arcs from the outside inward. Their borders were either not decarburized at all, or in part only in spots, i.e. preferentially in the interior of the cracks; in contrast, the surface of the drawn bar was not decarburized at all. Wherever decarburization took place, it appeared in the previously mentioned complete or in part columnar form characteristic for a low decarburization temperature and weakly oxidizing atmosphere **(Fig. 6.104).** From this it may be concluded that the cracks were formed during quench hardening and decarburized during tempering. No defects were found that could have promoted crack formation. Therefore quenching was too drastic.

Frequently occurring stress cracks in drop forged and heat treated **billets for drive shafts** of a type 1045 steel with a strength of 850 to 880 MPa (123-128 ksi) may be due to unfavorable mass distribution as shown in **Fig. 6.105 and 6.106** in exterior view and in section. The cracks in the thin forged dishes were located preferentially in the transitions to the more massive cross sections of shaft and flange. Their surfaces were not decarburized. Therefore these were quench cracks. Their formation was favored by a high manganese content of 0.85 % that exceeded the upper limit of the German standard's specification.

An example for the particular crack sensitivity of edges with acute angles is the following failure: **Six cylinder crankshafts** of chromium steel with an inductively hardened bearing face broke in several instances prematurely during testing. They disclosed far advanced fatigue cracks that had initiated in an oil bore and penetrated through an arm **(Fig. 6.107).** It was possible to expose by longitudinal sectioning of the oil bore the fracture origin of one shaft that showed an incipient crack but had not yet broken **(Fig. 6.108).** It was located closely below the bearing surface. Here in the low angle between bearing face and bore, several incipient cracks were found from which the fatigue fractures propagated **(Fig. 6.109 and 6.110).** Cracks of this type also were present in the shaft that had not yet been put into operation. Therefore these must have been quench cracks. Their formation was favored by the rough machining of the bore. **Figure 6.111** shows quench cracks in longitudinal section through a bore. They were located exclusively in the hardened part of the acutely angled tongue where quenching was especially drastic because the quenching medium could act from the bearing face as well as from the bore. Therefore it was recommended to the crankshaft manufacturer that the bore surface should be smoothed in the future and that all the openings of the channels be plugged prior to hardening. The manufacturer preferred to use a milder quenching medium because of less disrupting manufacturing conditions, which also achieved the same purpose.

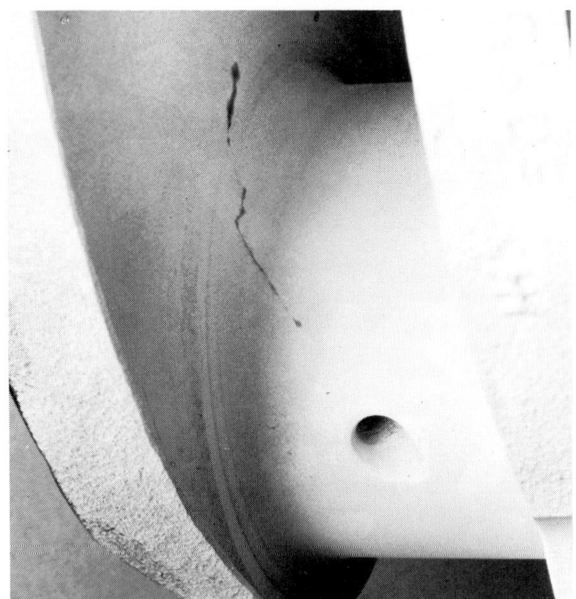

6.108

6.107

Fig. 6.107 to 6.111. Crankshaft with fatigue fractures that propagated from quench cracks in oil bore

Fig. 6.107. Fracture of shaft. Fracture origin designated by arrow (--cut)

Fig. 6.108. Incipient crack of another shaft, shown better by dye-penetrant inspection. 1 ×

Fig. 6.109. Quench crack broken open in rough grooves of bore. 5 ×

Fig. 6.110. Broken-open quench cracks and fatigue cracks propagating from them in oil bore. 5 ×

Fig. 6.111. Longitudinal section through oil bore. Etch: Picral. 10 × Top: Bore. Bottom: Bearing surface

6.109

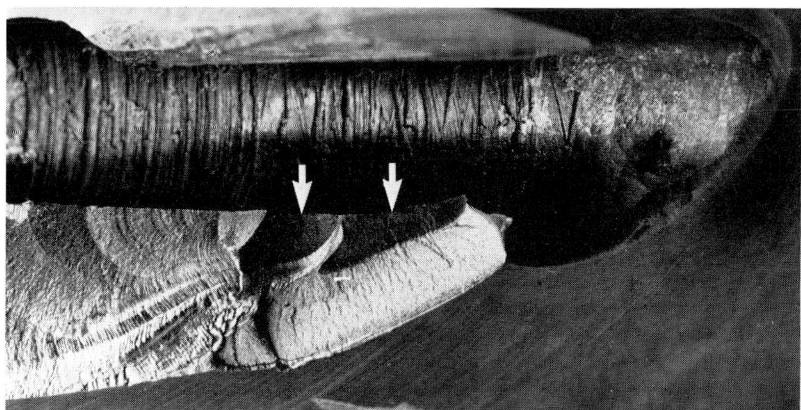

6.110

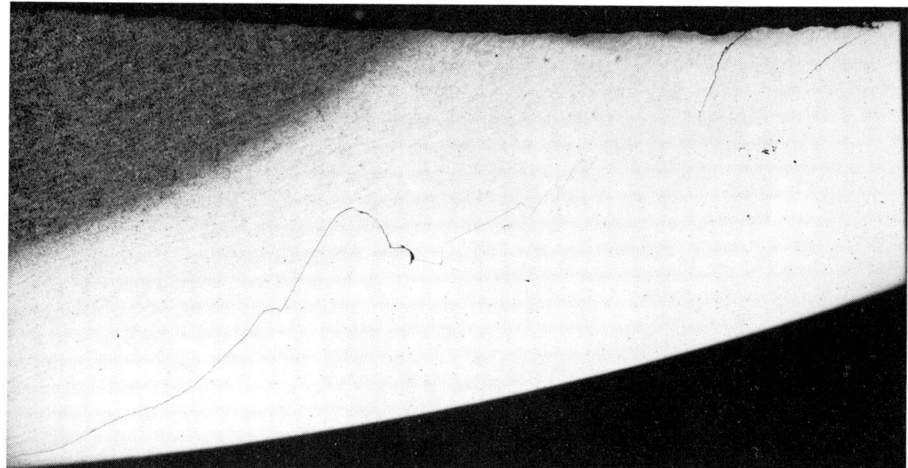

6.111

6.112

6.113

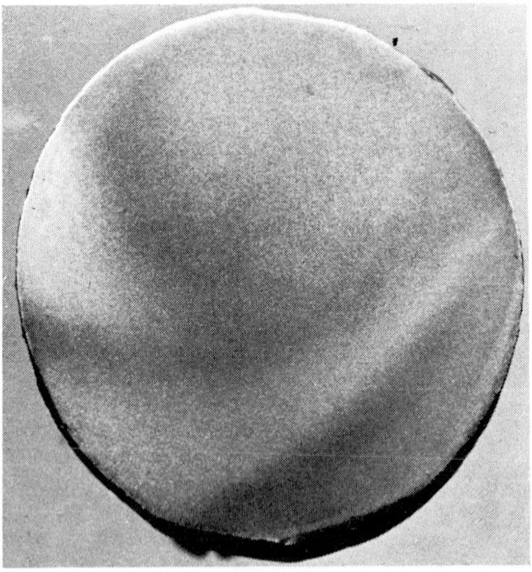

6.114

Fig. 6.112 to 6.114. Driveshaft of 1045 steel that cracked during quenching

Fig. 6.112. Circumferential quench crack in collar. Approx. 1.3 ×

Fig. 6.113. Longitudinal section through collar. Etch: Nital. 1.3 ×

Fig. 6.114. Transverse section through shaft. Etch: Nital. 3 ×

The formation of a circumferential quench crack in the collar of a **drive shaft** of steel grade 1045 was undoubtedly favored by oblique dipping into the quench bath **(Fig. 6.112).** In the longitudinal section through the collar **(Fig. 6.113),** as well as the transverse section through the shaft **(Fig. 6.114),** it became apparent after etching that the hardened zone was displaced to one side just like the crack.

6.2.4. **Errors During Case Hardening**

Even if wear resistance can be increased by well formed spherical carbides, surface carbon content in carburized cases as a rule should not exceed, or only slightly so, the eutectoid. This is so for two reasons: Firstly, the carbon-rich austenite decomposes incompletely during quenching and the high fraction of residual austenite reduces wear resistance and surface hardness; and secondly, surface layers with too high a carbon content are brittle and shock sensitive, especially if the hypereutectoid cementite is precipitated in the form of a closed network onto the austenitic grain boundaries. Such surface layers are also sensitive to the formation of grinding checks and pickling cracks (see also sections 8.2 and 9.1).

Figure 6.115 is an example of a microstructure of a **half-shell** that was too high in carbon and overheated before quenching; it was rejected because of low surface hardness. The structure of the case is very rich in austenite **(Fig. 6.116).** A heavy, closed cementite network can be seen at unquenched spots. The network is surrounded by ferrite and is separated from the pearlite matrix **(Fig. 6.117),** i.e. formed abnormally. The surface hardness of the quenched part was only 735 HV 3.

Chromium- and molybdenum-containing case hardened steels are particularly sensitive to excess carbon concentration. Reentrant angle and rectangular edges and corners are also jeopardized in this respect because the carburizing medium may penetrate from various sides. An example is a **hinge link** of an automobile with front wheel drive that failed due to fatigue cracks at the sharp link edge between two bores after a comparatively short period of operation. The hinge link consisted of chromium-manganese case hardened AISI grade 5120 steel. **Figure 6.118** shows a section parallel to the fracture plane. The edge from which the fracture propagated is designated by B. **Figure 6.119** shows this area at high magnification. It can be seen that hypereutectoid carbides are precipitated immediately below the surface in granular form and further down as an interconnected network. The fatigue fractures in this subsurface zone probably originated at incipient cracks caused by impact loading.

Excess carbon potential can be counteracted by the use of a mildly carburizing medium.

Under certain conditions, oxidation of easily oxidized alloying elements such as silicon[20]) may take place during case hardening (internal oxidation). The following failure may serve as an example. Several teeth had broken out of each gear of a pair of **bevel gears (Fig. 6.120).** All the final fractures originated from more or less advanced fatigue fractures **(Fig. 6.121 and 6.122).** Several teeth also showed incipient cracks in the outer part of the teeth **(Fig. 6.123).** Metallographic examination showed that the gears had been case hardened to a depth of approximately 0.9 mm. In the section, short incipient cracks that had propagated from oxidized surface defects **(Fig. 6.124)** could also be seen in addition to the fractures. Evidently oxidation preferentially followed the grain boundaries and penetrated from them into the interior of the grains **(Fig. 6.125).** The oxidized regions were also decarburized which could be recognized from the dissolution of the hypereutectoid carbides. Surface hardness was a little above 700 HV 0.1. The defects in part attained a depth of 0.07 mm. This is low in comparison to the carburized case depth and normally the remainder of the martensitic layer under residual compressive stress suffices to prevent fatigue fractures. But in this case a normal stress situation was out of the question. The onesided appearance of the fractures indicated the effect of local overloading – perhaps through oblique mounting of the gears (see section 10). Under these circumstances surface oxidation may have at least either favored or accelerated the fracture formation process.

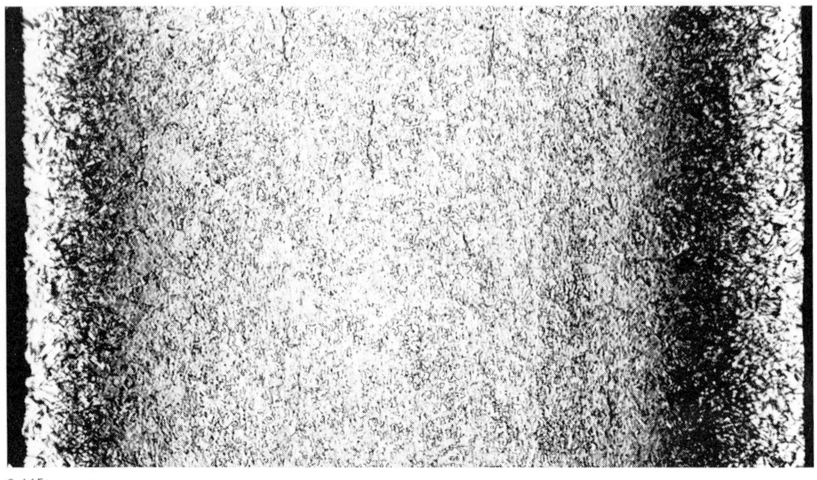

6.115

6.116

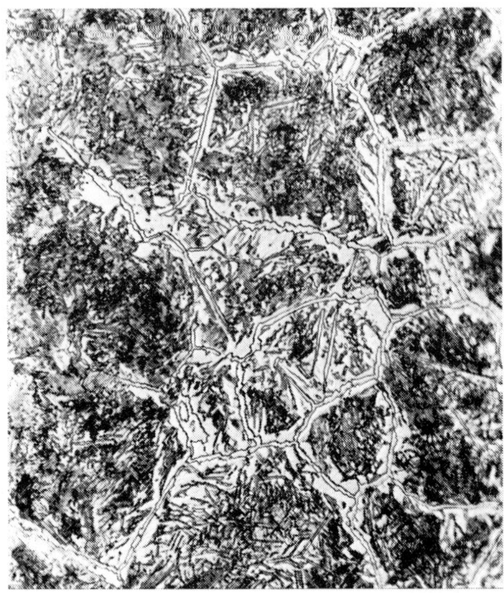

6.117

Fig. 6.115 to 6.117. Case hardened half-shell too high in carbon and overheated before quenching from carburizing temperature. Transverse section. Etch: Picral

Fig. 6.115. Transverse section. 100 ×

Fig. 6.116. Microstructure of the surface layer of the hardened part. 500 ×

Fig. 6.117. Structure near surface in unhardened part. 500 ×

6.2.5. **Errors During Nitriding**

During nitriding it is important that the parts are introduced into the furnace with clean and not decarburized surfaces and are free of working and machining stresses. They have to be positioned or hung in such a way that no distortion is possible. Nitriding itself causes but minor stresses because of the low temperature involved and the absence of quenching. These stresses find an expression only in minor calculable growth. If highest hardness and wear resistance are desired, the temperature for normal nitrided steels should be kept constant at around 500 °C, and for austenitic steels at approximately 550 °C, within narrow limits. The depth of penetration is a parabolic function of time. Because of the low nitriding temperature, attainable depth is shallow in comparison with other surface hardening processes. This should be considered in the selection of an appropriate surface hardening method. A somewhat deeper, though likewise hard nitride layer can be applied by double-stage nitriding first at low and subsequently at more elevated temperature[21]).

It has already been mentioned that aluminum-alloyed nitrided steels, because of their extended $\alpha + \gamma$ region, have a tendency to form completely decarburized ferritic surface layers due to the effect of decarburizing gases. **Figure 6.126** shows the microstructure of such a layer of a **plunger** made unusable by spallation of the surface.

The following failure constitutes an example of a cracked nitride layer that was overstressed by a high specific surface pressure.

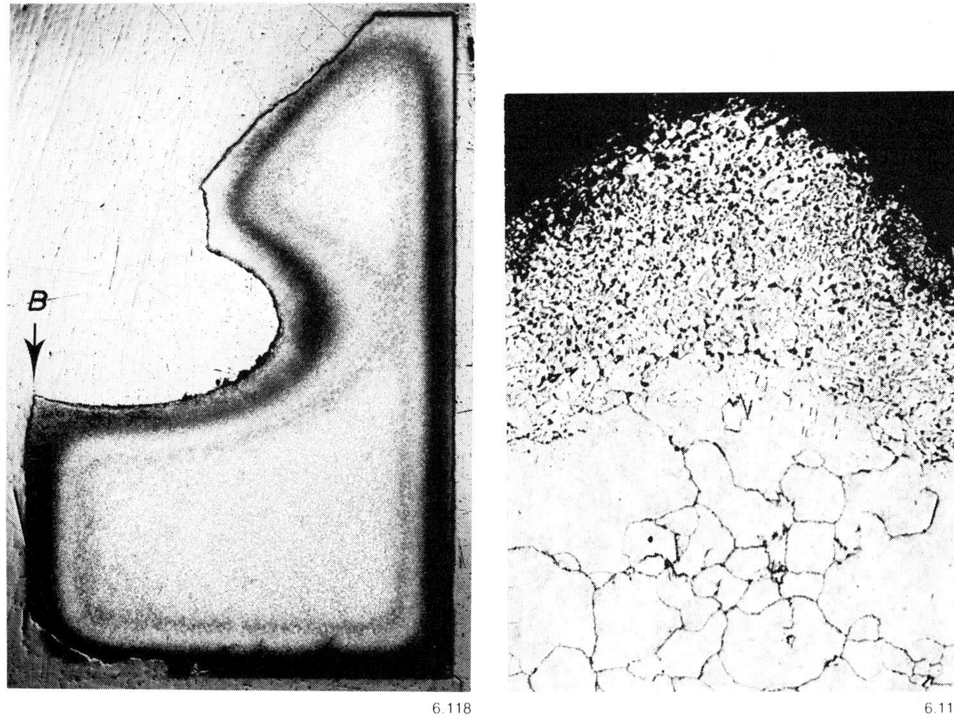

Fig. 6.118 and 6.119. Excessively carburized hinge link of an automobile

Fig. 6.118. Section parallel to fracture. Etch: Nital. 4.5 ×
B: Edge from which fracture propagated

Fig. 6.119. Edge B in Fig. 6.118. Etch: Alkaline sodium picrate solution. 200 ×

6.120

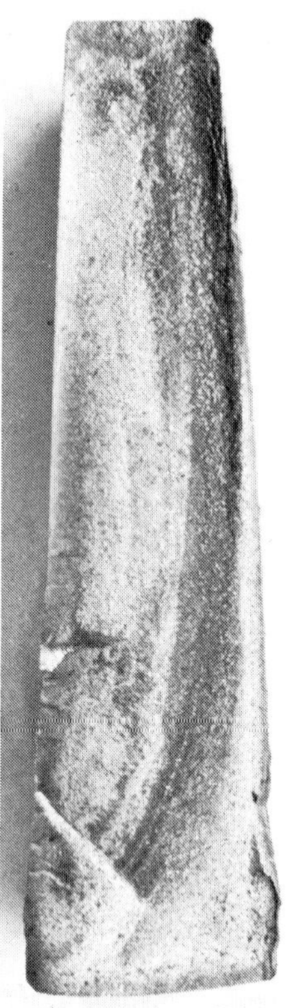

6.121

6.122

6.123

6.124

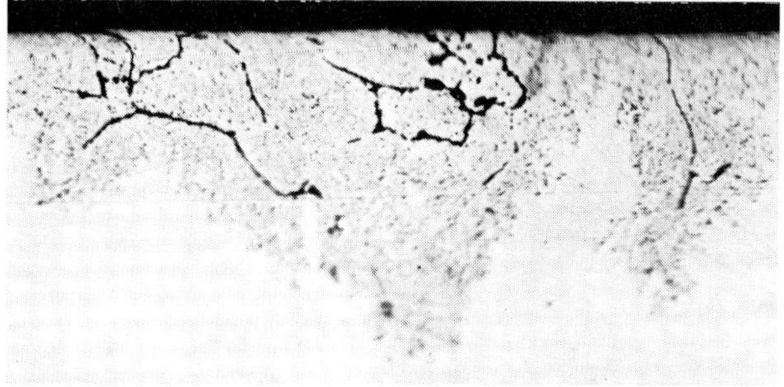

6.125

Fig. 6.120 to 6.125. Case hardened bevel gears with broken teeth due to surface oxidation

Fig. 6.120. Bevel gear pair with broken teeth. Approx. 0.4 ×

Fig. 6.121 and 6.122. Tooth break-outs that propagated from variously progressed fatigue fractures

Fig. 6.122. 10 ×

Fig. 6.121. 3 ×

Fig. 6.123. One-sided incipient fractures accentuated by dye-penetrant. 1 ×

Fig. 6.124 and 6.125. Internal oxidation under surface and incipient cracks propagating from them. Transverse section through a tooth flank. Unetched. 500 ×

A **worm gear** of a nitriding steel containing approx. 0.35 % C, 1.4 % Cr and 1 % Al had worn prematurely **(Fig. 6.127).** The gear was part of a triple-jaw chuck. The nitrided layer was depressed at the unworn edges and had broken out. This probably constituted the first stage of failure; the strong wear was the result of secondary grinding by the hard fragments. Surface hardness at the non-worn places was 1003 to 1097 HV 10, and accordingly was satisfactory. But core hardness was too low with 208 to 217 HV 10, corresponding to a strength of 700 to 720 MPa (100–105 ksi). **Figure 6.128** shows an etched section through the gear. It illustrates that the teeth were badly worn not only on one side, but they were in part also bent and subjected to shear. One tooth also showed an incipient crack below the engagement zone. It could be concluded that the teeth had been subjected to substantial overload. The nitrided layer was only 0.3 mm deep and had cracked and broken out in part. The microstructure of the heat treated core was composed of bands of ferrite and carbides that had agglomerated due to a high temperature. The worm gear accordingly was hardened at too low a temperature and tempered at too high a temperature. This explains the low core strength. According to the sender the worm gear was said to consist of molybdenum-containing nitrided steel (composition approx. 0.35 % C, 1.2 % Cr, 1 % Al and 0.2 % Mo) and had been nitrided for 62 h at 500 °C in ammonia. If this had been the case, the nitrided depth should have been approximately 0.4 mm and core strength 780 to 980 MPa (113 to 142 ksi). Accordingly, three mistakes had occurred. Firstly, a steel with low strength after heat treatment was selected; secondly, it was nitrided for too short a time or at too low a temperature; and thirdly, it was hardened at too low a temperature or tempered at too high a temperature. After correct tempering even the molybdenum-free steel used should have had a strength of 740 to 930 MPa (107–135 ksi). However, the question remains unanswered whether the avoidance of these errors would have caused the gear to withstand the very high stress applied, or whether a steel with higher core strength should have been selected; also, there is the other open question whether the nitriding process with its limited hardness depth and core strength was suitable in the first place for such highly stressed gears.

Fig. 6.126. Structure at surface of nitrided plunger. Nitrided layer spalled off as consequence of decarburization. Transverse section. Etch: Picral. 100 ×

In thin-walled parts for which nitriding is particularly suitable because of its freedom from distortion, occasionally the reverse situation takes place, namely surface hardening that is too deep or penetrates the core. This, however, is less frequent than occurs with other surface hardening methods. An example can be seen from the following failure analysis of **gears** having

6.127

6.128

Fig. 6.127 and 6.128. Worn nitrided worm gear of three-jaw chuck

Fig. 6.127. View. 2 ×

Fig. 6.128. Longitudinal section through teeth. Etch: Picral. 6 ×

a module of 0.8. In several of these gears, teeth had broken out during operation while others subjected to the same stress performed without trouble. As is proved by the **Fig. 6.129 and 6.130,** the damaged gears had teeth that were nitrided through, while the good quality gears were so lightly nitrided that the teeth still had a soft, deformable core.

Cutting tools made of high speed steel are often nitrided in order to increase wear resistance. But nitriding is recommended only for finishing tools because of shock sensitivity of the nitrided layers. This recommendation apparently was not observed in the following case. Two **spur gear hobs** with modules of 7.5 and 8.5, respectively, and made of high speed steel grade T 4, were rejected because shell-like spalling originating at the cutting edges occurred during operation. The hobs had been quenched and were also tempered in a cyanide bath. Experience has shown that this is unfavorable and does not correspond to the recommendation by the German Steel Industry*, which states that as far as feasable, nitriding should take place after tempering in a special bath at a temperature 10–20 °C lower than the tempering temperature; this should result in having a nitrided layer of 0.02 to 0.03 mm.

Figure 6.131 shows one of the hobs. Even the undamaged teeth (top) showed dark heat tinted zones along the cutting edges that indicated high stress. Metallographic examination proved that the hobs were nitrided exceptionally deeply to a depth of 0.05 to 0.06 mm **(Fig. 6.132).** This was due to the higher temperature of the salt baths and as a rule also to the longer holding time in the tempering bath as contrasted to a special nitriding bath. The core structure was normal. Core hardness was 65 HRC, while surface hardness was approximately 1100 HV 1. The fragmentation of the tooth edges was undoubtedly facilitated by the unusual depth of the very hard surface layer. Surface hardening that was not suitable in this case, was also done incorrectly.

* **Stahl-Eisen-Werkstoffblatt** 320

6.129

6.130

Fig. 6.129 and 6.130. Nitrided gears, module 0.8. Transverse sections. Etch according to Heyn. 4 ×

Fig. 6.129. Through-nitrided gear

Fig. 6.130. Good quality gear with correct nitrided depth

In conjunction with this section some hints for the examination of nitrided parts should be given. An exact metallographic determination of hardening depth is sometimes difficult. The nitrided layer consists of two layers in addition to the thin white zone of nitrides at the very surface. The outer one can be easily etched and probably is characteristic of the depth to which special nitride-forming alloying elements such as chromium and aluminum are precipitated from the iron lattice as nitride; and a second, inner zone that is attacked more slowly by the etching medium. Etching is conducted generally with picric or nitric acid in alcohol as is customary. If etching is not conducted for a sufficiently long period, the danger exists that the inner zone is either not etched at all, or incompletely, so that too small a penetration depth is measured. In case of doubt, this error may be avoided best by macro-etching with copper ammonium chloride solution according to Heyn (compare the photograph 6.129 with 6.130). The fraction of the deeply etched and darker appearing outer zone of the total layer thickness is the larger, the higher the nitriding temperature[21]). With the higher temperature the tendency for this zone to precipitate a network of iron nitrides onto the grain boundaries also increases.

6.131

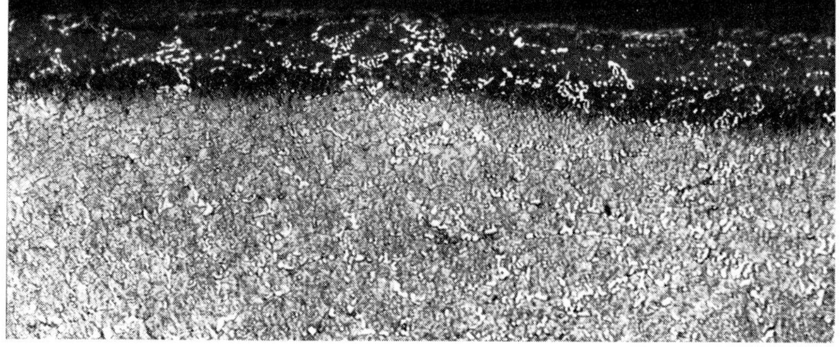

6.132

Fig. 6.131 and 6.132. Spur gear hobs, module 8.5, made of grade T 4 high speed steel with spalled cutting edges

Fig. 6.131. View. Approx. 1 ×

Fig. 6.132. Surface structure in transverse section through tooth. Etch: Nital. 200 ×

More and better information about nitriding than is possible by metallographic examination is provided by hardness measurements, and in particular by the plotting of hardness traverses from surface to core in the shape of hardness gradient curves. Measurements can be done on sections perpendicular to the surface or obliquely under a certain angle to it. In view of the shallow depth of the layers, it is recommended to use light test loads. The region of highest hardness, usually found close below the surface, is a function of the nitriding temperature. Correspondingly, the hardness gradient plots can be indications of the temperature profile used during nitriding. Low temperatures around 500 °C cause thin layers with high surface hardness, while high temperatures, i.e. around 550 °C, show softer but deeper case hardness layers for equal nitriding periods. The advantages of high surface hardness and deep nitriding depth can be combined if nitriding temperature is kept low initially and raised later. Nitriding then is deeper, but the hard wear resistant outermost zone is thinner than if nitriding had continued at the low initial temperature. From such a ,,narrow chested" but ,,thick bellied" hardness gradient curve, a temperature profile in the above sense can be assumed with confidence. Errors and mistakes can be explained in this way.

A section of **nitrided pipe** was shipped by a foreign institute because it had proven superior under compressive wear loading in contrast to prior tests of similar pipes. The questions were in what way did the nitrided case hardness differ from the normal one used previously, and what method of hardening had been used. Microstructural examination did not give any indication for the improved behavior. The hardness gradient plot **(Fig. 6.133)** showed the type of double stage nitriding with initially lower and subsequently higher temperature (approximately 24 h 500 + 24 h 550 °C). A comparison curve is drawn with broken line that shows the nitriding of a specimen of the same pipe for 48 h at 500 °C. Such a nitrided layer is certainly more suitable for parts that

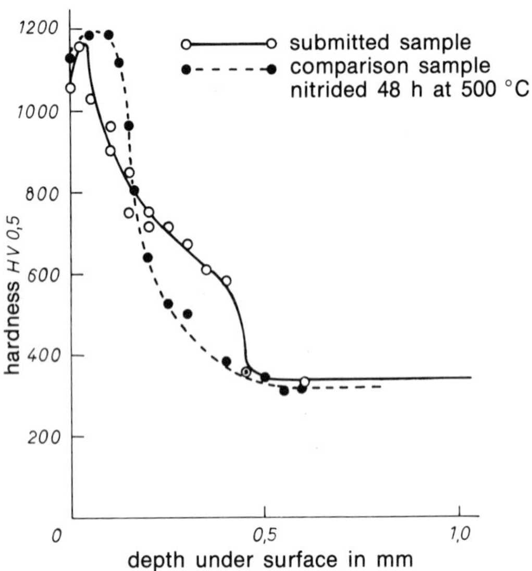

Fig. 6.133. Hardness gradient of nitrided pipe

are predominantly subjected to wear while the solid line curve represents a nitrided layer of higher support capability. During service of these pipes it was apparently more important to have resistance against high compressive load than continuous wear resistance. Based upon the preceding examination a simple method could be recommended to the inquiring party that was sure to bring the desired success without undue effort.

Literature Section 6

1) F. K. Naumann u. F. Spies: Rissiger Zentrifugendeckel und Extruderschnecke mit Oberflächenausbrechungen. Prakt. Metallographie 5 (1968) S. 397/401

2) G. Bandel: Verzundern und Entkohlen unlegierter Stähle. Stahl u. Eisen 58 (1938) S. 1317/26

3) F. Nehl: Die Oberflächenempfindlichkeit von Stählen gegen bestimmte Heizgase. Stahl u. Eisen 58 (1938), S. 779/84

4) H. Buchholtz u. R. Pusch: Ursachen feiner Oberflächenfehler bei der Warmverarbeitung von unlegiertem Stahl. Stahl u. Eisen 73 (1953) S. 204/12

5) M. Feller-Kniepmeier u. M. Thiele: Einfluß von Kupfer- und Zinnverunreinigungen auf die Verarbeitbarkeit von weichem unlegiertem Stahl. Stahl u. Eisen 82 (1962) S. 1432/36

6) W. Oelsen, K.-H. Sauer u. H. Brockmann: Ein einfacher Weg, das Zundern und Randentkohlen zu verfolgen. Härtereitechn. Mitt. 21 (1966) S. 47/54

7) H. Schrader: Empfindlichkeit legierter Stähle gegen Verbrennungserscheinungen Techn. Mitt. Krupp 2 (1934) S. 136/42

8) F. K. Naumann u. F. Spies: Gebrochenes Kettenglied. Prakt. Metallographie 9 (1972) S. 706/07

9) Dieselben: Beim Biegen angerissenes Kesselrohr. Prakt. Metallographie 10 (1973) S. 227/29

10) Dieselben: Entkohlung. Prakt. Metallographie 8 (1971) S. 375/84

11) W. Oelsen, K.-H. Sauer u. G. Naumann: Zur Entkohlung von Stählen in zundernden und nichtzundernden Gasen. Jahrbuch 1969 des Landesamtes für Forschung des Landes Nordrhein-Westfalen. Westdeutscher Verlag. Köln u. Opladen, S. 411/67

12) H. Berns: Entkohlung in Manganhartstahl. Zschr. f. wissenschaftl. Fertigung 63 (1968) S. 437/41

13) F. K. Naumann u. F. Spies: Gebrochene Federn. Prakt. Metallographie 7 (1970) S. 155/59

14) H. Schumann: Metallographie, 2. Aufl. Leipzig 1958, S. 367

15) F. K. Naumann u. F. Spies: Schlecht ziehbarer Kugellagerstahldraht. Prakt. Metallographie 9 (1972) S. 42/46

16) Dieselben: Untersuchung von Drähten für die Herstellung vergüteter Schrauben. Prakt. Metallographie 8 (1971) S. 437/42

17) Dieselben: Durch Kornzerfall zerstörte Siebstäbe. Prakt. Metallographie 9 (1972) S. 597/602

18) F. K. Naumann: Beitrag zum Nachweis der σ-Phase und zur Kinetik ihrer Bildung und Auflösung in Eisen-Chrom- und Eisen-Chrom-Nickel-Legierungen. Arch. Eisenhüttenwes. 34 (1963) S. 187/94

19) F. K. Naumann u. F. Spies: Gebrochener Rekuperator aus hitzebeständigem Stahlguß. Prakt. Metallographie 9 (1972) S. 709/10

20) C. Albrecht: Die Randoxydation bei Einsatzstählen, Härtereitechn. Mitt. 9 (1955) S. 9/26

21) O. Hengstenberg u. F. K. Naumann: Doppelnitrierung. Arch. Eisenhüttenwes. 7 (1933/34) S. 61/66

7. Failures Caused by Errors in Welding, Brazing and Riveting

During welding, defects may occur that are based solely on welding technique, as well as those in which the steel and additions play a contributory part[1]). Welding technique defects are mainly joining errors due either to contaminated edges[2]), insufficient heating, inadequate penetration, wrong torch or electrode manipulation, or unsuitable pass sequence, and also weld undercutting that may have a deleterious effect, particularly in the presence of hardened transition zones.

Also among these failures are **gas bubbles** of carbon oxides that are formed by combustion of the carbon in the steel or of hydrogen from the moisture content in the welding wire covering or the welding powder.

Oxide inclusions are formed by overheating and burning, especially during welding of steels with alloying elements that are readily oxidized and form high melting oxides with a large excess of oxygen. On the other hand, carburizing and hardening of the weld bead may occur through incorrect torch manipulation[3]). Hydrogen is also the cause of spots sometimes called **„fish eyes"** in the fracture of welding seams. Even **weld spatter** has caused failures in easily hardenable steels.

All these errors can be proved with the aid of various radiographic techniques and also can usually be identified easily by metallographic examination. Some illustrations will clarify this.

7.1 Joining Errors, Root Defects, Weld Undercuts

In section 1.1 (Fig. 1.1 and 1.2), a chain link which had been resistance butt welded, was found to be poorly bonded because the welding temperature was too low. **Figure 7.1** shows the fracture of a **tie rod** weld that was porous due to burning. The fracture was characterized by a blue stain and therefore was hot short. The lock had been welded together from a forged piece (left in the illustration) and a rolled bar, as was proved by the fiber orientation **(Fig. 7.2). Figure 7.3** shows the overheated microstructure permeated by oxides at the seam, as well as the intergranular propagation of the hot cracks.

A typical **root failure** characterized by insufficient weld penetration is shown in **Fig. 7.4.** Here a **pipe coil of acid-resistant steel** of type AISI 316 had been attacked by sulfur-containing oils.

Another root failure was found in a **high pressure pipeline** by radiography **(Fig. 7.5).** A crack propagated from the weld root, as could be seen by subsequent metallographic examination **(Fig. 7.6).**

Joint defects in a **gas welded pipe** made of a steel having an ultimate strength of about 450 MPa (65 ksi) are illustrated in the radiograph of **Fig. 7.7a** and a heavily etched section of **Fig. 7.7b.** According to the microscopic findings these defects were oxide slag inclusions.

Figure 7.8 shows a joint defect which was the result of oxidation at the edges of the welding bead, while **Fig. 7.9** shows a **stress crack** characterized by its intergranular propagation as a hot crack in the root of the same seam.

Further examples for joint failures are cited under 4) and 5).

Fig. 7.1 to 7.3 Butt welded seam of tie rod made porous and ruptured by oxidation

Fig. 7.1. Fracture, in part stained blue and intergranular. 1 ×

Fig. 7.2. Longitudinal section through fracture. Etch according to Oberhoffer. 1 ×

Fig. 7.3. Microstructure of seam. Longitudinal section. Etch: Nital. 100 ×

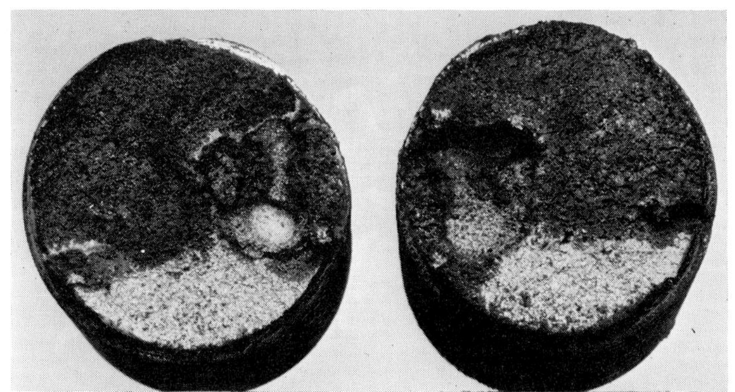

7.1

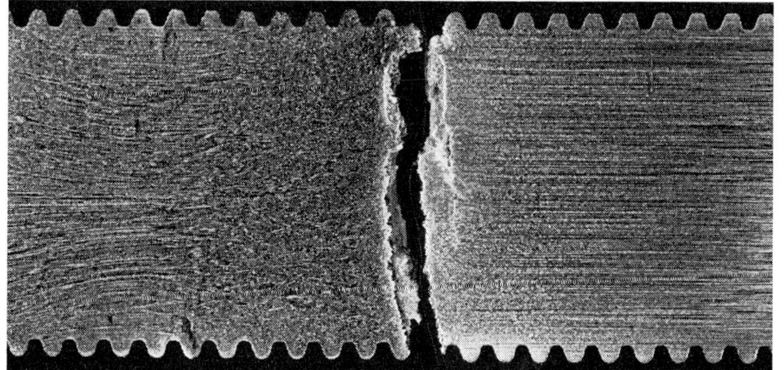

7.2

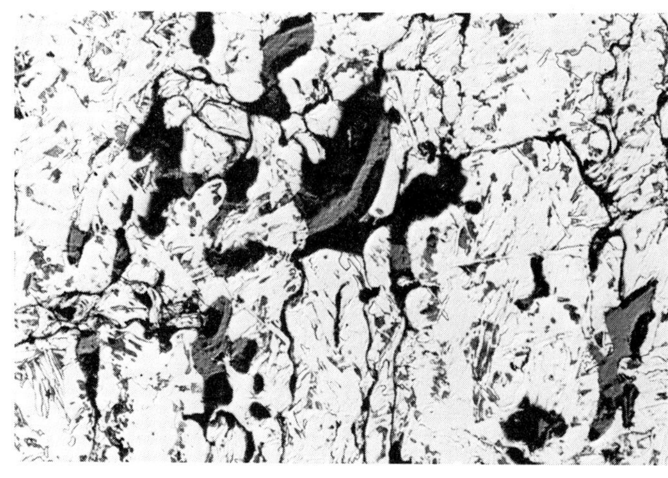

7.3

The effect of **weld undercutting** – especially for cyclic stresses – may be illustrated by the failure of a **fan rotor** made of a stainless steel containing approx. 18 % Cr, 12 % Ni, 2 % Mo, 0.5 % Ti min., and 0.1 % C max. The rotor is shown in **Fig. 7.10.** The upper face plate had cracked in several places in the flanged lip **(Fig. 7.11).** The cracks started in the interior from the weld at the intake edge of the blade foils **(Fig. 7.12,** arrow). **Figure 7.13** illustrates a crack in the initial stage. After breaking them open, the cracks displayed a structure resembling fatigue fractures **(Fig. 7.14).** Metallographic examination confirmed that the cracks propagated exactly from the weld undercuts **(Fig. 7.15).** In this austenitic material they could not have been caused by transformation stresses as in high hardenability steels (see the following section). Since the wheel was said to have conducted hot caustic soda, stress corrosion should not be ruled out as playing a contributory role (see section 15.3.4).

Weld undercutting also contributed in part to the failures described in 6).

7.2 **Hot Cracking, Weld Crack Susceptibility**

A defect that has caused many failures and continues to do so is associated with **"weld crack susceptibility"***. It occurs mainly during gas welding of thin sheets and thin-walled pipes adjacent to the welding seam. Judging by their shape, the defects can be classified as austenitic grain boundary cracks. Their cause is not yet completely clear. It is known that the cracks are formed at temperatures around 1200 °C, i.e. immediately below the solidus line of the steel, and that steels with high sulfur contents are especially sensitive to these defects[1][7]. This type of failure therefore should belong to the above mentioned second group in which the steel composition plays a contributory part.

* see Appendix I

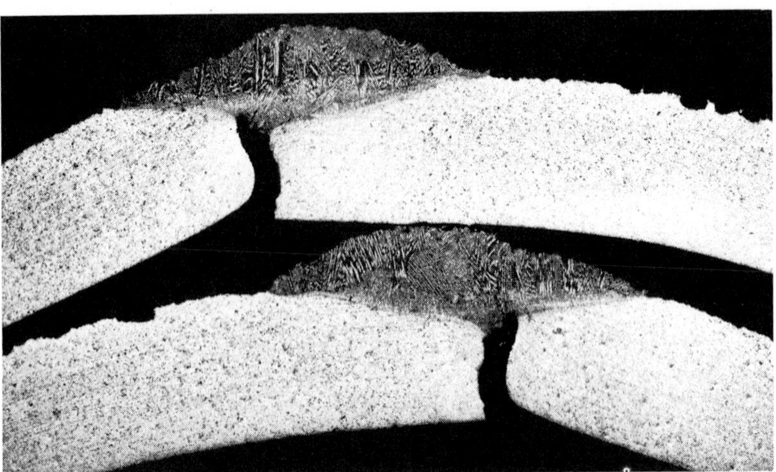

Fig. 7.4. Root defect in weld seam of pipe coil made of acid resistant steel. Transverse section. Etch: V2A-etching solution

In the following, two special cases of weld crack susceptibility are described. The first failure occurred during welding of a **bicycle frame** of seamless tubing made of a steel having about 450 MPa (65 ksi) ultimate strength. Magnetic particle inspection established that in a number of frames, cracks had occurred next to the tube seams. Several frames with and without cracks were examined. **Figure 7.16** shows a longitudinal section through a cracked joint and **Fig. 7.17** an unetched section at high magnification of one of the cracks. The crack is permeated by oxide inclusions and the surface zones are oxidized and decarburized. Therefore it is a typical hot crack. In the crack branches, fine precipitates were found **(Fig. 7.18).** They were probably manganese sulfides as will be described in a subsequent failure case. Phosphorus and sulfur contents were determined on two cracked frames and one comparison frame that was free of cracks. They are shown in **Table 1** together with the findings of the metallographic examination. The frame that was free of cracks and the crack-free tube of a frame that was rejected had a distinctly lower sulfur content than the welded cracked tubes in which sulfur contents substantially exceeded the highest value permissible for this tube alloy. It should be noted that during welding of the steering head tubes no cracks occurred. This may be explained by the fact, that these seams were welded first and therefore not exposed to such high stresses as the connecting tubes welded later, in which deformation was inhibited by the rigid clamping.

Table 1 Chemical Composition and Results of Metallographic Investigation for Cracked and Crack-Free Bicycle Frame Tubes

Frame	Tube	P %	S %	Weld at Top Tube	Weld at Head Tube
I	1	0.042	0.062	major cracks	no cracks
Rejected	2	0.043	0.049	minor cracks	no cracks
II	1	0.031	0.036	no cracks	no cracks
Rejected	2	0.042	0.044	cracks	no cracks
III	1	0.020	0.025	no cracks	no cracks
Good comparison frame	2	0.032	0.034	no cracks	no cracks

The second failure analysis concerned **pipe nipples** of 70 mm diameter and 3.5 mm wall thickness made of a steel having about 350 MPa (50 ksi) tensile strength that cracked next to the seam during welding to a manifold with thicker walls[8]). Welding was done horizontally. After tacking at two opposite spots, one-half the pipe circumference was welded from a point at the bottom between the two tack-welds, and subsequently the other half of the circumference was welded on top in the opposite direction. At the beginning and end, i.e. top and bottom, the two weld bead halves overlapped for a short length. The cracks found, using radiography, were always located in the vicinity of the end crater on the side of the first bead.

Figure 7.19 shows a longitudinal section that cuts through a crack. The crack originated next to the weld seam in the base metal and ended at its interface with the seam. The material was noticeably fine-grained and heavily ferritic in the cracked zone, while it showed a coarse-grained Widmannstätten structure lower in ferrite at some distance from the seam. This microstructure

7.5

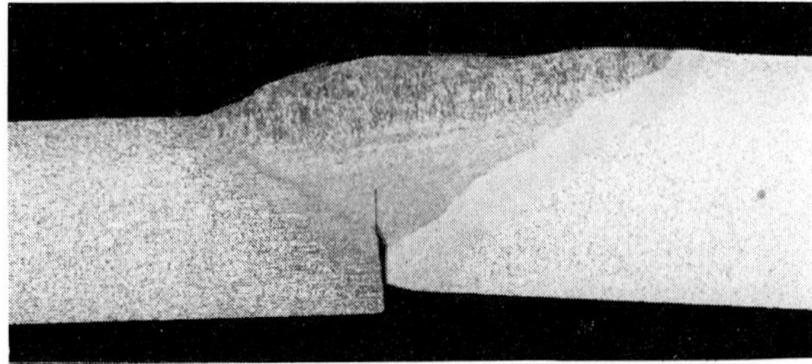

7.6

Fig. 7.5 and 7.6. Root defect with crack in weld seam of high pressure pipeline

Fig. 7.5. Radiograph
Fig. 7.6. Longitudinal section. Etch according to Heyn. 2 ×

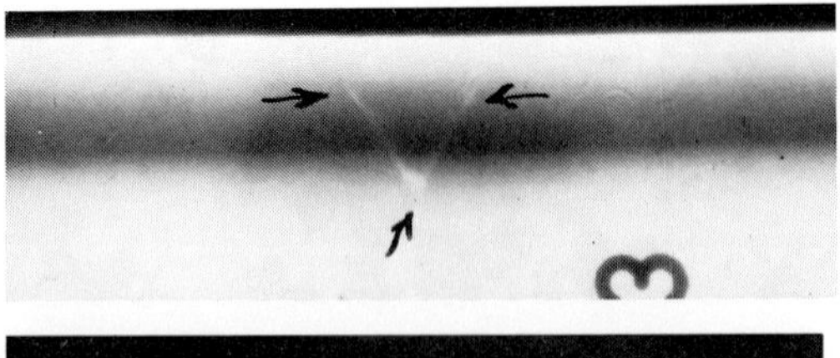

a

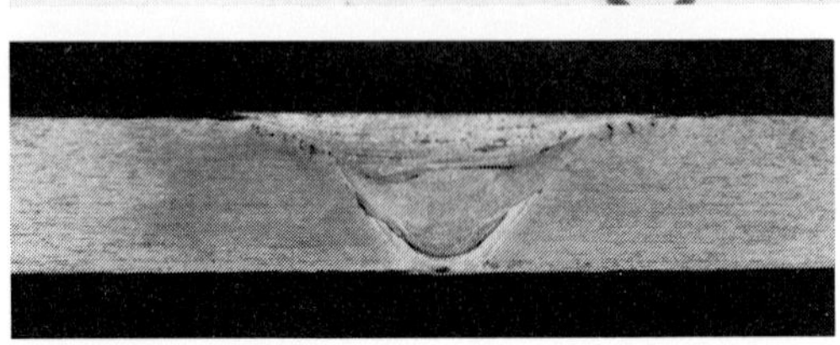

b

Fig. 7.7a and b. Joint defect caused by slag inclusions in a welded pipe a. Radiograph b. Longitudinal section, etched with diluted hydrochloric acid. 1 ×

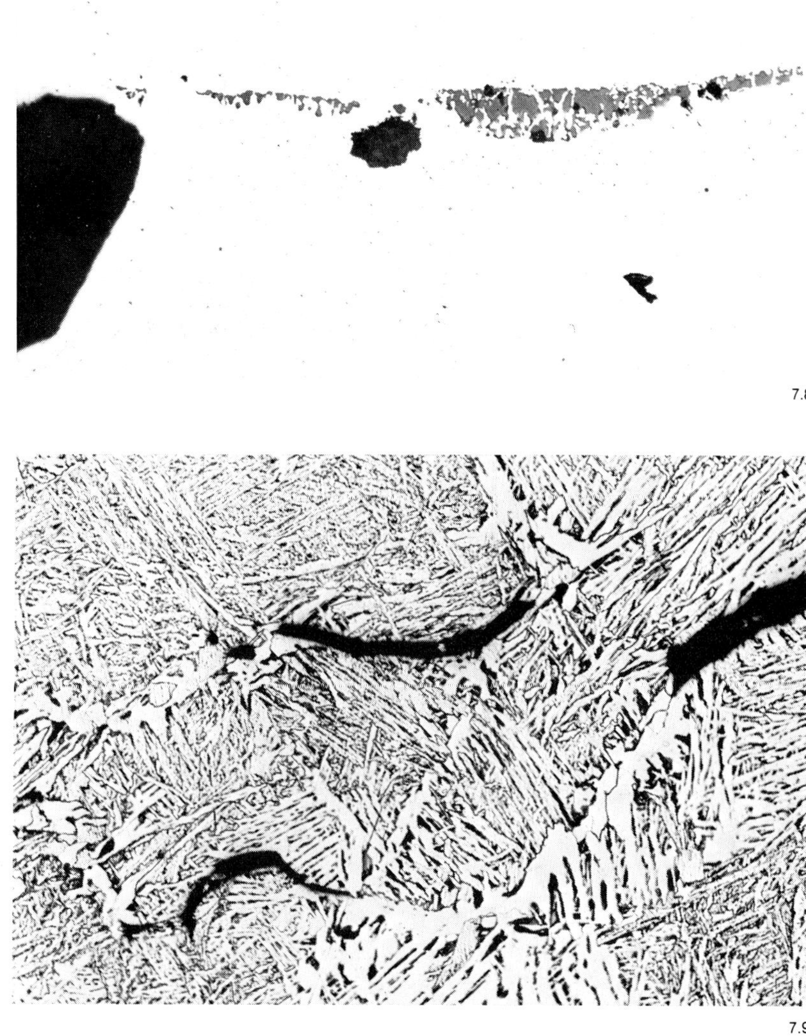

7.8

7.9

Fig. 7.8 and 7.9. Joint defect and hot cracks in weld seam of steel pipe

Fig. 7.8. Joint defect by oxidation at edge of a weld bead. Unetched section. 100 ×
Fig. 7.9. Hot cracking in root of weld seam. Etch: Nital. 100 ×

7.10

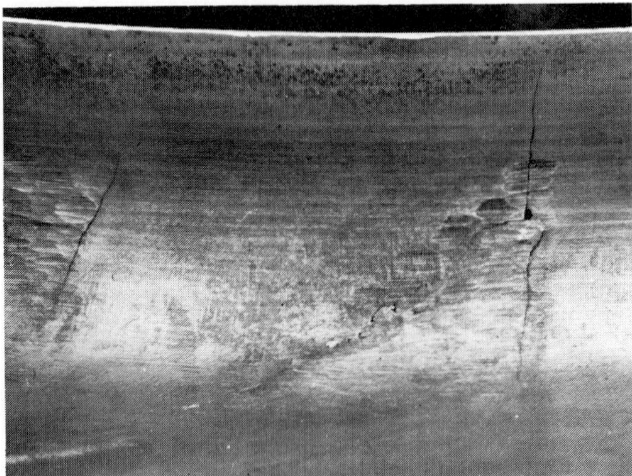

7.11

Fig. 7.10 to 7.15. Fan rotor of stainless steel with incipient cracks that originated at undercuts of weld seams

Fig. 7.10. Fan rotor. 0.167 ×
Fig. 7.11. Incipient cracks in flanged lip of upper face plate

Fig. 7.12. Cracks originating at weld seam at intake edge of blade foils (arrows). 1 ×
Fig. 7.13. Incipient crack formation at weld seam. 5 ×
Fig. 7.14. Opened crack with fatigue fracture. 5 ×

Fig. 7.15. Crack at weld undercut. Transverse section. Etch: V2A-etchant

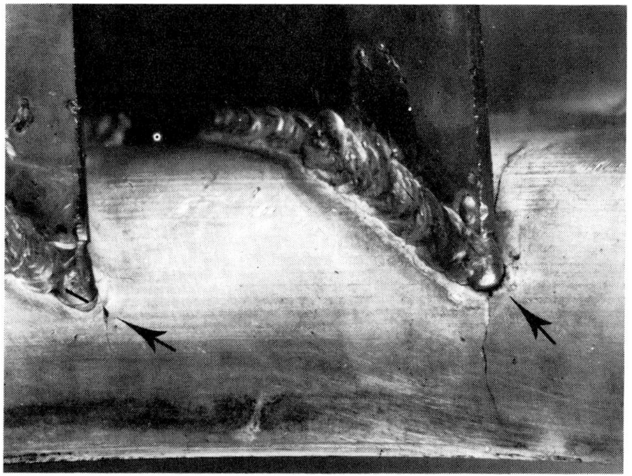

7.12

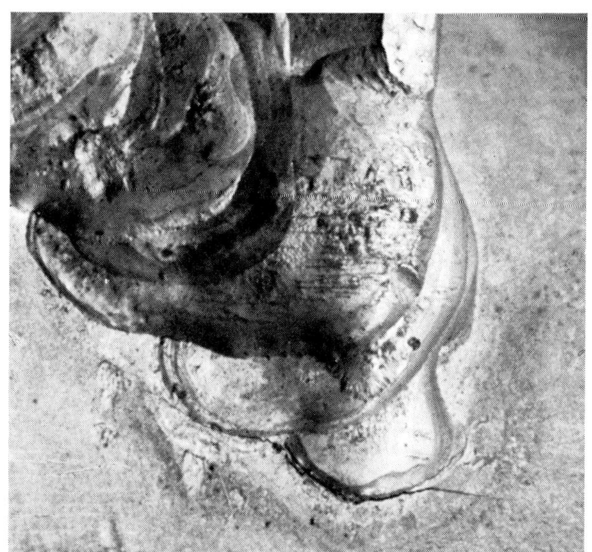

7.13

7.14

7.15 p 246

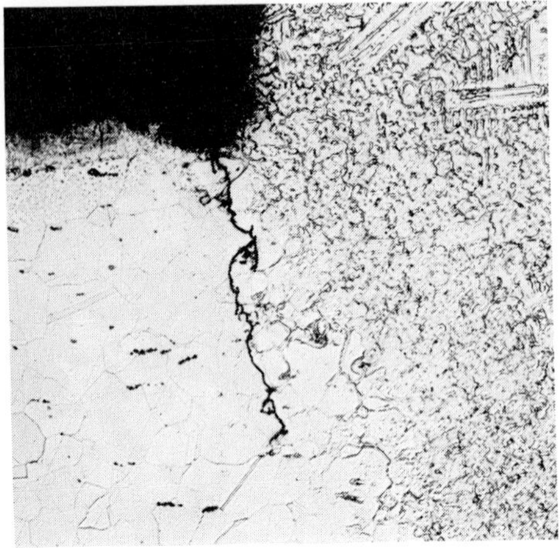

7.15

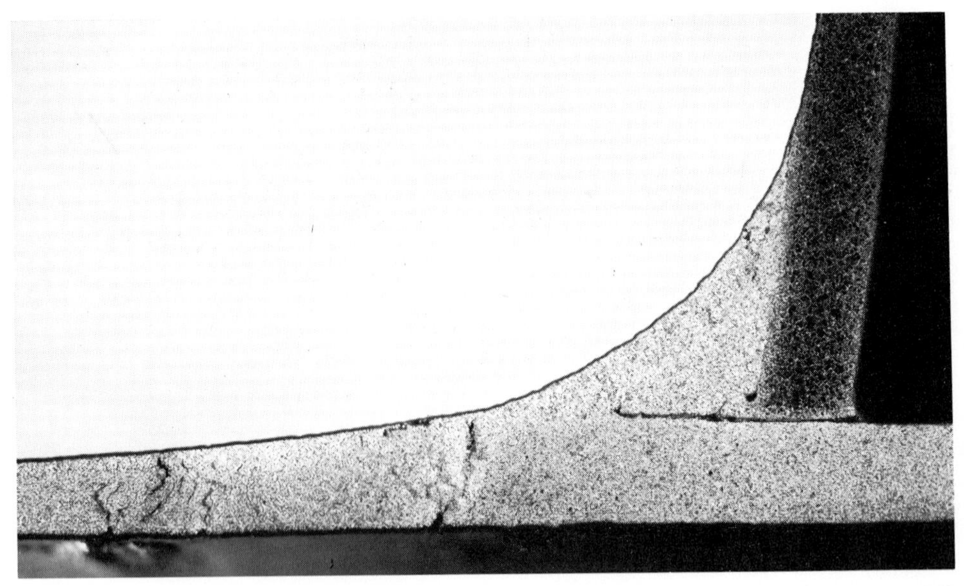

7.16

Fig. 7.16 to 7.18. Weld crack susceptibility in bicycle frame made from tubes of a steel having approx. 450 MPa (65 ksi) UTS

Fig. 7.16. Longitudinal section through weld at top of tube. Etch: Nital. 10 ×

Fig. 7.17. Hot cracking in tube adjacent to weld seam. Unetched. 200 ×

Fig. 7.18. Crack branches in austenitic grain boundary weakened by precipitates. Unetched. 200 ×

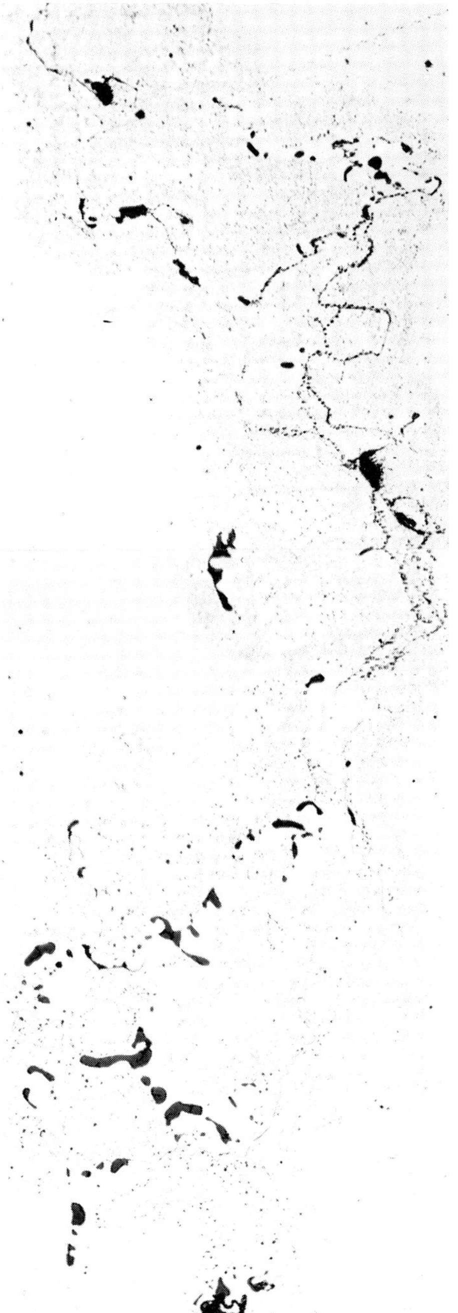

7.17

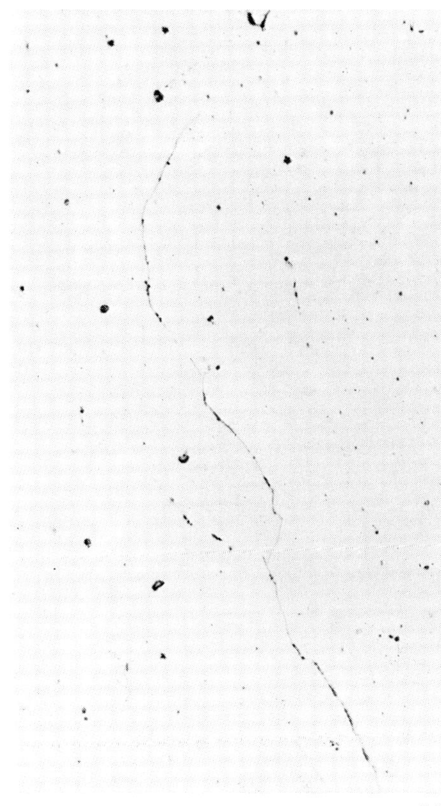

7.18

was the consequence of overheating during the welding. As could be seen from a macroetch according to Oberhoffer **(Fig. 7.20),** the raw material had a well defined banded structure that seemed to dissappear in the vicinity of the weld seam. The microstructure showed that the elongated sulfide inclusions were agglomerated in this zone, in part dissolved, and then precipitated onto the austenitic grain boundaries during cooling **(Fig. 7.21 and 7.22).** The relatively fine grain of this zone may be due to this redistribution of the non-metallic inclusions. The cracks followed the weakened grain boundaries in this process (Fig. 7.22). Their contours were oxidized and decarburized, and the decarburized zones were permeated with glassy precipitates of silicia-rich oxides (internal oxidation) **(Fig. 7.23 and 7.24).** Cracks occurred only in those places where the first deposited bead was remelted in the overlap zone.

The formation of the cracks may be visualized in this way: During deposition of the weld bead in the zones heated to just below the melting point next to the seam, mangnese sulfide has gone into solution and been precipitated during cooling at the austenitic grain boundaries. In this same region, there was a possibility of oxygen absorption. The grain boundaries weakened in this way tore open during welding of the opposite side under the effect of thermal stresses. The sulfur content of the steel for the nipple was comparatively high with 0.045 % to 0.050 % as determined by chemical analysis.

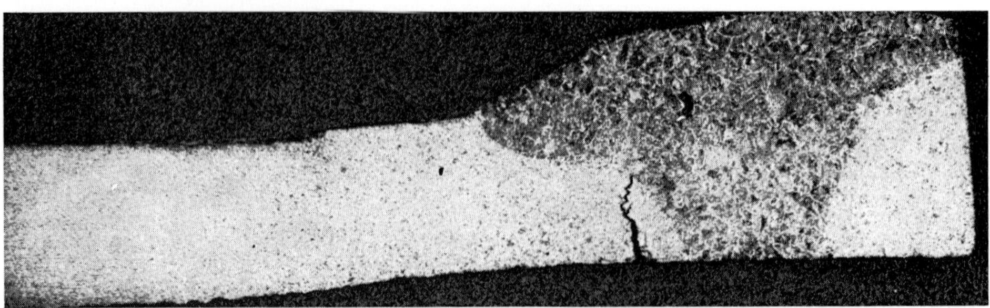

7.19

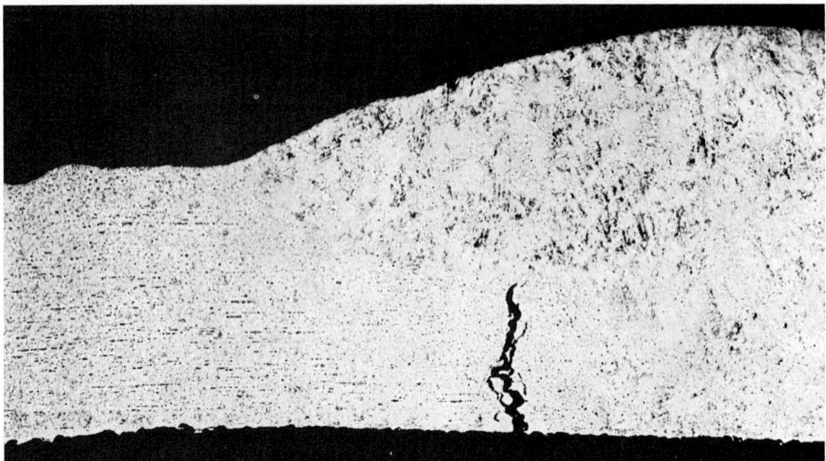

7.20

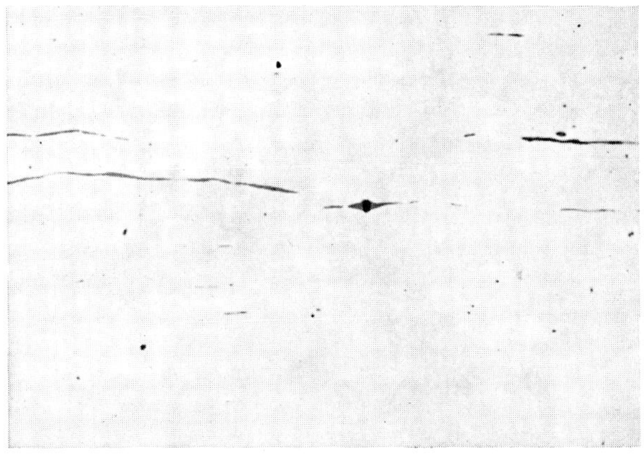

7.21

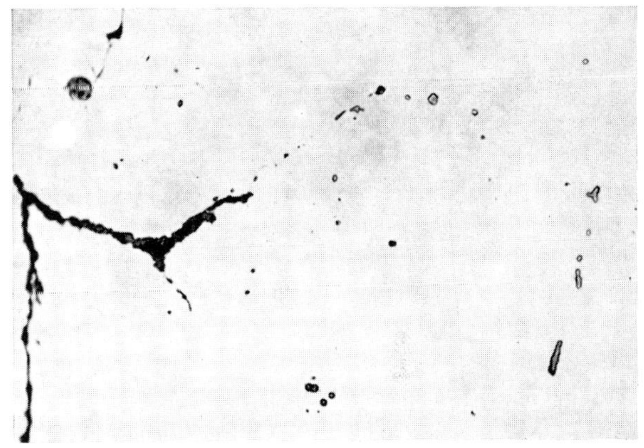

7.22

Fig. 7.19 to 7.24. Weld crack susceptibility in pipe nipples made of a steel with approx. 350 MPa (50 ksi) UTS

Fig. 7.19 and 7.20. Longitudinal section through cracks in circumferential seam of pipe nipple

Fig. 7.19. Etch: Nital. Approx. 6 ×
Fig. 7.20. Etch according to Oberhoffer. 10 ×

Fig. 7.21 and 7.22. Distribution of manganese sulfides in pipe nipples. Unetched longitudinal section. 500 ×

Fig. 7.21. In unaffected part
Fig. 7.22. In crack zone

Fig. 7.23 and 7.24. Incipient crack formation in pipe base metal adjacent to weld seam. Longitudinal section. Etch: Nital

Fig. 7.23. 50 ×
Fig. 7.24. 500 ×

p 250

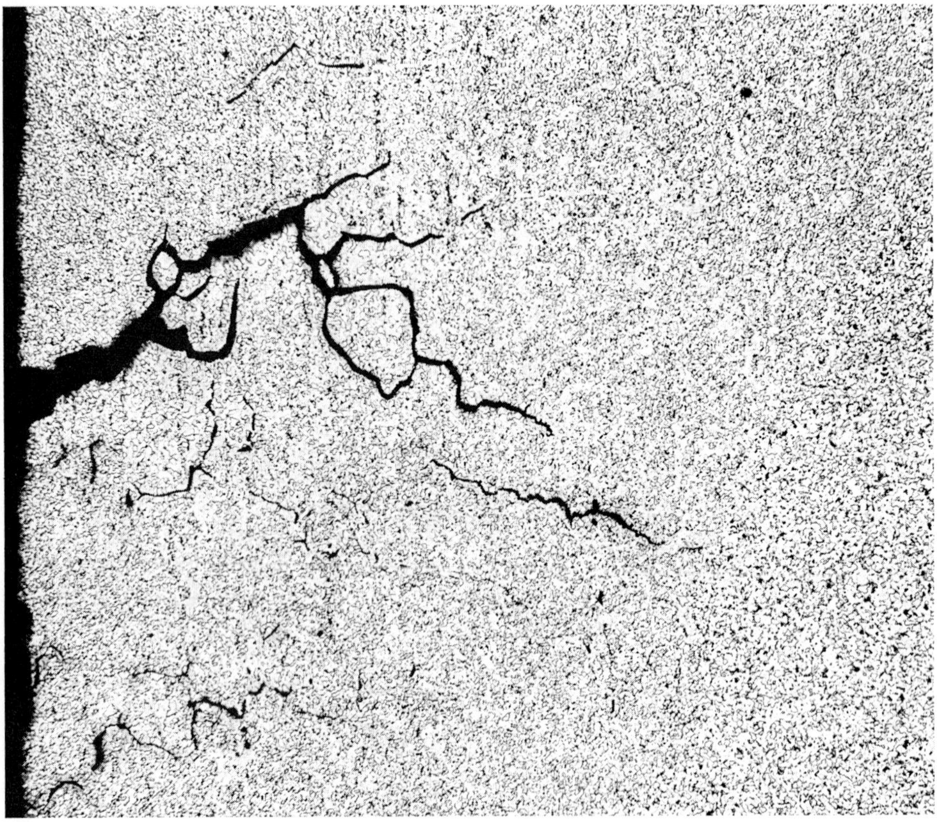

7.23

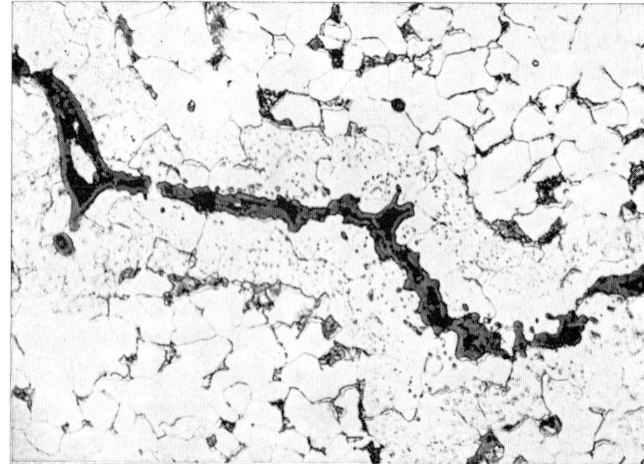

7.24

Accordingly, this investigation not only confirmed the contributory effect of sulfur to the crack formation during welding, but also permitted an insight into the mechanism of this cracking by examination of the microstructure. The weld crack suceptibility may be lowered[9]) by the addition of rare earth metals to the steel; this leads to the formation of high melting and harder-to-dissolve sulfides.

In this connection the hot crack sensitivity of niobium-stabilized austenitic steels with approximately 16 % Cr and 16 % Ni should be mentioned[10]). This cracking sensitivity is attributed to the incipient fusion of niobium-rich segregation zones at the primary grain boundaries.

7.3 **Hardening, Sensitivity to Welding Cracks**

Another kind of crack caused by welding[1]) should preferably be designated as "weld sensitivity"*; it occurs mainly in arc welding of thick sections. This failure is characterized by hardening of the weldment or the heat-affected zone next to the seam. The hardening may lead to the formation of fine transgranular cracks either during cooling from welding or later under service stresses. All strongly hardenable steels are weld sensitive in this sense. A manganese alloy steel with a yield point of about 355 MPa (52 ksi) was for a long time the highest strength grade steel licensed for use in welded construction. Even this steel caused major difficulties during welding of thicker cross sections[11]). Later, additions of aluminum with a concurrent increase of nitrogen contents produced **fine-grained structural steels** with higher yield points and a lower tendency toward brittle fracture[12]). An optimum with regard to yield point and brittle fracture susceptibility in conjunction with a moderate hardening tendency was finally achieved with further additions of a little nickel and vanadium[13]); this raised the yield point to at least 460–500 MPa (67–73 ksi). Even these steels should not be welded without precautionary meansures, especially where thick cross sections are involved; here the danger of hardening is particularly great due to a major heat loss caused by the large cold mass of the base metal and the fast cooling of the weld. Under all conditions the surface zones along the joint edges should be thoroughly preheated according to the wall thickness; temperatures of 100–300 °C should be used in order to lower the cooling rate.

In section 3.2 (Fig. 3.30 to 3.32) a failure was described that was due to tensile cracks in the weld seam of a pipe made of steel of high hardenability.

Another failure concerns a **pipe post** made of fine-grained structural steel with a 660 MPa (95 ksi) ultimate strength in which the edges were not preheated during or subsequent to welding into a ship's deck[14]). The post broke after a short time at a comparatively low load next to the weld seam. **Figure 7.25** shows the surface of the lower pipe section with the crack in the transition to the seam. A longitudinal section through the post is shown in **Fig. 7.26.** The failure originated in the heat-affected zone of the pipe metal, probably below the surface at the point where the crack opened the widest. From there it entered the weld seam and penetrated deeper into the pipe material. The base metal that elsewhere displayed a fine-grained structure of ferrite and pearlite **(Fig. 7.27)** and a hardness of 199 to 206 HV 5, had been transformed into martensite and bainite in the heat-affected zone **(Fig. 7.28),** and hardness here had increased to 376–401 HV 5. When such differences in structure occur in a narrow space, the formation of high stresses must be assumed that may lead to crack formation either during transformation or later during additional stresses in service.

---

* see Appendix I

The effect of preheating can be seen from the following experiment conducted jointly by the plant and steel producer[13]). A length of a cold formed **container** made from a 28 mm thick normalized steel plate, ultimate strength of approx. 700 MPa (100 ksi), exhibited fine transverse cracks in the longitudinal weld seam during qualification testing. The weld was originally made under powder. This welding must not have been conducted according to specifications, and in particular, the joint edges had either not been preheated at all, or insufficiently so. In order to prove that cracks of the type described occur when preheating is omitted, while they can be avoided with proper preheating, a 2500 mm long container section of the same material and the same dimensions as the one rejected was welded under the same conditions in five internal and two external passes under powder, i.e. once without preheating, and once with preheating of the edge zones to 200 °C.

Fine transverse cracks could be observed in the weld seam of the nonpreheated container during ultrasonic and radiographic testing. Two 250 mm lengths of the seam, that were found to have some cracks, were examined metallographically. Transverse sections were cut open through single cracks, while longitudinal sections were cut through many short cracks **(Fig. 7.29 and 7.30).** They were located preferentially in the fourth inner pass that was made after a two-hour pause so that the previous beads and plate edges could cool. The cracks were predominantly transgranular **(Fig. 7.31).** On the fracture they resembled flake cracks **(Fig. 7.32).** This indicates the possibility that during formation of this type of welding cracks, absorption of hydrogen from the welding powder may have played a contributory role.

Repetition of the welding on the same container length under controlled preheating of the edges to 200 °C, assured by accurate temperature measurement, resulted in a completely defect-free weld.

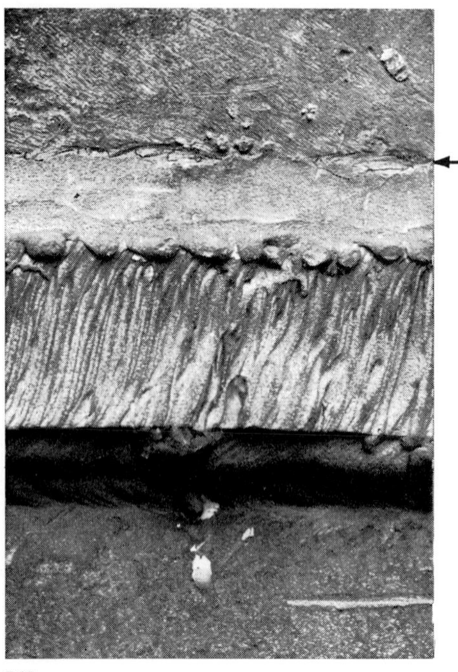

7.25

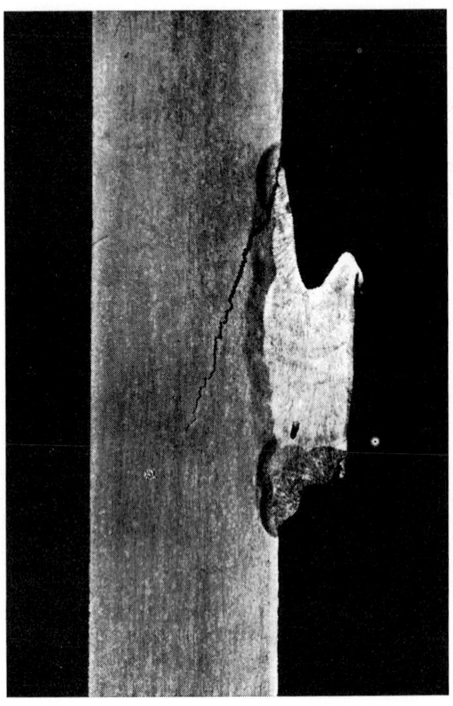

7.26

Without taking precautionary measures, no permanent solutions can be obtained by **weld build-up** of damaged or unusually worn parts that have experienced cyclic loading conditions[15]). Failures nevertheless still occur that are caused by this technique. **Figure 7.33** shows a far advanced bend fatigue fracture of a 77 mm diameter **driveshaft.** As could be noticed externally, it was located in a weldment zone that stretched along the entire circumference of the shaft. This fact can be seen better in **Fig. 7.34** after etching of the surface with nital. Metallographic examination showed a built-up weld structure of 3 to 4 mm depth **(Fig. 7.35).** The seam as well as the adjacent zone of the base metal show hardened structures with stress cracks **(Fig. 7.36).** The fatigue fracture probably propagated from one of these cracks. According to the microstructure, the shaft apparently consisted of an unalloyed steel with 0.3 to 0.4 % C. Special precautions should have been observed with such a high hardenability steel.

Further failures with respect to repair and build-up welding will be cited in section 11.

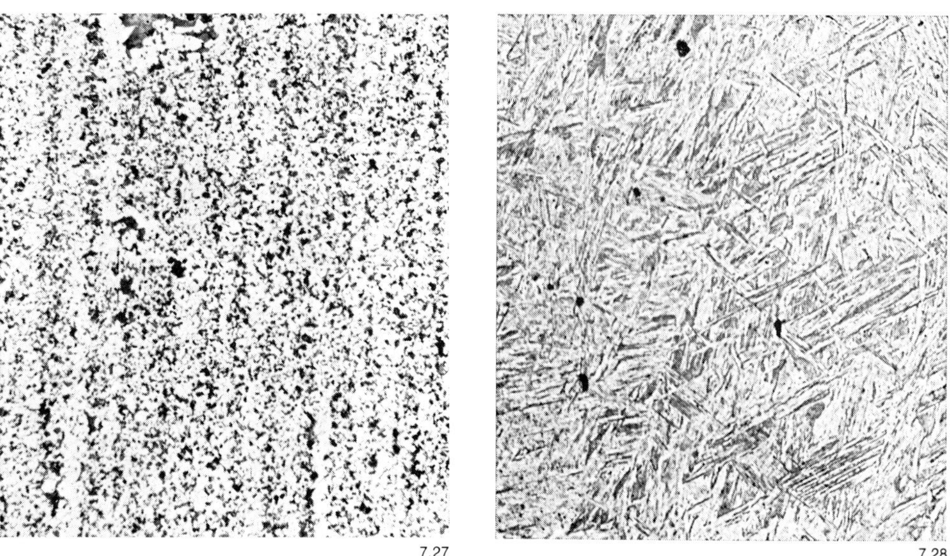

7.27

7.28

Fig. 7.25 to 7.28. Loading gear pipe post cracked next to weld seam due to hardening

Fig. 7.25. Surface of cracked pipe length (arrow) flame-cut from ship's deck. 1 × ← Crack

Fig. 7.26. Longitudinal section across weld seam. Etch according to Heyn. 1 ×

Fig. 7.27 and 7.28. Microstructure of pipe base metal. Longitudinal section. Etch: Picral

Fig. 7.27. Away from weldment. 100 ×

Fig. 7.28. In transition to weld seam. 500 ×

7.30

7.29

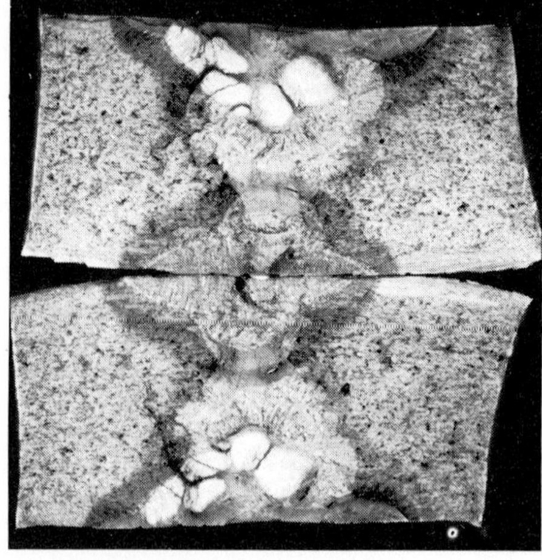

7.32

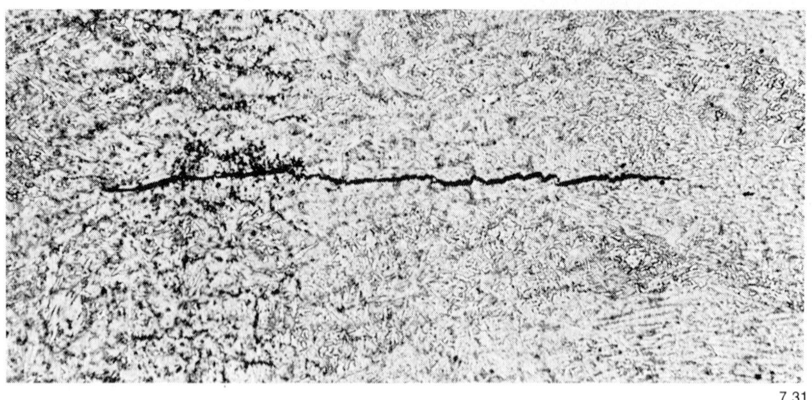

7.31

Fig. 7.29 to 7.32. Seam of container length of steel having approx. 700 MPa (100 ksi) UTS that was welded without preheating

Fig. 7.30. Longitudinal section through center of seam. Heyn etch. 1 ×

Fig. 7.29. Transverse section through center of seam. Heyn etch. 1 ×

Fig. 7.31. As Fig. 7.30. Etch: Picral. 100 ×

Fig. 7.32. Fracture across weld seam. Etch: Nital. 1 ×

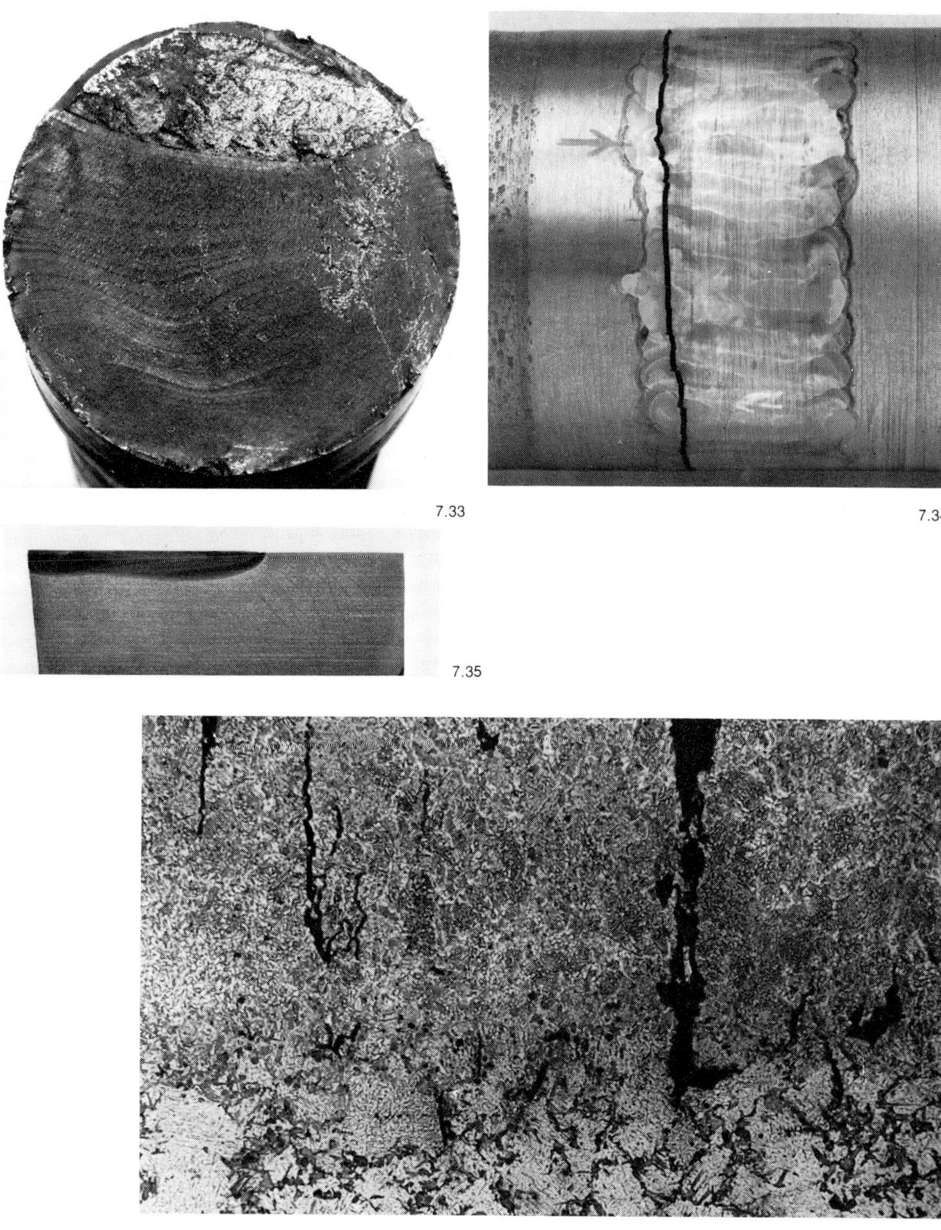

7.33

7.34

7.35

7.36

Fig. 7.33 to 7.36. Fracture of driveshaft propagating from built-up weld

Fig. 7.33. Fracture of shaft. 0.8 ×

Fig. 7.34. Surface at fracture. Etch: Nital. 0.8 ×

Fig. 7.35. Longitudinal section. Adler etch. 1 ×. Fracture surface left

Fig. 7.36. Microstructure at transition from weld (top) to base metal (bottom). Longitudinal section. Etch: Nital. 100 ×

7.37

7.38

7.39

7.4 **Errors During Flame Cutting**

What has been said about welding holds equally true for flame cutting. An example for a failure of this type was cited already in section 5.1 (Fig. 5.37 to 5.40). In another case a crack occurred during flame cutting of a **disk** out of 70 mm thick plate made of structural steel having a tensile strength of at least 700 MPa (100 ksi). The crack was located where the cuts met each other in the center **(Fig. 7.37 and 7.38).** Examination showed the cut edge was carburized by wrong torch manipulation and had hardened due to fast cooling **(Fig. 7.39),** just as in the case mentioned previously.

7.5 **Heat Treating Errors**

Subsequent tempering or soft annealing can generally relieve or reduce welding stresses. Vanadium or niobium alloy steels, such as are used for the construction of thick-walled welded pressure vessels, are subject to stiffening and embrittlement in the temperature range of 500 to 700 °C through the precipitation of finely distributed carbides or carbonitrides[17]). If high weld stresses are present, intergranular cracks may form in the heat-affected zones of the sheet material next to the weld. The danger of crack formation is said to be highest around 640 °C[16]). Welded pressure vessels of these steels therefore are stress relieved at temperatures below 600 °C. But even then crack formation apparently cannot be completely avoided. In this case fracture mechanics may give indications whether a crack of a given length will be stable or lead to the fracture of the part[18])[19]).

During the welding of alloy steels, the special carbides corresponding to equilibrium conditions are not always formed due to rapid cooling. Instead, transition phases of alloyed lower stability iron carbides are formed. In this case subsequent annealing may be absolutely essential for the restoration of equilibrium. For instance, the following failure is due to the fact that this precaution was not observed. It concerns a **heat exchanger** of hydrogen resistant steel of the type AISI 501 containing approx. 5 % Cr, 0.5 % Mo and less than 0.15 % C, that began to leak after five years' service in an ammonia synthesis plant[20]). Radiography showed a fine crack in the circumferential weld seam. **Figure 7.40** shows an etched section through this seam. The crack propagated from the outside, first penetrating the transition zone between the austenitic weld metal and the container wall, and then, following the highest stress, was deflected into the latter. The container alloy had been attacked by hydrogen in the transition zone transformed by the welding heat, even though the steel was supposed to be completely stable under the prevailing conditions, 400 °C and hydrogen partial pressure of 600 atm (see also section 15.2). The attack was evidenced by decarburization and intergranular crack formation **(Fig. 7.41).** The microstructure of this zone was purely martensitic **(Fig. 7.42).** The weld area therefore had not been post-annealed, and consequently the hydrogen-resistant chromium carbide could not form.

Fig. 7.37 to 7.39. Stress crack in carburized and fast cooled disk that was flame-cut out of a 70 mm structural steel plate

Fig. 7.37. Flame-cut surface. 0.9 ×

Fig. 7.38. Opened crack. 0.9 ×

Fig. 7.39. Surface structure at cut. Etch: Picral. 100 ×

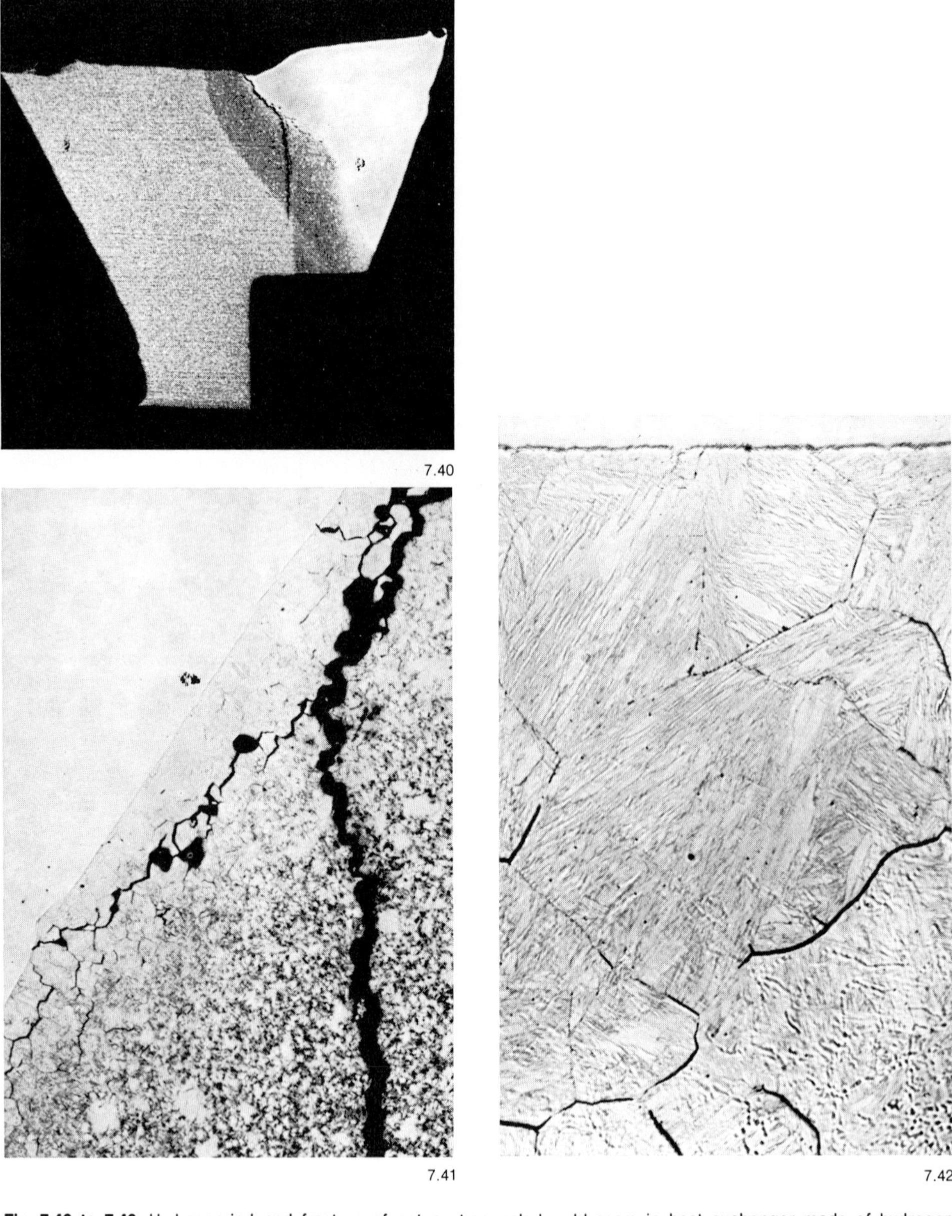

7.40

7.41

7.42

Fig. 7.40 to 7.42. Hydrogen-induced fracture of not post-annealed weld seam in heat exchanger made of hydrogen resistant steel

Fig. 7.40. Longitudinal section through seam. Etch: Picral. 4 ×

Fig. 7.41 and 7.42. Martensitic zone adjacent to austenitic weld seam. Etch: Picral

Fig. 7.41. 100 ×

Fig. 7.42. 500 ×

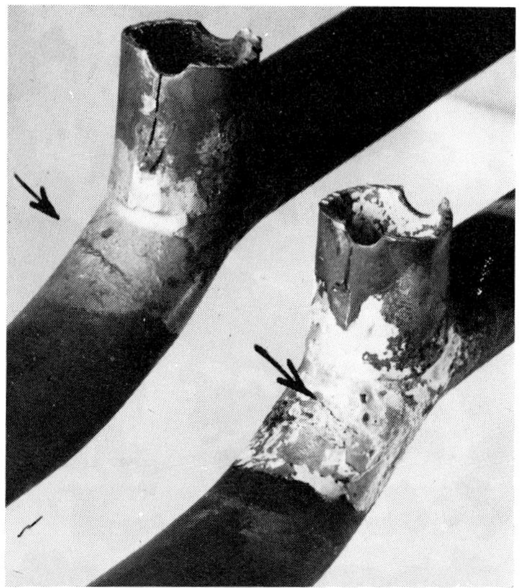

7.43

7.44

Fig. 7.43 and 7.44. Cold profiled bicycle tubes bent at right angles that fractured during brazing

Fig. 7.43. View. 1 ×
Fig. 7.44. Surface structure. Etch: Nital. 500 ×

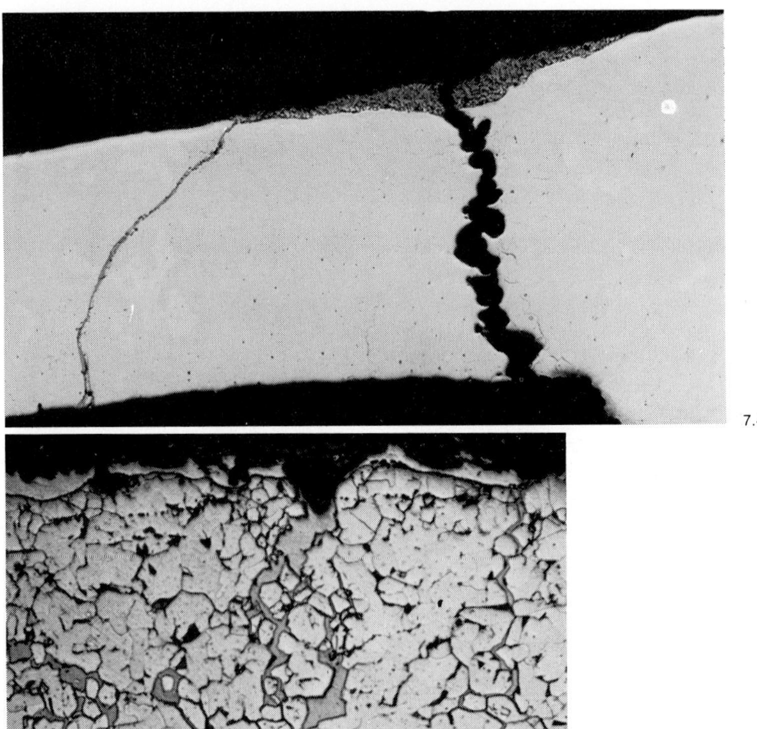

7.45

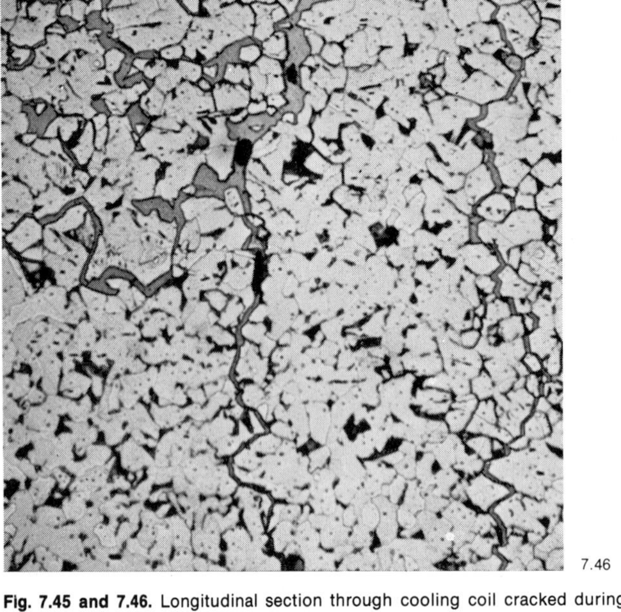

7.46

Fig. 7.45 and 7.46. Longitudinal section through cooling coil cracked during repair of weld crack by brazing

Fig. 7.45. Etch: Basic ammonium copper chloride solution. 25 ×

Fig. 7.46. Microstructure. Etch: Nital. 100 ×

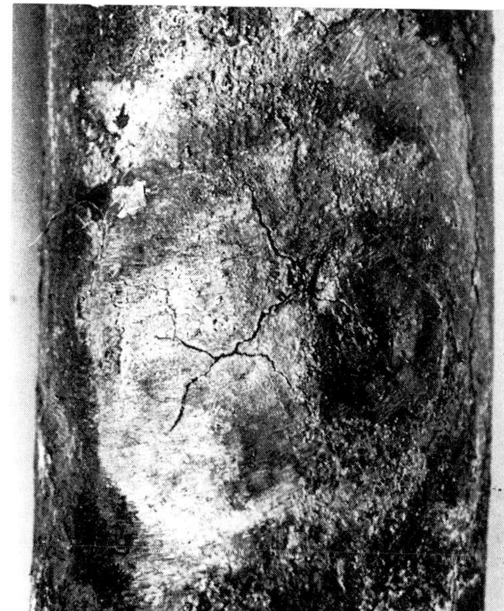

7.47

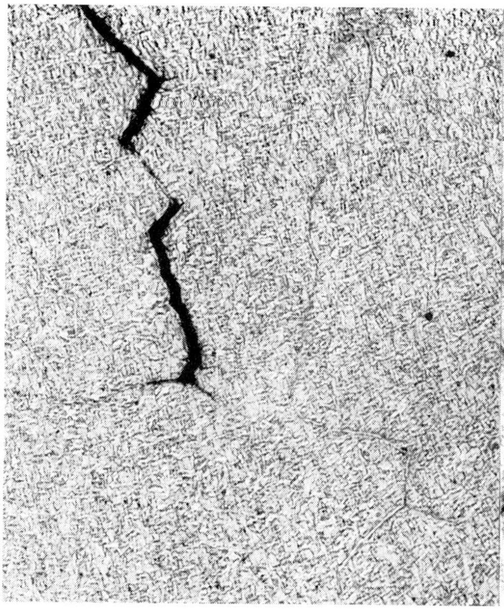

7.48

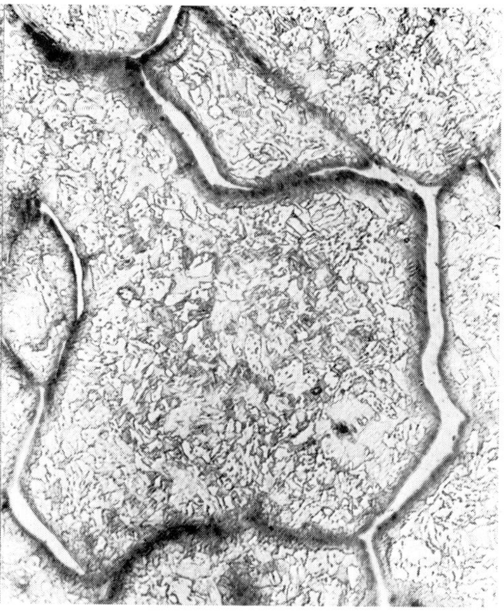

7.49

Fig. 7.47 to 7.49. Superheater pipe cracked by welding a thermocouple onto it

Fig. 7.47. External surface. 2 ×

Fig. 7.48 and 7.49. Microstructure of pipe at weld spot with intergranular cracks and precipitates of a yellow metal. Transverse section. Etch: Nital

Fig. 7.48. 100 ×

Fig. 7.49. 500 ×

7.6 **Brazing Errors**

If steel that is under tensile stress – external or residual – comes in contact with a liquid metal that is soluble in iron, such as copper, tin, or zinc, this metal may penetrate the grain boundaries of the steel and lead to crack formation[21]).

Figures 7.43 and 7.44 show examples of "**braze cracking**"* or "**braze fracturing**" in **bicycle tubes.** These had been cold formed and bent at right angles prior to brazing and therefore contained residual stresses. The tubes should have been annealed prior to brazing to relieve the stresses.

In the **spiral cooling coil** shown in **Fig. 7.45,** an attempt was made to repair a hot crack, produced during welding and characterized by oxide inclusions, by brazing with brass braze. The result was that the crack was glued together only superficially; adjacent to it some fine intergranular brazing cracks filled with brass appeared **(Fig. 7.46).**

A **superheater pipe** with 32 mm O.D. made from a heat resistant chromium-molybdenum steel became porous after 3500 hours of service through crack formation at the spot where a nickel-nickel chromium thermocouple was supposedly gas welded onto it. **Figure 7.47** shows the failure. Further small incipient cracks could be seen at the outer surface in transverse section through the cracked spot. They were intergranular. A metal of yellow color adhered to their surfaces. Metal of the same type also had been precipiated onto the extension of the cracks as well as onto the austenitic grain boundaries at crack-free places **(Fig. 7.48 and 7.49).** Accordingly, this was a typical case of liquid metal embrittlement. Since the penetrated metal had the color of copper or a copper alloy, the thermocouple could no have been a nickel-nickel chromium element, but rather an iron- or copper-constantan couple. In order to avoid stresses, it would have been better to fasten the thermocouple to a ground spot of the pipe surface either by tacking or through clamping.

7.7 **Riveting Errors**

Failures of **riveting compositions** have been described already in section 4 under the heading Aging. A more detailed description will be given in section 15.3.4 in conjunction with stress corrosion. But a special case may be mentioned here. In the highly stressed **face plate of a fan rotor** that served to convey coking plant gas, riveting hole cracks were formed, as was shown by magnetic particle inspection **(Fig. 7.50).** They propagated radially, i.e. perpendicular to the principal stress. As could be seen from etched flat sections, these cracks were located in zones with altered microstructure; they had formed around the riveting holes by heating and terminated precisely where the zones ended **(Fig. 7.51 and 7.52).** The zones had a hardened structure of martensite **(Fig. 7.53)** in contrast to the basic ferritic-pearlitic microstructure of the plate. Crack formation accordingly was caused or facilitated by transformation stresses with a possible contributory effect of corrosion.

* see Appendix I

Fig. 7.50 to 7.53. Cover plate of gas suction fan with rivet hole cracks

Fig. 7.50. Surface after magnetic particle inspection. 1 ×

Fig. 7.51. Flat section. Etch: Nital. 1.6 ×

Fig. 7.52. Crack branches. Flat section. Etch: Nital. 100 ×

Fig. 7.53. Microstructure at edge of rivet hole. Etch: Nital. 500 ×

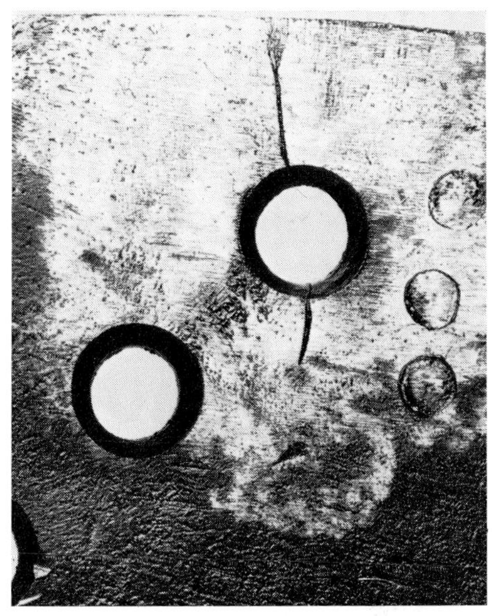

7.50

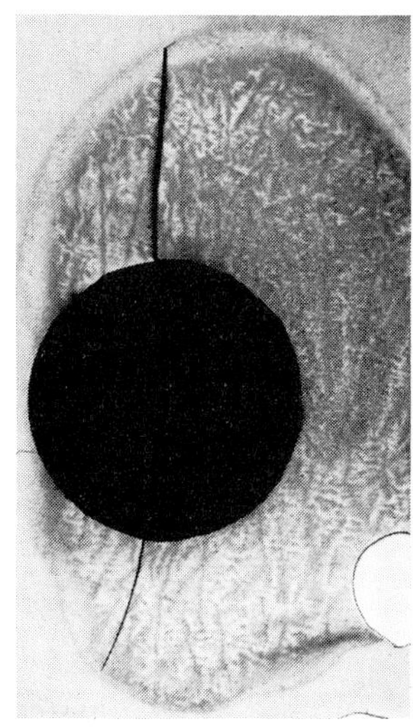

7.51

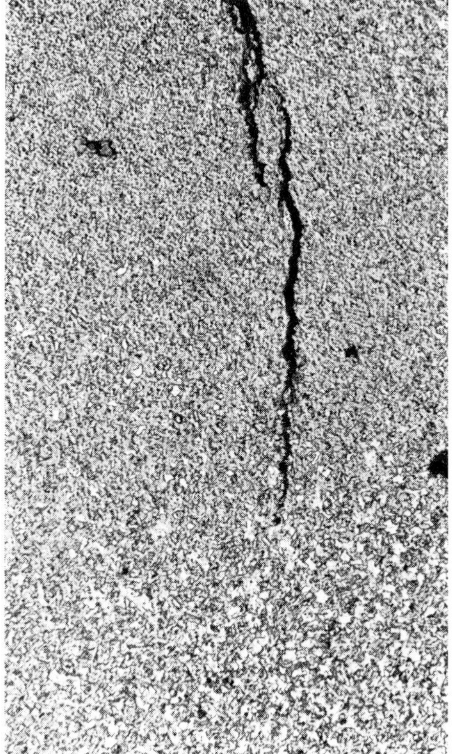

7.52

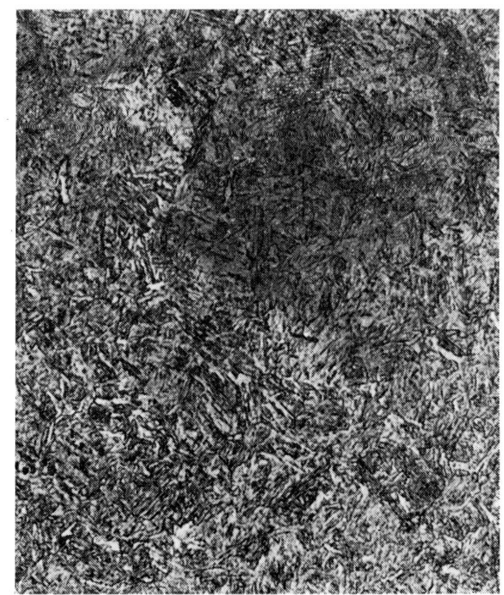

7.53

Literature Section 7

1) K. L. Zeyen u. W. Lohmann: Schweißen der Eisenwerkstoffe. Stahleisen-Bücher Bd. 6, 2. Aufl. 1948. Verlag Stahleisen mbH., Düsseldorf

2) E. Kauczor: Bindefehler in Schweißnähten an Stahlrohren durch nichtentfernten Schutzanstrich. Prakt. Metallographie 6 (1969) S. 69/70

3) F. K. Naumann u. F. Spies: Geschweißte Rohre mit harten Stellen. Prakt. Metallographie 10 (1973) S. 46/49

4) Dieselben: Gebrochene Spannbolzen. Prakt. Metallographie 11 (1974) S. 616/20

5) Dieselben: Untersuchung von Heißdampfschieberspindeln. Prakt. Metallographie 11 (1974) S. 621/27

6) Dieselben: Dauerbruch und Schweißung. Prakt. Metallographie 8 (1971) S. 551/59

7) A. Antonioli: Die Ursachen der Schweißrissigkeit von Chrom-Molybdän-Baustählen. Stahl u. Eisen 62 (1942) S. 540/45

8) F. K. Naumann u. F. Spies: Beim Schweißen gerissene Rohrnippel. Prakt. Metallographie 8 (1971) S. 667/74

9) W. G. Wilson: Weld. Res. 1971, S. 42/46

10) W. Dahl, C. Düren u. H. Müsch: Ursache der Heißrißbildung in Schweißverbindungen eines niobstabilisierten Stahls mit 16 % Cr und 16 % Ni. Mitt. a. d. Mannesmann-Forsch.-Inst. Stahl u. Eisen 93 (1973) S. 805/12

11) W. Grosse: Untersuchungen über Bruchbildungen an geschweißten Bauwerken aus Stahl St 52. Stahl u. Eisen 60 (1940) S. 441/52

12) H.-J. Wiester, W. Bading, H. Riedel u. W. Scholz: Einfluß der Stickstoffabbindung durch Aluminium auf die Eigenschaften von Baustählen. Stahl u. Eisen 77 (1957) S. 773/84

13) C. Düren, H. Müsch u. H. Adrian: Schweißen von höherfesten Feinkornbaustählen unter besonderer Berücksichtigung der Behälterfertigung. DVS – Ber. 22 (1971) S. 73/81

14) F. K. Naumann u. F. Spies: Gerissener Pfosten von einem Ladegeschirr. Prakt. Metallographie 11 (1974) S. 421/23

15) H. Schottky: Schmelzschweißung und Dauerbruch. Kruppsche Monatsh. 7 (1926) S. 213/16

16) K.-H. Piel: Verhalten der Druckbehälterstähle beim Schweißen. Stahl u. Eisen 93 (1973) S. 568/77

17) K. Forch, U. Forch u. K.-H. Piehl: Die metallkundliche Deutung der Kerbschlagzähigkeitsänderung in der Wärmeeinflußzone schweißbarer Baustähle. Stahl u. Eisen 98 (1978) S. 641/51

18) G. Schöne u. W. Fischer: Das Betreiben von Bauteilen mit Rissen. Stahl u. Eisen 98 (1978) S. 97/101

19) R. W. Nichols: Einige technische Anwendungen der Bruchmechanik. Stahl u. Eisen 93 (1973) S. 263/70

20) F. K. Naumann u. F. Spies: Bruch einer Schweißnaht von einem Wärmeaustauscher einer Ammoniak-Synthese-Anlage. Prakt. Metallographie 8 (1971) S. 503/08

21) P. Schlafmeister u. H. Schottky: Das Löten legierter Stähle. Metallwirtschaft 18 (1939) S. 43/47

8. Failures Caused by Machining Errors

Here a distinction should be made between groove formation and friction heat developed during grinding. Grooves and scratches are mechanical surface defects that act as notches. During service, stress concentrations are formed at such notches that may lead to premature fractures, especially during impact and cyclic loading. Surface defects caused by machining with dull tools lead to similar consequences. Heating of the zone near the surface may occur through rough grinding under too much pressure. This heating in turn may cause tempering, transformation, or even incipient melting of the material. Hardened parts are especially sensitive to such stresses. The stresses may lead to the formation of cracks that should be considered **grinding checks** based upon their origin[1]). These cracks are preferentially oriented transversely to the direction of grinding or they are arranged in a network-like formation; they are but a few millimeters deep. Steels with high carbon content or overcarburized case layers with a cementite network have a particularly strong tendency to grinding check formation. If the temperature reaches the austenitic region, hardening may occur, especially in thick parts with a large cold mass; if they have been hardened previously, grain size change and renewed hardening may occur. The transition from the hardened surface layer containing finely acicular martensite to the unchanged core structure occurs through an intermediate zone having a tempered or annealed microstructure. In this situation, the danger of cracking is particularly great because in addition to the purely thermal stresses, high transformation stresses occur with the martensite formation.

8.1

8.2

Fig. 8.1 and 8.2 Roller bearing housing that was machined with a dull lathe tool

Fig. 8.1 External surface with rough spot (arrow). 1 ×
Fig. 8.2 Transverse section through rough spot. Etch: Nital. 200 ×

8.1 **Machining Marks and Grooves**

Figure 8.1 shows a surface defect on the turned exterior face of a **roller bearing housing** that was thought to be a material defect by the machinist. Metallographic examination, however, did not confirm any material defect at the rough spot, but only deformation and tensile tears in the surface region. These may have occurred during machining with a dull lathe tool **(Fig. 8.2).**

A **Rush pin,** with which the shinbone of a patient was nailed, broke almost without deformation after several months with a renewed fracture of the leg. The pin consisted of a cold drawn rod of austenitic chromium-nickel-molybdenum steel and had an ultimate strength of 1200 MPa (175 ksi) at 17 % elongation and 51 % reduction in area. The fracture was a torsion fatigue type **(Fig. 8.3).** Its formation was facilitated by short circumferential grinding grooves **(Fig. 8.4).**

8.3

8.4

Fig. 8.3 and 8.4 Broken bone pin of austenitic chromium-nickel-molybdenum steel
Fig. 8.3 Fracture. 5 ×
Fig. 8.4 Surface with circumferential grinding grooves. 3 ×

8.5

8.6

Fig. 8.5 and 8.6 Fracture of piston pin from grooves in bore. Fatigue fracture origins in different planes
Fig. 8.5 Fracture. 1 ×
Fig. 8.6 Internal view of bore. 1 ×

A 140 horsepower engine failed because its piston hit against the cylinder head. During disassembly it was found that the **piston pin** had broken. A subsequent analysis was to determine whether the piston pin fracture was the cause or consequence of the failure. The pin showed three fatigue fractures that had propagated in different transverse planes from the longitudinal bore **(Fig. 8.5).** Breaking of the pin was facilitated by rough turning marks in the bore **(Fig. 8.6).** One of the fatigue fractures was contained entirely in one of the grooves. The failure of the piston therefore could not have been caused by a violent impact, but only by cyclic stresses.

A heat treated **crankshaft** of steel type AISI 5132 showed a crack in the bearing surface. After opening the crack, the failure was found to be a fatigue fracture originating in an oil bore **(Fig. 8.7).** The bore had a deformed surface full of machining marks in contrast to the flawless polished fillet. This was noticeable also in a section through the fracture origin **(Fig. 8.8).** These bores should be either cut with a sharp drill or subsequently reamed.

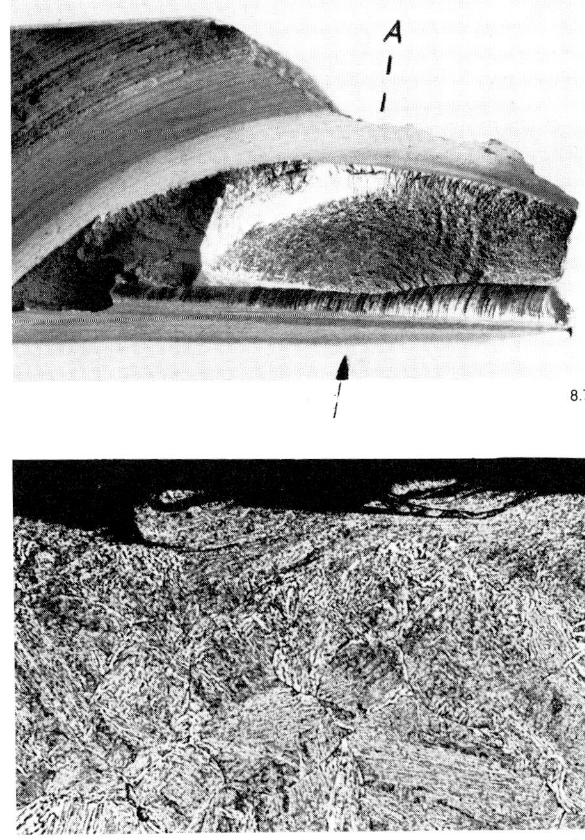

Fig. 8.7 and 8.8 Cracked crankshaft

Fig. 8.7 Fatigue fracture originating in oil bore. 2 ×
Fig. 8.8 Microstructure at edge of bore. Section A--A through fracture origin. Etch: Picral. 200 ×

8.2 **Heat Caused by Grinding**

During the flash grinding of a drop forged **connecting rod** of grade 1035 steel, the edges of the I-profile were inadvertently ground as well **(Fig. 8.9).** A fatique fracture developed from the grinding grooves that ran transversely to the principal direction of stress. This lead to the breaking of the rod after approximately 1000 hours of service **(Fig. 8.10).** A thin martensitic layer had formed in part under the ground areas **(Fig. 8.11)** which indicated excessive heating caused by rough grinding.

The torsion fatique fracture shown in Fig. 2.4 and 2.5 of the **helical spring** was also caused by a grinding spot at the fracture origin (arrow). Without this stress raiser the spring would have broken at the highest stressed inner fiber during overloading. But in this case the fracture origin was offset against the inner fiber by about 90° in the cross section. The heat treatable microstructure had been transformed under the surface into martensite to a depth of 30 μm maximum through heat caused by the grinding (Fig. 2.6).

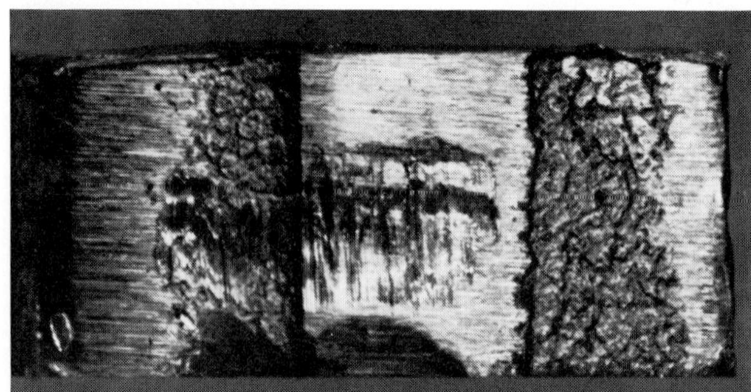

8.9

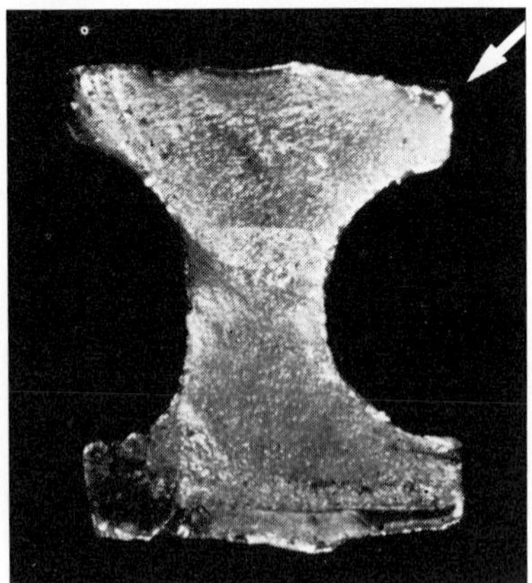

8.10

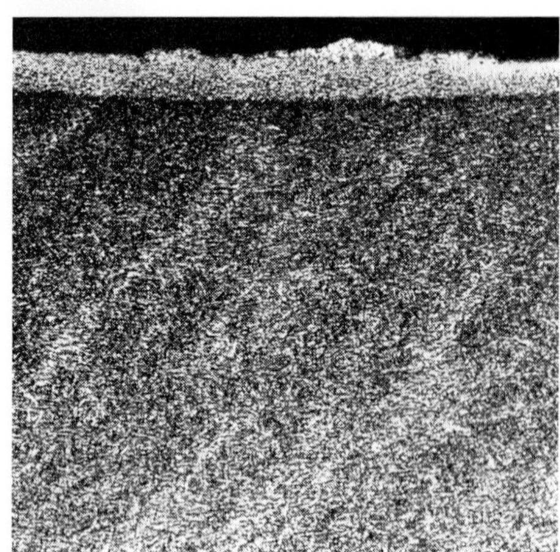

8.11

Fig. 8.9 to 8.11. Broken connecting rod of tractor engine

Fig. 8.9 Flash and grinding traces at edge (right) at which fatigue fracture originated. 5 ×
Fig. 8.10 Fracture plane. 2 ×. Fracture origin designated by arrow

Fig. 8.11 Microstructure of surface under ground area. Transverse section. Etch: Picral. 200 ×

Grinding checks are often mistaken for quench cracks. But as a rule they are easily distinguished. Cracks in the hardened **pressure plate** face planes (225 mm diameter and 30 mm thickness, **Fig. 8.12)** of a chromium-tungsten tool steel with 1 % C, were thought by the user to be quench cracks. Several factors contradicted this opinion: 1) a number of fine, tightly closed cracks appeared in addition to some penetrating cracks during magnetic particle inspection; 2) these fine cracks did not originate at the bore holes of the plates as had to be expected in the case of quench cracks, but from random surface spots, not particularly subject to quench stresses; also 3) the cracks had a shallow depth as could be seen in the fracture **(Fig. 8.13);** and 4) the fracture grain was velvety fine, indicating that hardening temperature had not been too high. These indications justified the conclusions that the fine cracks were grinding checks, and that the penetrating cracks could also have been caused by grinding stresses.

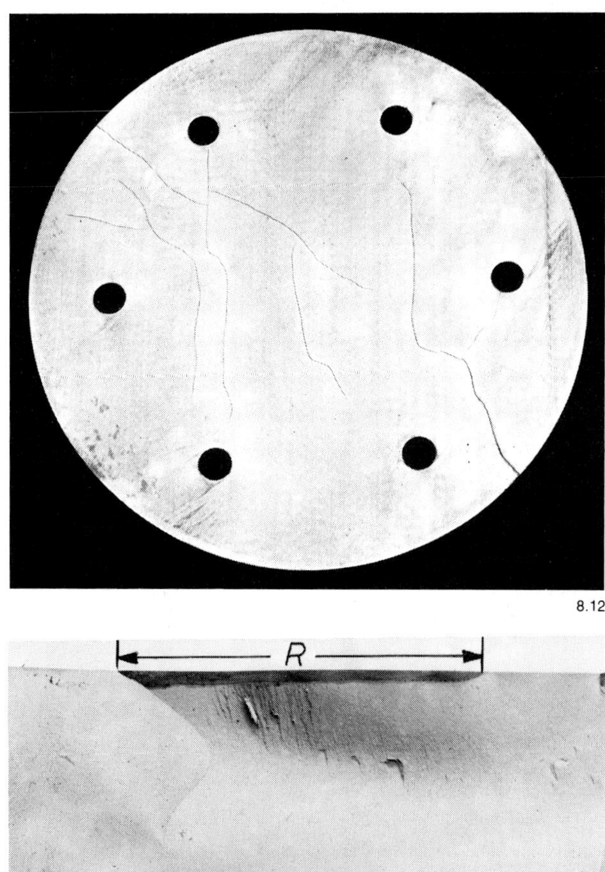

8.12

8.13

Fig. 8.12 and 8.13. Grinding checks in hardened pressure plate of chromium-tungsten tool steel

Fig. 8.12 Surface after magnetic particle inspection. 0.33 ×
Fig. 8.13 Fracture. 1 ×. R = grinding check

Figure 8.14 shows the fracture of a cylindrical **milling cutter** made of high speed steel that cracked in service. The fracture propagated from various spots on the teeth. It was velvety fine, which indicated that quenching had taken place from the correct temperature. The teeth were ground roughly and also had become hot as could be seen from the temper colors **(Fig. 8.15).** In spots some cracks could be found that ran perpendicular to the direction of grinding. These were typical grinding checks. They had caused the cracking of the milling cutter during service. Carbide distribution and hardened microstructure were in order.

8.14

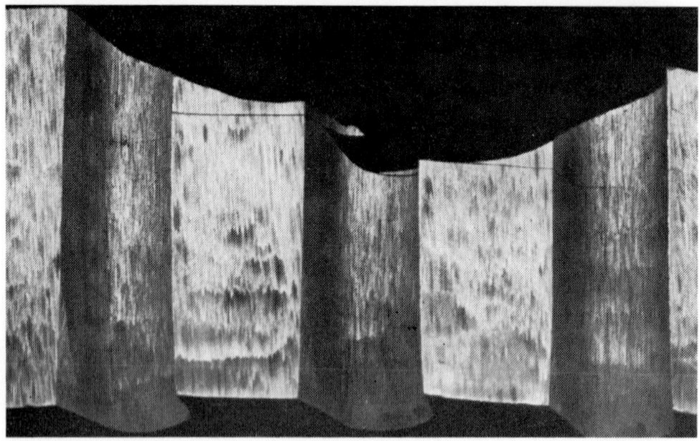

8.15

Fig. 8.14 and 8.15. Milling cutter of high speed steel cracked in service

Fig. 8.14. Fracture. 0.6 ×
Fig. 8.15. Rough grooves, temper colors, and grinding checks on teeth. 2 ×

Fig. 8.16 to 8.18. Hardened bearing ring of ball and roller bearing steel with grinding checks

Fig. 8.16. Ground bearing surface. 1 ×
Fig. 8.17. Grinding check and temper zone under bearing surface. Radial section. Etch: Nital. 20 ×
Fig. 8.18. As Fig. 8.17. 100 ×

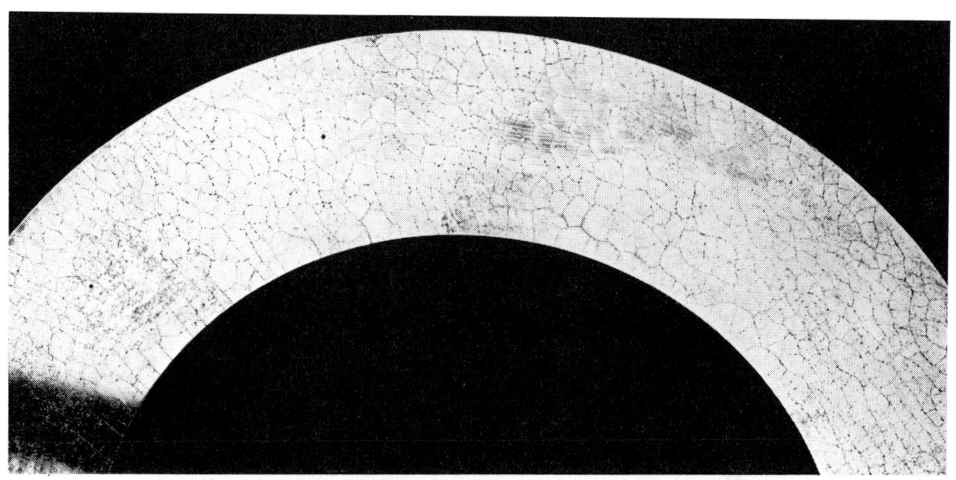
8.16

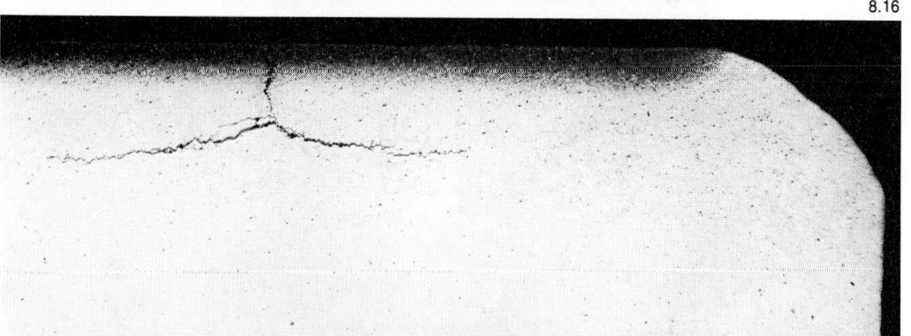
8.17

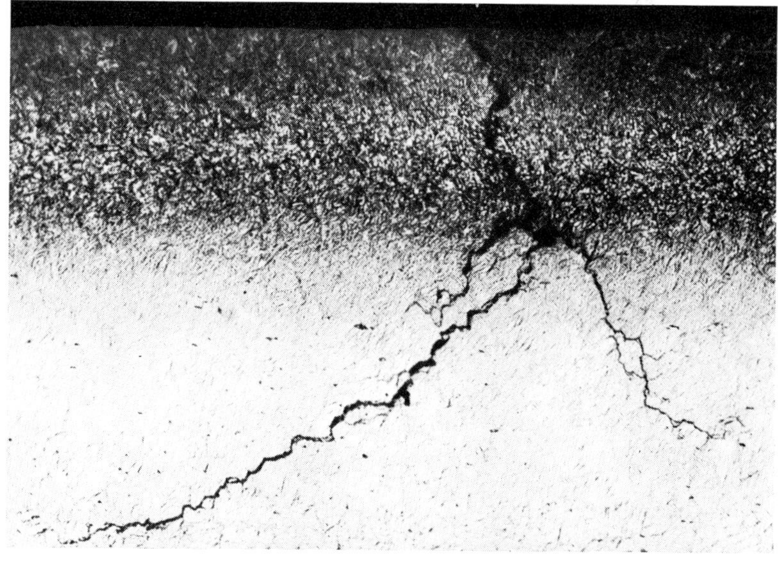
8.18

8.19

8.20

Fig. 8.19 to 8.23. Curved mold of wear resistant chromium-manganese steel damaged by grinding checks

Fig. 8.19. View. 0.2 ×
Fig. 8.20. Rear face after dye penetration test. 1 ×

Fig. 8.21. Surface structure below side face. Etch: Picral. 100 ×
Fig. 8.22. As Fig. 8.21. 500 ×
Fig. 8.23. Core structure. Etch: Picral. 500 ×

8.21

8.22

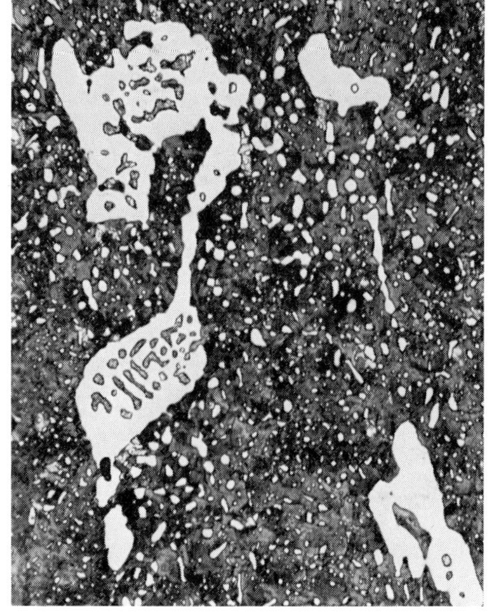

8.23

Grinding checks of a characteristic network pattern were found on the ground **bearing ring** surface made of alloy steel for ball and roller bearings, shown in **Fig. 8.16.** The quenched microstructure of the ring under the bearing surface, which was darkened by etching, revealed a tempered zone of 0.2 to 0.3 mm depth **(Fig. 8.17).** The cracks evidently propagated from a spot below the surface **(Fig. 8.18).**

In **molds,** used for briguetting soft coal, the sharp edges of the curved parts broke off either prior to service or shortly thereafter. They were made of wear resistant chromium-manganese steel containing approx. 2 % C, 2 % Cr and 1 % Mn. **Figure 8.19** shows a factory-new mold. The side faces were permeated by a great many cracks and the edges that were at an acute angle to the working surface were fragmented at several places. **Figure 8.20** shows the opposing side face. The cracks were made more noticeable here by using a dye penetrant detection method. The network pattern and likewise their shallow depth **(Fig. 8.21)** proved that these were grinding checks. The surface structure had become martensitic through heating by the grinding and rapid cooling **(Fig. 8.22).** In the transition to the ledeburitic-pearlitic core structure **(Fig. 8.23),** a zone with a tempered microstructure had formed.

Fine cracks were found at regular intervals of 214 mm in strips of hollow ground, but as yet unsharpened, **razor blades (Fig. 8.24).** The origin of the cracks was located parallel to the cutting edge and had penetrated in part through the entire blade thickness. The cracks always

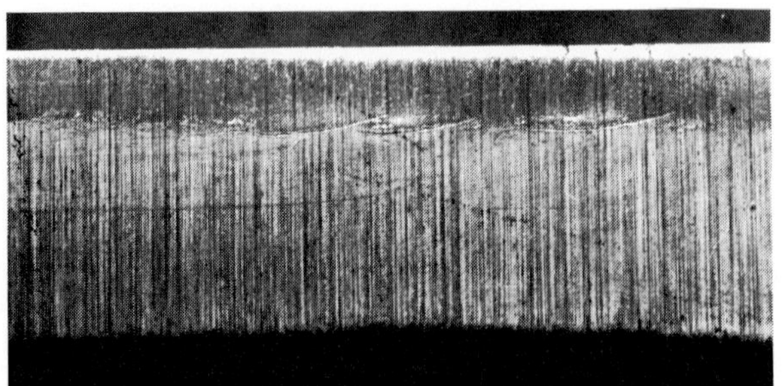

Fig. 8.24. Grinding checks in ground hollow of razor blade. 10 ×

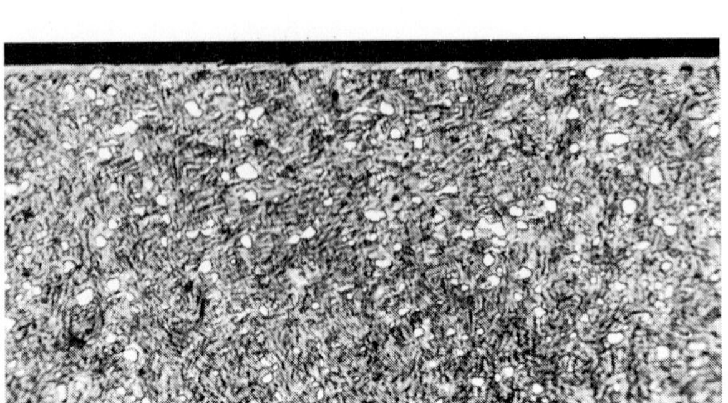

Fig. 8.25. Surface structure with grinding martensite. Transverse section. Etch: Nital. 1000 ×

appeared at the same edge and originated in the same face that had the trademark stamp; therefore they could have occurred only during processing of the strip into the blade. Their shape and location in the ground hollow indicated that they were grinding checks. After opening, the cracks showed a nonoxidized, velvety fine fracture. Therefore they could have occurred only after hardening. Correspondingly their surfaces were not decarburized. The flawless microstructure of the strips consisted of finely acicular martensite with spherical inclusions of hypereutectoid cementite. A very thin layer of newly formed martensite adhered to the ground face **(Fig. 8.25)**; it had flaked off in part. Therefore the cracks undoubtedly had occurred during grinding. The regular reappearance of the defect in the strip indicated deficiencies in the operational mechanics of the grinding process.

The formation of grinding checks may be avoided by the use of grinding wheels with suitable grains and binder, by a lower wheel r.p.m., by slower advance, and by adequate cooling. In case hardened camshafts, grinding check formation was avoided by raising the tempering temperature from 160 to 200 °C with corresponding reduction in hardness.

Literature Section 8

1) F. K. Naumann u. F. Spies. Schleifrisse. Prakt. Metallographie 5 (1968) S. 291/98

9. Failures Caused by Pickling and Surface Treatment Deficiencies

Pickling is often used to remove oxide scale following processing or machining, and electrolytic metal plating is then applied for corrosion protection. During pickling, failures may occur by anodic iron dissolution as well as by cathodic hydrogen reduction. Pitting may occur during dissolution of iron under unfavorable circumstances. Such conditions are given, for instance, if spent liquor is used or acid without addition of an inhibitor, or if an electrolytic potential is formed locally at firmly adhering or rolled-in scale particles[1]).

Failures through cathodic partial reaction may occur during pickling[2]) with the common acids as well as during galvanic depositions of metals onto iron. During this reaction atomic hydrogen is formed. The hydrogen atoms combine in part to molecules and rise as a gas, and in part they are dissolved in the iron, embrittle the steel, and precipitate as molecules at discontinuities such as inclusions, grain boundaries, and phase boundaries. Corresponding to the high concentration of absorbed hydrogen, very high pressures are created. In soft, well deformable steels, blisters may be generated under this effect, while in high strength steels with low deformation capacity cracks may be formed. The penetration of hydrogen into the steel may be prevented or diminished by addition of inhibitors (pickling oils) to the acid or by the use of alkaline baths. A large part of the absorbed hydrogen escapes in time during storage. This evolution process may be accelerated by heating to 100–200 °C. However, it may also promote blister formation.

9.1 Pickling Errors

Pits which occur as the result of pickling are generally randomly distributed, but sometimes are present in rows, stringers, or streaks. If they are formed after the last heat treatment, they have a bright and unoxidized interior. **Figure 9.1** shows a typical profile of a pickling pit. Such pits that have formed at a previous stage of production prior to the last heat treatment may sometimes not be differentiated clearly from other surface defects, such as those occurring in the material or during rolling.

Fig. 9.1. Pickling pits in surface of bright cold drawn pipe. Transverse section. Etch: Nital. 100 ×

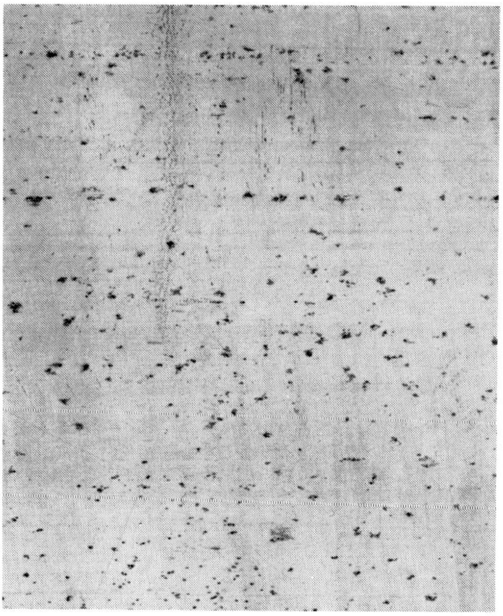

9.2

9.3

Rolling direction

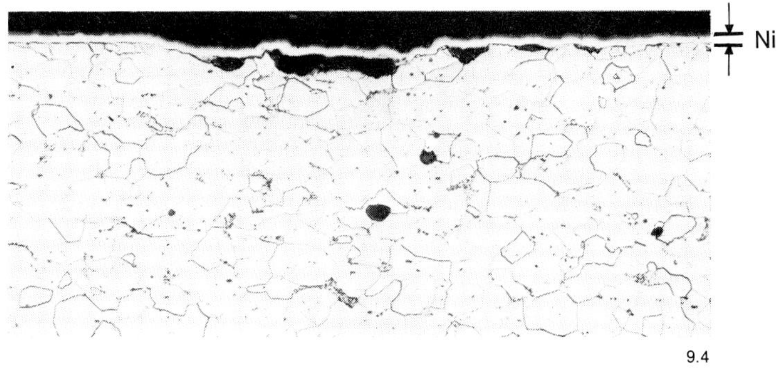

9.4

Fig. 9.2 to 9.4. Pickling pits and patches in row formation on cold rolled strip

Fig. 9.2 and 9.3. Surfaces. 1 ×

Fig. 9.2. Strip 1

Fig. 9.3. Strip 2

Fig. 9.4 Transverse section through strip 2. Etch: Nital. 100 ×. Nickel plated before sectioning to prevent breakout

If the pit holes occur in rows parallel to the rolling direction or preferentially at one side of the sheet, one may suspect that rolling defects may be involved. But this is not always the case. For instance, **Figures 9.2 and 9.3** show two **cold rolled strips** on which the pits of one were arranged at one side in streaks in the direction of rolling (strip 1), while on the other (strip 2) pitholes were aligned on both sides transverse to the direction of rolling. In at least one of these cases, and probably in both, this pit corrosion was not caused by rolling deficiencies. Metallographic examination showed that these defects were open pinholes without any traces of slag **(Fig. 9.4).** Evidently they were pickling patches that were pressed flat and stretched by cold rolling.

Streak formation of the type shown in **Fig. 9.5** is occasionally found in pickled **wires.** Pit corrosion occurs preferentially along well defined surface lines, probably in those places where individual turns of the wire coil were located adjacent to each other in the annealing furnace and in the pickling bath. **Figure 9.6** shows the pits in cross section as deep flat troughs.

The special pattern of the speckled surface defects on the **cold rolled strip steel** according to **Fig. 9.7** is in all probability due to the extended adhesion of an oxide coating in the pickling bath. The defects were pickling patches and in part still closed **pickling blisters (Fig. 9.8);** formation of the latter was aided by streaks of slag inclusions near the surface that were rich in aluminum oxide **(Fig. 9.9).**

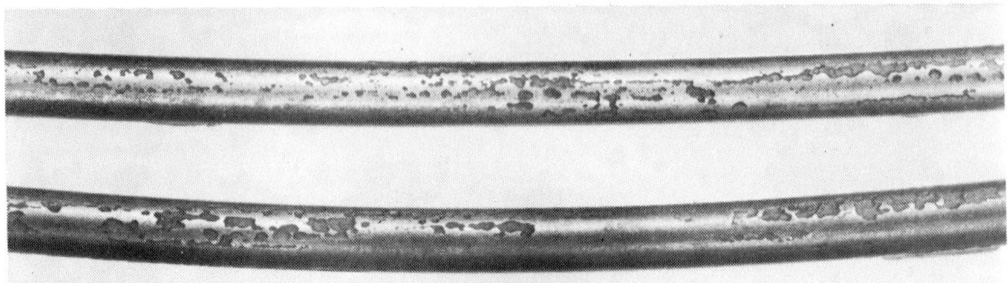

9.5

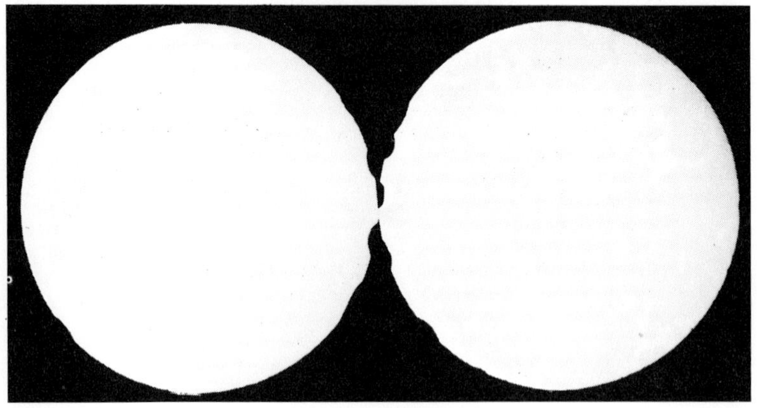

9.6

Fig. 9.5 and 9.6. Pickling defects in drawn high speed steel wire

Fig. 9.5. Surface. 1 ×
Fig. 9.6. Unetched transverse sections. 7 ×

9.7

9.8

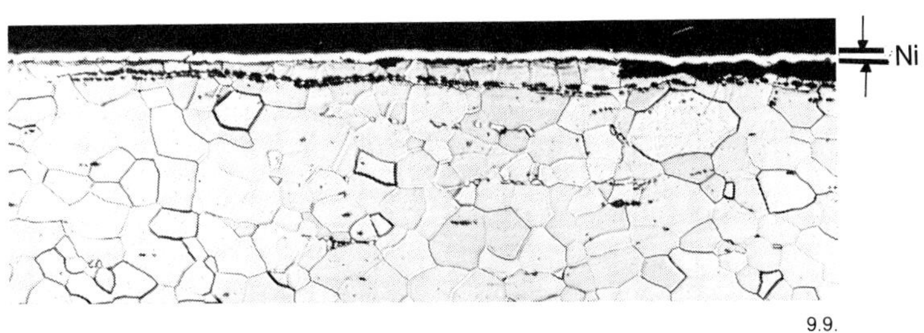

9.9.

Fig. 9.7 to 9.9. Pickling patches and blisters in cold rolled steel strip

Fig. 9.7. Pattern of pickling patches which may be due to extended adhesion of oxide coating. 1 ×
Fig. 9.8. Transverse section through a pickling blister. 500 ×. Surface nickel plated
Fig. 9.9. Inclusions near surface in longitudinal section. Etch: Nital. 200 × Surface nickel plated

9.10

9.11

Fig. 9.10 and 9.11. Surface of tensile specimens of rolled free cutting steel wire. 2 ×

Fig. 9.10. Pickled wire (24 % reduction in area, 2.8 ml H/100 g steel)

Fig. 9.11. Non-pickled wire (62 % reduction in area, 0.36 ml H/100 g steel)

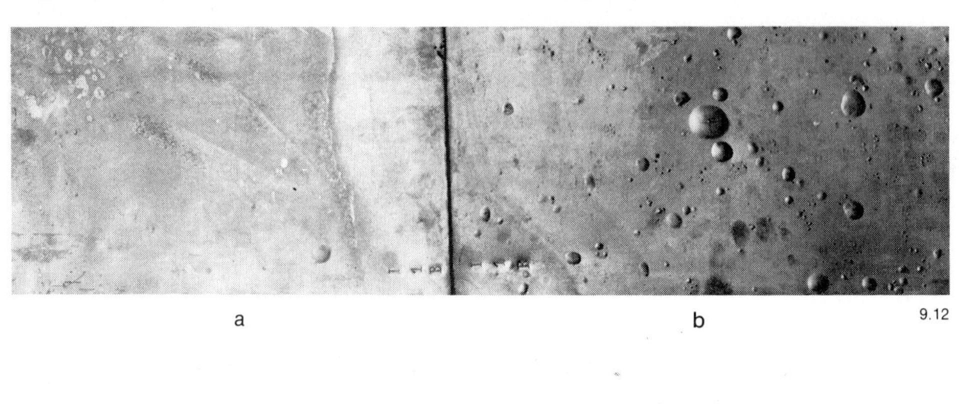

a b 9.12

9.13

Fig. 9.12 and 9.13. Steel strip of rimmed cast steel after prolonged pickling in sulfuric acid

Fig. 9.12. a. from ingot periphery, b. from segregation zone. Surface views. Approx. 0.15 ×

Fig. 9.13. Pickling blister (left) and slag inclusion stringers (right) in unetched longitudinal section. 100 ×

Another example of the pickling blister formation at stringers of nonmetallic inclusions that lie close to the surface has already been shown in section 4.1 (Fig. 4.12 to 4.14).

The following case shows what damage can be caused by extended and strong pickling. Rolled **free cutting 15 mm diameter steel wire** had been pickled deeply in sulfuric acid. This was evident from the rough surface which showed a tendency toward tearing and spallation of the surface during subsequent machining. This behavior also became apparent during tensile testing. In the pickled wires, the surface spalled off in the deformed zones **(Fig. 9.10).** Reduction in area was only 24 %. Comparison specimens of unpickled rolled wire remained smooth in contrast **(Fig. 9.11)** and had a reduction in area of 62 % before breaking. By heating the pickled

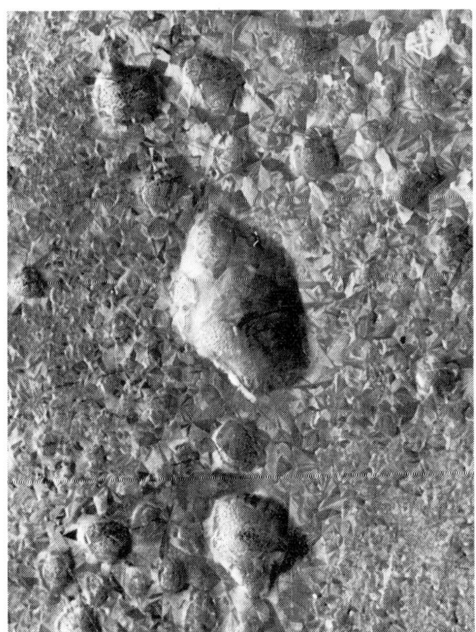

9.14

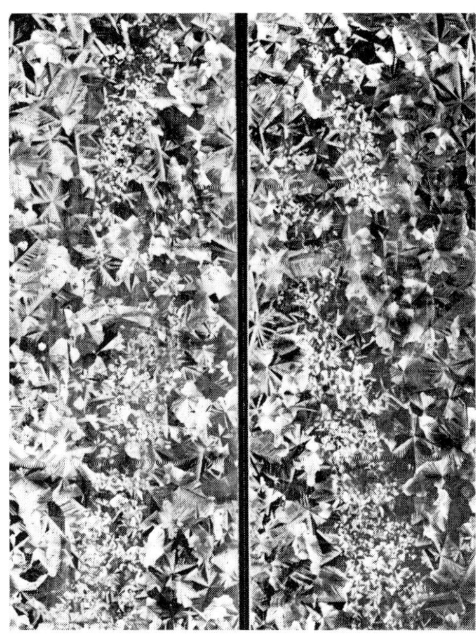

9.15

Fig. 9.14 and 9.15. Surface defects on hot dip galvanized sheets through discharge of hydrogen that had been absorbed during pickling

Fig. 9.14. Pickling blisters. 1 ×

Fig. 9.15. Fine grained "flash" on both sides of galvanized sheet. Surface and opposing face put side-by-side in mirror image. 1 ×

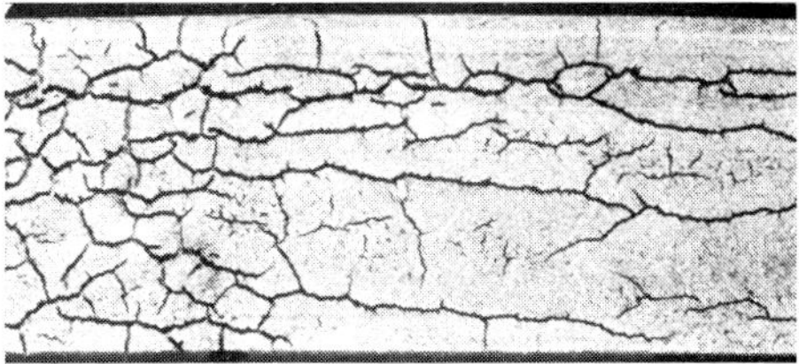

Fig. 9.16. Pickling cracks from deep etching of tool made of stainless knife steel. 1 ×

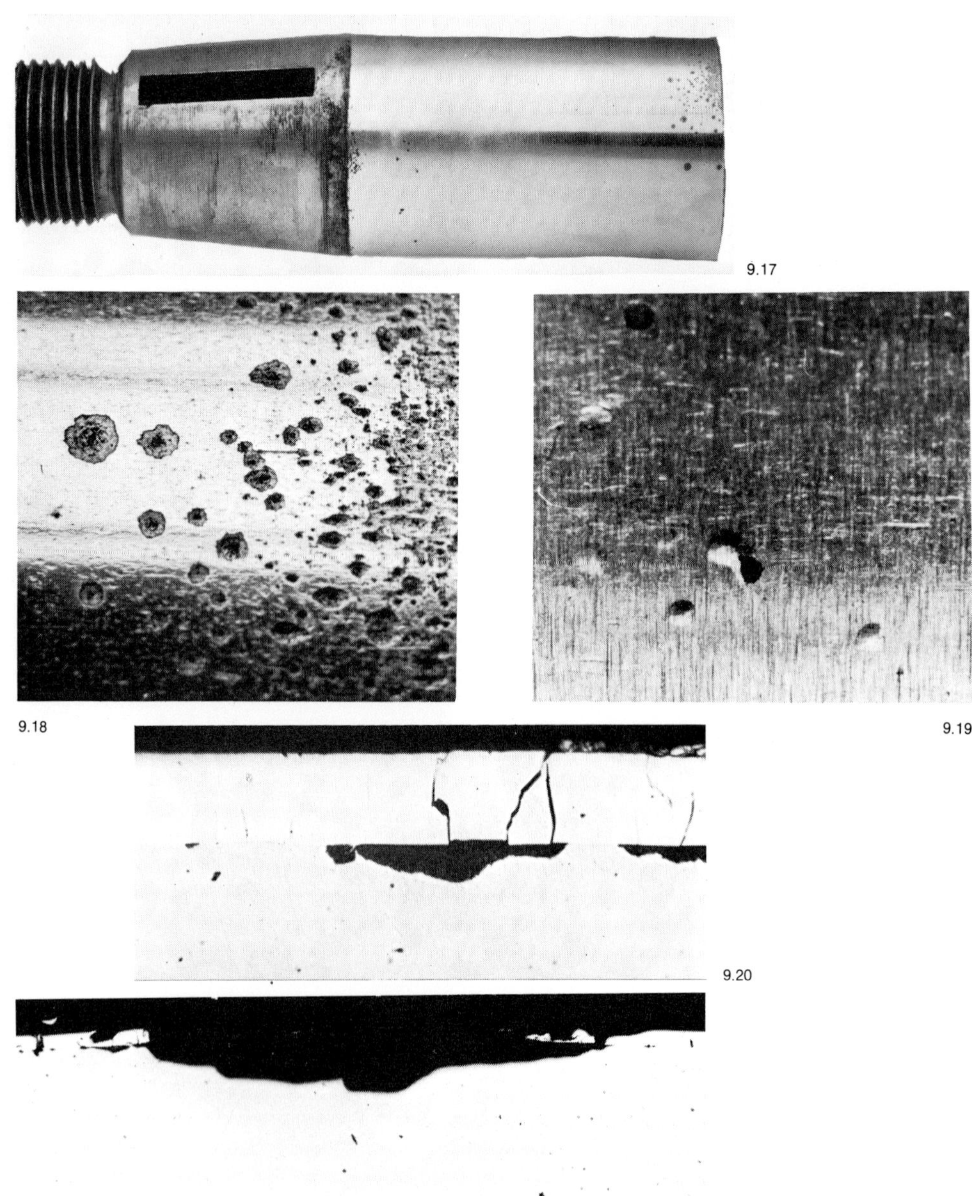

Fig. 9.17 to 9.21. Surface defects on hard chrome plated stub axle bolt through precipitation of hydrogen that was absorbed during plating

Fig. 9.17. Stub axle bolt with wear marks. 1 ×
Fig. 9.18. Highly stressed spot. 10 ×
Fig. 9.19. Less highly stressed spot. 10 ×
Fig. 9.20. Unetched longitudinal section through blister in formation. 500 ×
Fig. 9.21. Unetched longitudinal section through blister with torn-off top. 100 ×

9.22

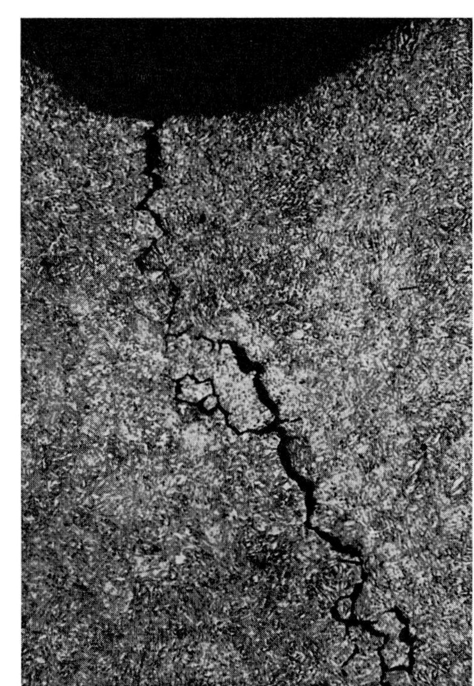

9.23

Fig. 9.22 and 9.23. Zinc plated cylinder head screws that broke after short service time

Fig. 9.22. Unetched longitudinal section through thread part. 7 × Arrow: Beginning of fracture

Fig. 9.23. Crack origin in thread base. Longitudinal section. Etch: Nital. 200 ×

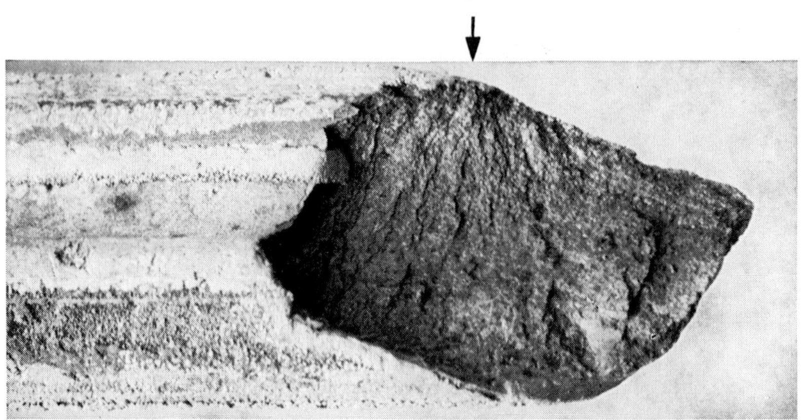

9.24

9.25

9.26

Fig. 9.24 to 9.27. Fracture of a zinc plated Z-profile wire strand from suspension cable of river bridge

Fig. 9.24. Fracture. 5 ×. Arrow: Fracture origin
Fig. 9.25. Fracture edge of wire, zinc plating removed. 10 ×
Fig. 9.26. Longitudinal section through edge of inner wire that was not zinc plated, unetched. 100 ×

Fig. 9.27. Fissure contiguous with stress crack in a broken zinc plated wire. Longitudinal section through edge from which fracture propagated. Unetched. 100 ×

wires to 400 °C, the reduction in area could be increased to 65 %. Even though the wires had already been stored for several weeks, 2.8 ml hydrogen per each 100 g of steel could be withdrawn from them at 400 °C, but only 0.36 ml/100 g steel from the unpickled wires.

The connection of pickling blisters with nonmetallic inclusions at which the hydrogen can precipitate becomes especially clear in the **sheet of rimmed cast steel** shown in **Fig. 9.12.** After strong pickling, blisters appeared predominantly in the parts of the sheets that corresponded to the segregated center of the ingot (right part b in the figure) and less so in the clean peripheral zone. **Figure 9.13** shows incipient blister formation on a silicate inclusion that was rolled flat. The presence of such coarse inclusions, however, is not essential for the formation of pickling blisters; after all they are not prevented by the use of high purity specialty steels, but are averted by reasonable pickling.

Blisters form in many cases when the parts are heated after pickling. This happens, for instance, when the acid traces are washed off in boiling water, or during hot dip galvanizing. **Figure 9.14** shows pickling blisters in a **galvanized sheet.** The so-called fine grained "flashes"* **(Fig. 9.15)**[3]) are another defect that occurs through evolution of hydrogen and the concomitant increased nucleation.

Figure 9.16 gives an example of pickling cracks in the surface of an overly pickled **tool made of a stainless knife steel.** The steel is more sensitive to the formation of hydrogen cracks the higher its strength. Martensitically hardened steels are particularly vulnerable due to the high residual stresses. Pickling in acids is therefore not a suitable means of investigating cracking in hardened steels since this may well become the cause of cracks that lateron cannot be distinguished from those that were present prior to pickling.

Pickle defects can be prevented by shortening the pickling period to the minimum possible; by renewing the acid in time, to keep it free from hydrogen sulfide, arsenic acid, hydrocyanic acid and thiocyanic acid; and by adding a suitable inhibitor (pickling oil or restrainer) to the bath.

* see Appendix I

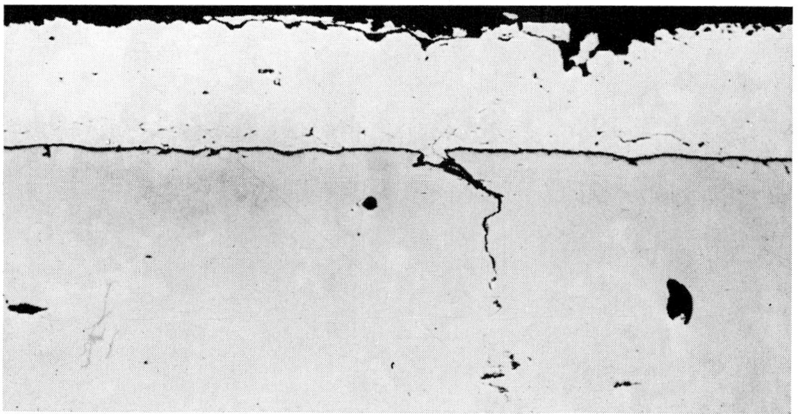

9.27

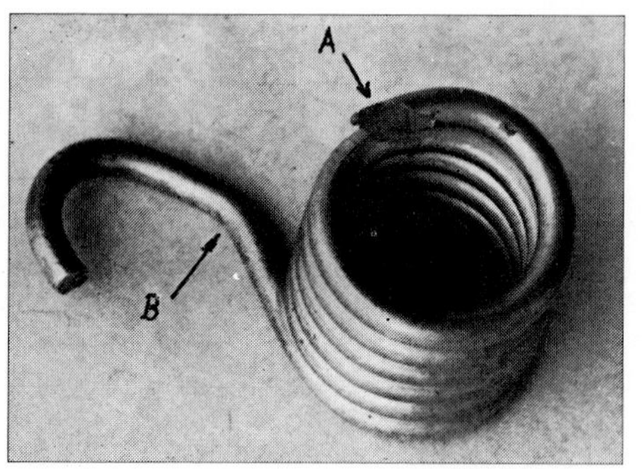

9.28

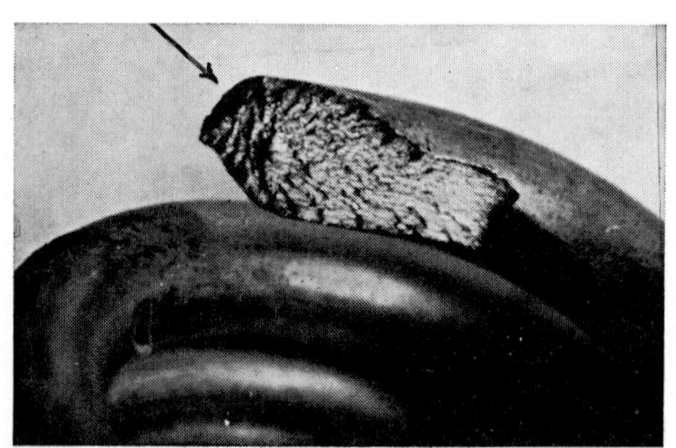

9.29

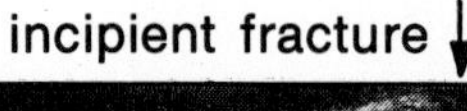

9.30

incipient fracture

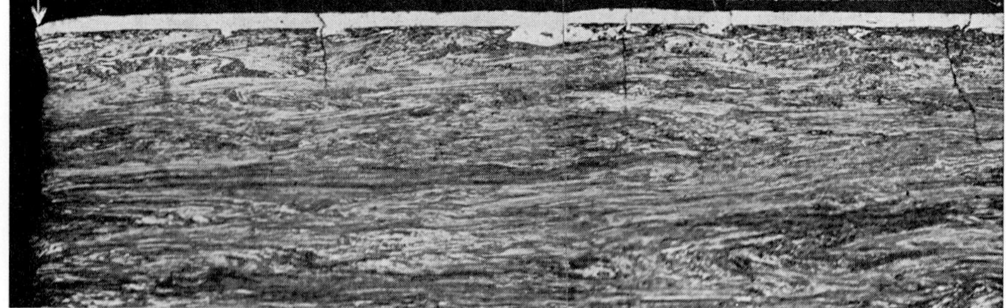

9.31

9.32

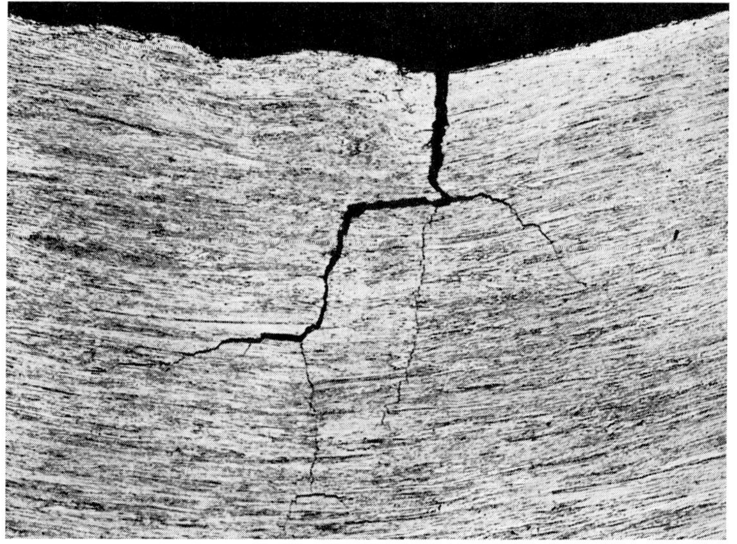

9.33

Fig. 9.28 to 9.33. Fracture of coil springs for door mountings that were electroplated with brass

Fig. 9.28. View. 3 ×. A and B = fracture and crack sites

Fig. 9.29. Fracture A. 10 ×. Arrow: Fracture origin

Fig. 9.30. Longitudinal section through a fracture. Etch: Picral. 50 ×

Fig. 9.31. Incipient cracks adjacent to fracture. Longitudinal section as Fig. 9.30

Fig. 9.32. Residual fracture. Longitudinal section as Fig. 9.30. 500 ×

Fig. 9.33. Crack at B according to Fig. 9.28. Longitudinal section. Etch: Picral. 500 ×

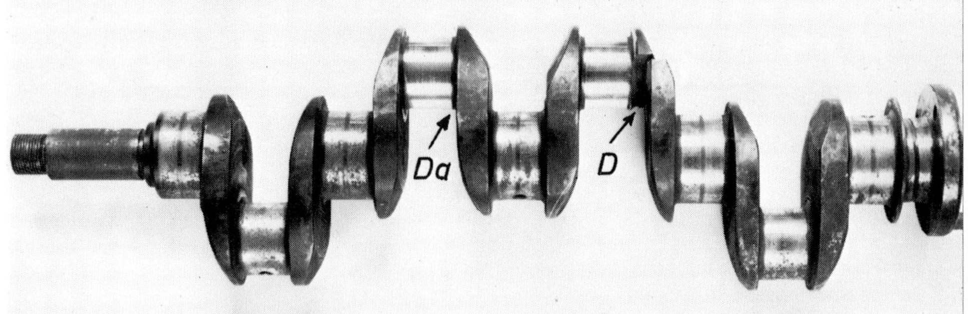

9.34

9.35

9.36

Fig. 9.34 to 9.37. Fracture of crankshaft at chrome plated connecting rod pin

Fig. 9.34. Crankshaft with fatigue fracture (D) and incipient fatigue fracture (Da). 0.25 ×
Fig. 9.35. Fatigue fracture. 1 ×
Fig. 9.36. Incipient fatigue fractures in fillet, made visible by magnetic particle inspection. 1 ×

Fig. 9.37. Bearing pins etched with 10 % diluted aqueous nitric acid. Lower left main bearing, upper right (arrow) connecting rod bearing with arm fracture. 0.5 ×

9.2 **Surface Treatment Errors**

In the following, examples are given of the blister and crack formation caused by hydrogen precipitation during metal electrodeposition coating. **Figure 9.17** shows pit-like wear marks on the surface of a **hard chromium plated stub axle bolt** at its highest stressed point. Under a magnifying glass they showed up as funnel-like fragmentations or erosions **(Fig. 9.18).** At less highly stressed sites it could be seen that the defects had originally been blisters whose tops later tore under service loads **(Fig. 9.19). Figure 9.20** shows the incipient blister formation in longitudinal section. Such protrusions may also occur from oxidation directly beneath the porous chromium plate (see also Fig. 12.47). But in this case no traces of a corrosion product could be found, and it is therefore more probable that these are hydrogen blisters that had formed at previously present voids or pickling pits. **Figure 9.21** shows the section of a blister whose top was torn off in service.

Zinc plated cylinder head screws, with rolled-in threads, broke after a brief service period. They were made with a steel having 0.45 % C and 1 % Cr that was heat treated to an ultimate strength of 1100 MPa (160 ksi). The fractures originated at the thread and had a fine grained structure. **Figure 9.22** shows a longitudinal section through the fracture of a screw. In addition to the crack that lead to rupture, incipient cracks could be observed in various threads; there were even several incipient cracks next to each other in the same thread. The crack pattern and crack propagation mode seemed to rule out fatigue cracks **(Fig. 9.23).** Rather the screw fractures were more likely of the so-called **"delayed fracture"** type in view of the fact that the screws were electroplated. This type of fracture occurs in high strength materials that have absorbed hydrogen during pickling or electrolytic surface plating. Such parts rupture under fatigue stresses below the yield point, but often only after a certain period of time. They break preferentially at notches, such as sharp cross sectional transitions or threads[4]).

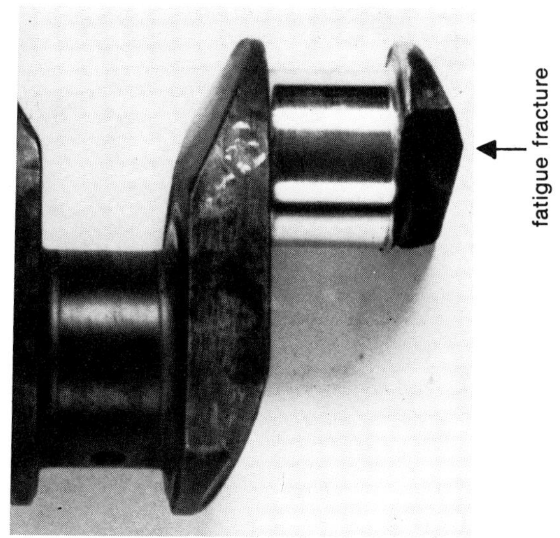

9.37

During construction of a **river suspension bridge**[5]), fractures were found in a number of wires of several ropes. They were located at the outside wrapping whose Z-profile wires were zinc plated. All the fractures originated at an edge of the Z-profile **(Fig. 9.24).** All wires, cracked or not, showed tears at the edges, regardless of whether or not they were located in the outside or inside wrapping **(Fig. 9.25).** According to their appearance in the longitudinal section **(Fig. 9.26),** they could be rolling defects[6]). Experience has shown that during static loading, defects of this type barely affect the strength of the wire, and only to a limited extent its ductility. Therefore it is improbable that these defects alone could be the cause of the wire failures. However, in many cases the cracked wires had an almost perpendicular penetrating jagged crack extending into their interior from the obliquely propagating fissure. This crack was oriented transversely to the fiber and had the marks of a stress crack **(Fig. 9.27).** The suspicion was justified, though it could not be proved, that hydrogen played a part in this fracture formation. The gas may have been absorbed during pickling or zinc plating. The fact that these wire fractures did not occur during cable stranding but were observed only later and therefore probably occurred after some delay, confirmed this suspicion[4]).

9.38

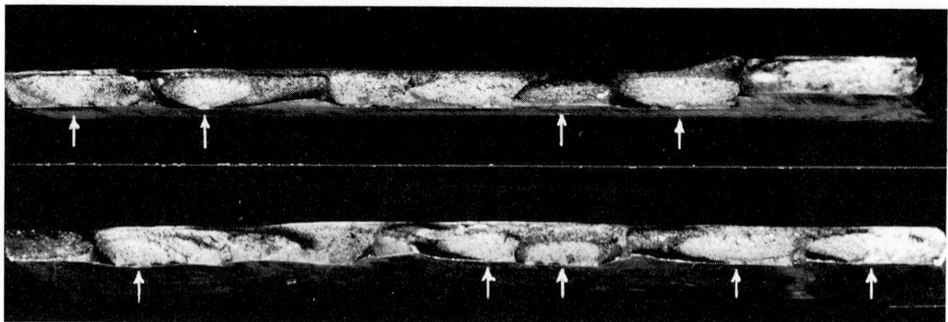

9.39

Fig. 9.38 to 9.40. Fracture of cadmium plated leaf spring

Fig. 9.38. Surface with melt beads and craters at fracture. 5 ×

Fig. 9.39. Fracture. Origins designated by arrows

Fig. 9.40. Microstructure in section through point of origin. Etch: Picral. 200 ×. Zone A: molten. Zone B: transformed, 733 to 773 HV 0.2. Zone C: unaffected, 447 to 463 HV 0.2

Brass plated spiral springs for door mountings broke frequently during service[7]). They were produced from hard drawn spring steel. An examination of thirty-four broken springs showed the following results: The fractures were located in two places, A or B, in the transition to the bent-open support ends **(Fig. 9.28).** They consisted of a short, dark incipient fracture that ran perpendicular to the surface and was located in the external fiber at point A, and the kink to the hook at point B. The remaining fracture propagated diagonally and often ended in a short transverse break **(Fig. 9.29).** This type of failure is often characteristic of fractures caused by hydrogen. It could be seen even more distinctly in the longitudinal section **(Fig. 9.30).** The incipient fracture cut across the fiber in almost a straight line transversely to the drawing direction. It could be seen at higher magnification that incipient cracks had formed in addition to the individual fractures **(Fig. 9.31).** The incipient fractures and cracks represented deformationless separations while the final fractures showed signs of deformation **(Fig. 9.32).** In the kink of the springs these cracks sometimes were deflected into the fiber under the effect of shear stresses and therefore underwent a step-like popagation **(Fig. 9.33).**

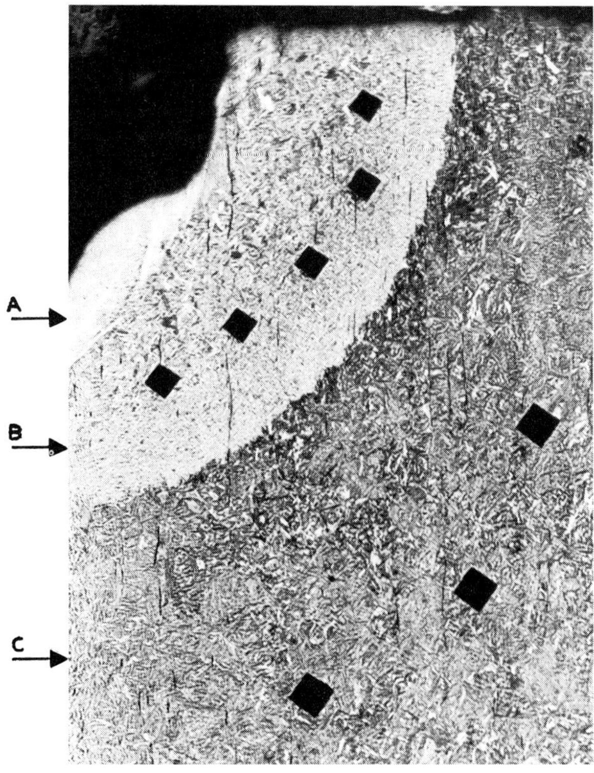

9.40

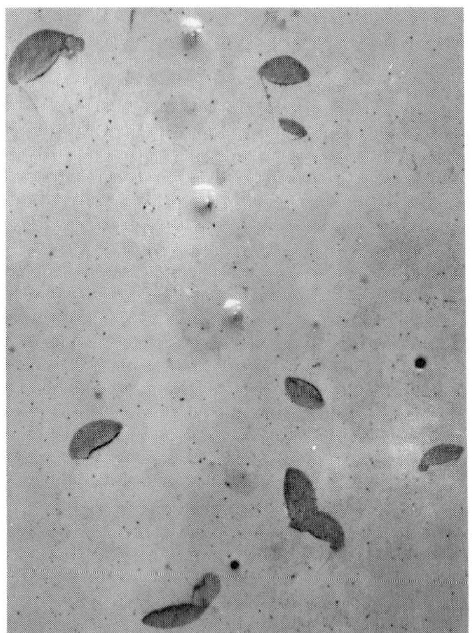

9.41

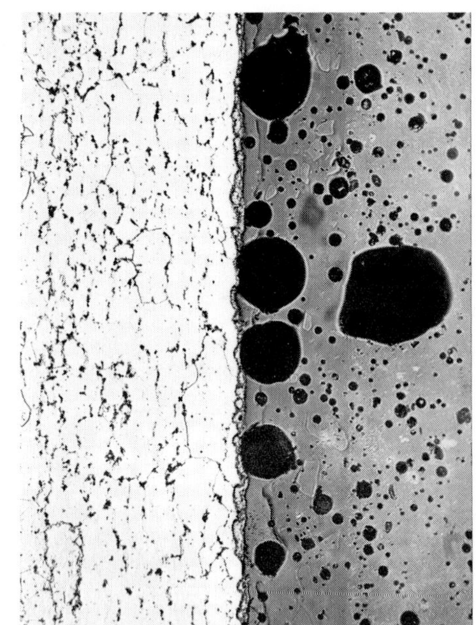

9.42

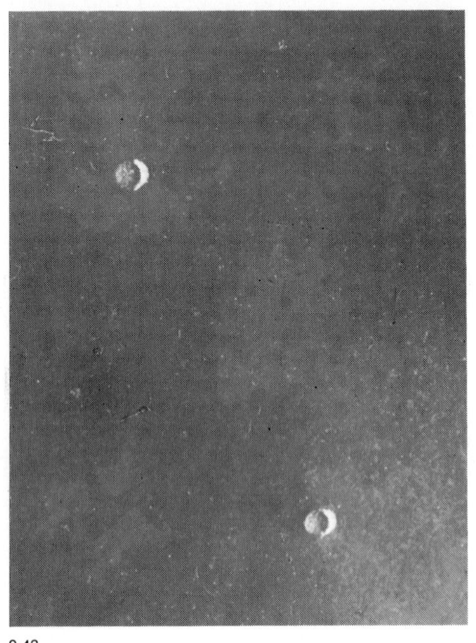

9.43

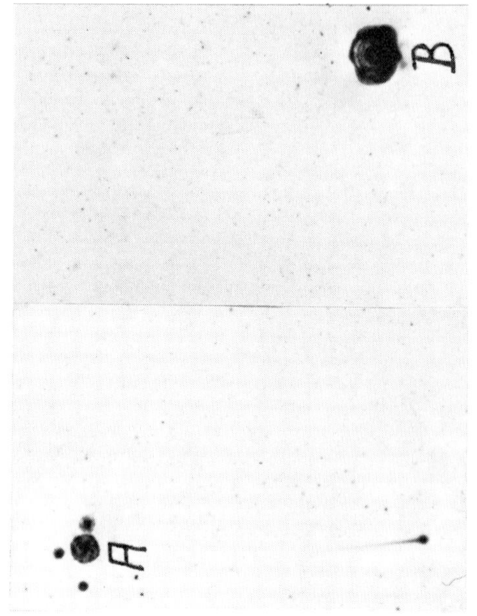

9.44

Cyclic fatigue strength, too, may be lowered by stresses produced by hydrogen that has penetrated and then precipitated during chrome plating[8][9]. **Figures 9.34 to 9.36** show, for instance, a fatigue fracture and incipient fatigue cracks in a **four-cylinder crankshaft**[10]. They propagated from the fillet between one arm and one connecting rod pin each. Only these two pins were chrome plated. This was evident in their bright appearance during delivery (Fig. 9.34) and was confirmed by superficial etching of all pins with 10 % diluted nitric acid. Etching attacked the main bearing pin and the two outer connecting rod pins, but not the two pins located in the center **(Fig. 9.37).**

In order to prevent this type of failure caused by the absorption of hydrogen, the pH-value of the electrolyte should be as high as possible, and the reverse applied to current density and voltage. A two-step procedure is also recommended with intermediate heating to remove the hydrogen, as well as the addition of inhibitores to the bath and the use of cyanide-free plating solutions[11]. The electrolyte should be renewed with time. Storage of the electroplated parts for at least a few days before stressing or boiling them for a few hours in water or oil is also useful.

The rupture of a **leaf spring** made of steel with 0.84 % Cr and electroplated with cadmium was due to another error[12]. The fracture propagated from a number of small cracks that were connected with melt beads and craters on the surface **(Fig. 9.38 and 9.39).** Metallographic examination confirmed that the material had been heated beneath the beads and craters to fusion at the points of fracture origin and had been transformed into martensite **(Fig. 9.40).** The transformation structure had a hardness of 733 to 773 HV 0.2 as compared to 447 to 463 for the unchanged heat treated structure (see impressions in Fig. 9.40). The fracture of the springs therefore was caused by stress cracks due to local quench hardening. The melt beads could have been formed by flying sparks from nearby welding operations or by arcing across a bad contact between spring and current leads. Experiments showed that melting of the type observed did not take place in the first mentioned case, but that localized fusion and martensite formation occurred during brief contact of the steel surface with the stylus of an electric recorder at 4 V potential. Therefore the quench cracks and spring fractures were probably due to errors committed during cadmium plating.

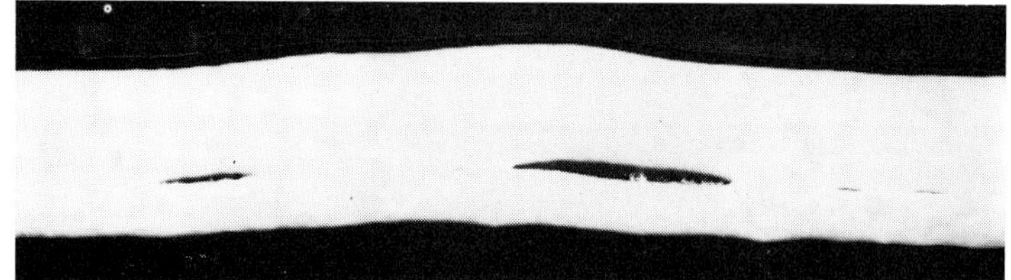

9.45

Fig. 9.41 to 9.42. Enameled bathtub with fish scales

Fig. 9.41. Surface. 1 ×
Fig. 9.42. Etched section. 100 ×

Fig. 9.43 to 9.45. Enameled steel sheet with hydrogen blisters in the enamel and the sheet

Fig. 9.43. Surface with blisters in enamel base coat. 2 ×
Fig. 9.44. Surface with inclusions of "penetration holes" in cover enamel. 2 ×
Fig. 9.45. Unetched section. 12 ×

9.46

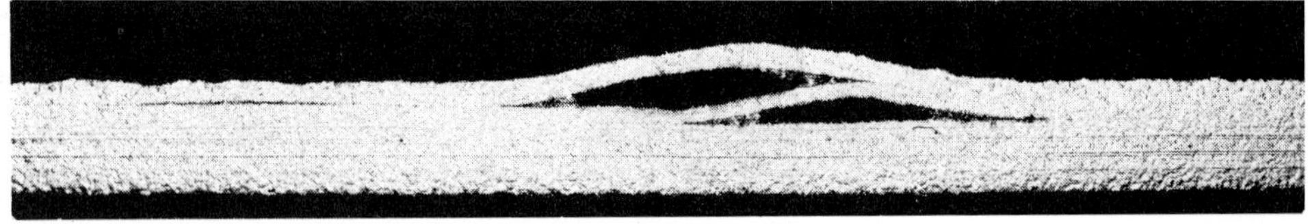

9.47

Fig. 9.46 and 9.47. Enameled coffeepot with hydrogen blisters

Fig. 9.46. View
Fig. 9.47. Section through blistered spot. Etch: Nital. 8 ×

In this connection, some failures should be reported briefly that may occur through the formation of gases during **enameling** of steel. Carbon monoxide may be formed by reaction of carbon in the steel or of carbon-containing contaminants with the oxides of the enamel, with oxygen from the environment, or with scale residues. These may cause blisters in the enamel. Furthermore, hydrogen may be generated from the crystal water of the clay in the slip by contact with iron; the gas not only produces blisters in the enamel layer or causes it to spall, but it can also penetrate into the steel, precipitate at inclusion sites and tear the material apart.

Figure 9.41 shows a section of an enameled surface of a **bathtub** with blisters and local breakouts, the so-called fish scales. **Figure 9.42** illustrates the transition from sheet to enamel in a section perpendicular to the surface of the same tub.

Figure 9.43 shows blisters in the enamel base coat, and **Fig. 9.44** demonstrates "penetration holes"* in the cover enamel of a sheet. In sections through such defects separations in the sheet became evident **(Fig. 9.45);** these originated in silicate inclusion stringers. Undoubtedly in this case hydrogen formation and precipitation was the cause of failure[2]).

Such failures may be reduced to a minimum by the use of high purity low carbon steel; by deoxidizing, cleaning, and degreasing of the steel surface; as well as by using suitable enameling procedures[13])[14])[15]).

A drastic example of blister formation on enameled containers is the **coffeepot** shown in **Fig. 9.46.** The unusually high blister formation **(Fig. 9.47)** gives rise to the suspicion that in this case hydrogen could have been absorbed during pickling, whereas the blisters were formed only during heating for enameling.

---

* see Appendix I

Literature Section 9

1) F. Wever u. H.-J. Engell: Elektrochemische Untersuchungen über den Einfluß des Walz- und Glühzunders von Stahl auf die Korrosion und über den Beizvorgang. Arch. Eisenhüttenwes. 27 (1956) S. 475/86

2) F. K. Naumann u. F. Spies: Beizblasen und Beizbrüchigkeit. Prakt. Metallographie 4 (1967) S. 663/70

3 H.-J. Wiester u. D. Horstmann: Blitze auf verzinkten Blechen. Stahl u. Eisen 74 (1954) S. 835/38

4 H. Krainer u. W. Reich: Einfluß von Spurenelementen auf die Eigenschaften von Stahl und Eisen. Werkstoff-Handbuch Stahl und Eisen 4. Aufl. Verlag Stahleisen mbH., Düsseldorf, Blatt G 12

5) F. K. Naumann u. F. Spies: Drahtbrüche in Tragseilen einer Schrägseilbrücke. Prakt. Metallographie 10 (1973) S. 588/91

6) P. Funke, H. Krautmacher u. R. Ohler: Die Entstehung von Oberflächenfehlern beim Warmwalzen von Stahldrähten mit 5.5 mm Dmr. und deren Bedeutung für die Verarbeitung des Drahtes. Stahl u. Eisen 87 (1967) S. 318/31

7) F. K. Naumann u. F. Spies: Gebrochene Schenkelfedern. Prakt. Metallographie 11 (1974) S. 223/26

8) H. Wiegand: Innere Kerbwirkung und Dauerfestigkeit. Metallwirtschaft 18 (1939) S. 83/85

9) R. J. Love: The influence of surface condition on the fatigue strength of steel. Inst. Metals, Monograph and Report Series No. 13 (1953) S. 161/96

10) F. K. Naumann u. F. Spies: Gerissene Achsschenkelbolzen und gebrochene Kurbelwelle. Prakt. Metallographie 7 (1970) S. 707/12

11) T. R. Croucher: Delayed static failure. Source Book in Failure Analysis, herausgegeben v. d. Amer. Soc. for Metals, 1974. S.20/25

12) F. K. Naumann u. F. Spies: Bruch vergüteter Blattfedern. Prakt. Metallographie 5 (1968) S.219/21

13) L. Vielhaber: Email-Technik. Berlin, Göttingen, Heidelberg. 1939

14) A. Petzold: Email. Berlin 1955

15) D. Horstmann: Die Wechselwirkung zwischen Kohlenstoff des Stahlblechs, dem Email und der Einbrenntemperatur beim Emaillieren. Stahl u. Eisen 81 (1961) S. 629/40; vgl. a. Mitt. Ver. Dt. Emailfachl. 9 (1961) S. 77/87

10. Failures Caused by Errors in Assembly and Shipping

During assembly, many errors occur that may lead to failures. Unnecessary, unsuitable, and faulty welding, crooked fastening or overtightening of screws or nuts, and untrue mounting of gears are some examples.

A failure caused by an obliquely screwed-on **ring of a safety valve** had been reported already in section 1 (see Fig. 1.3 to 1.5).

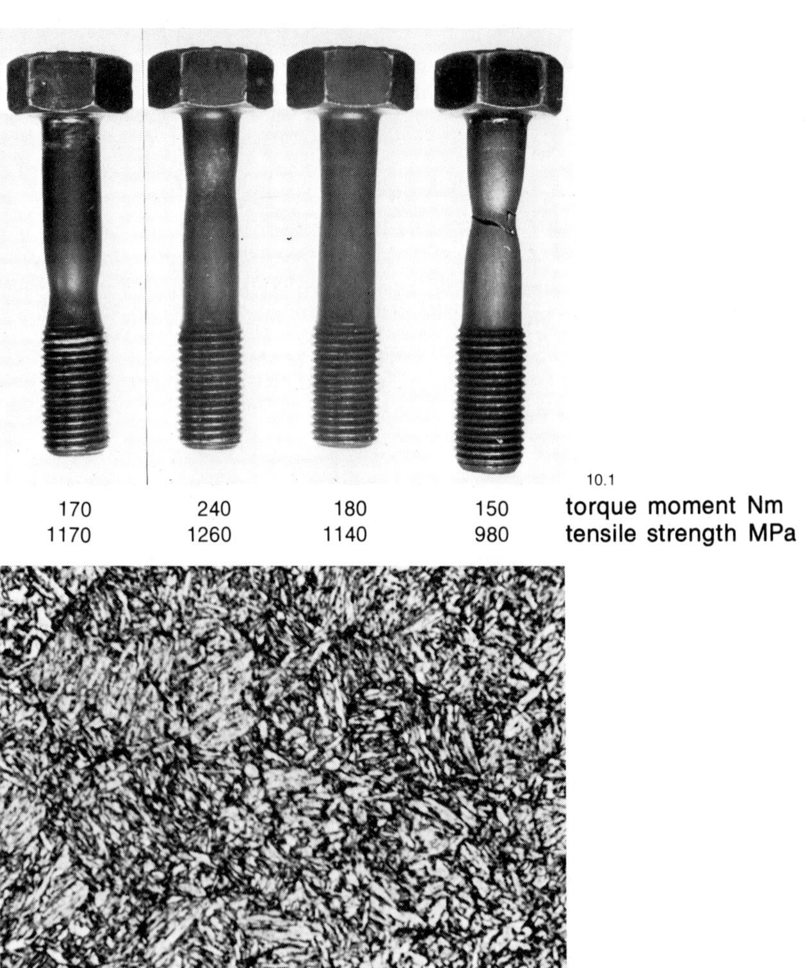

Fig. 10.1 and 10.2. Connecting rod bearing screws stretched or broken during tightening

Fig. 10.1. Views. 0.75 ×

Fig. 10.2 Microstructure of cracked screw. Transverse section. Etch: Picral + nital. 1000 ×

Heat treated **connecting rod bearing screws** having an ultimate strength of 980 to 1180 MPa (142–172 ksi) cracked or stretched during tightening according to specifications at a torque moment of 150 to 240 Nm **(Fig. 10.1).** All screws had the specified strength, but the cracked screws had values bordering on the lower limit set by the specifications. The screws had a finely acicular heat treated microstructure **(Fig. 10.2)** and also good elongation as illustrated in Fig. 10.1. It could not be determined whether the strength specifications in this case were too low, the specified torque moment was too high, or the measurement of the moment was wrong.

The **hook of a crane** broke almost without deformation during service. The fracture originated at the root of the first supporting thread. It was designated a catastrophic fracture based on the grainy microstructure. The hooks had been welded to the nut at the upper end to prevent turning. As could be seen from a longitudinal section through the threaded part and nut **(Fig. 10.3),** the weld was tight and flawless, and the heat-affected zone was not deep and did not reach the fracture area. An examination of the material from which the hook was made showed that it consisted of a low carbon steel, and no cause for embrittlement could be found. Accordingly, weld stresses must have contributed to the fracture. They lead to a tilt of the hook in the nut. Hence, a bend stress was superimposed on the presumed impact-type tensile stress. Welding therefore was an unsuitable method to secure the nut against turning.

A case hardened **bevel gear** of an automotive gear unit, that had to be dismantled after less than 1000 km service, showed strong wear and breakout of the teeth at the base of the truncated cone. The fractures were located all at one side **(Fig. 10.4).** Rupturing was induced by fatique fractures that originated below the case in the core material **(Fig. 10.5). Figure 10.6** presents an etched section of a heavily worn spot. The case hardened layer was in order, judging by its microstructure, depth, and hardness, while core strength was manifestly high. Accordingly, the strong local attack could have been caused only by high overloads that probably were concentrated in a comparatively small part of the tooth flanks due to careless assembly.

A similar case was already reported in section 6 (Fig. 6.123).

The fracture of tensile wires during construction of a viaduct described in section 3.3 (Fig. 3.82 to 3.85) should also be regarded as the result of an assembly error.

Fig. 10.3. Fracture of a loading hook secured by welding against turning in the nut. Longitudinal section through low-deformation fracture (below). Etch according to Heyn. 1.33 ×

10.4

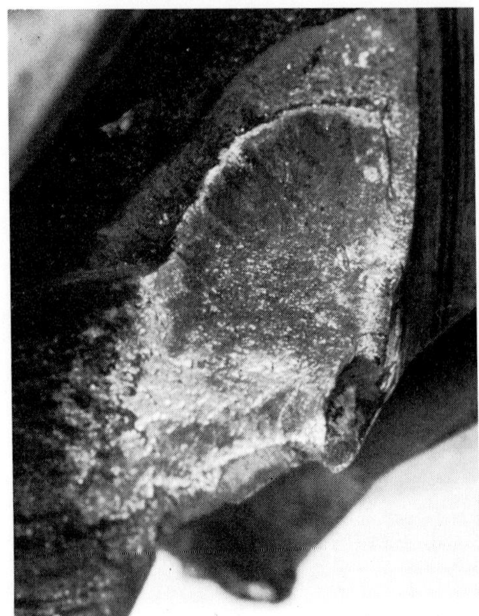

10.5

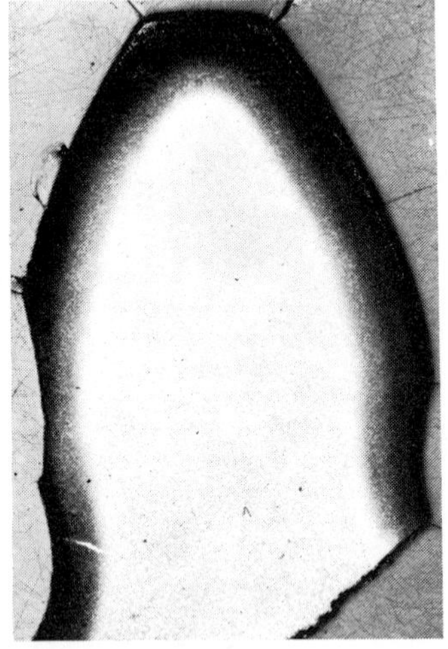

10.6

Fig. 10.4 to 10.6. Tooth failures in case hardened bevel gear of automotive gear unit

Fig. 10.4. Side view. 1 ×

Fig. 10.5. Fatigue fracture with origin below case

Fig. 10.6. Section through knocked-off worn tooth. Etch: Nital. 5 ×

Shipping failures are almost always corrosion failures. The question then usually arises what caused the corrosion and when did it occur. The last part of the question can never be answered precisely by the materials testing engineer. Not even the question whether the steel had rusted before, during, or after shipping, can be reliably determined. With respect to the corrosive medium it may be assumed as a rule – with certain exceptions – that in a normal corrosion process the cause was oxygen-containing water that was deposited in drops on the shipped material either directly or through condensation from supersaturated air. In an attempt to protect a given shipment from corrosion, the supplier often wraps the steel in crepe or oil-soaked paper, or even solders it into sheet metal containers. During loading into cold railroad cars or storage areas moisture may then precipitate onto the metal. Just that type of condensation corrosion is particularly damaging and dangerous because of the formation of local electrochemical cells and the pitting type of corrosion caused by it (see section 15.3.2.2). Usually it is better to ship and store in well ventilated areas in which no moisture can precipitate, rather than packing in air-tight and poorly ventilated containers. These points are of particular importance during shipping and storage of high strength structural steels, such as reinforcement rods for prestressed concrete structures, because creep resistance can be considerably reduced by corrosion pits.

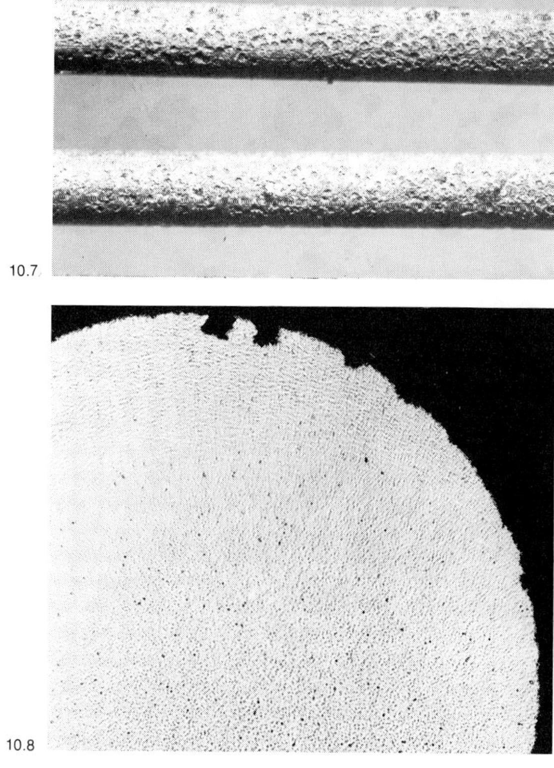

10.7

10.8

Fig. 10.7 and 10.8. High speed steel wire rejected because of rust formation during overseas shipping

Fig. 10.7. Surface after dissolving of rust with 10 % diluted ammonium citrate solution. 10 ×

Fig. 10.8. Unetched transverse section. 100 ×

Since it is well known that **chlorides** are particularly aggressive, it is often thought that corrosion by seawater which contains about 3 % chlorides causes or intensifies rust damage in overseas shipments. This possibility certainly exists during storage on deck, but shipments that are located below deck are in much greater danger of condensation corrosion. But it is in most cases not possible to prove this with certainty.

Figures 10.7 and 10.8 show as example the pitted surface and the profile characteristic for pitting corrosion of a **high speed steel wire** that was rejected due to rusting during overseas shipping. No chlorine ions could be found in the rust with the exception of the very minor amount that is always present everywhere.

Findings were different in the following case. One of the two 200-ton shipments of 100 x 0.65 mm AISI 1095 band **saw steel strips** was rejected because the strip had arrived in a heavily rusted condition at an East Asian port, whereas the strip of the other shipment remained bright. **Figure 10.9** shows a coil of the rejected shipment. Only some scraps of crepe paper and jute rope remained from the packaging. **Figure 10.10** shows the outer (79th), two middle (54th and 8th) and the inner (1st) turns of the coil. Accordingly, corrosion was heavier in the outer lying turns than in those in the middle, and also more pronounced near the edges than in the center of the strip. Therefore corrosion was caused by a liquid penetrating from the outside.

Rust that was brushed off from the surface of a coil and then was dissolved in nitric acid had the following composition:

Cl' %	So_4'' %	Na^+ %	K^+ %	Ca^{++} %	Mg^{++} %
5.74	0.10	0.14	0.01	0.58	0.04

The high contents of chlorine ions could point to seawater as the corrosive agent but the necessary cations, especially sodium, were present in much too small amounts. A part of the rust could easily be dissolved in water. It contained 1.30 % Fe and 1.95 % Cl. This corresponds approximately to the composition of ferrous chloride. The remaining part of the rust presumably contained the chlorine in the form of the hard-to-dissolve iron hydroxychloride ($Fe\ (OH)_2\ Cl$). Also the brown crepe paper with which the strips were wrapped contained considerable amounts of chlorine without the corresponding cations of the seawater.

In search of the source of the chlorine – whereby the oil with which the coils were greased was also taken into consideration – it was finally established that containers with carbon tetrachloride were stored in the same hold on the freighter. Carbon tetrachloride, a usually very stable compound, is hydrolized at room temperature in the presence of iron under formation of hydrochloric acid[1]). Corrosion tests in a water vapor-air mixture at 40 °C confirmed that the amount of iron removed in the presence of carbon tetrachloride vapors was more than ten times as high as in a test without carbon tetrachloride. The rust dissolved in boiling water contained 1.28 % Cl.

This proved that the failure could have been caused by the effect of carbon tetrachloride that escaped from one or more leaky containers stored together with the band saw steel strips. The failure may be visualized in such a way that water or moisture-saturated air together with vapors of carbon tetrachloride penetrated into wrappings of the rings. The iron then went into solution as a bivalent ion and formed ferrous chloride with the chlorine ion of the hydrochloric acid produced by dissociation of the carbon tetrachloride. The ferrous chloride was partially oxidized to hard-to-dissolve iron hydroxychloride and precipitated next to the ferric hydroxide. This enabled the chlorine to accumulate heavily on the iron.

Literature Section 10

1) H. Remy: Lehrbuch der Anorganischen Chemie. 8. Aufl. 1955, Bd. I, S. 372

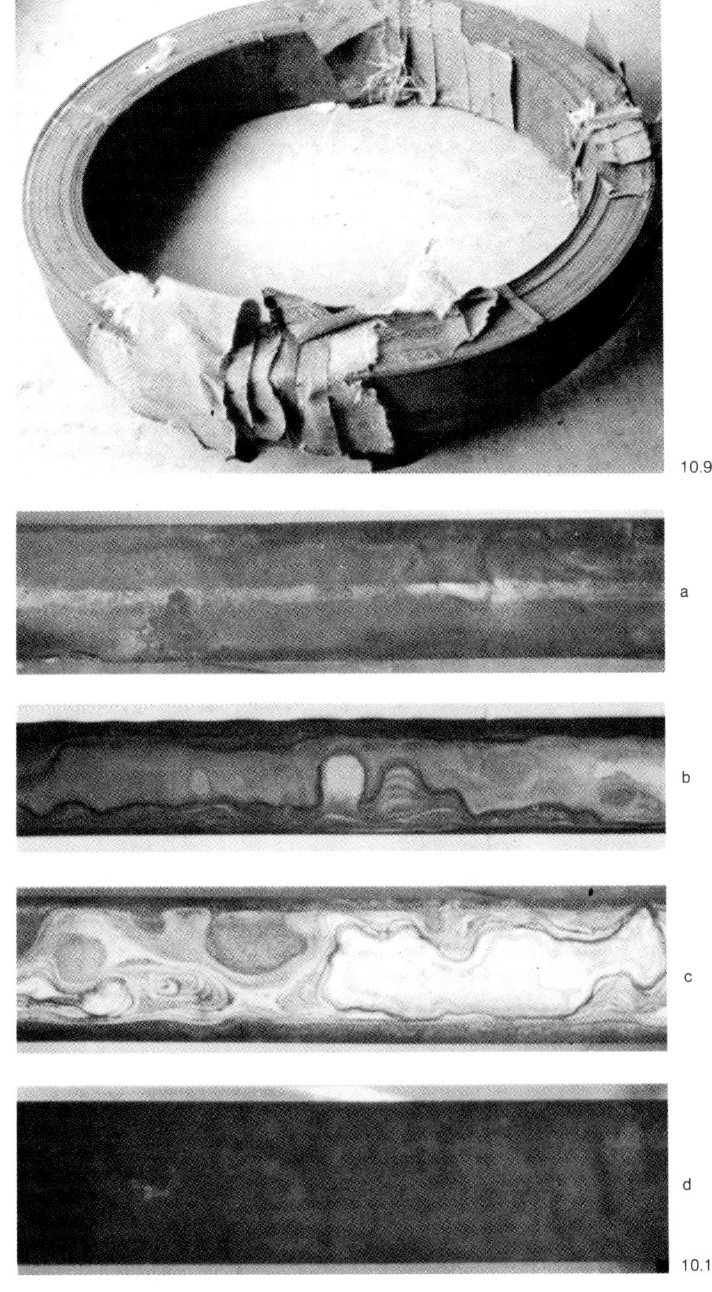

Fig. 10.9 and 10.10. Rust formation on band saw steel during shipping overseas

Fig. 10.9. Rusted coil of steel strip 100 × 0.65 mm

Fig. 10.10. Surface view of various turns of coil according to Fig. 10.9. a. outer (79th) turn, b. 54th turn, c. 8th turn, d. inner (1st) turn

11. Failures Caused by Repair Errors

Errors in producing castings are often corrected by welding. If this is done properly and expertly, it is not objectionable. But sometimes welding is attempted to restore broken and worn machine parts with inadequate home remedies. But here, in particular, proper preparation and a superb technique are of the utmost importance. The reason is that, especially under cyclic stresses, undercutting and stress cracks lead to rapid failure of the patched parts. Thus the repair job often only postpones final replacement[1)][2)][3)].

11.1. **Patch Welding, Weld Build-up**

An example of a poorly and carelessly made repair by welding is the rupture of a **pull rod** of a truck trailer, shown in **Fig. 11.1.** The fact that the examination was requested by the office of the district attorney permitted the assumption that the break had caused an accident. The fracture was discolored, partly intergranular, and covered with melt beads **(Fig. 11.2.)**; this characterized the fracture site as a weldment. From the longitudinal section of **Fig. 11.3** the joint was seen to be only superficial and apparently had remained open in the root.

A patch weld also lead to the rupture of a forged **chromium steel roll** after a brief period of service **(Fig. 11.4).** Confirmation of a weld failure was obtained after purposely breaking open the steel roll. Failure originated in an old longitudinal crack in one of the four drive pin keyways. Once initiation had occurred, the crack was deflected in a transverse direction. In the fresh final fracture **(Fig. 11.5),** fine-grained fracture zones, a few millimeters in depth, were noticeable in the fracture origin (A) as well as at a point offset by 60° (B). In cross section **(Fig. 11.6)** these zones proved to be welds that were deposited in a width of 40 to 50 mm at five uniformly distributed points over the perimeter. The crack originated in one of these points; in two other places additional cracks formed as the result of the weld stresses. The purpose of these weld beads could not be determined; it was assumed that they were used to correct planning and processing errors.

Fortunately a serious accident did not occur when the **rear wheel suspension** of a racing sports car broke at high speed on the highway[4)]. Examination showed that the circular break **(Fig. 11.7)** was induced by several fatigue fractures that originated in the fillet at the transition from bushing to disk. Four paired weld build-ups were made in the torus at opposite sides. One of these spots in etched longitudinal section is shown in **Fig. 11.8.** The weldment was quench hardened to bainite. The base metal in the remaining torus was martensite which was permeated by quench cracks **(Fig. 11.9).** The fatigue fractures probably propagated from such cracks. The suspension of the rear wheel of the car that had not yet ruptured also showed the same characteristics.

Fig. 11.1 to 11.3. Pull rod of truck trailer broken at weld which contained an open root
Fig. 11.1. Fracture. 1 ×
Fig. 11.2. Fracture planes, cleaned with gasoline. 1 ×
Fig. 11.3. Longitudinal section. Adler etch. 1 ×

11.1

11.2

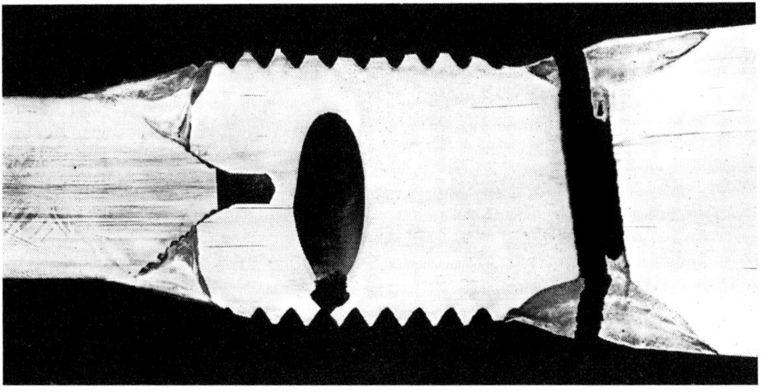

11.3

11.4

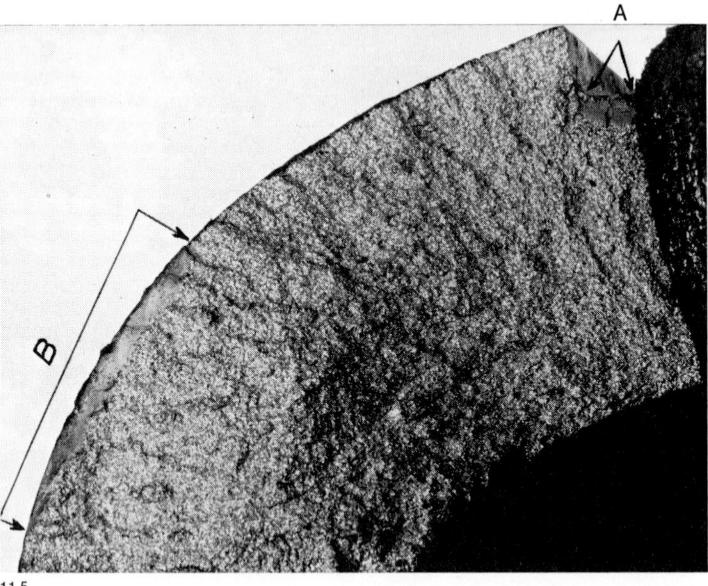

11 5

An attempt was made to restore the original dimensions of the **rear axle tube of a bus** by weld build-up of a gradually worn bearing face[3]). The tube broke shortly thereafter **(Fig. 11.10)** due to several fatigue fractures that were located in part at (a) and in part adjacent to the sharp edged transition of the cross section (b). A longitudinal section through the fracture site showed that the built-up layer (light in **Fig. 11.11)** was permeated by stress cracks and the base metal below it (dark in Fig. 11.11) was hardened. The cracks originated at the transition of the weld to the base metal where the residual tensile stress was highest **(Fig. 11.12).** The tube consisted of a steel with 0.5 to 0.6 % C. Welding of such a high carbon steel would have required uniform preheating of the part, a fact which obviously had been disregarded.

The fracture of a **driveshaft** that had been repaired by weld build-up has already been described in section 7 (Fig. 7.33 to 7.36).

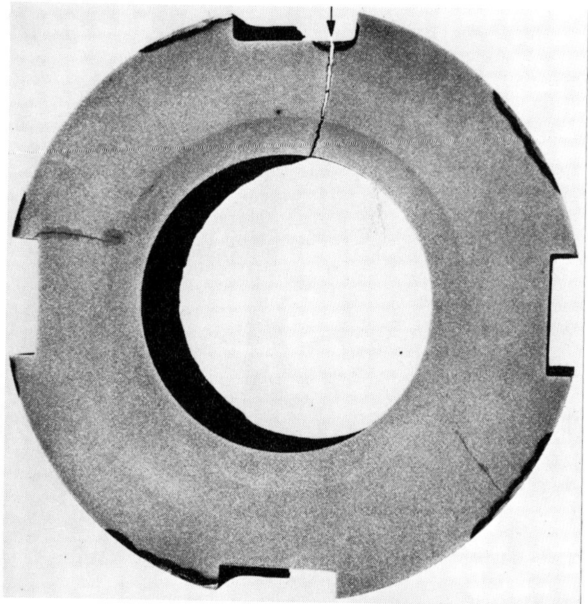

11.6

Fig. 11.4 to 11.6. Patch weld-induced break in the drive pin keyway of a chromium steel roll

Fig. 11.4. Fracture. 0.2 ×
Fig. 11.5. Old crack (right) and artificially produced fracture in drive pin. 1 ×. Fine-grained weld areas delineated by arrows

Fig. 11.6. Transverse section adjacent to pin fracture, etched according to Heyn. 0.5 ×. ↓ Crack initiation point

11.7

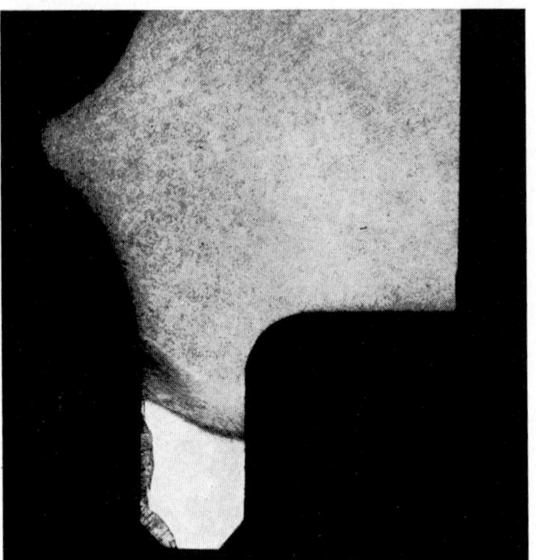

11.8

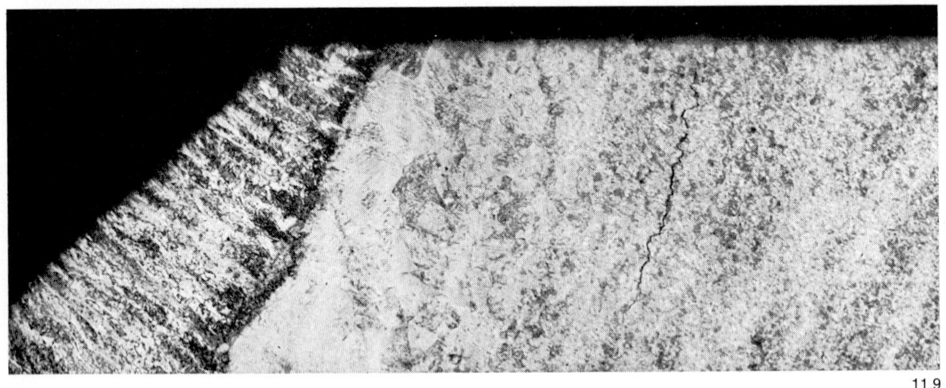

11.9

Fig. 11.7 to 11.9. Broken rear wheel suspension of racing sports car

Fig. 11.7. Fracture. 0.6 ×

Fig. 11.8. One of four sites with weld build-ups at torus. Longitudinal section. Etch: Picral. 3 ×

Fig. 11.9. As Fig. 11.8. 50 ×. Left: Weld metal. Right: Hardened zone with crack

11.10

11.11

11.12

Fig. 11.10 to 11.12. Rear axle tube of autobus which broke as a result of weld build-up

Fig. 11.10. Fracture. 0.3 ×

Fig. 11.11. Longitudinal section through fracture. Etch according to Heyn. 2 ×

Fig. 11.12. Stress crack in transition from weld to base metal. Longitudinal section. Etch: Nital. 100 ×

11.13

11.14

11.15

The **drum of a centrifuge** that had been in service for three years while operating at 4800 r.p.m. broke after repairing the cover seat with a deposit of austenitic chromium-nickel-molybdenum steel. During opening of the drum housing for examination, the circular saw blade got stuck in the cut indicating the presence of high stresses. The fracture is shown in **Fig. 11.13.** Judging by the fiber orientation, the fracture originated in the welded-on seat. The fine-grained smooth structure beneath the austenitic weld metal suggested that the first incipient crack was formed in the zone that was hardened by the weld heat. The extent of the weld heat effect can be seen from the longitudinal section illustrated in **Fig. 11.14.** The microstructure below the austenitic deposit was transformed into a coarse acicular martensite **(Fig. 11.15)** and had a hardness in excess of 500 HV 10. The base metal that consisted of a chromium-vanadium steel with 0.43 % C had a finely acicular heat treated structure. However, this region had a notch toughness of only 14 to 21 J at a yield point of 990 MPa (144 ksi). Thus, it had only a limited capacity to reduce stresses through deformation. Therefore, it was possible that the entire drum had torn open from the seat which had been subjected to excessive welding stresses. In addition, the welded-on seat may have already had cracks.

Finally an example is given for a **patch weld on a casting** that contained the fracture origin. In this example, an **angle lever** broke during service in one of the I-cross section legs **(Fig. 11.16)**[7]). It was said to consist of spherulitic graphite cast iron with a tensile strength of at least 490 MPa (70 ksi). The failure originated at one of the flanges **(Fig. 11.17).** The fracture origin was intergranular and the fracture surface was stained blue by oxidation. Accordingly this was an old incipient crack that originated while the weld was hot. An etched transverse section that was cut from a region next to the fracture **(Fig. 11.18)** showed deposits of a foreign material at the fracture origin. It had the microstructure of a hypereutectoid steel with local precipitates of temper carbon **(Fig. 11.19),** while the base metal of the angle lever consisted of a ferritic cast iron with worm-shaped graphite (vermicular graphite) **(Fig. 11.20).** Therefore the lever did not consist of spherulitic graphite cast iron. The very low strength may have contributed to its breaking. But the principal cause of failure had to be attributed to hot cracking caused by the patch welding, and the notch effect produced thereby.

Fig. 11.13 to 11.15. Lower part of centrifuge drum which cracked as a result of weld build-up

Fig. 11.13. Fracture. Approx. 0.33 ×. Fracture origin upper right

Fig. 11.14. Longitudinal section through seat with weld build-up. Etch: Nital. 2 ×

Fig. 11.15. Microstructure below weld build-up. Longitudinal section. Etch: Nital. 500 ×

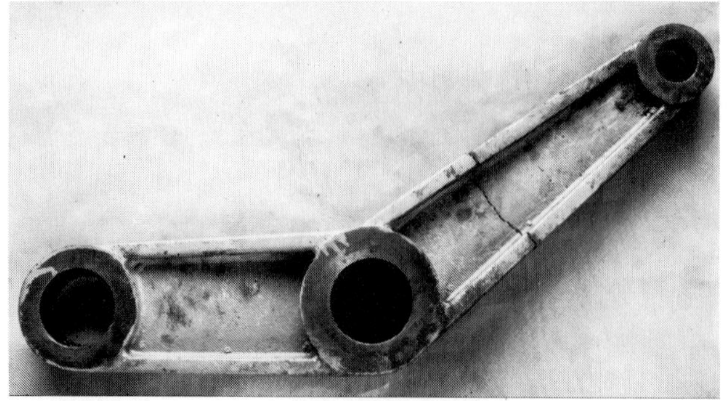

11.16

11.17

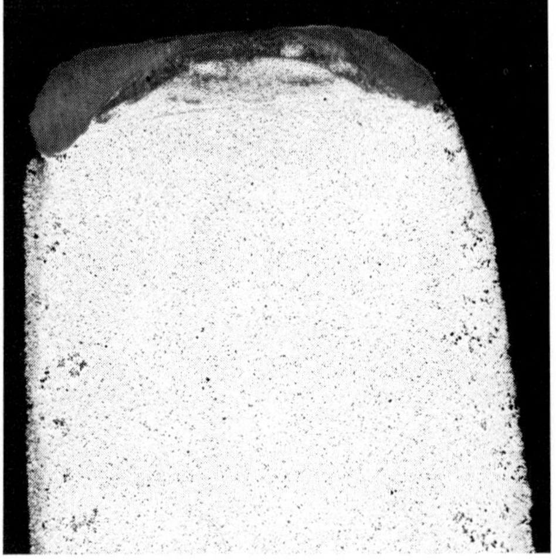

11.18

Fig. 11.16 to 11.20. Cast iron angle lever which broke as a result of patch weld

Fig. 11.16. View. Approx. 0.067 ×
Fig. 11.17. Fracture. 1 ×. ↓ Origin

Fig. 11.18. Transverse section through fracture origin. Etch: Nital. 3 ×

Fig. 11.19 and 11.20. Microstructure. Etch: Picral

Fig. 11.19. Built-up metal. 500 ×

Fig. 11.20. Base metal. 100 ×

11.2 **Spray Metallizing, Brazing**[5])

The described advantages of build-up welding, i.e. high heating with consequent danger of crack formation and hardening, may be avoided by spray coating of metals or alloys. This method may also lead to heating, oxidation, and melting of the coating metals, but not to high heating of the base metal. In this process the surface of the work piece is often rough machined in order to obtain better adherence of the coating. This decisively reduces impact and fatigue strength.

An example ist the fracture of an **engine shaft** that occurred at the seat of a V-belt pulley used for driving a centrifuge drum **(Fig. 11.21)**[6]). The fracture **(Fig. 11.22)** was strongly abraded, but could be identified by SEM observation as a fatigue type **(Fig. 11.23).** At the fracture edge parts of a surface layer had broken out. Evidently this was a deposit because beneath it turning grooves could be seen **(Fig. 11.24).** Probably the seat had been reconditioned earlier to the original dimensional tolerances after sustaining heavy wear. In the section it could be seen that a highly porous metal coating was present that was not attacked by picral, but dissolved quickly in alkaline potassium ferricyanide solution according to Murakami **(Fig. 11.25).** The hardness of the coating was 600 to 700 HV 0.1, and according to chemical analysis, it consisted principally of metallic molybdenum. The steel surface was very rough due to coarse turning. As can be seen from Fig. 11.22, several small flaky cracks were present in the interior of the shaft. Their presence was confirmed by an examination of the metallographic mount **(Fig. 11.26).** These flaky cracks may have accelerated propagation of the fracture, but were of no consequence for its genesis. Instead, the cause of the shaft fracture must be seen in the excessive surface roughness. If roughening is required for a good adherence of the coating, it should be done by sand blasting or shot peening. This makes the surface less notch sensitive and furthermore subjects it to compression stresses.

In section 7.6 an unsuccessful attempt was reported to patch a weld-induced hot crack by brazing (Fig. 7.45 and 7.46).

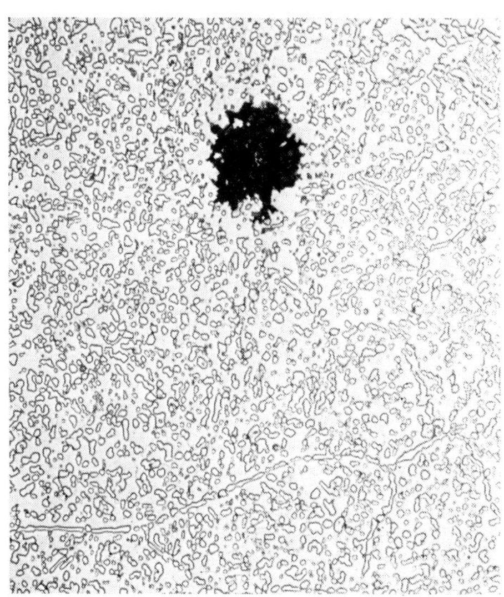

11.19

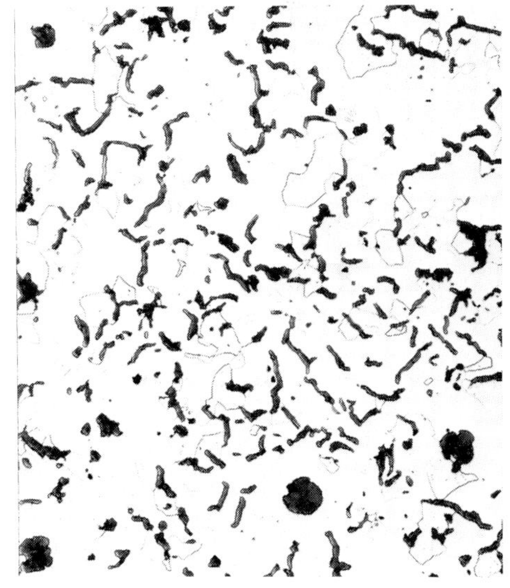

11.20

11.21

11.22

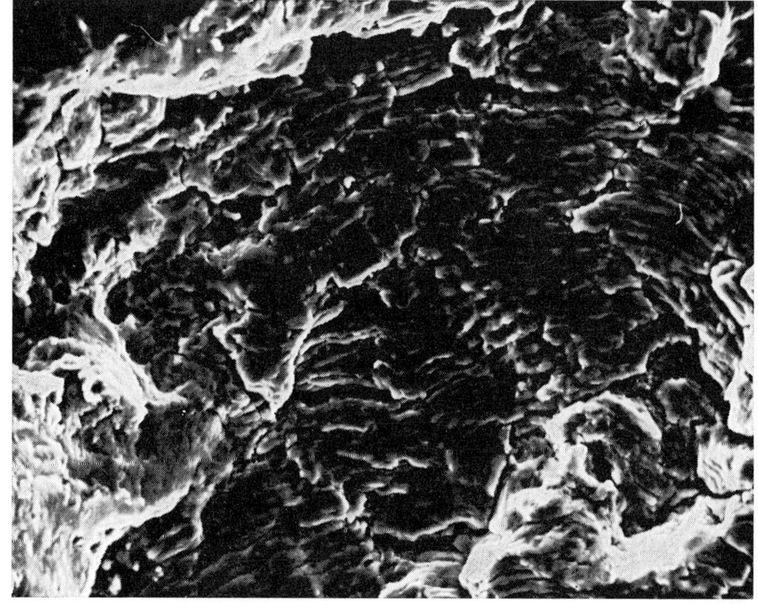

11.23

11.24

Fig. 11.21 to 11.26. Engine shaft broken because of rough turning grooves used to anchor spray coated metal layer

Fig. 11.21. Lateral view (V-belt pulley taken off). 0.8 ×. Bottom: Fracture plane with breakouts

Fig. 11.22. Fracture. 1 ×

Fig. 11.23. SEM of fracture. 600 ×

Fig. 11.24. Breakouts at fracture edge and turning grooves below spray coated metal layer. 5 ×

Fig. 11.25. Edge structure. Etch according to Murakami. 100 ×

Fig. 11.26. Transverse section next to fracture with flaky cracks. Etch according to Heyn. 1 ×. p 314

11.25

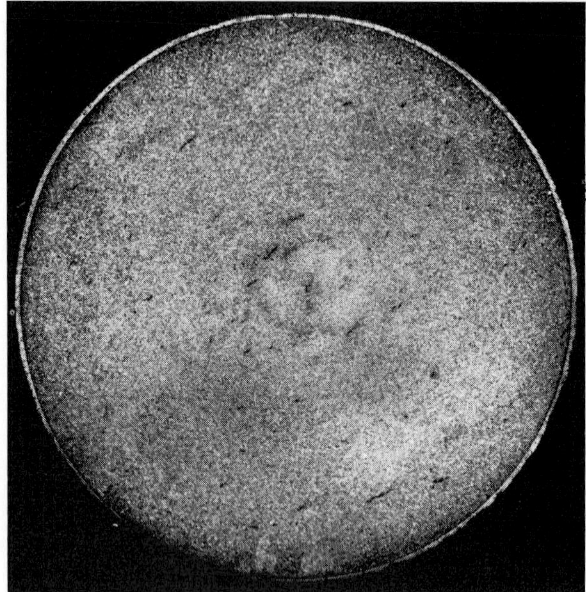

11.26

Literature Section 11

1) H. Schottky: Schmelzschweißung und Dauerbruch. Kruppsche Monatshefte 7 (1926) S. 213/16

2) G. Kühnelt: Der Einfluß einer Auftragschweißung auf die Dauerhaltbarkeit von Stahlwellen. Maschinenschaden 13 (1936) S. 57/64

3) F. K. Naumann u. F. Spies: Dauerbruch und Schweißung. Prakt. Metallographie 8 (1971) S. 551/59

4) Dieselben: Gebrochene Hinterradaufhängung. Prakt. Metallographie 9 (1972) S. 99/104

5) Chr. Kampmann u. K. Kirner: Metallographische Untersuchungen an Molybdänaufspritzungen. Prakt. Metallographie 9 (1972) S. 363/69

6) F. K. Naumann u. F. Spies: Gebrochene Motorwelle. Prakt. Metallographie 11 (1974) S. 293/97

7) **Nach einer Flickschweißung gebrochener Winkelhebel Dieselben: aus Gußeisen. Prakt. Metallographie 17 (1980) S. 35/38**

12. Failures in Iron Castings

In this section failures that are characteristic for cast iron will be described. Others that do not differ greatly in cause and appearance from steel defects have been or will be discussed in the appropriate chapters of this book.

Of special importance for cast iron, as well as for cast steel, is the design and construction of the casting with regard to formation of cavities. Defects in the solidified microstructure cannot be corrected because there is no further processing. The connection between strenght and chemical composition in cast iron is subject to specific laws because of the graphitic structure. Heat treatment has in part other purposes; processing errors lead to different failures than those in steel. Finally a particular form of corrosion failure characteristic for cast iron will be reported.

Internal cavities are formed by shrinkage and gas development, particularly in thick walled regions if the flow of iron is inhibited through narrowed cross sections that had solidified earlier. Their formation can be prevented by suitable design of molds, and especially by the correct placement of gates and risers. **Stresses** are caused by premature solidification and shrinkage in the thinner parts. They may lead to crack formation during cooling or later under service stresses. Therefore large differences in the casting wall thickness should preferably be avoided. If this is not possible, the cross sectional transition should at least be flattened or rounded off. A high **pouring rate** leads to large temperature differences within the casting and promotes the formation of cavities and stresses.

The **molds** should permit uninhibited shrinkage. After pouring, obstructions that prevent shrinkage, such as cores and core arbors, should be removed early.

Fig. 12.1. Pipe which occurred below the gate of a heavy cast iron chill used in a steel plant. 1 ×

The **cooling rate** is also of great importance after pouring, not only because stresses are raised by fast cooling, but also because supercooled solidification may occur, which in turn may lower the strength of the cast iron. In the presence of large cross sectional differences, the cooling rate and consequently the microstructure and strength within the casting may be very different. For this reason, too, large differences in wall thickness should preferably be avoided or care should be taken that cooling of thick-walled areas is accelerated.

Chemical composition of cast iron also has a substantial effect upon castability, solidification and supercooling capacity of transformations, and thus also upon microstructure and strength. Silicon facilitates solidification according to the stable system, i.e. the formation of grey cast iron. But manganese stabilizes the iron carbide; it increases the tendency toward cavities and shrinkage formation. Sulfur makes the iron viscous, promotes shrinkage and thus cooling stresses and crack formation. Phosphorus that appears in cast iron as a ternary eutectic with iron and carbon, steadite, increases fluidity and wear resistance; but in larger amounts it makes the

12.2

12.3

Fig. 12.2 and 12.3. Shrinkage cavities in a fan rotor that failed

Fig. 12.2. Fracture. 2 ×
Fig. 12.3. Section. 2 ×

12.4

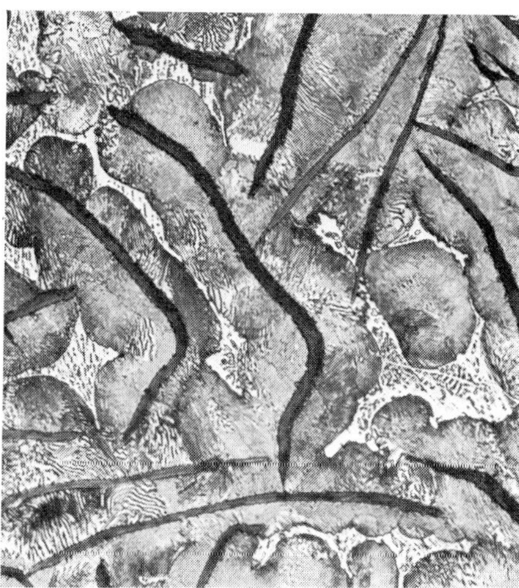

12.5

12.6

12.7

Fig. 12.4. to 12.7. Broken watermain of 700 mm diameter

Fig. 12.4 Defective site
Fig. 12.5. Microstructure of pipe. Etch: Picral. 200 ×
Fig. 12.6. Fracture with blow holes. 1 ×
Fig. 12.7. Bend specimens with blow holes. 1 ×

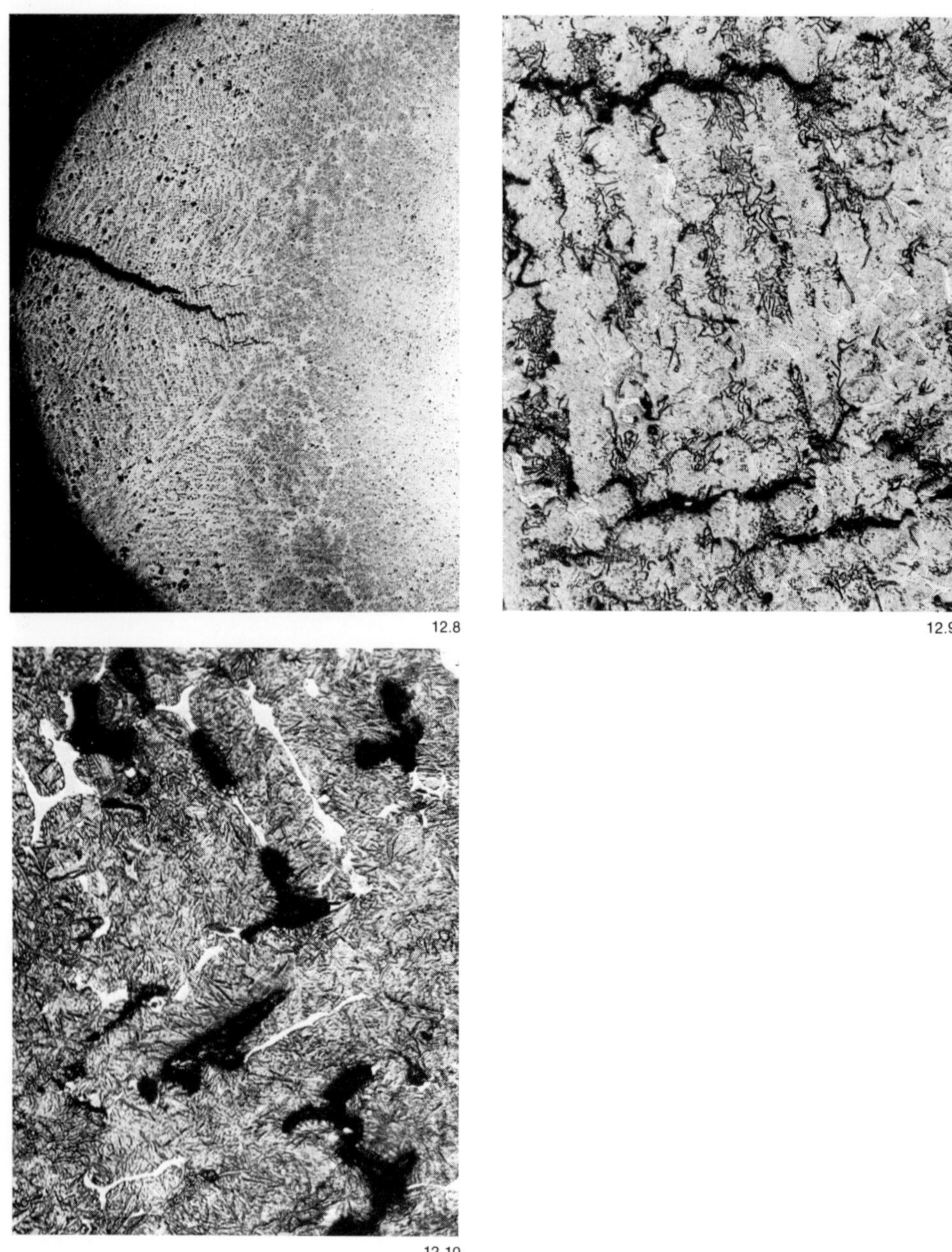

Fig. 12.8 to 12.10. Cam of grey cast iron camshaft that cracked during induction hardening. Micrographic section. Etch: Picral

Fig. 12.8 and 12.9. Crack region
Fig. 12.8. 7 ×

Fig. 12.9. 100 ×

Fig. 12.10. Microstructure with microcavities. 200 ×

iron sensitive to impact blows and abrupt temperature changes. Carbon content has a special effect upon the strength of the iron. It increases strength as in steel, provided that it is present as a component of pearlite; but it decreases strength due to a notch effect, if it is present in the form of lamellar graphite. Therefore the strength level of cast iron is affected by the amount of primary solid solution formed and decreases with increasing graphite content. In addition to the amount of carbon, the eutectic composition is also dependent upon silicon and phosphorus contents. The relation of carbon content to the true eutectic composition of the alloy is expressed by the carbon equivalent S_C that is approximated by the following equation:

$$S_C = \frac{C}{4.23 - 0.275 \cdot P - 0.312 \cdot Si}$$

All alloys with $S_C < 1$ are hypoeutectic and all with $S_C > 1$ are hypereutectic. The lower the carbon equivalent the higher is the fraction of primary solid solution, and also the strength of the cast iron.

An extensive presentation and discussion of casting defects is given in an atlas issued by the Verein Deutscher Giessereifachleute[1]).

W. Jähnig[2]) reported on the metallography of cast iron alloys and prepared a special section on defect phenomena of the individual alloys. Therefore this chapter can be brief and limited to a few special cases.

12.1 **Casting Defects: Shrinkage Cavities, Pipes, Blowholes**

Figure 12.1 shows a pipe that occurred under the gate of a heavy cast **iron chill** for steel making. The feeding of the mother liquor was interrupted when only the skeleton of the primary crystals had formed.

Figures 12.2 and 12.3 show in the fractures and metallographic sections, respectively, typical shrinkage cavities – judging from their location – in a **fan rotor** that burst at 1500 r.p.m.

A **waterline main** of cast iron socket pipes having a 700 mm diameter and 18 mm wall thickness broke some 50 to 60 years after it was put into service. The failed piece, 3200 mm in length and approximately one-third of the pipe circumference in width at the center, that broke out, can be seen in **Fig. 12.4.** As in two other previous breaks of the same type, the fracture had occurred at the downward located sector of the pipe. The pipe consisted of a hypereutectic pearlitic cast iron with high phosphorus content (S_C = 1.07). Its microstructure is illustrated in **Fig. 12.5.**

The following relatively low mean strength values resulted from five tensile tests:

Tensile strength MPa (ksi)	Bend strength MPa (ksi)	Hardness HB
135 (20)	2892 (420)	169

The strength should have been sufficient to withstand the internal pressure of 6 atm. But an examination of fracture and micrographic sections showed that the pipe wall was permeated by many small blow holes **(Fig. 12.6)**; this also was apparent during machining of tensile and bend specimens **(Fig. 12.7).** From previous experience pipes cast vertically are known to leak when pressurized water finds its way through the walls of such pipes, even though they had been

12.11

12.12

Fig. 12.11 to 12.19. Crankcase with shrinkage cracks

Fig. 12.11. to 12.13. First crankcase
Fig. 12.11. Fragments of front wall. 1 ×
Fig. 12.12. Fracture in front wall. Incipient fracture colored dark. 1 ×

Fig. 12.13. Microstructure. Etch: Picral. 200 ×

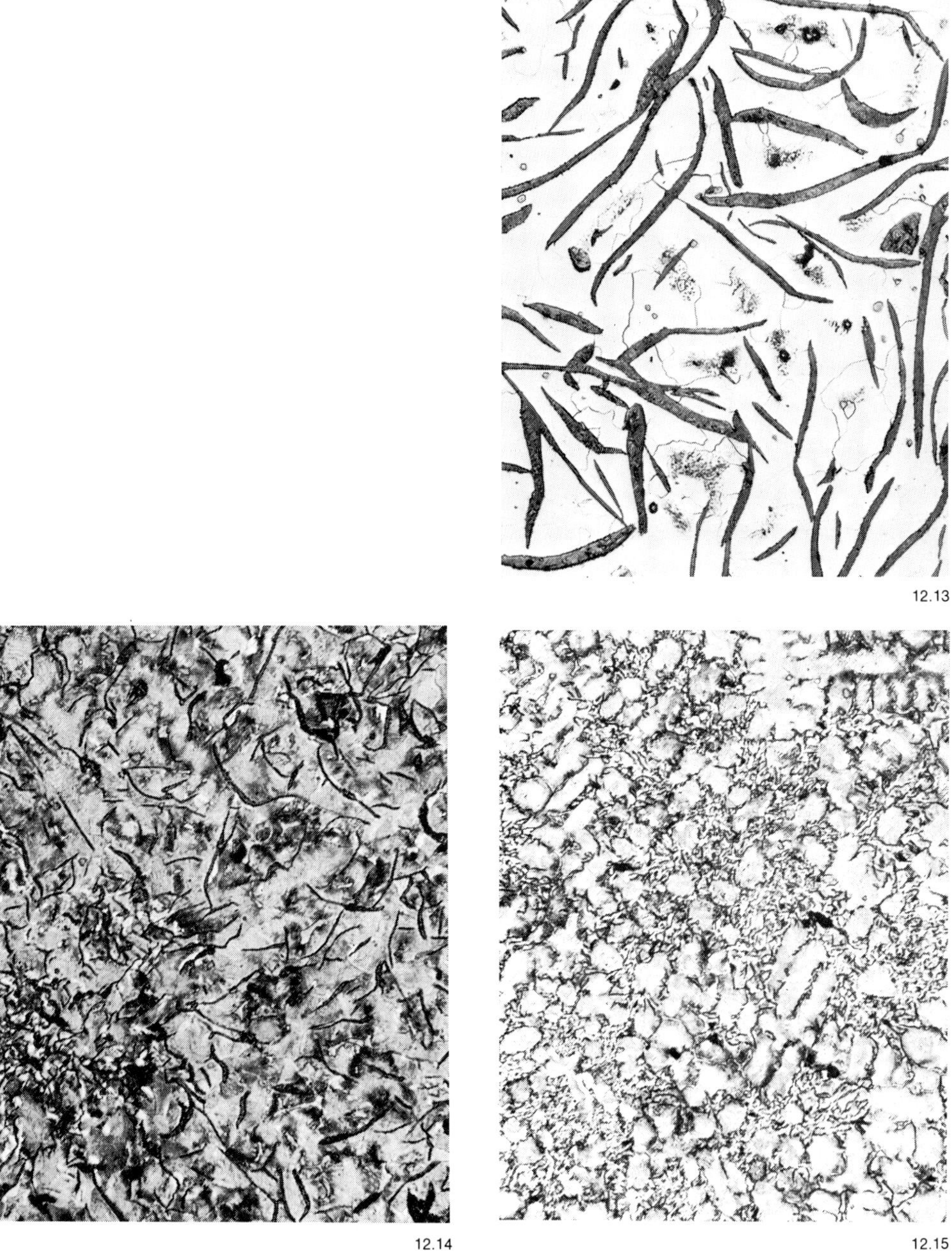

12.13

12.14

12.15

Fig. 12.14 and 12.15. Microstructure of second crankcase. Etch: Picral. 100 ×

Fig. 12.14. In thick-walled part

Fig. 12.15. In thin-walled part

12.16

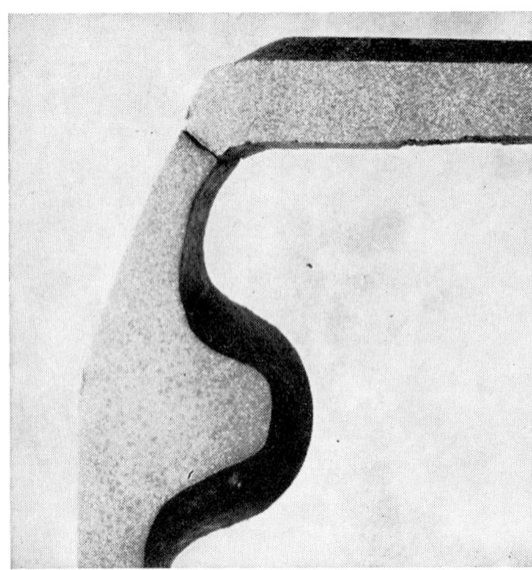

12.17

Fig. 12.16 to 12.19. Third crankcase

Fig. 12.16. View. 0.25 ×. Cracks at a and b
Fig. 12.17. Section through crack region a in Fig. 12.16. Etch: Picral. 1 ×

water pressure tested for tightness. Such holes enlarge quickly by erosion. It may be assumed that the same occurrence took place in the case at hand. This caused the pipe to be underwashed and rupture under the additional bend stress of its own weight, the load of the water, and of the soil. Nowadays these pipes are cast centrifugally, and therefore are completely tight and have higher strength levels.

A last example for shrinkage cavities is presented by the fracture of a grey cast iron **camshaft** during induction hardening of a cam. It could be seen in the metallographic section that the microstructure of the cam was very porous **(Fig. 12.8).** These microcavities were located in the dendritic interstices just like graphite and phosphide eutectic **(Fig. 12.9 and 12.10).** The subsequent quench crack followed these microcavity zones and was caused or facilitated by them.

12.2 **Cooling Defects: Shrinkage Stresses, Supercooling**

The crack of the cast iron **crankcase** shown in **Fig. 12.11** was primarily caused by shrinkage stresses[3]). According to the dark temper coloration on the fracture **(Fig. 12.12)**, it evidently originated in the thin front wall and then continued into the transition to the thick bosses. These bosses are subsequently used for drilling the bolt holes. The cracking, however, was favored by low strength characterized by an almost purely ferritic matrix **(Fig. 12.13).**

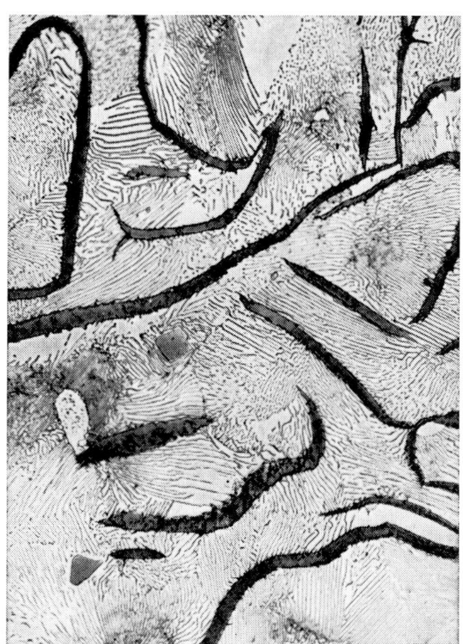

12.18

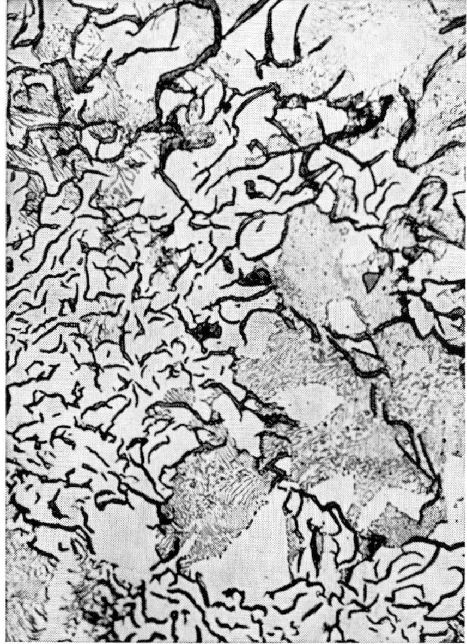

12.19

Fig. 12.18 and 12.19. Microstructure of third crankcase. Etch: Picral. 500 ×

Fig. 12.18. In thick-walled part
Fig. 12.19. In thin-walled part

In a second crankcase that had also cracked, the microstructure in the thick-walled part was formed normally **(Fig. 12.14)**, but in the thin-walled part it consisted of vermicular graphite in a ferritic matrix **(Fig. 12.15).** This vermicular graphite may be formed by fast cooling of the melt through supercooling of the solidification of the residual eutectic melt which becomes richer in carbon[4])[5]).

This type of structure, too imparts only low strength to the cast iron. In this case therefore the cracking was facilitated by low strength in the thin-walled parts.

A third crankcase **(Fig. 12.16)** showed two cracks, of which one (a) was located in an thin-walled area of the front end and the other (b) in the thinnest place between two cylinder bores which was only 6 mm thick in the raw casting. **Figure 12.17** shows the crack a in a transverse section. A comparison of the microstructure in the thick-walled and thin-walled parts of the front face shows again that the first mentioned was formed normally **(Fig. 12.18),** while the latter consisted predominantly of vermicular graphite in a ferritic matrix **(Fig 12.19).** In this case, too, the crack formation was induced by a combination of shrinkage and service stresses. This was unquestionably favored by the anomalous microstructure in the thin cross sections caused by supercooling.

The same holds true for the following failure cases that occurred during service of tractor engines with a cantilevered body. The side wall of the **cylinder head blocks**[6]) of these engines tore open in several instances **(Fig 12.20).** The crack was always located near the center of the side wall and propagated horizontally, namely parallel to a cast-on internal reinforcement rib. This can be seen even more clearly in the larger magnification of the section shown in **Fig 12.21.** This again is an area where high shrinkage stresses are to be expected. These already appear to have caused a constriction of the wall cross section in the plastic state during cooling of the casting. According to an analysis of three blocks, the carbon equivalent was 0.95 to 0.96. Therefore the chemical composition was satisfactory. Nevertheless, the microstructure in the thin wall showed again the well known supercooling phenomena **(Fig. 12.24).** Accordingly, cooling had taken place very fast. Considerable improvement could be obtained by a reinforcement of the case wall as shown in **Fig. 12.23,** but the formation of the microstructure was still not entirely satisfactory **(Fig. 12.25).**

12.3 **Unsuitable Chemical Composition**

In some other engines, cast iron **bearing caps**[7]) broke in numerous cases after a short service period. The fractures always originated in a cast-in groove and ran approximately radially to the shaft axis **(Fig. 12.26).** This point had not only the smallest cross section, but also had a stress peak that should have been expected during loading due to the notch effect of the groove which had a rounded off radius of only 2 mm. Casting stresses should not have played any decisive role in this case because the pieces had a simple shape and no large cross sectional variations.

Micrographic sections through the notch base parallel to the fracture showed ferrite clusters with vermicular graphite under the surface **(Fig. 12.27 to 12.29).** The core structure was improved but not free of ferrite **(Fig. 12.30).** The same phenomena were observed in four caps. The inherent low strength of the supercooled ferritic microstructure is assumed to have accelerated the fracture of the caps in these highly stressed areas. A comparison investigation of five other bearing caps that had given satisfactory service over longer operating periods confirmed this assumption. These caps showed an edge and core structure low in ferrite **(Fig. 12.31 and 12.32).** Chemical analyses of the broken and satisfactory bearing caps resulted in the following mean values:

Bearing caps	C %	Graphite %	Si %	Mn %	P %	S %	S_C
Fractured	3.55	3.05	2.74	0.65	0.109	0.064	1.06
Not fractured	3.25	2.73	2.62	0.66	0.312	0.124	0.97

Carbon, graphite, and silicon contents were significantly higher in the broken caps than for the satisfactory caps, while phosphorus and sulfur contents were lower. According to the carbon equivalent, all fractured caps had a hypereutectic composition, while all satisfactory caps showed a hypoeutectic composition. Based on the relation $\sigma_B = 100 - 80\ S_C$ postulated by P. A. Heller and H. Jungbluth[8]), the fractured caps should have a tensile strength of 150 MPa (22 ksi) and the satisfactory ones a strength of 220 MPa (32 ksi). This is expressed in the Brinell hardness as well, which resulted in a mean value of 164 HB for the broken caps, and 204 HB for the satisfactory ones.

Accordingly, the fracture of the bearing caps had been caused primarily by an unsuitable chemical composition. Rupturing was facilitated by the unnecessary weakening of the cross section at the groove and by insufficient rounding off of its bottom. In order to avoid such damage in the future the following precautions were successfully employed:

1. Analysis specifications were changed so that a hypoeutectic cast iron was obtained reliably.
2. The grooves were eliminated and the transition to the bent part was well rounded off.
3. Cooling of the pieces after casting was slowed.
4. In order to prevent supercooling, solidification was nucleated by inocculation with finely powdered ferrosilicon in accordance with the approach developed by P. Bardenheuer[9]).

A further example of unsuitable composition is presented by the failure of a **rotor of a turbocompressor.** The rotor consisted of a cylindrical cast iron body of 300 to 350 mm diameter into which slits were milled to a depth of 65 mm. The remaining ribs had a thickness of 33 mm on the outside, but only 12 mm on the inside. They had broken off in the small inside cross section. The cross sectional variation was sharply angular. Three ribs were examined. They had a dark grey coarse-grained fracture **(Fig. 12.33).** Analysis of one of the ribs showed the following composition:

C %	Graphite %	Si %	Mn %	P %	S %	S_C
3.67	3.14	2.30	0.46	0.324	0.077	1.07

Strong graphite precipitation had to be expected due to high carbon and silicon contents, so that only approximately 0.5 % C remained for pearlite formation. The solidification of the hypereutectic alloy had started with the graphite precipitation, and the graphite flakes could grow to unusual size due to the slow solidification of the heavy casting. In the macrosection **(Fig. 12.34)** they could be seen with the naked eye and had partially broken out. **Figure 12.35** shows the microstructure. The coarse graphite lamellae were surrounded by ferritic precipitates. Such microstructure could not be expected to have good strength properties. As a consequence, Brinell hardness was only 114 HB, whereas 180 to 200 HB are required in highly stressed cast iron parts. According to P. A. Heller and H. Jungbluth[8]), the following strength values may be expected as a function of rod cross section for highly stressed parts of a hypereutectic cast iron poured into round molds, as compared to a hypoeutectic alloy with $S_C = 0.9$:

12.20

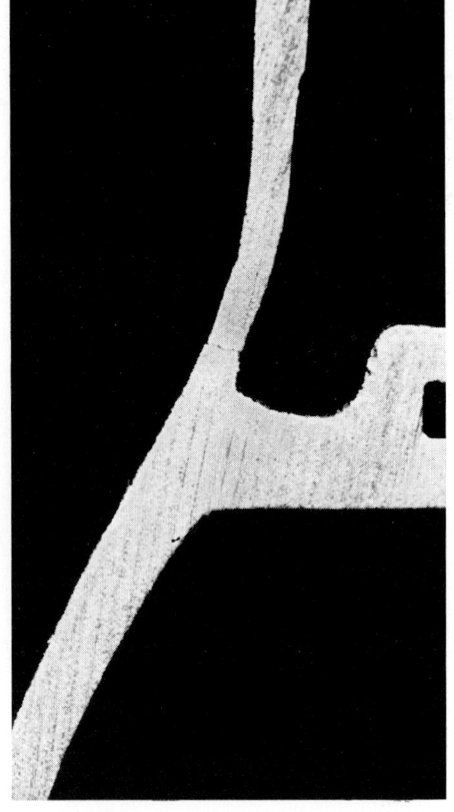

12.21

12.22

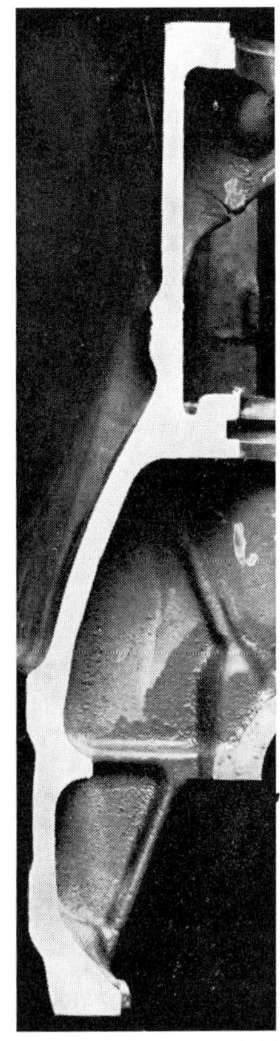

12.23

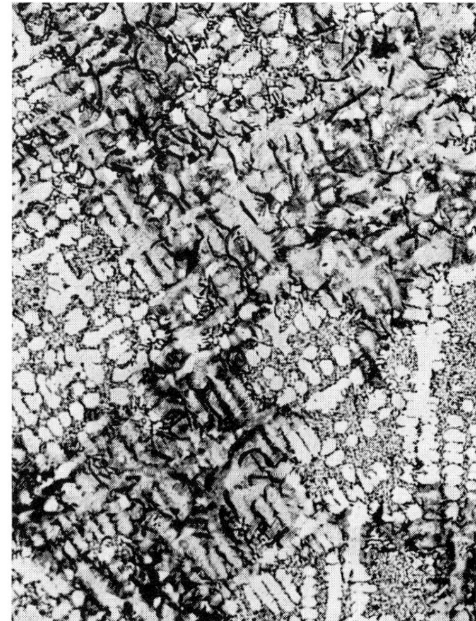

12.24

12.25

Fig. 12.20 to 12.25 Cracked cylinder head blocks of cast iron

Fig. 12.20. Internal view with horizontal crack (arrow) 0.2 ×
Fig. 12.21. Transverse section through crack. 1 ×

Fig. 12.22. and 12.23. Section through side wall. 0.25 ×

Fig. 12.22. Old design with crack
Fig. 12.23. New design, unused block

Fig. 12.24 and 12.25. Microstructure in thin part of side wall. Etch: Picral. 100 ×

Fig. 12.24. Block of old design
Fig. 12.25. Block of new design

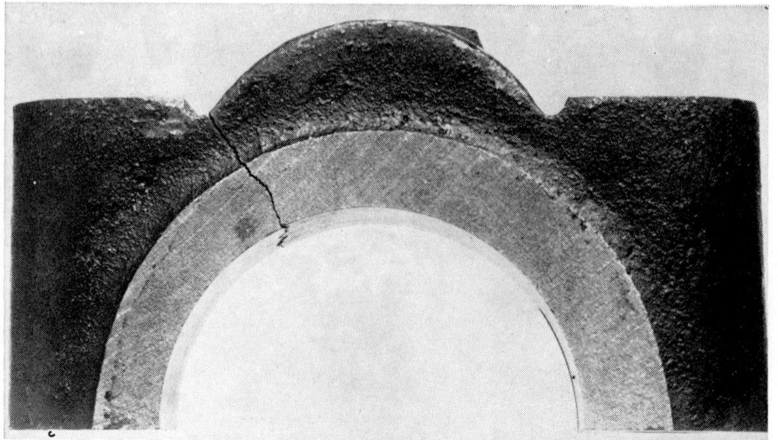

12.26

12.27

Fig. 12.26 to 12.32. Cracked bearing caps

Fig. 12.26. Side view of a cracked bearing cap. 1 ×
Fig. 12.27. Section through groove parallel to fracture. Etch: Picral. 3 ×

Fig. 12.28 and 12.29. Microstructure in edge zone. Etch: Picral

Fig. 12.28. 100 ×
Fig. 12.29. 500 ×

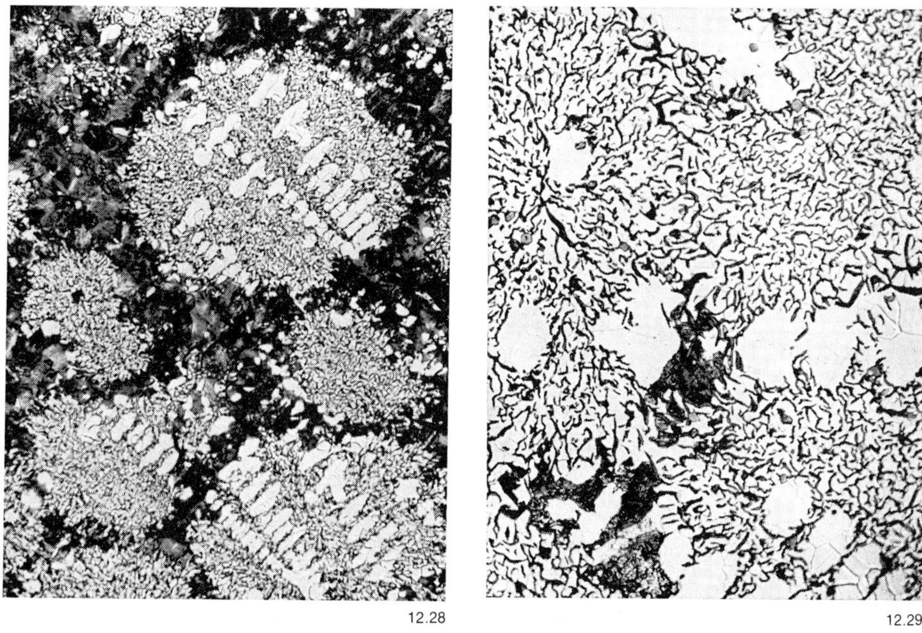

12.28

12.29

12.31

Fig. 12.31. Section through groove of cap of hypoeutectic cast iron Etch: Picral. 3 ×

Fig. 12.30 and 12.32. Core structure of a broken (left) and a satisfactory bearing cap. Etch: Picral. 200 ×

Fig. 12.30. Cap of hypereutectic cast iron

Fig. 12.32. Cap of hypoeutectic cast iron

p 330

Rod diameter mm	Tensile strength $S_C = 1.07$ Mpa (ksi)	Tensile strength $S_C = 0.9$ Mpa (ksi)
10	314 (45.5)	412 (59.7)
20	196 (28.4)	314 (45.5)
30	137 (19.9)	275 (39.9)
60	59 (8.55)	216 (31.3)
90	39 (5.65)	182 (26.4)

Therefore the strength of the fractured rotor must have been very low, but could have been increased substantially by the use of a hypoeutectic cast iron. A rounding off of the transition in the base of the slit could also extend service life.

12.4 **Heat Treating Errors**

Casting stresses may be prevented by subsequent **annealing.** The annealing temperature should however not exceed 550 to 600 °C depending on annealing time, as otherwise the cementite decomposes, which results in expansion. This heat treating error can be recognized metallographically by the disappearance of pearlite and the formation of temper carbon nodules at the graphite lamellae. During very slow cooling of the casting, this disappearance may have already

12.30

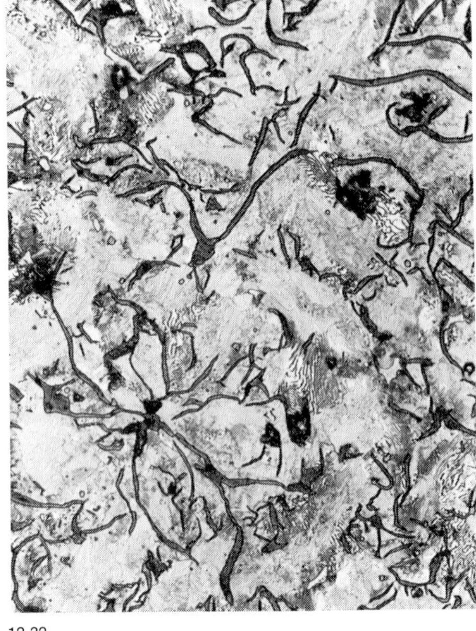

12.32

12.34

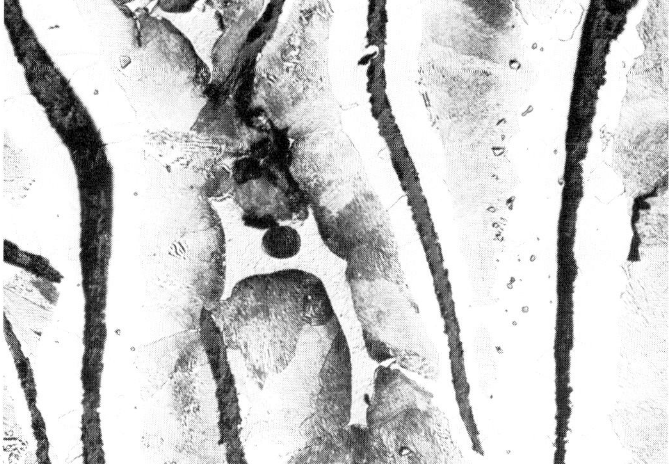

12.35

12.33

Fig. 12.33 to 12.35. Broken-out rib of cast iron turbocompressor

Fig. 12.33. Fracture of rib. 0.85 ×
Fig. 12.34. Unetched section transversely to longitudinal axis. 1 ×
Fig. 12.35. Microstructure in rib. Etch: Picral. 100 ×

taken place upon completion of solidification. If cast iron parts are exposed for extended periods to oxidizing atmosphere at elevated temperatures, oxidation of the silicon-containing ferrite along the graphite lamellae may also occur; this penetrates deeper for the coarser flaked graphite. This phenomenon, too, is connected with an increase in volume, the **"growth"** of the cast iron. It also can be easily confirmed metallographically. Both occurrences may cause stresses that lead to distortions or cracks. In the following, some examples of this are cited.

A **spinning plant** had some thin-walled cast iron parts annealed for stress relief in order to avoid distortion after machining. First, after eight hours of preheating the parts were annealed for eight hours at 400 °C and then cooled for fifty hours. Lateron the temperature was raised to 520 °C for the same time cycle. This caused considerable distortion of the parts so that they became unusable. The temperature was assumed to have exceeded the specified value. An investigation showed that this was indeed the case. While the part annealed at 400 °C had a purely pearlitic matrix structure **(Fig. 12.36),** much of the cementite in the pearlite of the piece annealed at the higher temperature had coalesced and decomposed; the carbon hat precipitated at the graphite lamellae **(Fig. 12.37).** Analysis showed **(Table 1)** that the combined carbon con-

Table 1. Free and combined carbon contents of two annealed castings

Annealing temperature °C	Total C %	Graphite %	Combined C %
400 °C	3.48	2.81	0.67
520 °C (?)	3.38	3.12	0.26

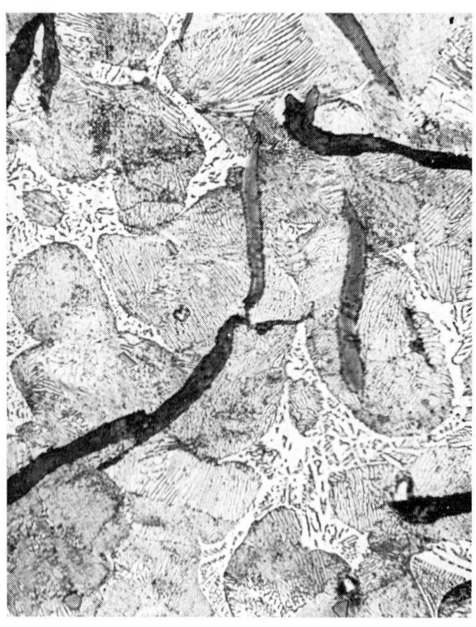

12.36

12.37

Fig. 12.36 and 12.37. Microstructure of cast iron machinery parts. Etch: Nital. 500 ×

Fig. 12.36. Correctly annealed

Fig. 12.37. Annealed at too high a temperature and heavily distorted

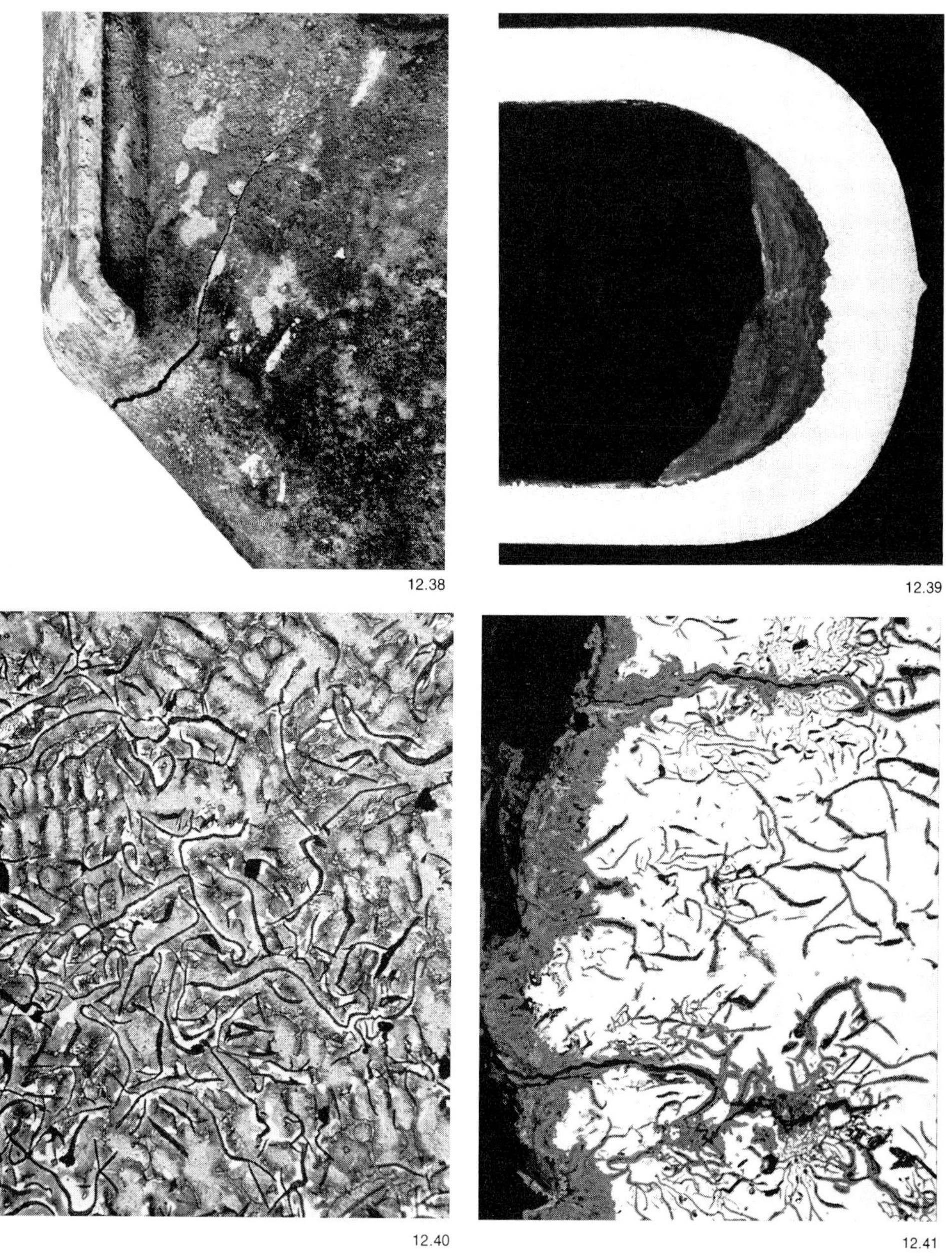

Fig. 12.38 to 12.41. Broken member of a central heating plant boiler

Fig. 12.38. Crack region. 0.5 ×

Fig. 12.39. Internal deposit of boiler scale. Unetched section parallel to crack. 1 ×

Fig. 12.40. Core structure. Etch: Picral. 100 ×

Fig. 12.41. Structure at water side, unetched. 100 ×

tent was substantially lower in the part annealed at higher temperature as opposed to the one annealed at 400 °C. Hardness had dropped correspondingly from 198 to 112 HB. The microstructure and hardness of piece 1 may have corresponded to the cast condition. But such a major change as that occurring in piece 2 could not have occurred only by annealing at 520 °C. After some annealing tests comparable microstructures were found in cast iron of a similar composition only after long term tempering above 600 °C.

12.5 **Service Errors**

Conditions that lead to the growth of cast iron parts also occur in service at steam and boiler installations. A **boiler section** of the central steam plant of a housing complex had to be dismantled after only 14 months due to the formation of a crack. This was located in the vicinity of the elbow at which the firebox of the boiler tapers to the top **(Fig. 12.38).** This is the region of highest thermal stress. At the interior of the section, an unusually thick layer of scale had been deposited considering the short time of service **(Fig. 12.39).** Analysis confirmed that this was indeed boiler scale and not remnants of the core. Such a thick thermally insulating layer hinders heat transfer and therefore leads to overheating. The ferrite in the structure **(Fig. 12.40)** had oxidized along the graphite lamellae at the side of the water **(Fig. 12.41),** a process that also takes place under volume expansion. The stresses enhanced by the growth had already lead to the formation of additional fine cracks. Fatigue stresses due to constant temperature changes may also have had a contributory effect. It was not clear, however, how thick a layer of scale could have formed in such a short time. Probably the feed water was renewed very often.

The guide shoulder of a **hot steam valve** that was supposed to consist of pearlitic cast iron was distorted so badly after a service period of only 3 months that the part had to be changed. The shoulder served as guide for a ground-in stainless steel stem and as steam chamber seal. The lower, thicker part was exposed to superheated steam of 425 to 450 °C at 20 atm, while the upper part, not submitted, was air cooled externally. **Figure 12.42** shows the lower part seen from the side. The distortion can be noticed by the outward bulging of the lower edge and by the elliptical deformation of the two originally circular transverse bore holes. In the longitudinal section of this spot, a loosening of the structure could already be seen in the finish-ground condition **(Fig. 12.43).** The microstructure in the non-deformed part consisted mostly of coarse-flaked graphite with nodules crystallized on its surface, and a matrix of ferrite and pearlite containing coalesced cementite lamellae **(Fig. 12.44).** The ferrite at the edge of the graphite lamellae was strongly oxidized at the bore for the stem **(Fig. 12.45).** In the distorted part between stem bore and cross bore, oxidation had progressed to the extent that little remained of the metallic matrix **(Fig. 12.46).** The distortion of the casting therefore was caused by internal oxidation and the subsequent growth. Furthermore, it could be determined from the composition of the casting (3.42 % C, 2.08 % Si) that the iron did not have a pearlitic microstructure to begin with.

Fig. 12.42 to 12.46. Guide shoulder of a hot steam valve distorted by oxidation (growth)

Fig. 12.42. Distorted lower part. Approx. 1 ×
Fig. 12.43. Longitudinal section through lower part, finish-ground. Approx. 1 ×

Fig. 12.44 and 12.45 Microstructure of colder upper part. Etch: Nital. 200 ×

Fig. 12.44. In core

Fig. 12.45. At stem bore

Fig. 12.46. Microstructure in distorted part between stem bore and cross bore. Etch: Nital. 200 ×

12.42

12.43

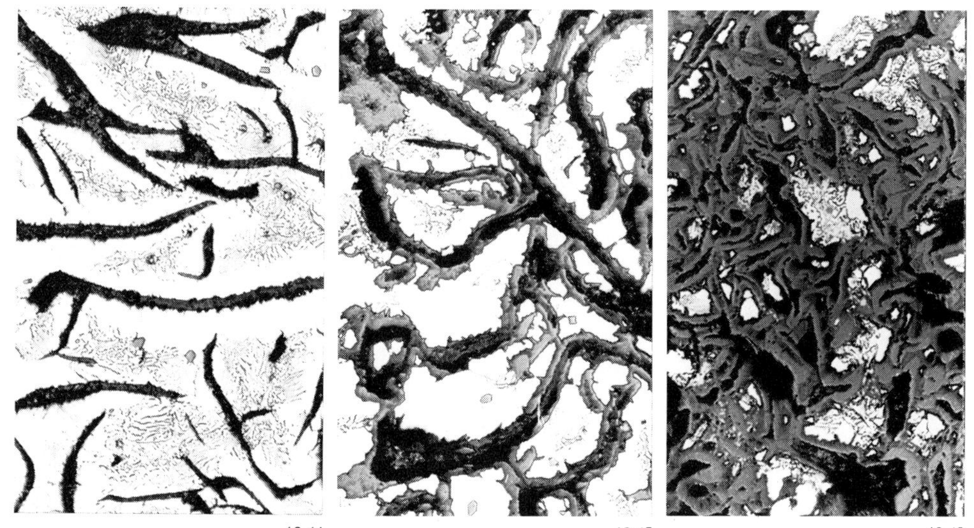

12.44 12.45 12.46

A **piston ring** with a hard chromium plated and ground bearing surface showed small protrusions at the surface after two years' service under absence of any substantial mechanical stresses; operating in a hot steam valve, it had come in contact only with saturated or superheated steam up to 520 °C. Whereas the chrome plated bearing surface remained shiny and had acquired only yellow to blue temper colors, the internal surface and front faces had scaled. **Figure 12.47** shows the unetched section through a protrusion. The unusually thick but probably not quite dense chromium plating – like most electrodeposits – was penetrated here by a corrosive agent. The corrosion, intensified by the electrochemical potential between iron and chromium, had bulged and torn the chromium layer due to the increase in volume associated with the interaction of the elements. Corrosion took place preferentially along the graphite flakes. It may have been caused by steam dissociation or by water.

In this connection a type of corrosion should be described that occurs in cast iron if it is exposed to the effect of salt solutions or weak acids, for instance, in acidic soil or in hydrogen sulfide-containing waters[10])[11]). It is called **"spongiosis"*** or graphitic corrosion and acts selectively in such a way that ferrite and cementite are dissolved successively leaving the phosphide eutectic and graphite behind. Without changing the external shape, this lowers strength to such an extent that the material may be cut with a pocket knife whereby a black powder is produced.

Pipe bursts in a cast iron **watermain,** that was laid in the year 1876, were caused by such graphitic corrosion. Even a freshly made fracture showed a deep corrosion layer which appeared dark **(Fig. 12.48).** It could be seen even more distinctly in the metallographic section **(Fig. 12.49).** The depth of the layer differed considerably. Next to almost unattacked areas there were some where the corrosion had destroyed almost the entire wall thickness. In these places the wall could be pierced with a dull crowbar. **Figure 12.50** illustrates the ferritic core structure. In the corroded surface region, the attack took place first along the graphite veins **(Fig. 12.51)** and then expanded across the entire pearlitic matrix **(Fig. 12.52).**

Additional mechanically or chemically caused service failures, that are not specifically characteristic of cast iron, are described in sections 13 to 15.

---

*) see Appendix I

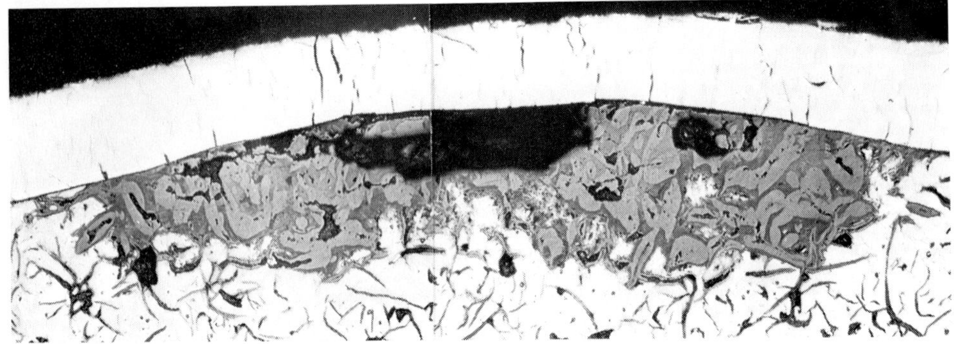

Fig. 12.47. Corrosion under surface of hard chromium plated piston ring of a hot steam valve. Unetched transverse section. 140 ×

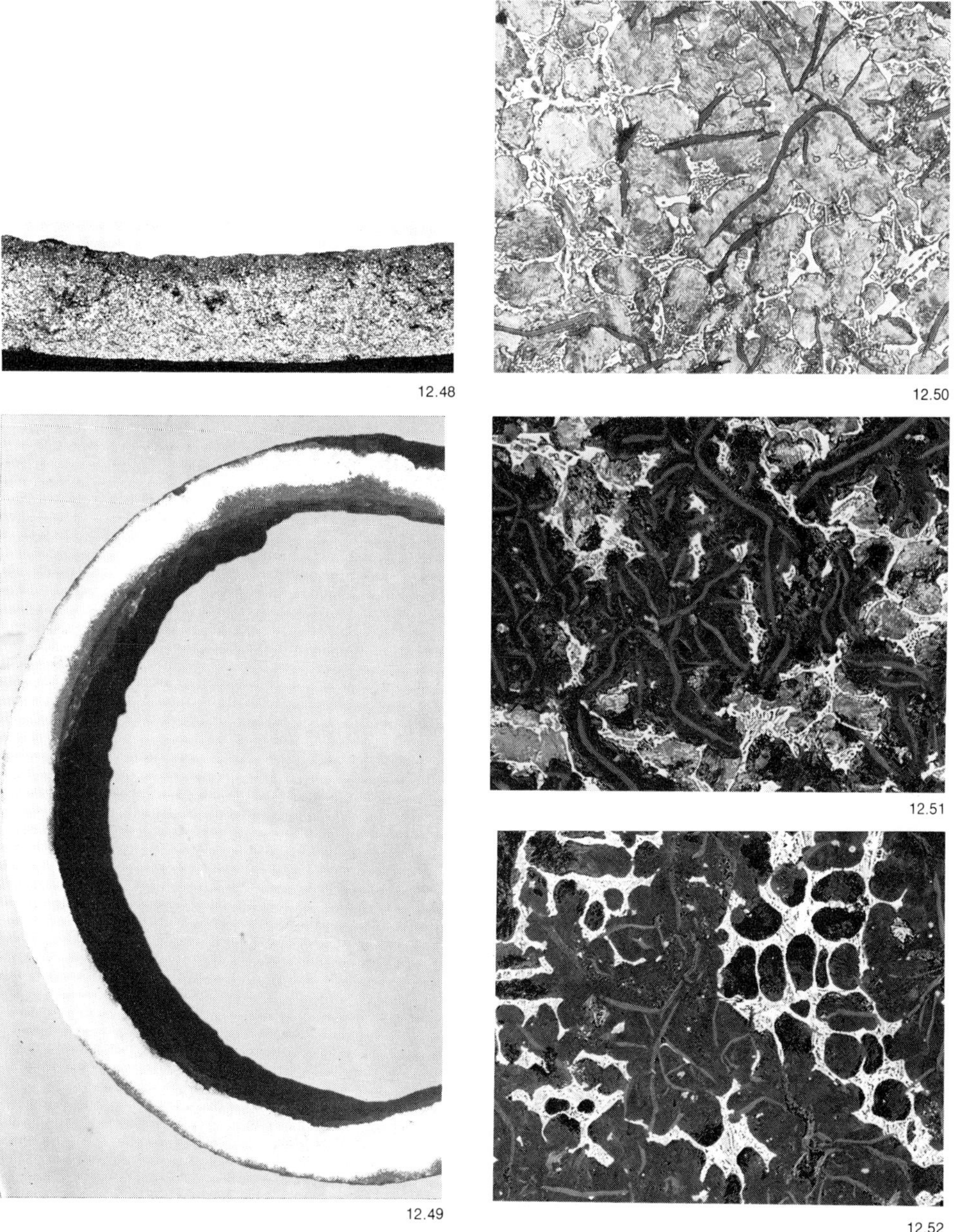

Fig. 12.48 to 12.52. Graphitic corrosion in a cast iron watermain pipe

Fig. 12.48. Fracture with dark corrosion layer. 1 ×
Fig. 12.49. Unetched transverse section. 1 ×

Fig. 12.50 to 12.52. Microstructure of pipe. 200 ×

Fig. 12.50. In core. Etch: Picral
Fig. 12.51. In transition zone. Etch: Picral
Fig. 12.52. In corrosion zone. Unetched.

Literature Section 12

1) Gußfehleratlas. Zweite völlig neu bearbeitete Auflage für die Kommission Metallurgie und Gießeigenschaften im Internationalen Komitee Gießereitechn. Vereinigungen unter Mitwirkung namhafter Fachleute bearbeitet von H. Reuter und Ph. Schneider. Gießerei-Verlag Düsseldorf 1971

2) W. Jähnich: Metallographie der Gußlegierungen. Herausgegeben vom VEB GISAG Kombinat für Gießereiausrüstungen und Gußerzeugnisse – Forschungszentrum – Leipzig. VEB Deutscher Verlag für Grundstoffindustrie. Leipzig 1971

3) F. K. Naumann u. F. Spies: Gerissene Kurbelgehäuse aus Gußeisen. Prakt. Metallographie 10 (1973) S. 290/97

4) P. Bardenheuer u. A. Reinhardt: Einfluß der Schmelzbehandlung durch eisenoxydulreiche und saure eisenoxydularme Schlacken auf die Kristallisation und die mechanischen Eigenschaften von grauem Gußeisen. Mitt. Kais.-Wilh.-Inst. Eisenforsch. 16 (1934) Abh. 250, S. 65/75

5) L. Sofroni, J. Riposan u. J. Chira: Some considerations on the crystallisation features of cast irons with intermediary shaped graphite (vermicular type). The metallurgy of cast iron Proceedings of the second international symposium on the metallurgy of cast iron, Geneva, 29.–31. Mai 1974, S. 177/95

6) F. K. Naumann u. F. Spies: Risse in Zylinderblöcken und in einem Zylinderkopf aus Gußeisen. Prakt. Metallographie 10 (1973) 159/65

7) Dieselben: Gerissene Lagerkappen aus Gußeisen. Prakt. Metallographie 10 (1973) S. 151/58

8) P. A. Heller u. H. Jungbluth: Die chemische Zusammensetzung des grauen Gußeisens und seine Zugfestigkeit. Gießerei 42 (1955) S. 255/57

9) Lehrgang: Das Impfen von Gußeisenschmelzen. Schulungsdienst des Vereins Deutscher Gießereifachleute.

10) K. G. Schmitt-Thomas u. G. Fenzl: Über Erscheinungsform und Bildungsbedingungen der Spongiose (Graphitierung) an grauem Gußeisen. Maschinenschaden 38 (1965) S. 516

11) F. K. Naumann u. F. Spies: Korrosion an Grauguß. Spongiose. Prakt. Metallographie 4 (1967) S. 367/70

13.-14.-15. Service Overload Failures

Service failures may occur due to mechanical, thermal, or chemical overloading. The latter is caused by the effect of hot oxidizing or decarburizing gases or through corrosion in electrolytes.

13. Mechanical Overload Failures

In failures of this type, a differentiation has to be made between those that are caused by excessive tensile, bend, or torsion stresses, those that occur as a consequence of static or dynamic compression or shear stresses, and those that are caused by friction loads.

Overloading may occur either if stress is too high or material strength too low. Generally the necessary purity and specified strength can be easily established for the procured material. However, deciding whether the specified strength will suffice to enable the material to withstand the normally expected loads is more difficult. Stress conditions are often hard to estimate for the materials specialist, especially in cases where complicated multi-axial stress patterns are present. Typically, determination of the magnitude and direction of the anticipated stresses is difficult, but the actual service stresses can almost never be accurately estimated. Therefore a decision of whether the failure was due to design or operational error, is often not an easy one to make for the materials testing engineer and sometimes cannot be made at all. Valuable hints for such cases are given in a handbook issued by the German insurance company Allianz Versicherungs A.G. which summarizes statistically long term experiences for numerous cases investigated by their engineers and materials specialists for the purpose of prevention of failures[1]).

Designers, plant engineers and materials testing engineers should cooperate as far as possible in handling such failure cases.

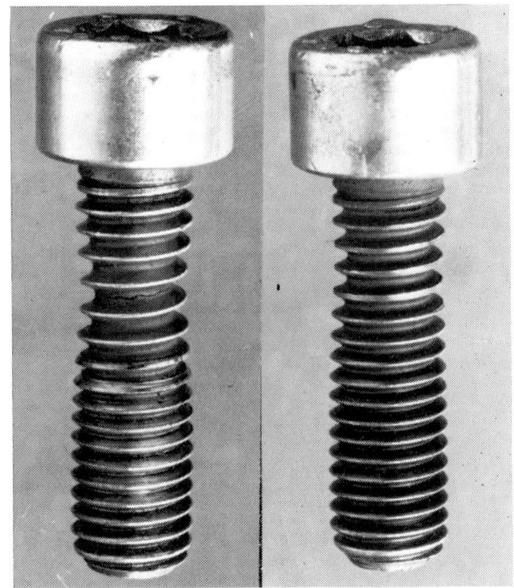

Fig. 13.1. Bolts of 930 MPa (135 ksi) strength overstretched due to overtightening. 1.5 ×

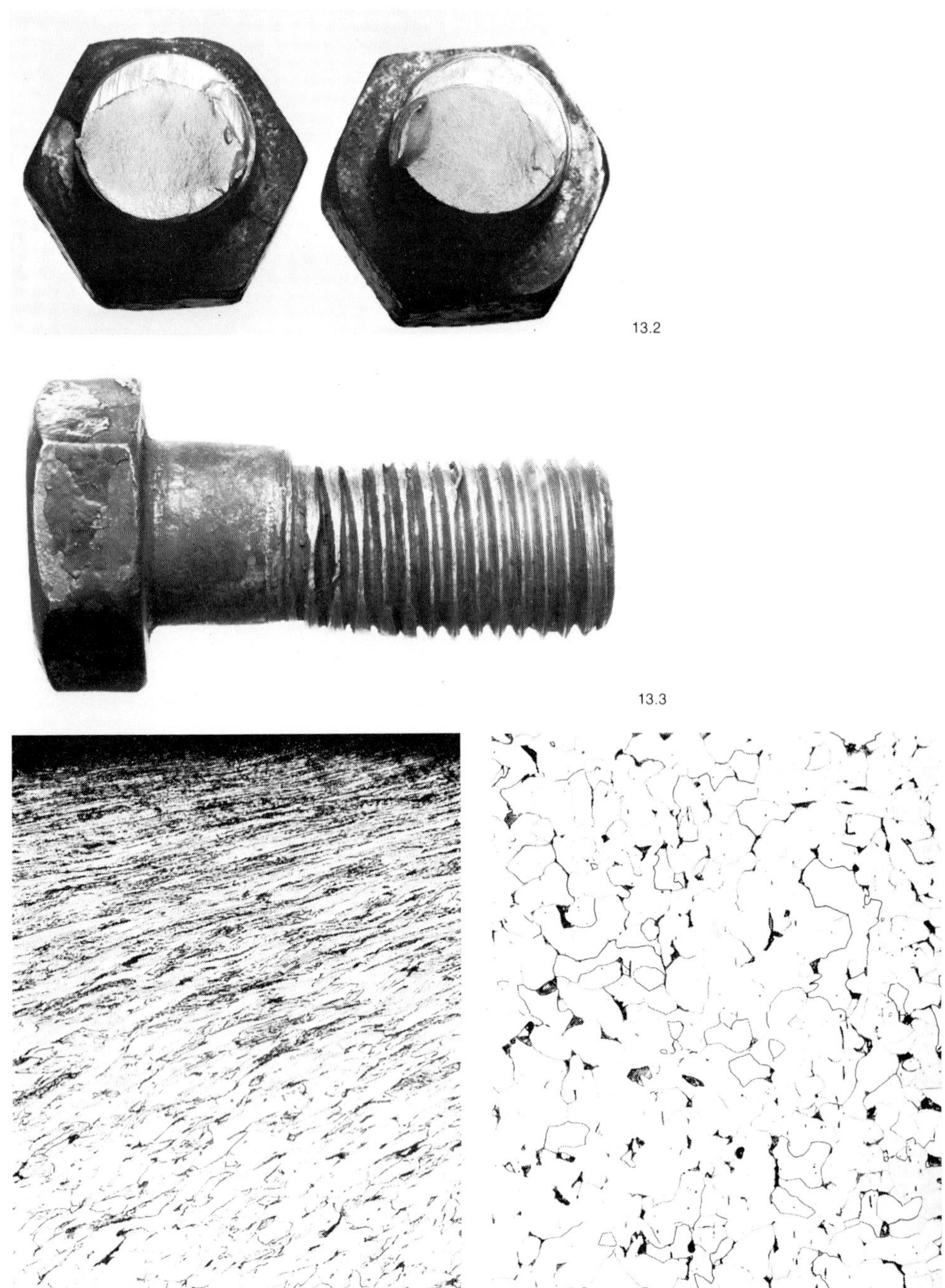

13.2

13.3

13.4

13.5

13.1 **Tensile, Bend, Tangential, and Torsion Stresses**

Bolts according to **Fig. 13.1** may serve as examples for overloading in tension. They consisted of defect-free steel and had a perfect heat treated microstructure, a fact that could be deduced from extensive elongation and reduction in area values under the strongly multi-axial stress caused by the notch effect of the thread. Their tensile strength was 930 MPa (135 ksi) and therefore they corresponded to the German specifications for this type of fastener*. The over-stretching of the bolts most probably was caused by uncontrolled tightening of the nuts.

Bolts of a revolving tower crane, whose fractures are shown in **Fig. 13.2,** broke in service due to excessive shear stresses. The fractures were of the deformation type which could be confirmed by a longitudinal section through the fracture plane **(Fig. 13.4).** Another bolt was only deformed by shear across the first thread **(Fig. 13.3).** i.e. where the others had broken. Therefore the steel had good deformability. It was fine-grained **(Fig. 13.5),** free of defects, and had a tensile strength of 390 to 490 MPa (56 -71 ksi).

* DIN 267, class 8 G

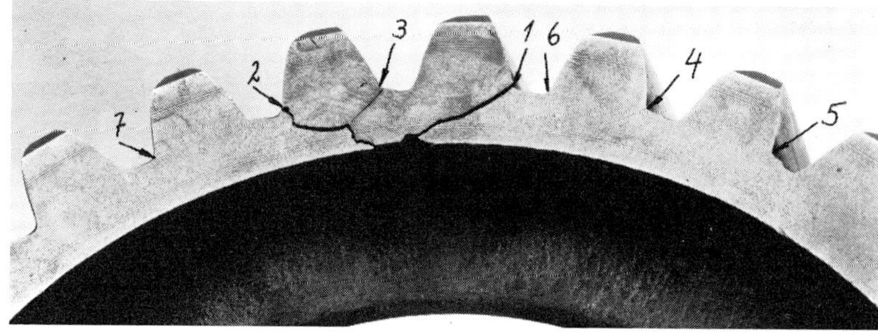

13.6

13.7

Fig. 13.6 and 13.7. Broken gear of a gear shift transmission

Fig. 13.6. Side view. 1 ×. 1 and 2 = fatigue fractures, 3 to 7 = incipient fatigue cracks
Fig. 13.7. View of fracture. 1 ×

Fig. 13.2 to 13.5. Sheared bolts of revolving tower crane

Fig. 13.2. Fractures. 1 ×
Fig. 13.3. Bolt deformed by shear stress. 1 ×
Fig. 13.4. and 13.5. Microstructure of a sheared bolt. Longitudinal section. Etch: Nital. 100 ×
Fig. 13.4. Below fracture
Fig. 13.5. In head

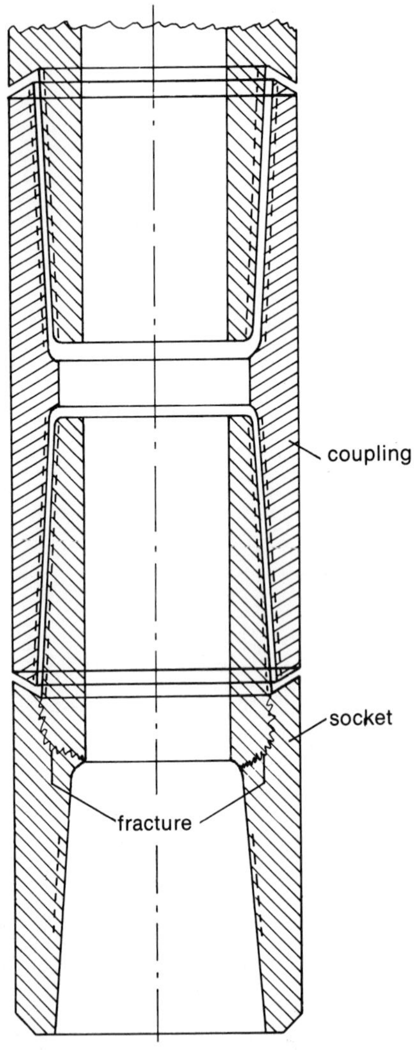

13.8

Fig. 13.8 to 13.11. Broken coupling of a pipe rod structure used in a crude oil drilling operation

Fig. 13.8. Transverse section through coupling and socket. Approx. 0.33 ×

Fig. 13.9a. Socket fragment in coupling. Approx. 0.6 ×
Fig. 13.9b. Opposing fracture in socket. Approx. 0.6 ×
Fig. 13.10. Incipient crack in socket of another coupling. Longitudinal section. Etch according to Heyn. Approx. 0.6 ×
Fig. 13.11. Quench structure in thread of socket according to Fig. 13.10. Etch: Nital. 5 ×

A failure due to bend fatigue stress of a defect-free overloaded **crankshaft** has already been described in section 2 (Fig. 2.24 and 2.25).

Impact-type overloading during shifting was probably responsible for the cracking of the 1 mm deep case of a **gear** that was part of a variable speed gear shift transmission of a heavy duty dump truck **(Fig. 13.6).** Fatigue cracks originated from the notches at several places at the root of the teeth on both sides **(Fig. 13.7),** of which two lead to breakout of the teeth. Material and case hardening were in order; core strength was even exceptionally high at about 1420 MPa (205 ksi). A failure due to corrosion stresses has been described already in section 2 (Fig. 2.7 to 2.11).

13.9a

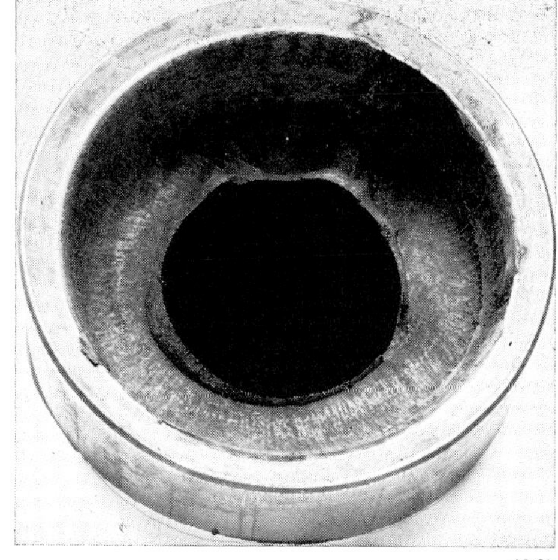

13.9b

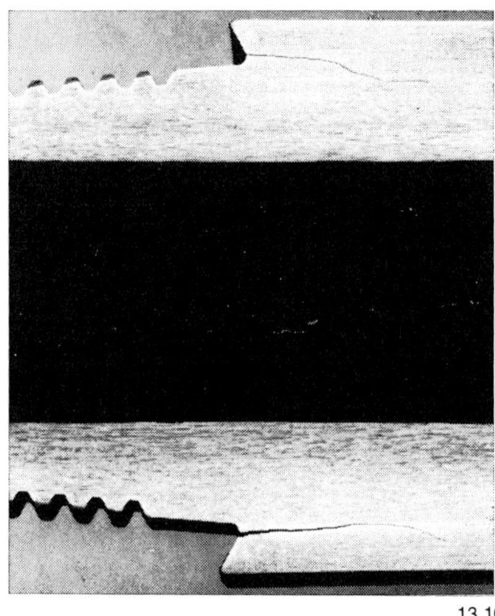

13.10

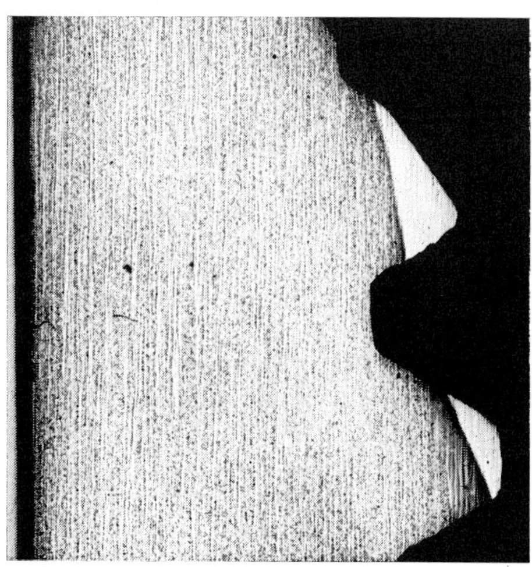

13.11

Impact-type torsion stress caused a failure in a **crude oil drilling** operation during recovery of a pipe rod structure[2]). The pipe, 3600 m in length, was used to catch lost drill rods that had broken during use. Such rod pipes are highly stressed by static tensile, torsion, and bend stresses, and therefore often break. The rods or couplings remaining in the bore hole, especially those of the lower part of the structure, are thus severely stressed through elastic spring back. They may develop incipient cracks, and fracture at a low torque moment when they are being retrieved. Two such broken couplings were examined. They were made of seamless tubing of AISI 9840 steel, and were heat treated to a strength of 1080 to 1180 MPa (157-171 ksi). The failures originated at the low angled fillet between thread and cylindrical part of the socket **(Fig. 13.8).** The fracture of a socket is illustrated for both pieces in **Figs. 13.9a and b.** At first, it propagated almost parallel to the cylinder surface and then penetrated inwards. The fracture in the cylindrical part with indications of a fiber structure probably occurred during the first rupturing of the pipe rod due to impact-type shear and torsion stresses. The final failure, located transversely to the fiber and spirally coiled around the longitudinal axis, occurred as a result of torsional stress developed during retrieval of the pipe rod. The remaining part got stuck in the threaded part of the coupling.

Another coupling showed an incipient crack of the same type **(Fig. 13.10).** It is noteworthy that the backs of the socket thread teeth had hardened in part martensitically **(Fig. 13.11).** The newly formed martensite could be clearly distinguished from the heat treated structure of the socket, both macroscopically and microscopically, e.g. by the fact, that martensite did not etch so darkly, by a finer grain size, and by the absence of carbide precipitates. The martensite may have been formed by frictional heat during rapid spring back of the rod pipe.

Microstructure, strength, and toughness of the socket were acceptable. The failure in this case was clearly due to overloading. A transverse tensile stress probably occurred in the wall of the socket due to the conical inclination of the seal plane. The recommendation was to change the design accordingly. A more favorable fiber orientation that would make the part more resistant to the unusual prevailing stresses could be obtained if the couplings were not manufactured anymore of seamless tubing, but were forged individually with intermediate upsetting.

In case of **fractures occurring at elevated temperature** the question usually arises whether the specified temperature or the calculated stress was exceeded.

Tubes of boiler plants occasionally crack longitudinally under the effect of tangential tensile stresses that exceed the hot strength[3]). The anticipated temperature can sometimes be shown to have been substantially exceeded but the cause of the rise in temperature can often not be found; for instance, the excessive temperature may have been due to an interruption of water or steam supply or to thermally insulating deposits from the feed water.

Figure 13.12 shows the fracture of a ruptured seamless **superheater tube** of high temperature structural steel containing about 0.15 % C and 0.3 % Mo. The tube had a fine-grained structure of ferrite and finely banded pearlite outside of the heating zone, as is normal for this steel **(Fig. 13.13).** However, in the fracture region, the pearlite was agglomerated **(Fig. 13.14).** Molybdenum carbide Mo_2C was precipitated in the ferrite grains and was fairly strongly coagulated. The pearlite had been partly transformed into martensite in the thin cross section of the constricted crack edges **(Fig. 13.15).** Accordingly the temperature had exceeded the A_{C1} point and had been at least by 200 °C higher than normal operating temperature. At this temperature, the low alloy molybdenum steel only has a very low creep strength left.

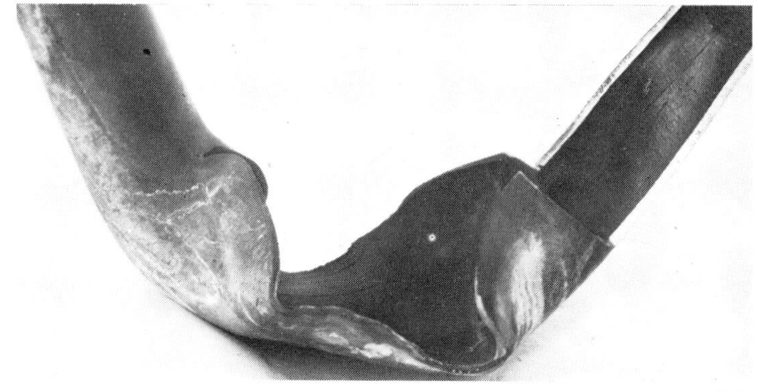

13.12

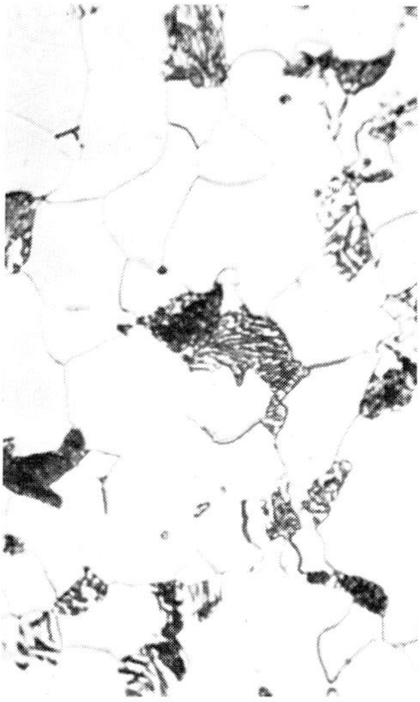

13.13

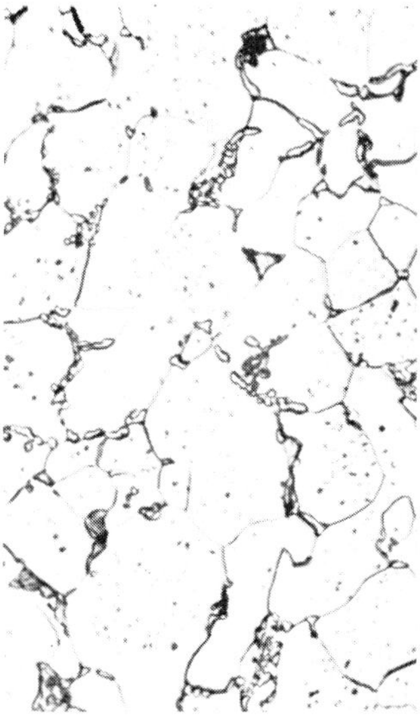

13.14

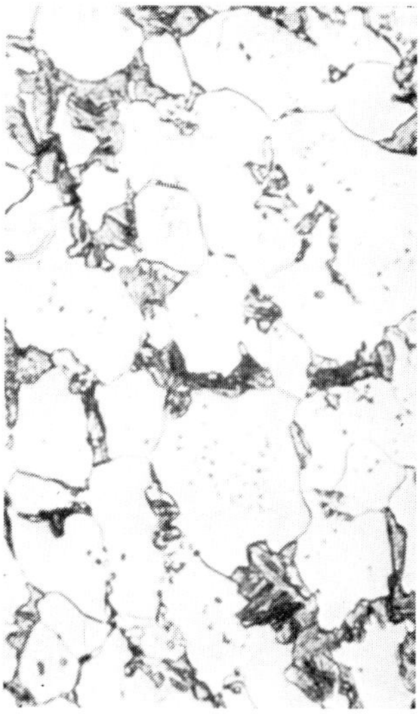

13.15

Fig. 13.12 to 13.15. Superheater tube torn open due to excessive temperature

Fig. 13.12. Ruptured spot. Approx. 0.33 ×

Fig. 13.13 to 13.15. Microstrucure in transverse sections. Etch: Picral. 1000 ×

Fig. 13.13. Outside the heating zone

Fig. 13.14. Inside the heating zone in the part not widened

Fig. 13.15. In thin wall at crack region

13.16

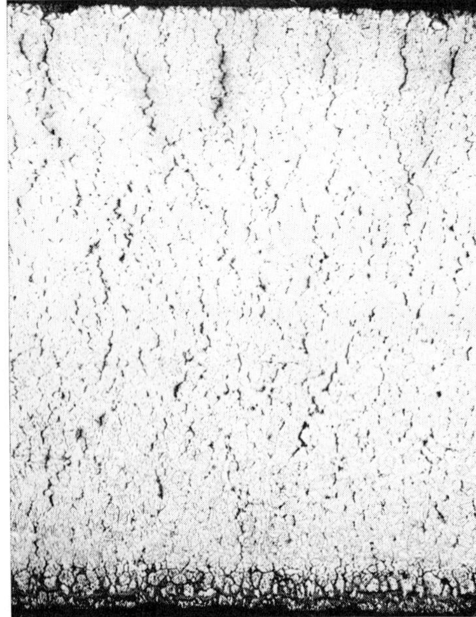

13.17

Fig. 13.16 to 13.20. Austenitic pipe from a crude oil refinery which broke when the rupture strength was exceeded as a result of the supplementary action of bending stresses

Fig. 13.16. Unetched longitudinal section through fracture of tube. 7 ×.

Fig. 13.17. Microstructure adjacent to fracture. Longitudinal section. Etch: V2A-etching solution. 20 ×

Fig. 13.18 and 13.19. Surface structure adjacent to fracture

Fig. 13.18. Outside, unetched. 100 ×

Fig. 13.19. Inside, etched with V2A-etchant. 200 ×

Fig. 13.20. Locally enhanced oxidation (Röschenformation) and carbon enrichment at inner surface. Longitudinal section. Etch: V2A-etchant. 50 ×

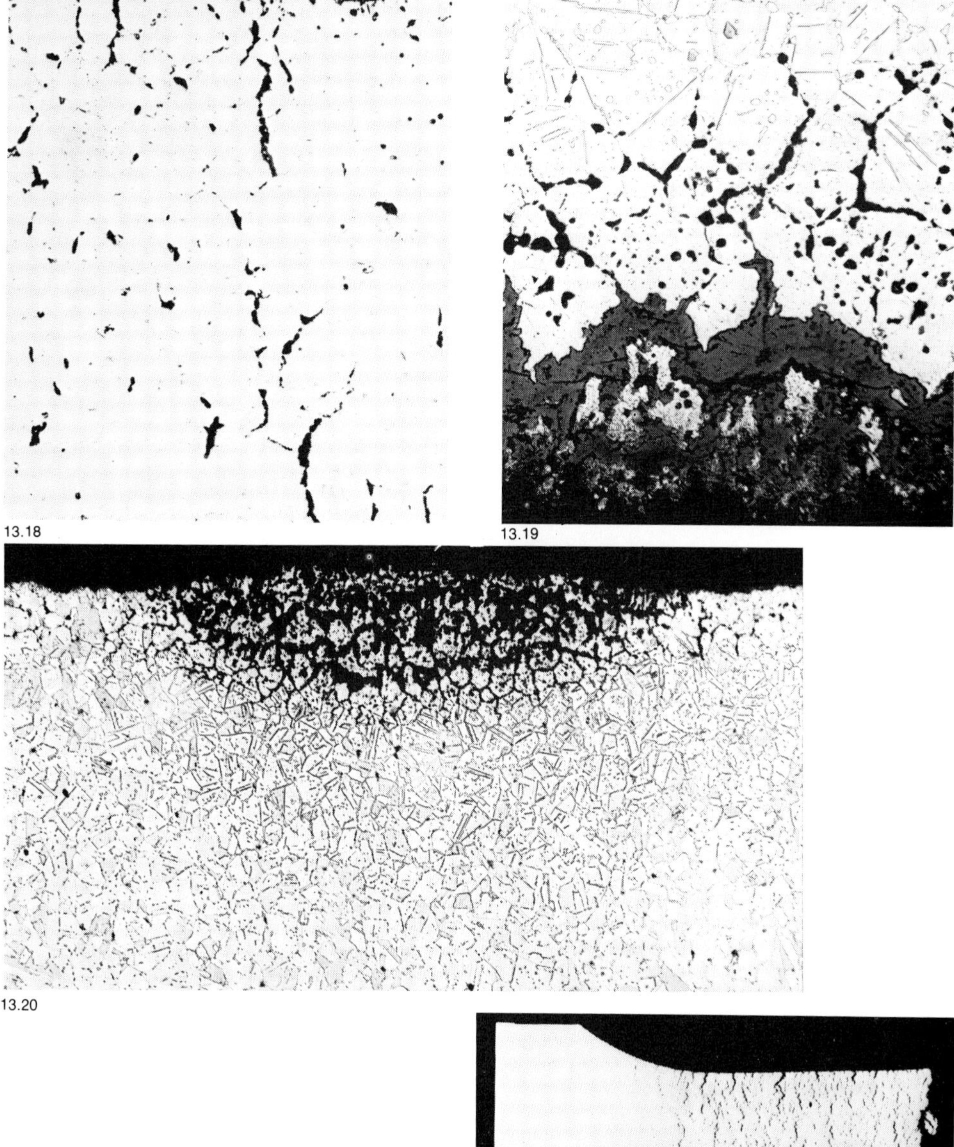
13.18

13.19

13.20

Fig. 13.21. Longitudinal section through stress-rupture specimen of steel containing 0.08 % C, 18.20 % Cr, 10.05 % Ni and 0.64 % Ti, that broke under a 20 MPa (2900 psi) tensile stress at 850 °C in air after 4398 h. Etch: V2A-etchant. 2 ×

In a **petroleum refinery** the tubular connectors (pigtails) between the cracking retort and accumulator fractured after approximately 7000 hours of service. They had a 32 mm O. D. and a 3.8 mm wall thickness and consisted of an austenitic steel with 0.07 % C, 19.7 % Cr, 35.0 % Ni and 0.29 % Ti. The temperature of the gas flow was said to be 820 to 870 °C at a gas pressure of 1.9 MPa (19 atm) The designer had calculated a stress of 11.3 MPa (1640 psi) under these conditions and believed that the safety factor was 1.8. The tubes were air colled on the outside.

The fractures were located immediately adjacent to the weld seam at the outlet valve. They originated at the side of the tubes that was pointed upwards **(Fig. 13.16).** In addition to the tangential and axial stresses, bend stresses would be caused at this point by internal pressure. In the vicinity of the fractures many intergranular cracks were found that permeated the entire tube wall thickness and had a preferred orientation transverse to the tube axis as in the case of the fracture **(Fig. 13.17 and 13.18).** The steel was attacked intergranularly below the inner wall of the tube **(Fig. 13.19).** The attack originated preferentially in some centers **(Fig. 13.20)** where protrusions ("Röschen") had formed. Evidently, at these spots the protective oxide layer that imparts corrosion stability to these steels was first penetrated. The scale at the surface and on the grain boundaries had the green color of chromium oxide. Accordingly, this was a case of "green rot"*; this is caused by a preferential oxidation of the chromium (see section 15.1). Precipitates of carbides could also be found in such areas (Fig. 13.19). Accordingly the atmosphere must have been so mildly oxidizing that combustion of chromium took place, whereas the nickel, iron, and carbon were not attacked.

---

* see Appendix I

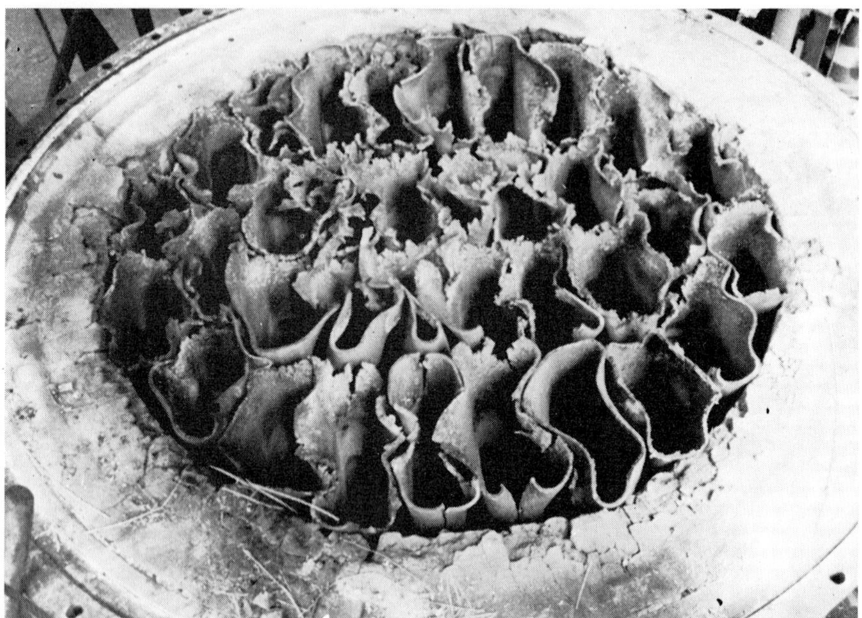

Fig. 13.22 to 13.26. Tuyeres of blast furnace air preheater (Cowper) which scaled and bulged through overheating and exceeding of high temperature strength

Fig. 13.22. View of combustion chamber

Fig. 13.23. Surface structure near tuyere mouth. Etch: V2A-etchant. 200 ×

Fig. 13.25. As in Fig. 13.23. Etch: Diluted ammonium hydroxide 1.5 V. 500 ×. Carbide lightly etched

Fig. 13.24. As in Fig. 13.23. Etch: 10N sodium hydroxide 1.5 V. 500 ×. Ferrite very lightly etched

Fig. 13.26. Edge structure at distance of 200 mm from mouth of tuyere. Etch: 10N sodium hydroxide 1.5 V. σ-phase lightly etched. 500 ×

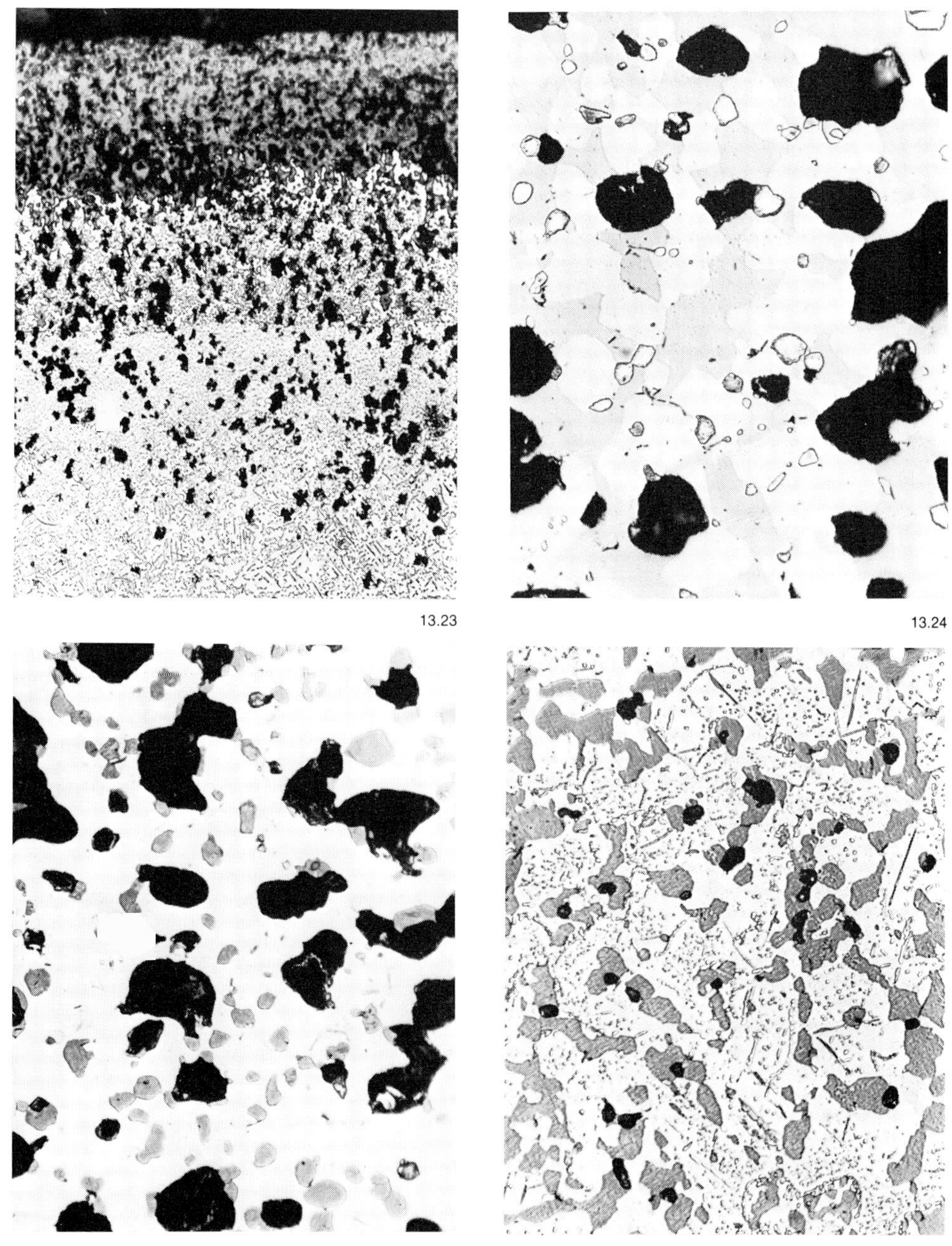

13.23

13.24

13.25

13.26

13.27

13.28

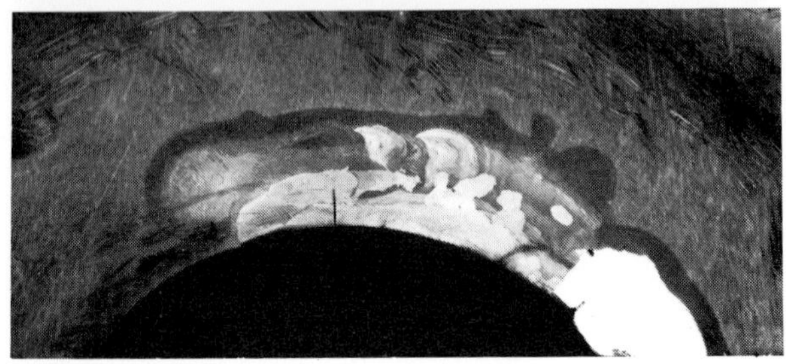

13.29

But this attack evidently was not the factual cause of fracture as was confirmed by the distribution of the intergranular cracks throughout the entire wall thickness, and especially their preferential occurrence in the upper part. Rather the cause must have been due to the fact that the creep strength, which is very low at this high operating temperature, was exceeded during service. **Figure 13.21** shows a longitudinal section through a stress-rupture test specimen of an austenitic chromium-nickel steel with 18 % Cr and 10 % Ni for comparison purposes. It ruptured after 4400 h at 850 °C under a 20 MPa (2900 psi) tensile stress in air and shows the same intergranular cracks distributed over the cross section as did the examined tubular connectors. In order to avoid this type of failure in the future, the elimination of bend stresses is recommended as a principal precautionary step.

The following case is characteristic of those failures where the high temperature strength is exceeded by substantial overheating. A burner of a **cowper recuperator,** that had been found entirely functional during inspection after 1½ years of service, collapsed ¼ year later. The tuyeres not only were heavily scaled, but were also badly dented **(Fig. 13.22).** The operating temperature was said to have risen as high as 800 °C during heating and dropped within 1 hour to the normal temperature of the environment during the blast period. The specific material was cast steel containing approx. 0.4 % C, 27 % Cr, and 4.5 % Ni. It also had been used properly as was shown by quality control analysis. Since this material is heat resistant up to 1150 °C under the strongly oxidizing conditions normally prevailing in such a burner, and since it also posseses high temperature strength as a semi-austenitic steel, the initial assumption was that completely abnormal circumstances must have prevailed during the last quarter year.

Metallographic examination of sections that were taken from the moutnpiece and the other end of the tuyeres, located about 200 mm away, showed that the microstructure of the tuyeres had an oxide scale that had flaked off to a large extent. Under the scale, the surface was enriched by carbon, and completely permeated by pores **(Fig. 13.23 to 13.25).** At the colder end the ferrite had decomposed to austenite and sigma phase **(Fig. 13.26).** Accordingly at this point, the operating temperature did not greatly exceed 800 °C[4]). The carbon enrichment at the mouthpiece evidently occurred through carbon accumulation from the burned austenite (see also section 6.1.3). The pores may have been a coalescence of vacancies; during annealing tests a comparable microstructure was obtained only after two hours of heating at temperatures exceeding 1200 °C. Hence the tuyeres were at least temporarily heated to 1200 °C at their mouthpieces. Perhaps this happened when the flame of the burner backfired.

13.2 **Compression and Shear Stresses**

Cracks in a characteristic pattern are formed by **shear stresses under high contact pressure**[5])[6]). An example that shows cracks originating from the bottom of the engraving in a **forging die** made of a steel containing approx. 0.45 % C, 1.5 % Cr, 4 % Ni and 0.3 % Mo, is presented in **Fig. 13.27.** As can be seen from the transverse section of **Fig. 13.28,** the cracks had penetrated into the material from the bottom as well as from the side walls of the engraving at an angle of approximately 45°. This mode of flaw propagation is characteristic of shear stresses and differs from the pattern produced by quenching. In contrast to the previous situation, the cracks at the upper edge of the engraving may have occurred due to tensile stresses during widening. They had already been patched partly by welding with austenitic electrodes **(Fig. 13.29).** Material and microstructure of the forgings did not have any defects. Accordingly, the failure of the die was not due to material or hardness errors, as the forge operator assumed, but solely due to overloading in service.

Fig. 13.27 to 13.29. Cracked forging die. 1 ×

Fig. 13.27. Partial view

Fig. 13.28. Section A--A according to Fig. 13.27. Etch according to Heyn

Fig. 13.29. Cracks in upper edge of engraving patched by welding with austenitic electrodes. Surface lightly ground and etched according to Heyn

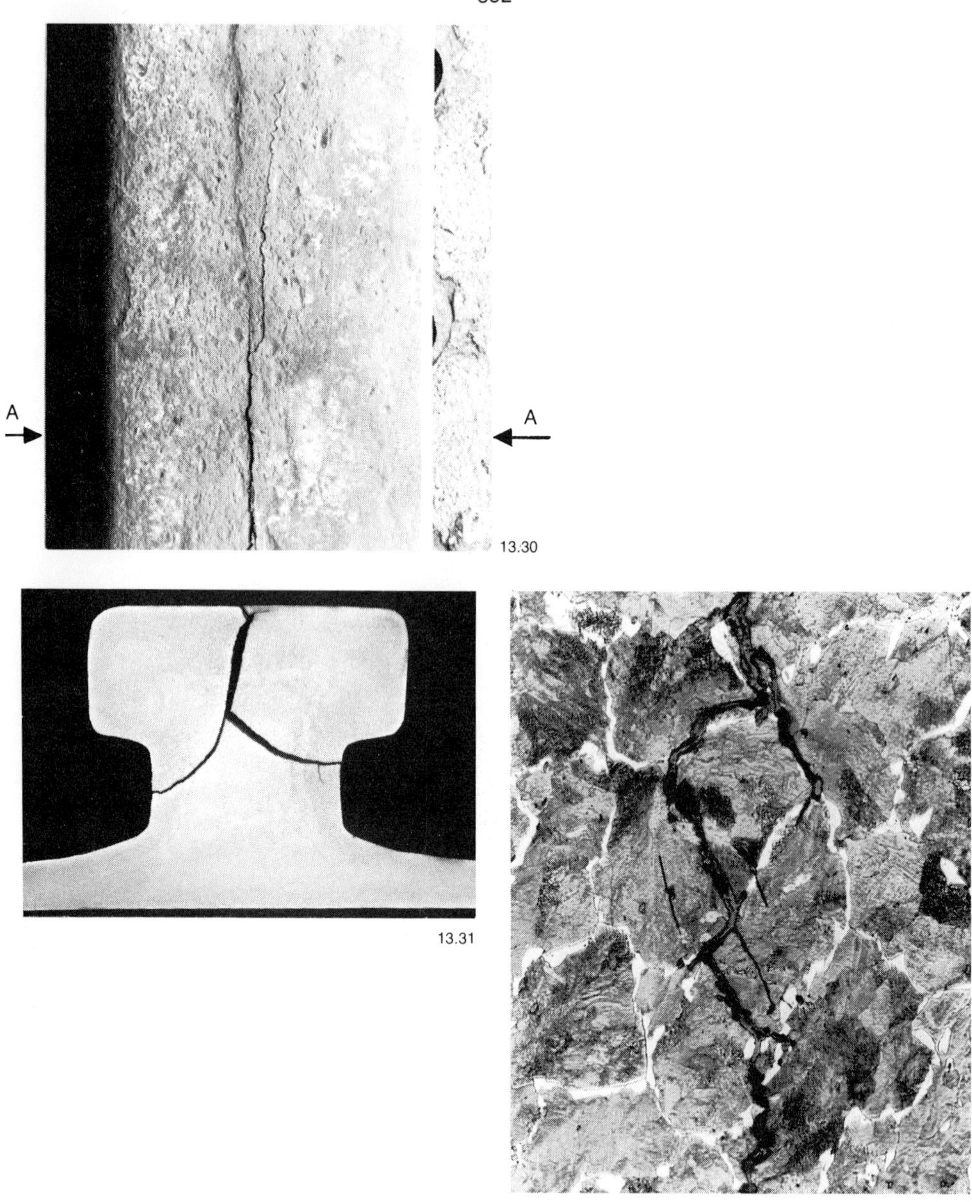

Fig. 13.30 to 13.32. Crane track rail of ore loading bridge, cracked longitudinally by overloading

Fig. 13.30. Bearing surface with longitudinal crack. 0.5 ×

Fig. 13.31. Transverse section A--A according to Fig. 13.30. Etch according to Heyn. 0.5 ×

Fig. 13.32. Transverse section showing crack propagation. Etch: Picral. 200 ×

13.33

13.34

Fig. 13.33 and 13.34. Damaged bridge supports

Fig. 13.33. View of damaged section. Approx. 0.2 ×
Fig. 13.34. Fragment of roll. 2 ×. Fracture origin directly below bearing surface

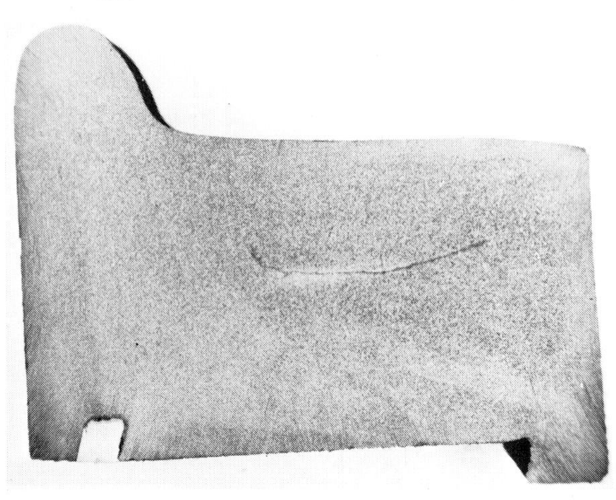

13.35

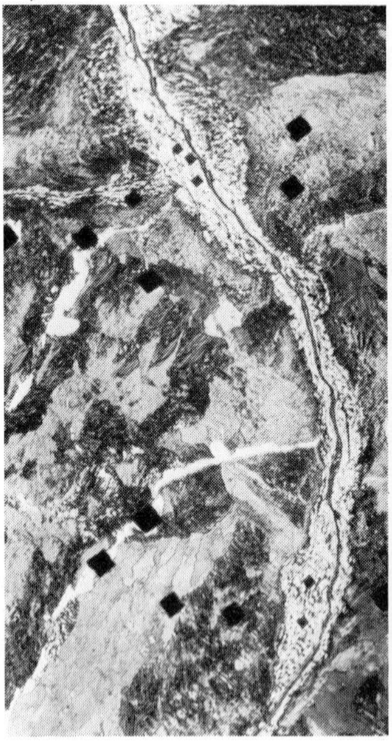

13.36

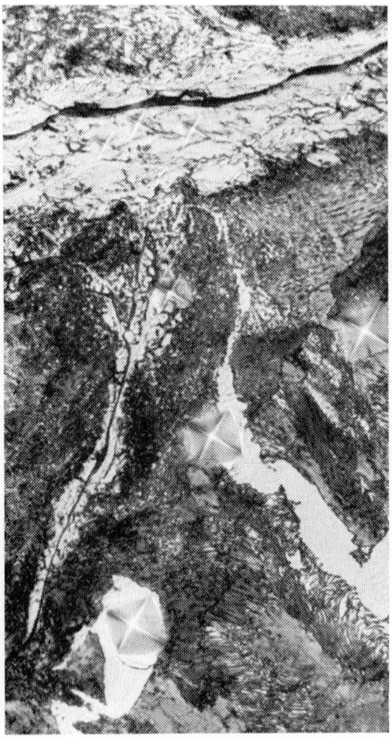

13.37

13.38

Fig. 13.35 to 13.38. Railway wheel rim with internal fatigue fracture

Fig. 13.35. Transverse section. Etch according to Heyn. 0.5 ×

Fig. 13.36 to 13.38. Crack propagation in microstructure. Etch: Picral

Fig. 13.36. 200 ×

Fig. 13.37. 500 ×

Fig. 13.38. 500 ×

Crane rails for ore loading bridges showed a similar crack propagation mode when they broke en masse during service[7]). The rails had the profile KS 75 and were made from a basic converter steel with at least a 590 MPa (85 ksi) strength level. The first indication of failure was seen in a widening of the track, and later by the formation of longitudinal cracks of 3 cm to 10 m in length on the rail heads. **Figure 13.30** shows such a longitudinal crack that branches off at the end and continues as a groove-like depression. From the transverse section **(Fig. 13.31),** the crack can be seen to gape wide open in the interior of the rail head and branch off into several individual cracks at the bearing surface and the web. Apparently it originated at a point below the surface as is often the case in shear stress cracks. This was confirmed by the fractures. They were darkly colored, probably by oxidation due to friction. Nevertheless they had the markings of a fatigue fracture. On the microscopic scale, too, the fracture showed the characteristics of a fatigue type **(Fig. 13.32;** see also Fig. 2.23). As can be seen from Fig. 13.31, the rails as a rule were free of material defects, a fact that was also confirmed by metallographic examination. According to G. Steinrück[8]) and H.J. Wiester[9]), the formation of this type of fracture is normal for overloaded rails. Therefore the profile was too weak or the strength too low for the stress encountered.

The impression and edge fragmentations in the lower third of the **rolls and** in the **plates of a bridge support,** illustrated in **Fig. 13.33,** were also caused by excessive contact pressure. The stainless knife steel parts had been heat treated to a hardness of approximately 700 HV 10. They had to be changed after three to four years because of failure. In this case, too, the fractures originated at a point immediately below the rolling surface, as can be seen clearly in **Fig. 13.34.** Because of the numerous failures, defects in the steel could be eliminated as a potential cause; in fact, no material defects were found during testing. Local stress raisers probably occurred

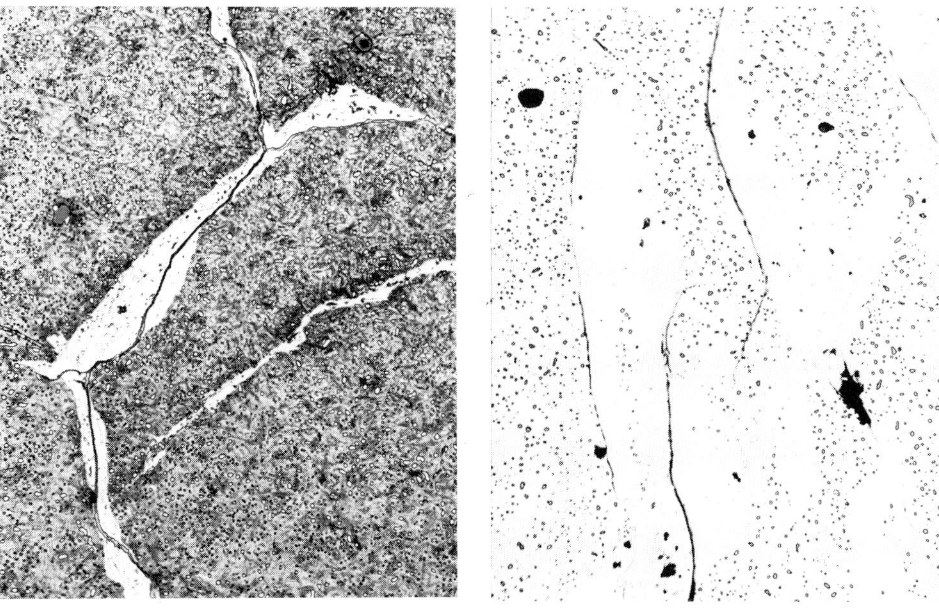

13.39 13.40

Fig. 13.39 and 13.40. Internal cracks (butterflies) with local carbide dissolution and martensite formation below surface of roller bearing ring. 500 ×

Fig. 13.39. Etch: Picral

Fig. 13.40. Etch: Alkaline potassium ferricyanide solution according to Murakami

due to rolling over foreign particles. These may have been accumulations of sand that might have been left behind after blasting of the supports. The sand collected preferentially in the angle between the roll barrels and the guide strips gripping the circumferential grooves (arrow in Fig. 13.33).

Sometimes internal cracks are formed in **railway wheel rims** under the effect of alternating compression and shear stresses. As a rule they originate in internal defects, such as coarse slag inclusions or flakes, and propagate almost parallel to the running surface[10]). **Figure 13.35** shows such an internal fatigue fracture of a railway wheel rim in transverse section. The predominantly intergranular crack propagation in the microstructure is demonstrated in **Fig. 13.36 to 13.38.** A yellowish constituent stretching along the main crack and some individual branches was found; it was much harder than the pearlite, as demonstrated by the diamond impressions (Fig. 13.36 and 13.37). After extensive examination, there was no doubt about it being very finely acicular martensite. This meant that a narrow region around the fatigue fracture had been heated to temperatures in the austenitic region under the influence of high frequency vibrations through friction. No material defect could be found in this case.

In this connection a failure must be mentioned that occurs in roller bearing elements. Sometimes places are found in them immediately below the surface in the zone of highest shear stress that are not etched by picric acid. These spots often start from non-metallic inclusions and are permeated or delineated by fine cracks[11]). In the Anglo-Saxon literature such areas are designated as **"butterflies"***. **Figure 13.39** shows this phenomenon in a **roller bearing ring.** The white phase around the crack is harder than its surroundings. This region is composed of fine

* see Appendix I

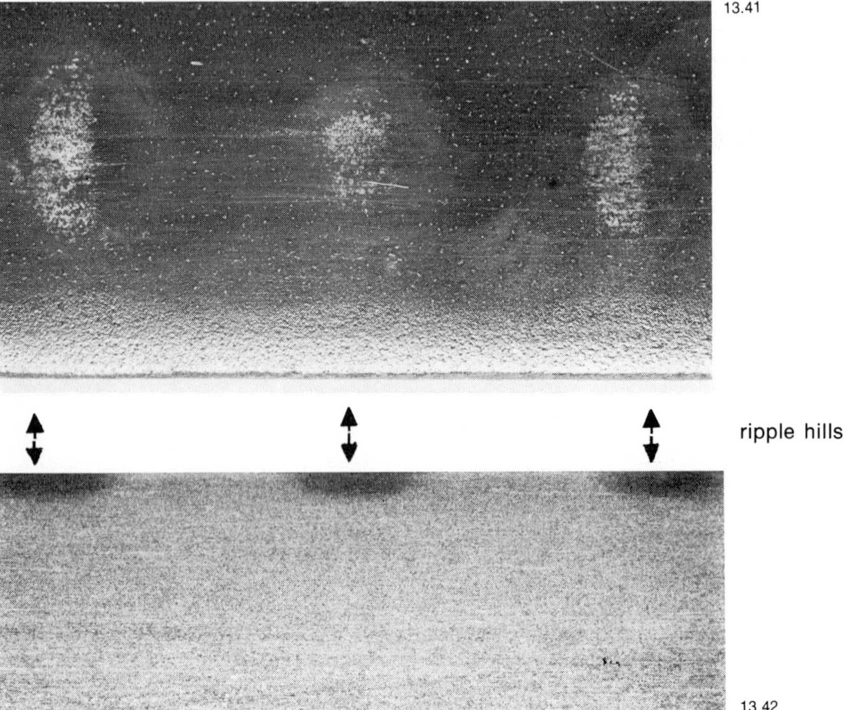

13.41

ripple hills

13.42

carbides which are dissolved in large part, as can be seen even more distinctly by Murakami etch **(Fig. 13.40).** In all probability this is martensite. H. Schlicht[12]) interpretes its genesis as follows: During the rotation of roller bearing elements, very high stresses may be induced under the surface by simultaneous hydrostatic pressure and shear; these stresses locally exceed the shear strength of the material. The flanks of the cracks propagating in the direction of highest shear stress may be heated up to the melting temperature by the plastic deformation of the structure in an extremely short period. The large cold mass of the surrounding material causes a fast heat transfer and has a quench effect.

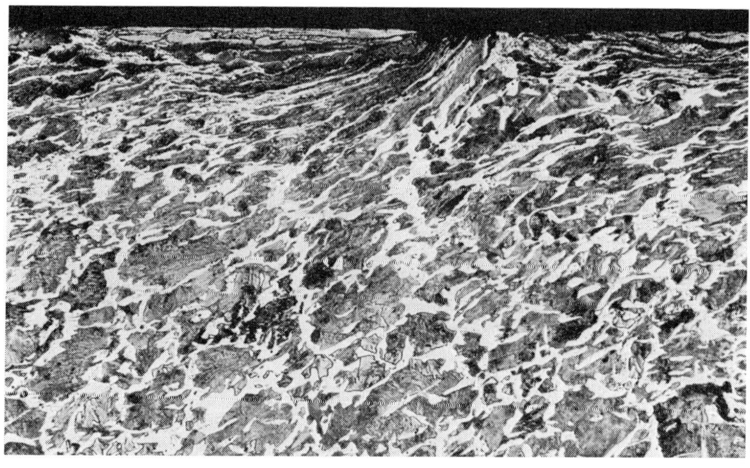

13.43

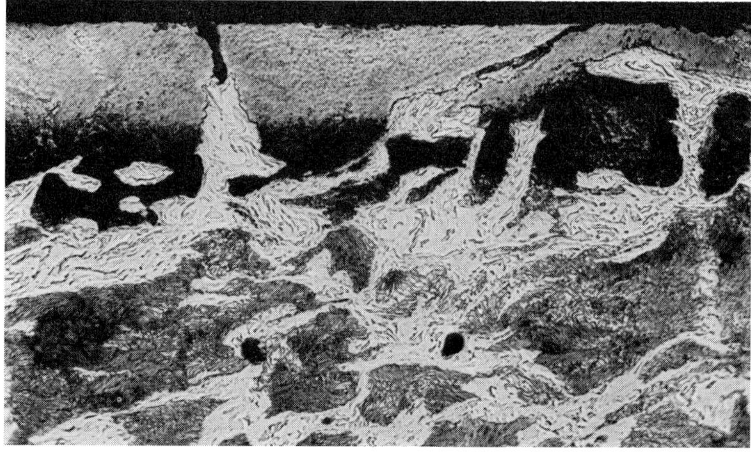

13.44

Fig. 13.41 to 13.44. Rail with ripples on bearing surface

Fig. 13.41. Bearing surface. 1 ×. Ripple hills light

Fig. 13.42. Longitudinal section through bearing surface. Etch according to Heyn. 1 ×. Ripple hills dark

Fig. 13.43. Microstructure of ripple hill surface. Etch: Picral. 100 ×

Fig. 13.44. As Fig. 13.43. 500 ×

13.3 **Rolling Friction, Compressive Wear**

A failure that results primarily from dynamic compressive shear stresses is the **rippling or corrugation on rails** that causes the well known rumbling noise, and that still poses problems and costs for the railroads[13]). Rippling is a wave-like sequence of "hills" and "valleys" that are present on the bearing surface of the rail heads at intervals of 30 to 60 mm. The difference in height between hills and valleys as a rule is some hundreths to tenths of a millimeter. The hills are rubbed bright, while the valleys are corroded a dark color. **Figure 13.41** shows a bearing surface of a rippled rail. In longitudinal section through the bearing surface, 3–4 mm deep cold deformation zones were found under the hills by macroetching **(Fig. 13.42)**[14]). Under the microscope, the deformation proved to be a tongue-like overlapping **(Fig. 13.43).** At the surface and the rubbed and torn open faces, martensite had been formed and below it a tempered microstructure could be seen **(Fig. 13.44).** Since the heating and cooling cycle may be repeated frequently and the temperature of the last heating is the deciding one, the microstructure may also change. Traces of slip lines could still be found in the ferrite up to a depth of 4 mm. Some areas leave the impression that cold deformation preceded martensite formation and heating occurred primarily through internal friction. Electron microscopic and X-ray analysis[14]) showed that the light phase was indeed martensite.

There is as yet no complete understanding of this rippling or means to avoid it, nor about the part played by the rail steel. From the changes in microstructure, hardenability and cold working capacity are known to be important for a tendency toward corrugation; however, no effective means of prevention can be anticipated from a materials point of view.

A broken-out piece from a roll of a **self-aligning roller bearing** is reproduced in **Fig. 13.45.** This failure, which occurred after 8000 h of operation in a coal dust mill, was the result of material overloading. The characteristic metallographic appearance of such defects is shown in **Fig. 13.46.** A fatigue fracture that originated in one of these regions is illustrated in **Fig. 13.47.** In these structures characteristic markings which may indicate that the bearing was running hot could also be observed **(Fig. 13.48);** therefore it is not clear whether the overheating was the cause or the consequence of the breakout. The formation of such defects in roller bearing elements is facilitated by nonmetallic inclusions. But inclusions of such a size that they could have had the effect described could not be found in this case.

Dimple formation or **pitting in gears** is due to excessive contact pressure. According to G. Niemann[15]), these are small flat tearouts in or below the circular pitch line of the tooth flanks. The formation of such pits is a fatigue phenomenon in which the shear-fatigue strength has been exceeded and the lubricant has been pressed into the fine cracks. The fractures may originate at the surface or inside the material in the region of highest shear stress. Overloading may have been caused by a local excess in rolling compressive strength as a consequence of nonuniform support of the tooth flanks. The overloaded **rim of a rolling mill transmission gear**[16]) is shown in **Fig. 13.49.**

The incipient pitting on a **bevel gear** made of a steel with about 0.6 % C and the corresponding **pinion** made of a steel containing approx. 0.35 % C, 1.25 % Si, and 1.25 % Mn is shown in a transverse section through the tooth flanks in **Fig. 13.50.** The cracks in the pearlitic-ferritic microstructure of the bevel gear followed a transcrystalline path; they appear to have formed without any prior deformation, and therefore may be interpreted as fatigue cracks. They were directed from the back of the teeth to the root. A piece had broken out along one of the crack planes. In contrast, in the softer pinion with an annealed structure, the cracks were oriented from the root of the teeth to the back with otherwise the same appearance. According to G. Niemann and H. Glaubitz[17]), this difference expresses the dissimilar perception of the stress in the driving and driven gear.

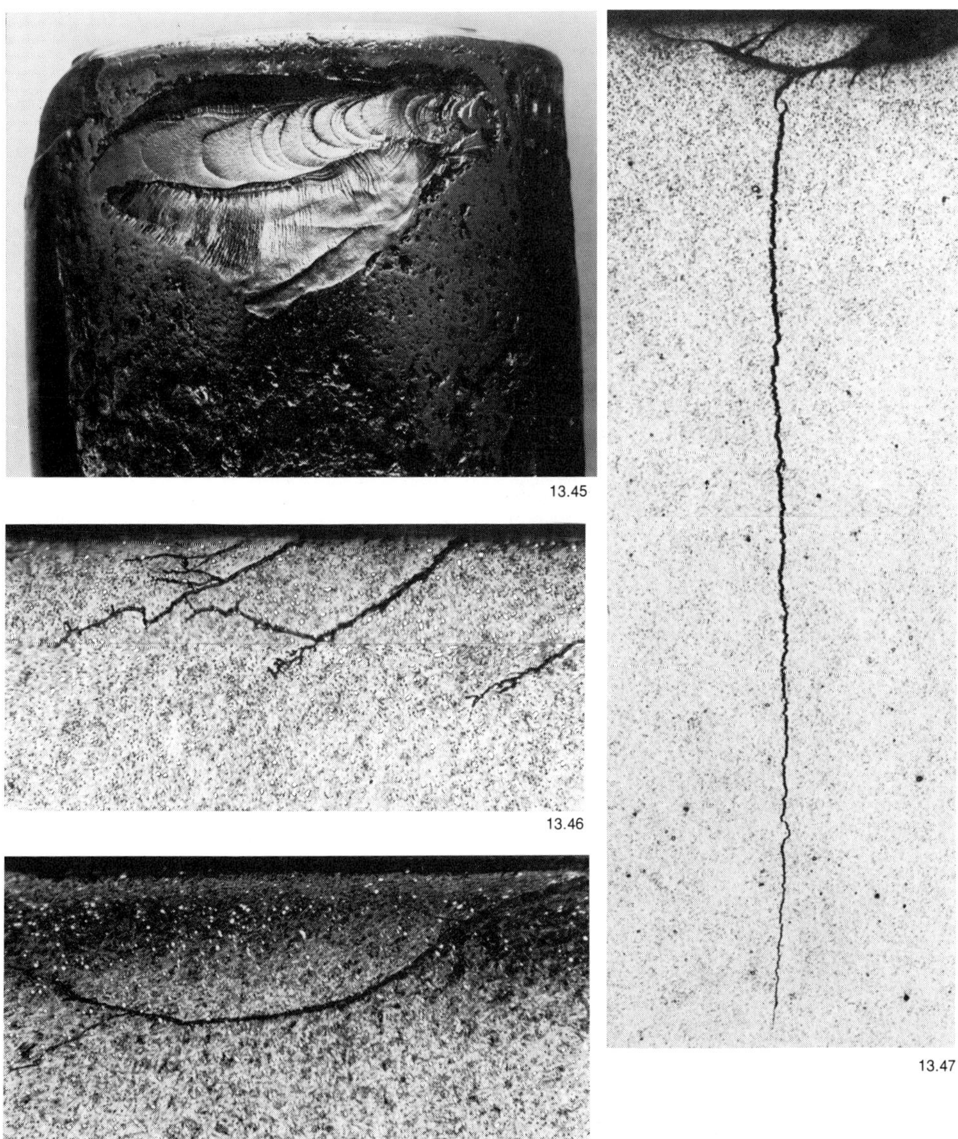

Fig. 13.45 to 13.48. Roll of self-aligning roller bearing destroyed by overloading

Fig. 13.45. Bearing surface with breakouts. 1.4 ×

Fig. 13.46. Incipient breakouts in section. Etch: Picral. 400 ×

Fig. 13.48. Heated area with tempered structure. Etch: Picral. 400 ×

Fig. 13.47. Fatigue fracture originating at a breakout of a surface. Etch: Picral. 100 ×

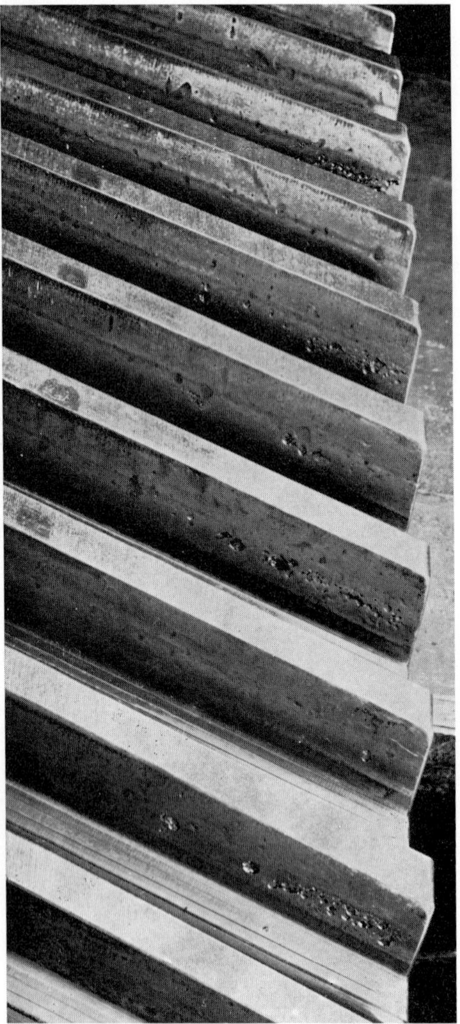

Fig. 13.49. One-sided pitting in teeth of rolling mill transmission gear. 0.3 ×

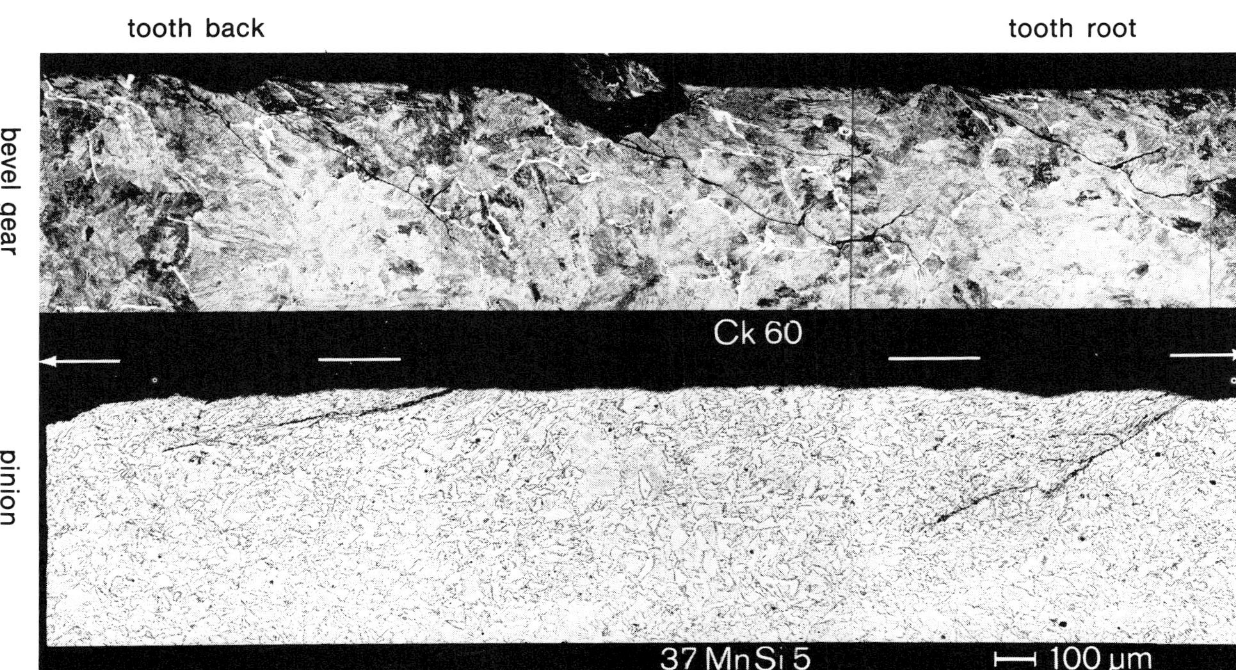

Fig. 13.50. Bevel gear and pinion with pitting. Transverse section through tooth flanks. Etch: Picral. 50 ×

In **cast iron gears,** too, pitting may occur as a result of overloading. For instance, a potential failure of this type in a worn and torn worm gear tooth is shown in **Fig. 13.51.** This gear had been made of a defect-free cast iron that had a tensile strength of 230 MPa (33.5 ksi).

13.4 **Sliding Friction, Hot Runs**

Friction and deformation between two surfaces may cause them to become so hot, that microstructural changes and stress cracks may occur. In addition, chips may be torn from the surface by non-lubricated friction or by insufficient or interrupted lubrication of sliding parts – whereby a deepseated cold deformation may take place first.

Braking is a process whereby friction is used to transform kinetic energy into heat. If failures are to be prevented, the heat must be conducted away so fast that the rubbing parts do not reach high temperatures. If this is not possible, hardening and the formation of quench cracks may result. **Figure 13.52** shows as an example the worn bearing surface of a **railway wheel rim** that was subjected to "**hard braking**"[18]). Significant surface damage including gaping grinding cracks (see also section 8.2) can be seen. A metallographic examination revealed that the surface damage consisted of tongue-shaped overlaps. In transverse section through another rim with similar defects **(Fig. 13.53),** the heated areas could be seen as darkened zones with a changed structure after macroetching. A longitudinal section through the zone which exhibited numerous cracks, Fig. 13.52, showed that some of them had penetrated far beyond this transformed region into the interior of the rim **(Fig. 13.54).** In the cracked zones the microstructure consisted of grainy pearlite or of featureless martensite in accordance with the respective temperature reached. Whether heating occurred in this case through friction between brake shoe and wheel rim or through slipping of the firmly braked wheel on the rail could not be determined. Rapid cooling may have taken place by conducting the heat into the large cold mass of the rim or by wind slip stream.

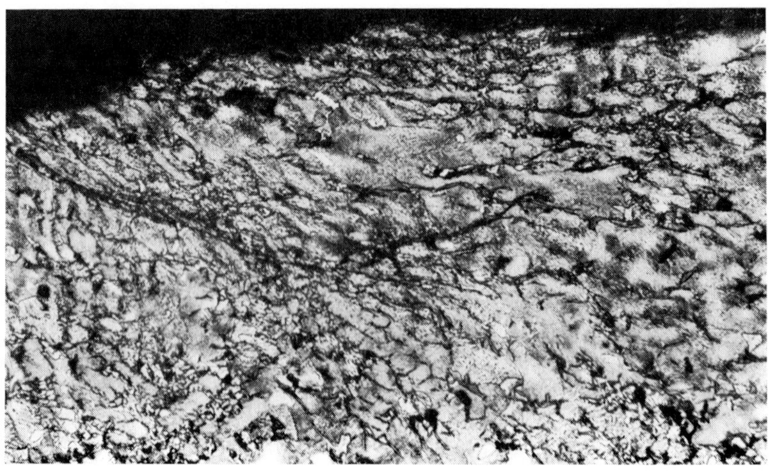

Fig. 13.51. Pitting in flank of broken-out tooth of cast iron worm gear. Surface structure. Etch: Picral. 100 ×

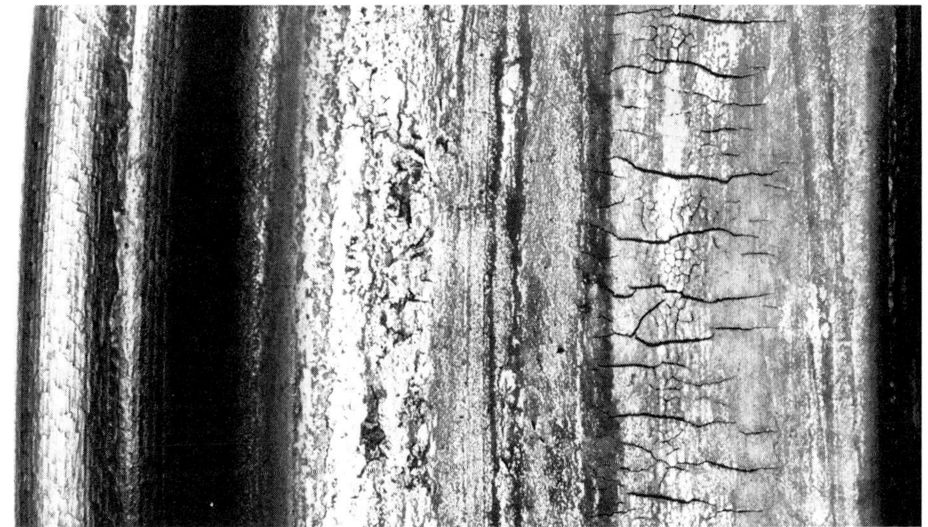

13.52

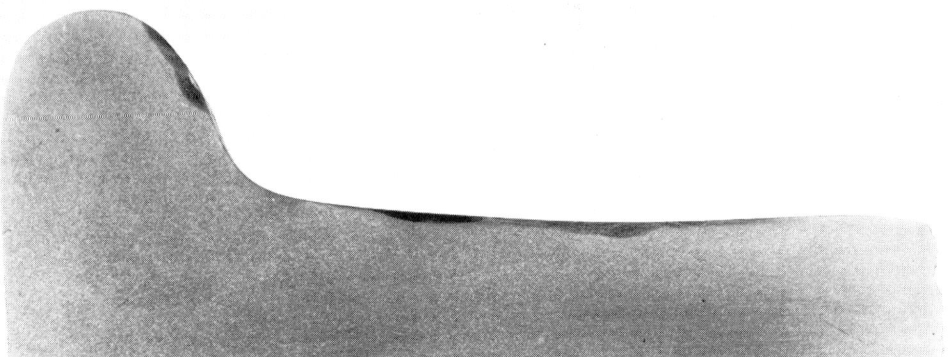

13.53

13.54

Fig. 13.52 to 13.54. Hard-braked railway wheel rim. 1 ×

Fig. 13.52. Bearing surface with cold deformation and grinding cracks

Fig. 13.53. Transverse section through another rim. Etch according to Heyn. 1 ×

Fig. 13.54. Longitudinal section through zone with cracks according to Fig. 13.52. Etch according to Heyn. 2 ×

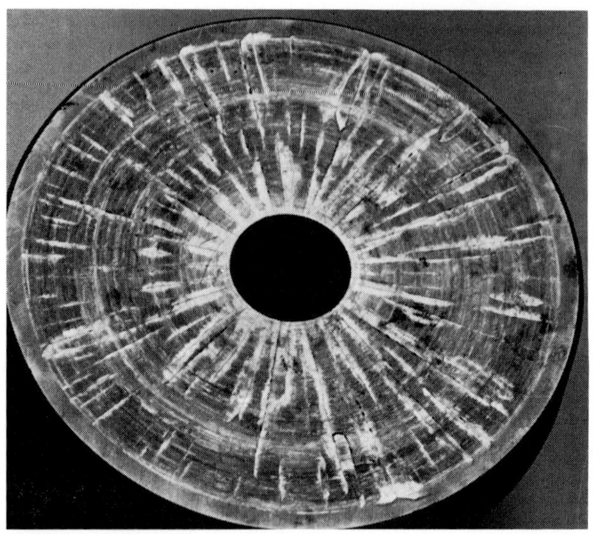

13.55

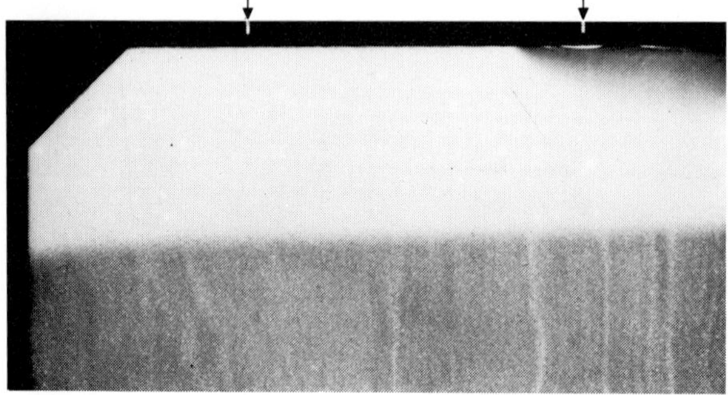

13.56

hardness test series 3

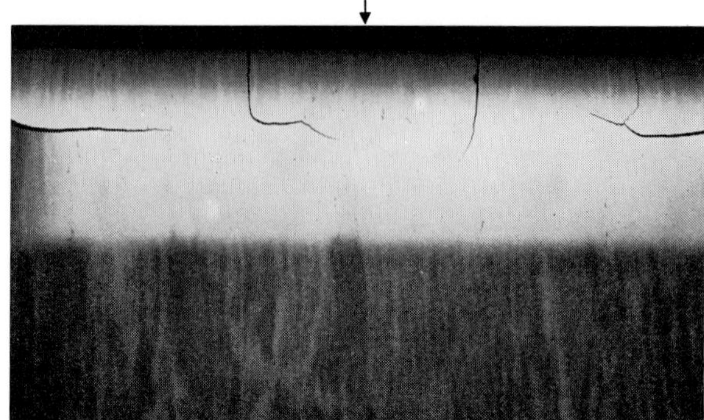

13.57

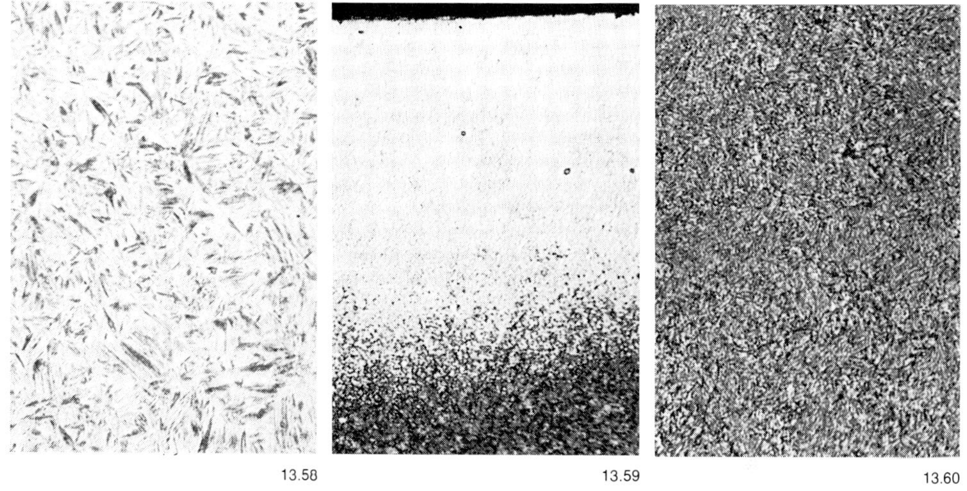

13.58 13.59 13.60

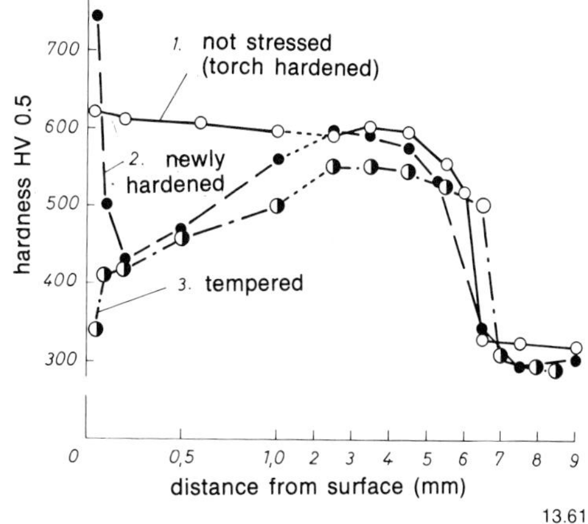

13.61

Fig. 13.55 to 13.61. Cracked pressure plate of upper part of blooming-mill stand

Fig. 13.55. Wear surface of pressure plate. 0.2 ×

Fig. 13.56. Radial section of edge zone not subjected to stress. Etch: Picral. 3 ×

Fig. 13.57. Tangential section. Etch: Picral. 3 ×

Fig. 13.58 to 13.60. Etch: Picral. 500 ×

Fig. 13.58. Microstructure of torch hardened outer edge of disk

Fig. 13.59 and 13.60. Microstructure in crack zone

Fig. 13.59. Newly hardened spot

Fig. 13.60. Tempered zone

Fig. 13.61. Hardness depth traverse in the test series 1, 2 and 3 according to Fig. 13.56 and 13.57

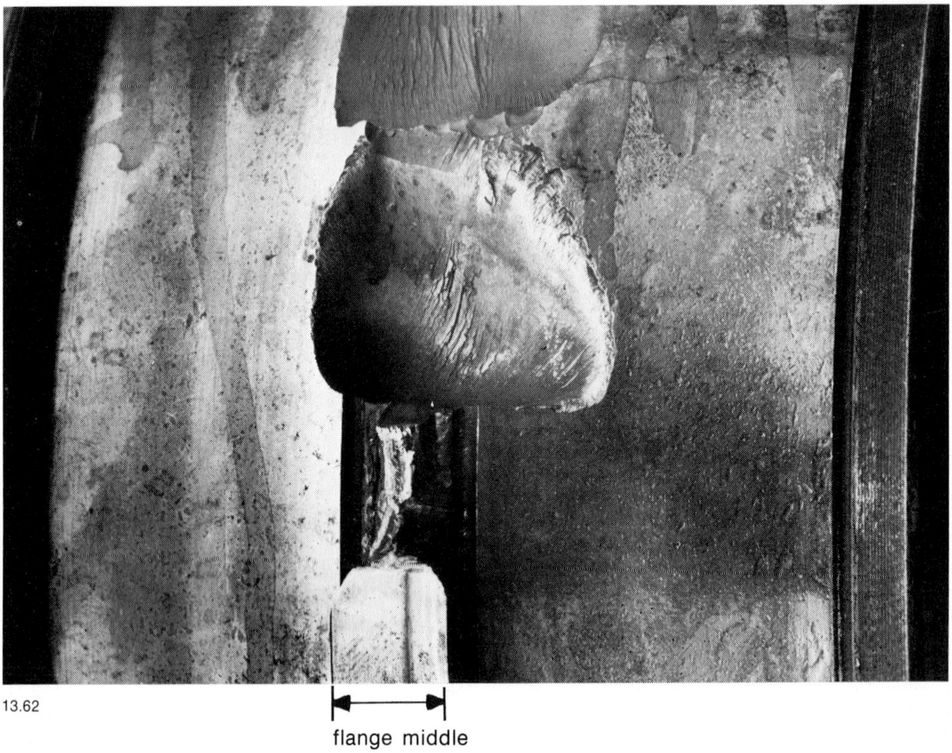
flange middle

13.62

flange middle

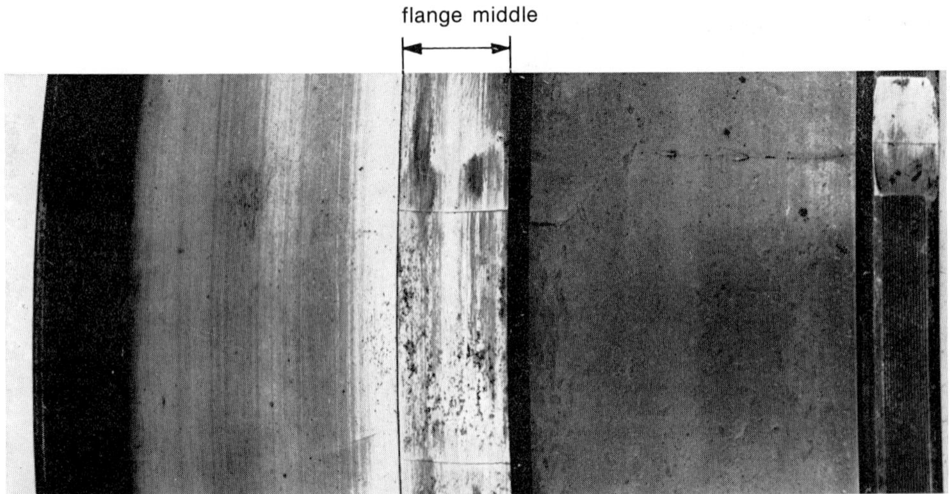

13.63

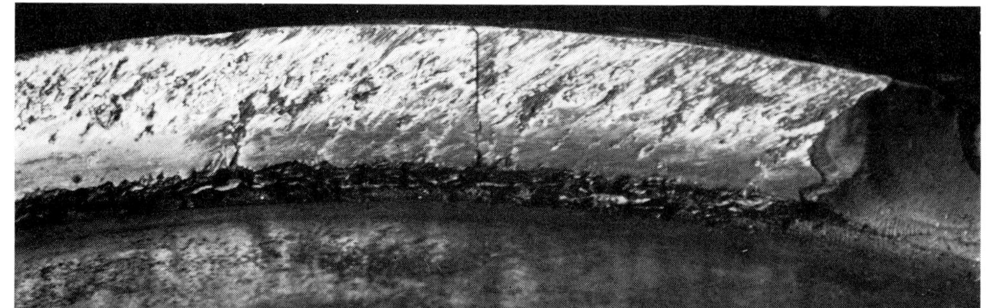

13.64

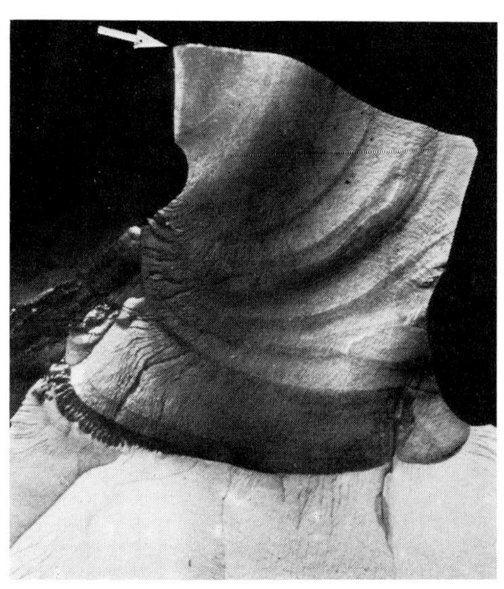

13.65

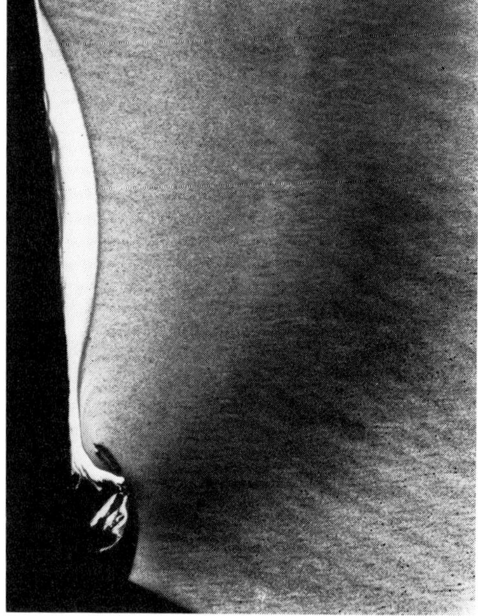

13.66

Fig. 13.62 to 13.66. Cracked self-aligning roller bearings

Fig. 13.62. Inner ring with breakouts in flange middle. 0.8 ×

Fig. 13.63. Bearing surface of a ring with cracks in flange middle. 0.8 ×

Fig. 13.64. Worn flank of flange middle. 2 ×

Fig. 13.65. Opened crack in flange middle. → Fracture origin. 2 ×

Fig. 13.66. Transverse section through worn flank. Etch: Nital. 8 ×

13.67

13.68

13.69

13.70

Fig. 13.67 to 13.71. Cold rolls that became hot locally due to slippage of rolling material

Fig. 13.67. Surface, etched with Nital. Approx. 0.33 ×

Fig. 13.68. Section of Fig. 13.67. Approx. 1 ×

Fig. 13.69. Surface peeling. Barrel surface etched with Nital. Approx. 0.2 ×

Fig. 13.70. Barrel surface of broken-off peeling after etching with Nital. Approx. 0.66 × ← Fracture origin

Fig. 13.71. Peel fracture according to Fig. 13.69. 1 × ↘ Fracture origin p 370

A **pressure plate** from the upper part of a blooming-mill stand developed cracks during service. The plate was made of heat treated chromium-nickel-molybdenum steel with 0.35 % C, 1.53 % Cr, 1.48 % Ni, and 0.30 % Mo, and had been surface hardened with a torch. The designer provided the following information about type of work and load: An opposing plate presses onto the pressure plates during service. It consists of a complex aluminum bronze which can turn against the torch-hardened plate. Three radially running oil grooves had been machined into the bronze plate; they had the purpose of keeping the wear-subjected surface immersed in oil. During the relative motion between pressure plate and opposing plate each point of the working surface of the pressure plate is stressed and relieved alternately in relation to its position to the groove. This results in a threshold compressive stress from contact pressure in addition to the shear stress caused by frictional forces. If the part moves under load, mean contact pressures of up to 98 MPa (14 ksi) may occur at speeds of 4 r.p.s., corresponding to a mean rubbing speed of approximately 3 m/s.

The plate showed a number of partially gaping radial cracks on the bearing surface, that did not quite extend to the outer edge **(Fig. 13.55).** The crack peripheries were offset against one another by stress relief; in addition, they had chipped off some bronze from the opposing plate. This in turn was deposited in the plate in circular bands. An etched radial section through the edge zone is shown in **Fig. 13.56,** and in **Fig. 13.57,** a tangential section through the crack zone. In the crack-free edge zone, the microstructure remained unchanged **(Fig. 13.58),** but under the working surface a dark zone with a tempered structure was present **(Fig. 13.60).** In some places under the surface, renewed hardening occurred **(Fig. 13.59).** The cracks propagated almost at a right angle through the tempered zone, which must have been under a tensile stress. Where a martensitic zone preceded the tempered zone, the cracks apparently originated at the transition between the two zones. At the transition into the compression-stressed zone of the original

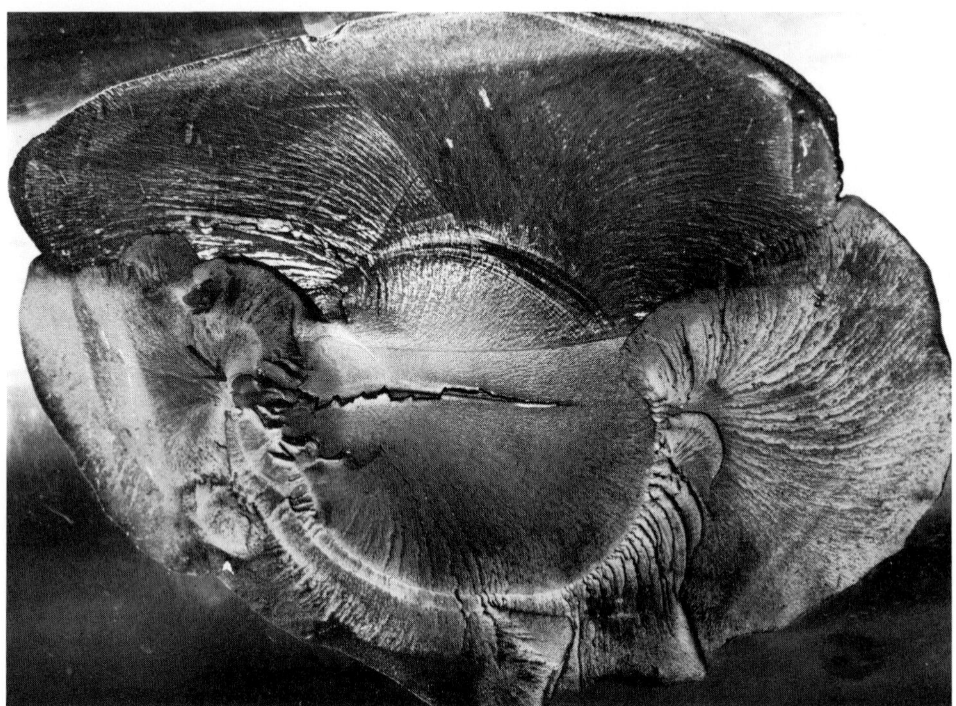

13.71

martensite, the cracks were bent by 90° into the horizontal plane. The dynamic compression stress from contact pressure mentioned before and the dynamic shear stress of the frictional forces may have been superimposed upon the residual stresses and thus widened or extended these cracks.

Since the unstressed edge zone remained crack-free, the cracks certainly had not occurred during grinding in the production of the plate. Furthermore an inquiry with the manufacturer established that the disk had not been ground circumferentially as would have been necessary to cause radial crack propagation.

A hardness traverse under the surface in the test series 1, 2, and 3 according to Fig. 13.56 and 13.57 is shown in **Fig. 13.61.** The effect of heating on hardness is expressed by a considerable drop during annealing and a marked increase through renewed hardening.

Remedial measures should first of all address a reduction of friction by improved lubrication.

Inner rings of several **self-aligning roller bearings** made of a chromium-manganese ball bearing steel containing about 1 % C, 1 % Mn, and 1.5 % Cr failed in service[19]). As can be seen from **Fig. 13.62,** whole parts had broken out of the flange middle. In addition, incipient cracks had formed in the flange middle in numerous places perpendicular to the rotational direction **(Fig. 13.63).** All the cracks and fractures originated at an edge of the flange middle, whose adjoining flank showed heavy wear and scouring **(Fig. 13.64).** In the forcibly opened crack **(Fig. 13.65)** its origin could be clearly recognized. The final fracture had a satin-like fine structure that indicated correct hardening. The abraded flank had gotten so hot that underneath a martensitic structure an adjacent tempered zone had formed **(Fig. 13.66),** and the edge that had originally been beveled, had been ground to a knife edge by wear. Accordingly, the fracture of the rings was caused by frictional heat as a consequence of the rolls or the roll cage rubbing on a flange middle flank. This lead at first to the formation of grinding cracks and finally to the breaking out of some parts.

Hardened **steel rolls** sometimes become unusable if the barrel surfaces become hot during slippage of the material to be rolled. The micro-structural changes caused by tempering or renewed hardening lead to the formation of grinding checks which result in pieces of the rolls breaking out. The fracture origin usually is located below the surface.

Cracks appeared after only three days' service on each one of a pair of steel rolls; these rolls were used to produce cold rolled brass and aluminum strip. The cracks were arranged in narrow bands around the circumference and ran transversely to the rolling direction **(Fig. 13.67)**[20]). Etching with nitric acid caused the bands in which the cracks were located to become darker than their surroundings. This suggested that the surfaces had become hot. In both rolls the bands appeared in corresponding places and were limited to an area in the central part of the barrel. Accordingly, they could not have occurred during final grinding of the rolls, but only during rolling of the strips as a result of slippage of the rolling material. **Figure 13.68** shows a section of Fig. 13.67 in natural size in which the characteristic pattern of grinding and friction cracks perpendicular to the grinding direction is distinctly expressed.

The stresses and cracks caused by frictional heat may in combination with shear stresses during rolling also lead to a breaking away of the barrel surface. **Figure 13.69** shows a hardened steel roll with such a peeling effect. Surface etching disclosed again some regions that had been heated. The relationship of the surface peeling to the heating effect becomes even clearer in **Fig. 13.70** which shows the etched surface of the broken-off layer. The point of fracture origin, designated by an arrow, was located exactly at a dark colored spot, i.e. the point of highest stress just below the surface **(Fig. 13.71).**

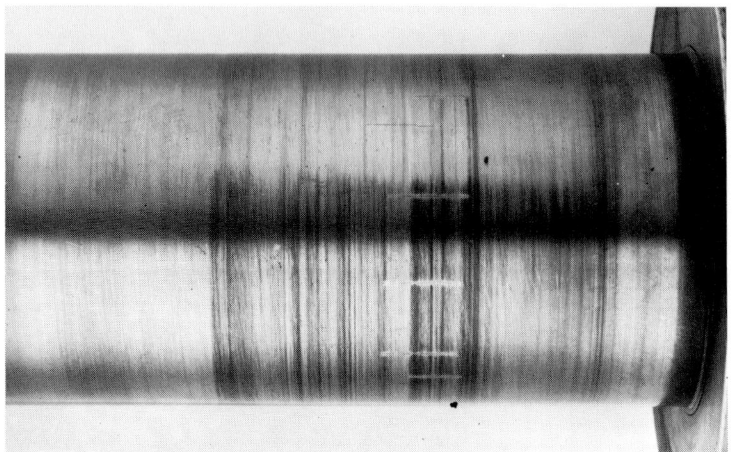

13.72

13.73

13.74

The darkly etched bands with a tempered structure have generally a shallow depth. If no deep-seated surface peeling has occurred, the rolls can usually be reconditioned by regrinding. Hence, metallographic sections can be dispensed with as a rule and the investigation confined to surface etching, supplemented by hardness measurements if indicated.

Failures as a result of frictional heat occasionally occur with fast rotating shafts as a consequence of locally increased contact pressure or insufficient lubrication. The following failure is an example, featuring a case hardened **lathe spindle**[21]). The bearing surface exhibited a group of short cracks parallel to one another and to the spindle axis **(Fig. 13.72).** Circumferential bands with temper colors in this region indicated that the spindle had run hot. This was confirmed by etching the surface with nital **(Fig. 13.73).** In the dark regions the surface hardness had dropped from approximately 700 to 420 HV 10. **Figure 13.74** shows an etched longitudinal section through the transition from a light to a dark band. The light areas illustrated the original hardened martensitic structure of the case hardened layer of approximately 0.5 mm depth; in the more strongly etched band, the structure had darkened due to tempering and had become martensitic again in spots by heating into the austenitic range **(Fig. 13.75).** The cracks had started from the dark zone that had been subjected to tensile stresses due to contraction resulting from the martensite decomposition **(Fig. 13.76).**

13.75

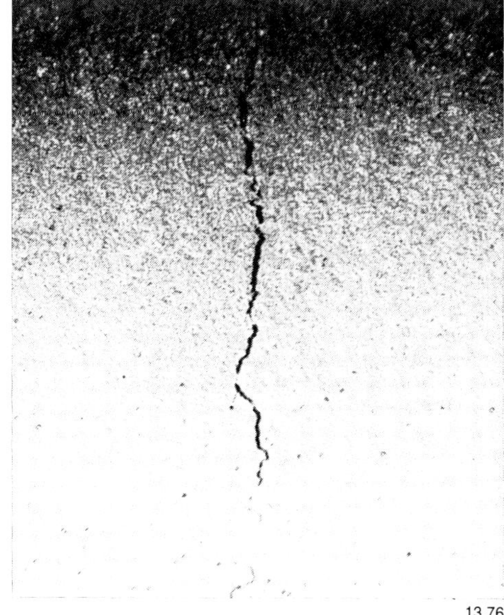

13.76

Fig. 13.72 to 13.76. Hot running bearing shaft of lathe spindle

Fig. 13.72. Surface view. 0.66 ×

Fig. 13.73. As Fig. 13.72. Surface etched with Nital. 0.66 ×

Fig. 13.74 to 13.76. Edge structure. Etch: Nital

Fig. 13.74. Longitudinal section through heated zone. Transition from not tempered (left) to tempered region (right). 10 ×

Fig. 13.75. Longitudinal section with quench and tempered microstructure. 200 ×

Fig. 13.76. Transverse section through a crack. 100 ×

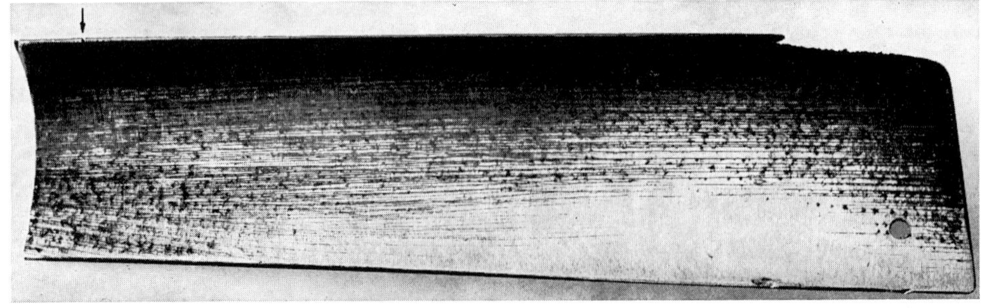

13.77

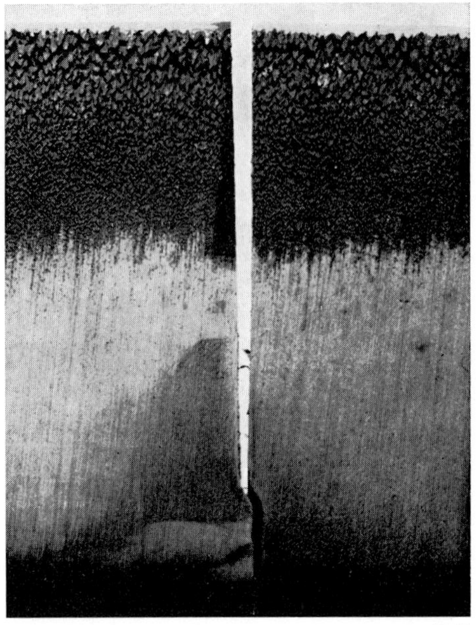

13.78

13.79

Fig. 13.77 to 13.81. Broken steam turbine blades

Fig. 13.77. Inner surface of broken-off outer part of blade. 0.5 ×. Fracture (left) and fatigue crack (↓) originated at eroded inlet edge

Fig. 13.78. Outer surface of the blade with erosion and incipient crack at the inlet edge. 3 ×

Fig. 13.79. Fracture of opened-up incipient crack. 2 x

Fig. 13.80 to 13.81. Eroded inlet edge. Transverse section. Etch: V2A-etchant

Fig. 13.80. 200 ×

Fig. 13.81. 500 ×

13.5 **Erosion**

Finally an example of failures of a special type of wear, i.e. **erosion,** should be cited. It is a surface attack that occurs in fast moving liquids and gases, especially if they carry along solid particles.

Seven out of 112 blades of the last low pressure stage disc of a **35 000 kW high pressure condensing turbine** fractured in an electric power plant 8 to 14 months after being put into service[22]). The milled blades without tie wires did not break at the fastening, i.e. the location of highest bending stress, but in a central region which was 165 to 235 mm away from the gripped end. The blades were produced from a stainless heat treatable chromium steel containing 0.2 % C and 13.9 % Cr. **Figure 13.77** reproduces the broken-off outer part of a blade with an incipient crack (arrow). The surface was covered with a dark oily layer which was soluble in toluene and could be wiped off easily. The inlet edge showed strong erosion characteristics which increased toward the end of the blade **(Fig. 13.78).** The crack originated at this edge. Forcing the crack open revealed a fatigue fracture **(Fig. 13.79).** The fracture surface was tarnished blue. **Figures 13.80 and 13.81** illustrate the surface structure in transverse section through the eroded inlet edge. The edge was full of fissures and incipient cracks. The microstructure showed signs of stress- and fatigue-induced corrosion cracking (see section 15.3.4 and 15.3.5). Otherwise it had an appearance normal for that steel. The cyclic fatigue strength of the blades was considerably reduced by the notch effect of the erosion pits – an effect that was possibly enhanced by corrosion; this may have lead to premature failure of the blades.

Another failure case in which erosion played a role will be described in section 15.3.2 (Fig. 15.57 to 15.60).

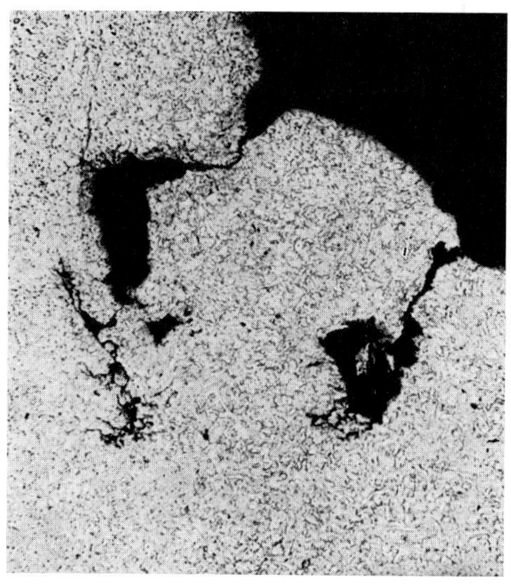

13.80

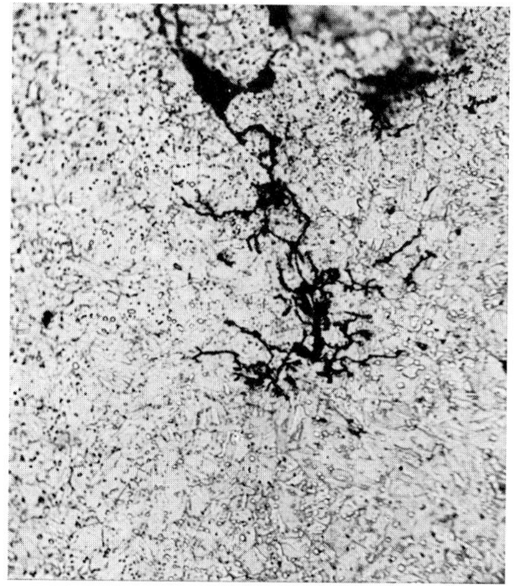

13.81

Literature Section 13

1) Allianz, Handbuch der Schadensverhütung. München u. Berlin 1972

2) F. K. Naumann u. F. Spies: Gebrochene Verbinder eines Fanggestänges. Prakt. Metallographie 11 (1974) S. 98/103

3) A. Grögli: Zeitstandbrüche an Rohren aus Dampferzeugeranlagen. Prakt. Metallographie 8 (1971) S. 179/85

4) F. K. Naumann: Beitrag zum Nachweis der σ-Phase und zur Kinetik ihrer Bildung und Auflösung in Eisen-Chrom- und Eisen-Chrom-Nickel-Legierungen. Arch. Eisenhüttenwes. 34 (1963) S. 187/94

5) E. Siebel: Die Formgebung im bildsamen Zustande. Verlag Stahleisen mbH., Düsseldorf, 1932

6) H. H. Heinemann: Formänderungsfähigkeit verschiedener Aluminium- und Kupferlegierungen bei hohen Formänderungsgeschwindigkeiten. Diss. T. H. Aachen, 1961

7) F. K. Naumann u. F. Spies: Gerissene Kranbahnschienen. Prakt. Metallographie 4 (1967) S. 495/98

8) G. Steinrück: Maschinenschaden 30 (1957) S. 79

9) H.-J. Wiester: Moderne Unfallverhütung, Essen 1958, S. 24

10) F. K. Naumann u. F. Spies: Radreifen mit Innenfehlern. Prakt. Metallographie 4 (1967) S. 541/46

11) H. Schlicht: Strukturelle Änderungen in Wälzelementen. Wear 12 (1968) S. 149/63

12) H. Schlicht: Der Überrollvorgang in Wälzelementen. Härterei-Techn. Mitt. 25 (1970) S. 47/54

13) W. Spieker, H. Köhler u. M. Kühlmeyer: Untersuchung über die Riffelbildung auf Schienen in Versuchsstrecken unter üblichen Bedingungen des Fahrbetriebes. Stahl u. Eisen 91 (1971) S. 1470/87

14) F. K. Naumann: Gefügeuntersuchungen an verriffelten Schienen. Arch. Eisenhüttenwes. 32 (1961) S. 617/26

15) G. Niemann: Maschinenelemente, 2. Bd. Getriebe. Springer-Verlag, Berlin, Göttingen, Heidelberg, 1960

16) F. K. Naumann u. F. Spies: Gebrochene Bandage von einem Walzwerksgetriebe. Prakt. Metallographie 8 (1971) S. 371/72

17) G. Niemann, H. Glaubitz: Zahnflankenfestigkeit geradverzahnter Stirnräder aus Stahl. Fachtagung Zahnradforschung, Hefte 1 bis 3 der Heftreihe Getriebeforschung, VDI-Verlag. Zeitschr. VDI 92 (1950) S. 741/42 u. 93 (1951) S. 121/26

18) F. K. Naumann u. F. Spies: Hartgebremste Eisenbahnradreifen. Prakt. Metallographie 6 (1969) S. 235/39

19) Dieselben: Gebrochene Innenringe von Pendelrollenlagern. Prakt. Metallographie 8 (1971) S. 247/51

20) Dieselben: Durch Rutschen des Walzgutes warm gewordene und gerissene Stahlwalzen. Prakt. Metallographie 5 (1968) S. 647/51

21) Dieselben: Durch Heißlaufen gerissene Drehbankspindel. Prakt. Metallographie 6 (1969) S. 125/28

22) Dieselben: Gebrochene Turbinenschaufeln. Prakt. Metallographie 9 (1972) S. 105/08

14. Thermal Overload Failures

In this section those failures that are caused by thermal overloading without any decisive participation of external mechanical stresses or chemical influences will be discussed. Such failures are caused by overheating and thermal fatigue stresses.

The following case may serve as an example for **failure by overheating.** A **heating coil** of 4 mm resistor alloy wire containing 30 % Ni and 20 % Cr, that was built into a bright annealing furnace, became brittle and broke at one end after a few weeks. The chemical composition corresponded to the specifications. The operating temperature was said to be around 930 °C. Since this alloy is stable up to 1100 °C in air under constant operating conditions, the temperature stated was subject to doubt from the start. The fracture was intergranular and very coarse-grained. Accordingly, the coil not only had an unusually coarse austenitic grain, but also showed incipient melting at the austenitic grain boundaries **(Fig. 14.1).** The same phenomena were also present, though to a smaller extent, at the other end that was not quite so brittle. In the surface zone, the originally fine grains had been stabilized by oxide inclusions **(Fig. 14.2).** The temperature of the coil therefore had substantially exceeded 930 °C, at least locally for short periods of time.

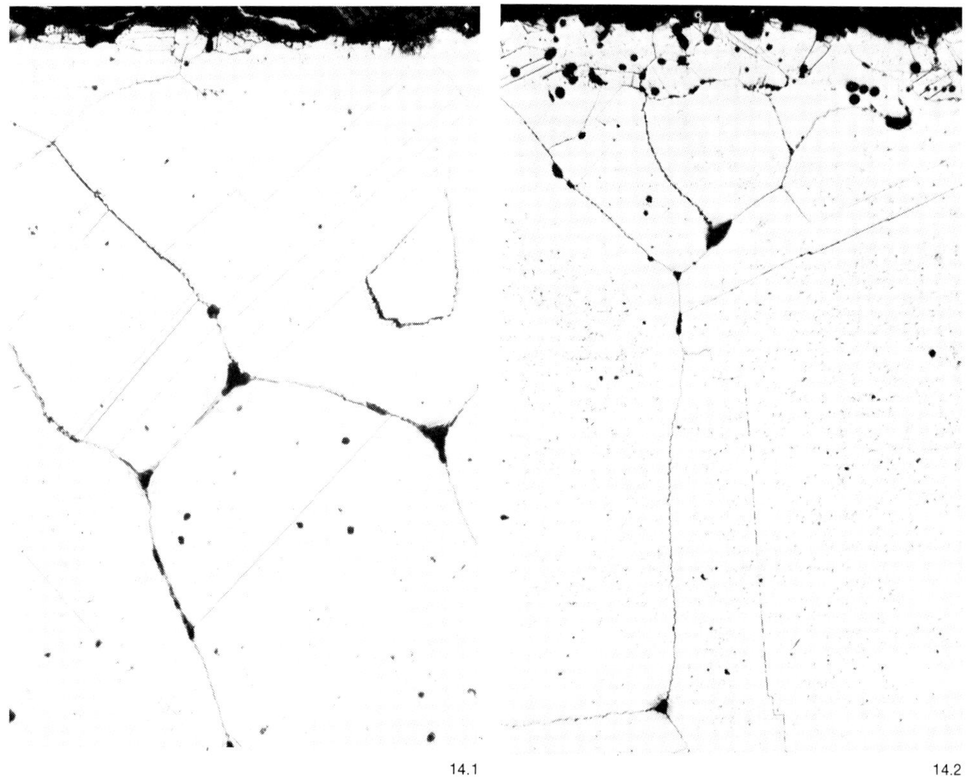

14.1 14.2

Fig. 14.1 and 14.2. Surface structure of a heating coil of 30–20 Ni-Cr alloy embrittled by overheating and incipient melting. Longitudinal sections. Etch: V2A-etching solution. 100 ×

Fig. 14.1. Adjacent to fracture

Fig. 14.2. At the other, less embrittled end

The failure of tuyeres of a blast furnace recuperator discussed in the last section (Fig. 13.22 to 13.26) could also have been included in this chapter.

In hot working tools and boilers, failures occur that have been designated as burn cracks, network cracks, or thermal shock cracks, but which are best designated as thermal fatigue **stress cracks**[1][2]. These are a group of cracks of shallow depth which frequently are arranged in a network-like pattern, occur due to continuously repeated rapid heating and cooling, and become wider and deeper with the number of cycles. They may lead to failures with severe consequences, if the parts affected by them are not repaired or removed in time.

Figure 14.3 shows such network of cracks at the internal wall of a **centrifugal casting chill mold** made of a hot working steel with 0.17 % C, 1.17 % Si, 0.76 % Cr, 0.32 % Mo, and 0.30 % V. After being artificially overloaded to fracture, the thermal fatigue cracks can be seen in **Fig. 14.4** to have a depth of less than 1 mm. Radiography gave no indication of their existence. A more severe crack which is much deeper, as is shown in **Fig. 14.5 and 14.6,** consists of an amalgamation of small cracks, all oriented in the same direction. This crack could clearly be seen in the radiograph. The cracks had a smooth transgranular path in the heat treated structure **(Fig. 14.7).** Their edges were oxidized but not decarburized. When the mold had been repaired by machining 2 mm from the inner wall, it later cracked longitudinally after barely 800 castings. Cooling water could then penetrate the interior and cause ejection of the liquid iron.

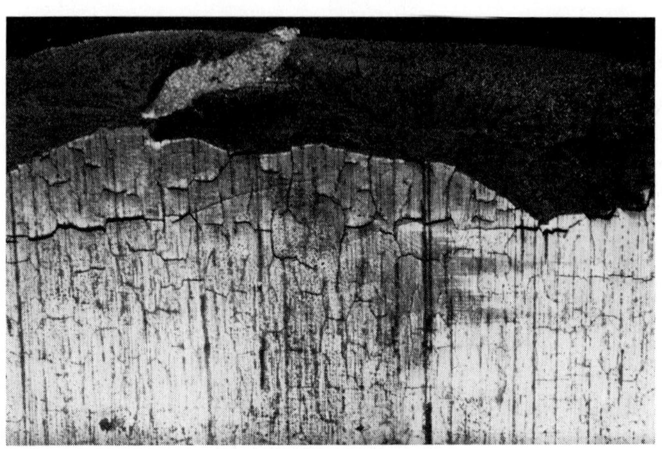

14.3

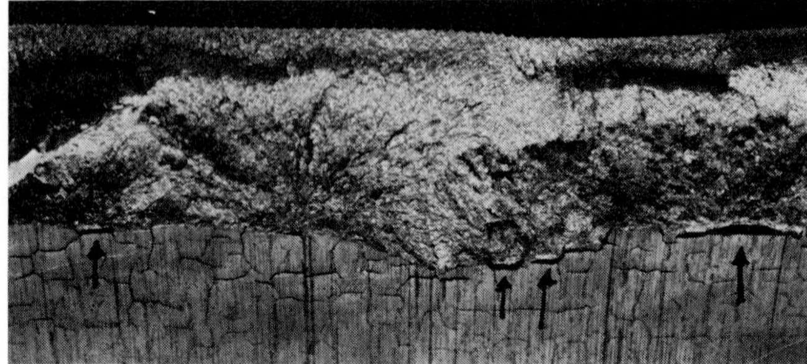

14.4

14.5

14.6

14.7

Fig. 14.3. to 14.7. Thermal fatigue stress cracks in centrifugal casting steel mold

Fig. 14.3. Network cracks at inside. 1 ×

Fig. 14.4. Network cracks in deliberately forced fracture. 1 ×

Fig. 14.5. and 14.6. Further advanced crack. 1 ×

Fig. 14.5. Seen from inner surface

Fig. 14.6. Fracture surfaces after deliberately inducec break

Fig. 14.7. Edge structure with crack. Transverse section. Etch: Picral. 100 ×

The process of crack formation may be visualized in such a way that during fast heating the thin surface layer is heated to a higher degree, but that the colder core zone is prevented from expanding and becomes subject to compressive stress; this may lead to upsetting when the yield point is exceeded. During cooling the outer zone is subjected to tensile stress. The repeated alternation of multi-axial compressive and tensile stresses leads to cracking when the endurance limit is exceeded.

Crack formation of this type can be counteracted by an appropriate design and operation that endeavors to avoid sharp temperature contrasts, as well as through the use of steel of high fatigue strength and good thermal conductivity.

14.8

14.9

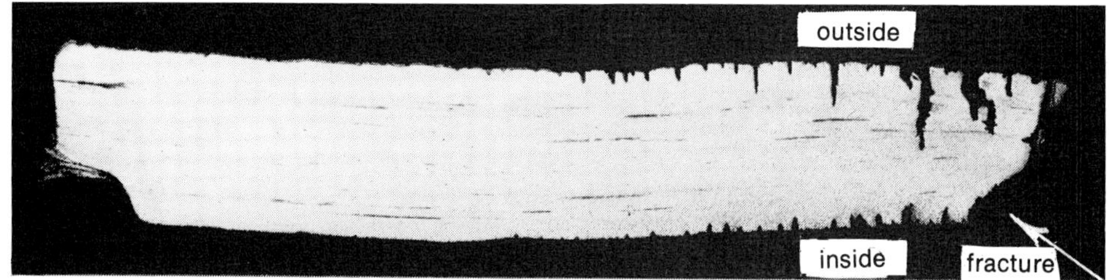

14.10

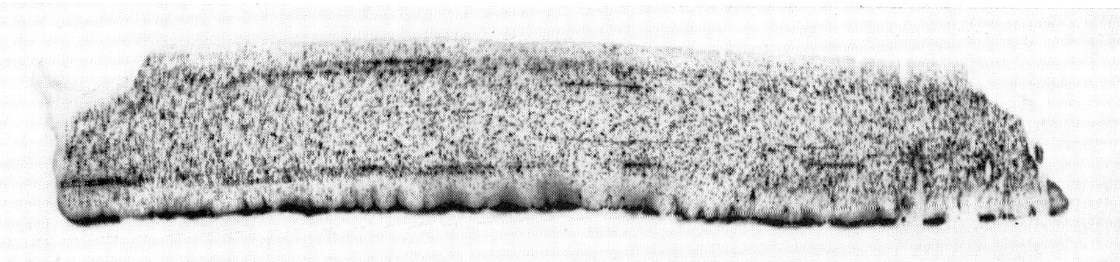

14.11

14.12

Fig. 14.8 to 14.12. Cracked drum of rotary furnace for calcination of soda

Fig. 14.8. Internal surface. 1 × --- Direction of retort axis. S--S metallographic section

Fig. 14.9. Opened crack. 1 ×

Fig. 14.10. Cut S--S according to Fig. 14.8. Etch according to Heyn. 1 ×

Fig. 14.11. Baumann sulfur print of cut S--S

Fig. 14.12. Unetched section. 3 ×

The **drum of a rotary kiln furnace** for the calcination of soda showed cracks inside and outside; it also developed a penetrating fracture below them at an angle auf 45° to the longitudinal axis of the drum. The drum was heated by oil burners from the outside. A section of the internal surface of the drum after dissolving the residual soda is shown in **Fig. 14.8.** The cracks propagated in part like the above fracture in a direction approximately 45° to the axis. Torsion stresses may have contributed to their formation. Other cracks, longer and flatter, were oriented parallel to the retort axis. But the outer cracks ran approximately at right angles to the longitudinal axis. Hence bend stresses would have been primarily responsible for their formation. After opening, these cracks were obviously much deeper than the internal cracks **(Fig. 14.9).** They had the appearance of fatigue failures. **Figure 14.10** reproduces a section that was cut perpendicular to the fracture, and **Fig. 14.11** is a Baumann sulfur print of it. The retort sheet consisted of a cast

14.13

14.14

14.15

Fig. 14.13 to 14.15. Coupling bar of streetcar broken at impact during shunting. Spot hardened by arcing shown in fracture origin

Fig. 14.13. Fracture. 1 ×

Fig. 14.14. Fracture. 5 × ↘ fracture origin

Fig. 14.15. Microstructure in fracture origin. 50 ×

rimmed steel as was proven by core segregation. The scale at the outside contained much sulfur. Therefore high sulfur content oil had probably been used to heat the furnace. The crack surfaces also were scaled **(Fig.14.12).** The cracks propagated across the fiber. From the straight line of the cracks that hat not yet gaped wide open due to scaling, the cracks could be seen to have taken a transgranular path. Judging by the microstructure, the steel contained approximately 0.1 % C and had been soft annealed. The strength level was correspondingly low.

Failure therefore occurred due to high alternating stresses acting in different directions. According to the established findings, thermal fatigue must also have had a contributory effect.

Among the failures due to thermal effects we may also count those that occur through sparking during cutting or welding and in the discharge of arcs in production or plant operation. Such failures have already been reported in section 1.2 (Fig. 1.16 to 1.18) and section 9.2 (Fig. 9.38 to 9.40). The fracture of a **coupling bar of a streetcar** could also have been due to such a cause[3]). The bar, consisting of a steel with approximately 0.6 % C, broke after an impact during shunting at a point that was not subjected to any particularly high stress **(Fig. 14.13).** In the fracture origin close to the edge of the rectangular section bar, a fine-grained area could be seen in the fracture **(Fig. 14.14)** that proved to be a martensitically hardened spot in a subsequently etched metallographic section **(Fig. 14.15).** A spot of the same type also appeared at exactly the corresponding point at the other end of the bar. Measurements showed that both burn marks were located just at the point where the bar left the coupling jaw. A flow of current between the motor driven car, which could be insulated, for example, by a snow blanket, and the well-grounded trailer could encounter a high contact resistance at this point, leading to the momentary formation of an arc. In view of the fact that the fracture propagated from this minimally stressed point, the conclusion was that the fracture of the bar was caused by local hardening.

Literature Section 14

1) W. Rädeker: Rißbildung in niedriglegierten Stählen durch schroffe Temperaturwechsel. Stahl u. Eisen 74 (1954) S. 929/43

2) Derselbe: Untersuchungen über die Wirkung schroffer Temperaturwechsel auf die Oberflächenbeschaffenheit von Stahl. Stahl u. Eisen 75 (1955) S. 1252/63

3) F. K. Naumann u. F. Spies: Bruch einer Kupplungsstange durch Funkenbildung. Prakt. Metallographie 5 (1968) S. 153/56

15. Failures caused by Excessive Chemical Attack (Corrosion)

Corrosion ist generally defined as the change of a material initiated at the surface due to undesirable chemical or electrochemical attack. Within this concept a differentiation should be made between corrosion by hot and primarily oxidizing gases, the special case of hydrogen attack, and the metal regression or selective attack by aggressive liquids (electrolytes).

15.1 High Temperature Corrosion

The normal case of scaling is the attack of the metal by the oxygen in the atmosphere. This oxidation is strongly temperature dependent and obeys a parabolic time law. The growth of oxide scale occurs in part through the inward migration of oxygen and in part through outward diffusion of the metal. In unalloyed steel the scale consists of the oxides of iron with an oxygen

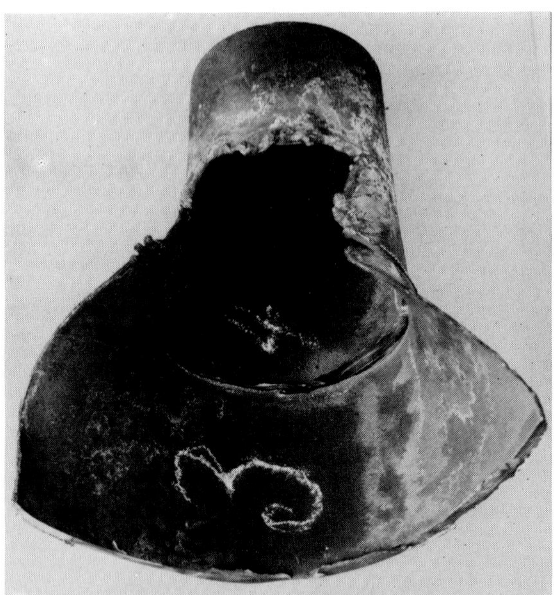

15.1

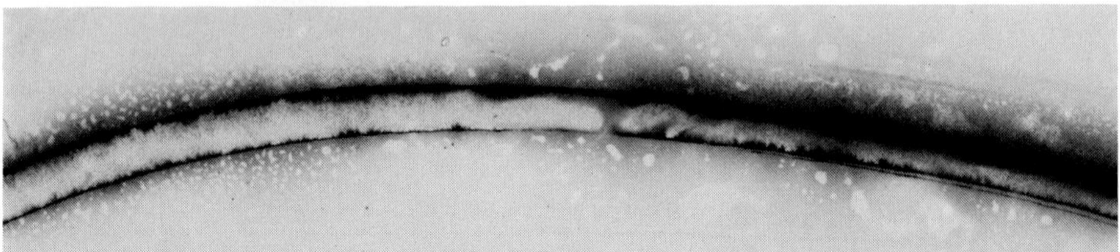

15.2

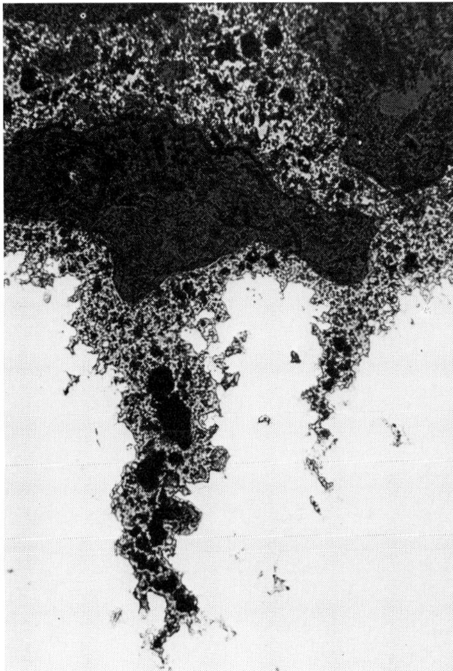

15.3

15.5

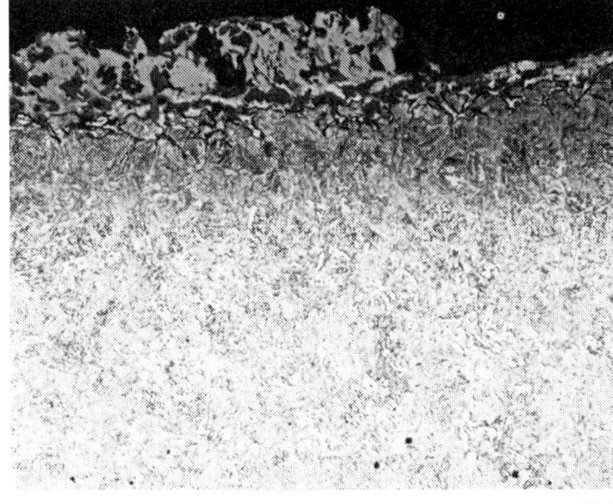

15.4

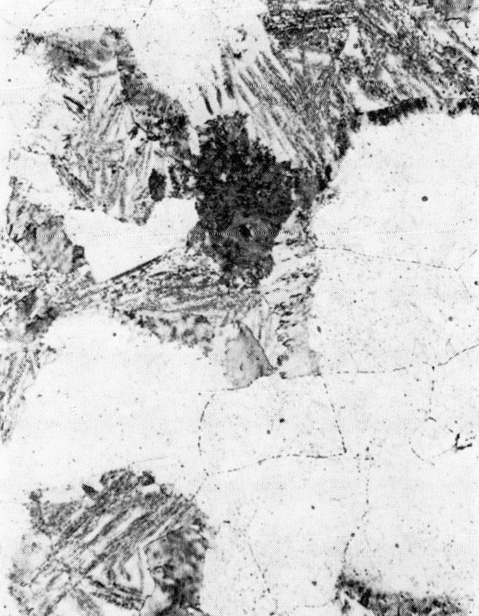

15.6

Fig. 15.1 to 15.6 Burst pipe of a naphtha desulfurization plant

Fig. 15.1. Damaged region. 0.25 ×

Fig. 15.2. Baumann sulfur print of transverse section next to rupture. Reproduction. 1 ×

Fig. 15.3. Outside edge structure in vicinity of crack. Unetched transverse section. 100 ×

Fig. 15.4. Inside edge structure. Transverse section. Etch: Picral. 100 ×

Fig. 15.5. Core structure of a burst pipe. Transverse section. Etch: Picral. 500 ×

Fig. 15.6. Core structure of an unused pipe. Transverse section. Etch: Picral. 500 ×

content that decreases from the outside to the inside. They are easily distinguished metallographically. Oxide scale layers are never fully dense. Sulfur in combustion gases reinforces the attack, especially in cases where partial pressure of oxygen is low. In those cases, metal sulfides constitute the corrosion products, while with high partial pressure of oxygen, oxide formation predominates. Hydrogen sulfide is already aggressive at temperatures of 400 to 600 °C[1]).

Resistance to scaling is increased if the steel is alloyed with chromium, silicon and/or aluminum, and also if rare earth metals and zirconium are added. All these alloy additions form non-penetrable and firmly adhering oxide scale layers. Nickel has an unfavorable effect in the presence of sulfur because of the formation of a low melting nickel-nickel sulfide eutectic. The protective effect may also be cancelled and oxidation intensified by the deposition of vanadium pentoxide-containing oil ashes and other dusts such as lead oxide and alkaline sulfates which react with the protective scale layer under formation of low melting reaction products[2])[3])[4])[5]).

In nitrogen-containing and carburizing gases, scale resistance can be affected by a decrease of chromium and aluminum in the steel matrix through formation of nitrides or carbides.

At low partial pressures of oxygen or water vapor, a dense protective oxide layer cannot be formed. As a consequence, in nickel-rich austenitic alloys internal oxidation only occurs with the more easily oxidized chromium while it leaves the nickel unattacked; this phenomenon is called "green rot" because of the green color of the precipitated chromium oxide.

An example of hydrogen sulfide attack scaling is given by the following failure in a **naphtha desulfurization oil refinery plant.** A pipe in which crude oil was heated by light refinery gases at 35 atm pressure from 290 to 370 °C ruptured longitudinally. The pipe wall temperature was said to heat to 600 °C. The pipe, which had a 114 mm O.D. and 6.4 mm wall thickness, consisted of a steel with 2 to 2.5 % Cr and 1 % Mo. The fracture, after knocking off a black deposit of 0.8 mm

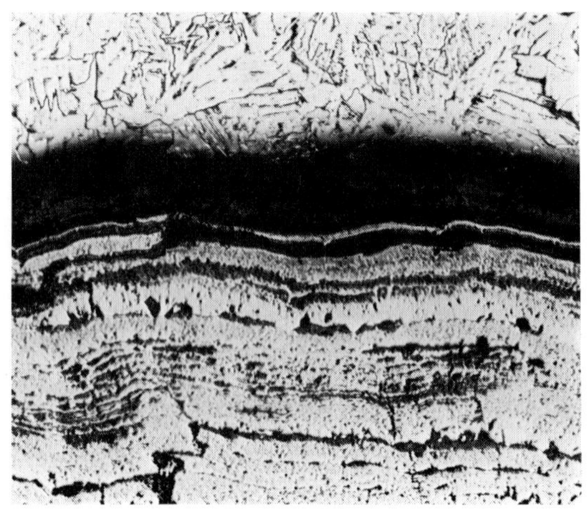

Fig. 15.7. Hydrogen sulfide deposit formed at 400–500 °C at inside of pipe from a gas generating plant. 100 ×

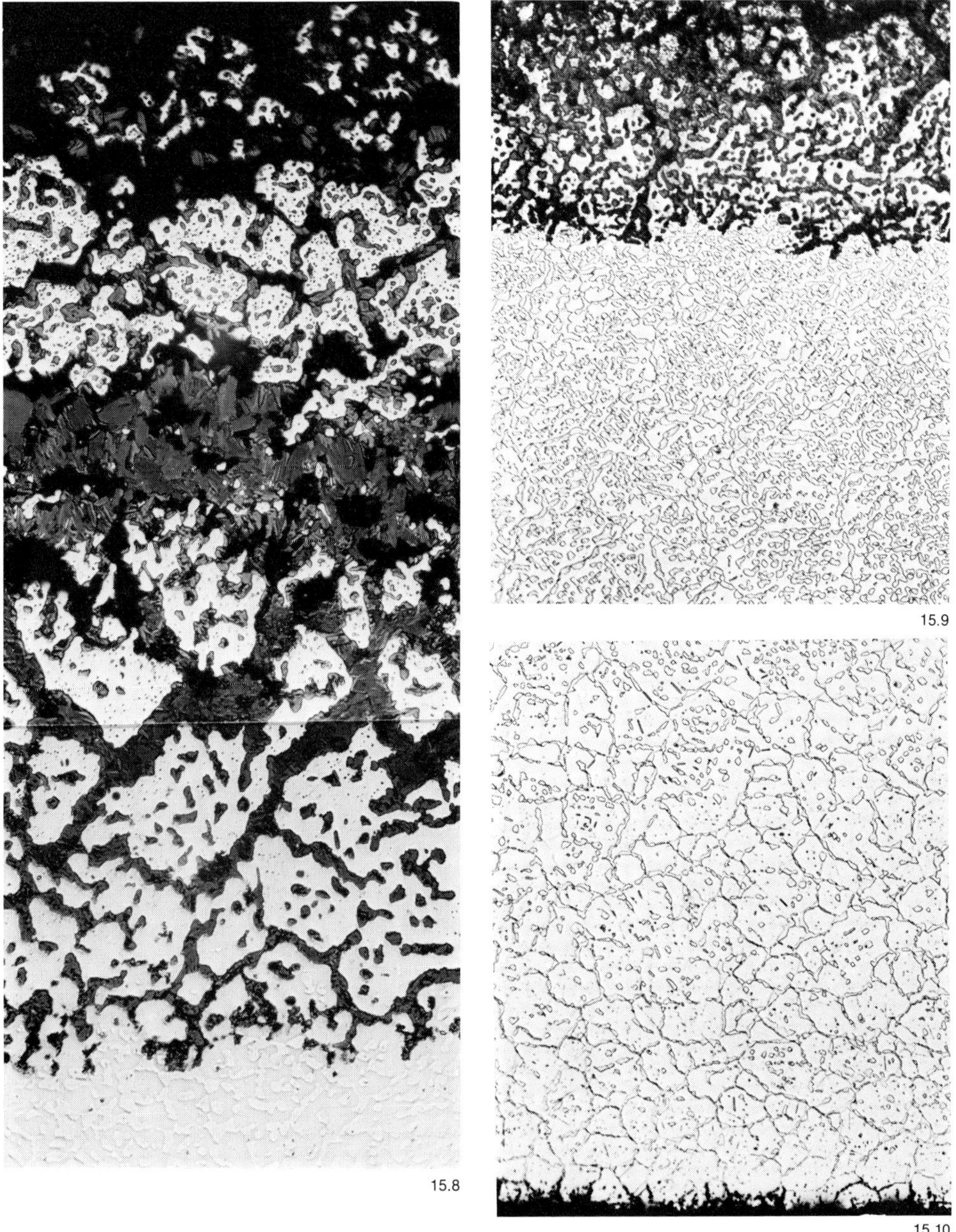

15.8

15.9

15.10

Fig. 15.8 to 15.10. Internally scaled tube of AISI type 310 steel damaged by moist hydrogen-rich protective gas

Fig. 15.8. Oxide scale at internal surface. Unetched. 500 ×

Fig. 15.9. Microstructure at internal surface. Etch: V2A-etchant. 100 ×

Fig. 15.10. Microstructure at external surface. Etch: V2A-etchant. 100 ×

thickness on the inside, is shown in **Fig. 15.1.** The pipe had bulged considerably prior to cracking. The wall thickness had decreased to about 5.8 mm in the fracture cross section and contracted to about 4 mm at the crack edges.

A Baumann sulfur print established that the pipe was strongly sulfurized on both the inside and outside **(Fig. 15.2).** The inner deposit contained 70 % iron sulfide as was confirmed by analysis. During microscopic examination of the section, several short incipient cracks were found in the direction parallel to the crack **(Fig. 15.3).** The surface and crack walls were covered with iron sulfide and slightly carburized **(Fig. 15.4).** The core structure of the ruptured pipe **(Fig. 15.5)** showed the fine intergranular cracks that form in steels of this type if the creep strength ist exceeded[6]). A comparison with the microstructure of an unused pipe **(Fig. 15.6)** indicated that the specified wall temperature of 600 °C had been substantially exceeded, provided that the original state was the same. This could be assumed already from the strong bulging of the pipe prior to cracking. Probably this high temperature was not the cause but the consequence of the formation of the thermally insulating deposit.

The cause of failure therefore was the attack of the steel by hydrogen sulfide that had formed from the crude oil. The failure could have been prevented or at least delayed if an alloy higher in chromium or a chromium-silicon steel would have been used as material for these pipes[1]).

Figure 15.7 shows a micrograph of the black deposit at the inner wall of a pipe used in a **gas generating plant** that had failed after 4 months service as a result of considerable scaling. The generator was used to produce gas with a composition of 6 % CO_2, 20 % CO, 8 to 12 % H_2, 0.5 to 1.5 % CH_4, remainder N_2. The temperature was said to be only 400° to 500 °C. An analysis of the deposit showed 5.7 % C, 26.5 % S, and 50.9 % Fe. Accordingly it consisted of iron sulfide that was mixed with soot and rust. This type of scale which is formed at low temperature can be differentiated by its sedimentary structure from that formed a higher temperature for which examples will be shown later.

Two examples for the scaling of heat resistant steels at high temperature in oxygen-lean gases are cited below.

Tubes of a wire annealing furnace made from austenitic type 310 steel became completely brittle after 1½ months and partly cracked in the longitudinal and girth welds. They had been heated from the outside with burner gases to 1100 °C and were fed a hydrogen-rich gas mixture on the inside for protection from oxidation of the wire that passed through them.

The tubes were not attacked on the outside, but considerably scaled on the inner surface. The scale contained only 0.002 % S and therefore consisted primarily of oxide. In the remaining base metal, 1.46 % and 0.46 % N were found, while otherwise the composition corresponded to specifications except for the deficiency in chromium due to selective oxidation.

Metallographic examination of the tubes indicated that the external surfaces had only a thin, dense and firmly adhering protective oxide layer that is characteristic for this type of steel under the influence of strongly oxidizing atmospheres. However, in the interior of the tube a thick oxide scale had formed that had largely flaked off. Oxidation had progressed preferentially along the austenitic grain boundaries **(Fig. 15.8).** After etching of the metallographic sections, the metal that was lower in chromium and richer in nickel was found to be free of precipitates in the oxide network. But underneath, carbon had accumulated in a considerable concentration **(Fig. 15.9).** The amount of precipitates decreased continuosly from the inner to the outer surface **(Fig. 15.10).**

The embrittlement of the tubes accordingly was caused by precipitation of carbides and nitrides beneath the oxide scale. Water vapor probably was the oxidizing agent.

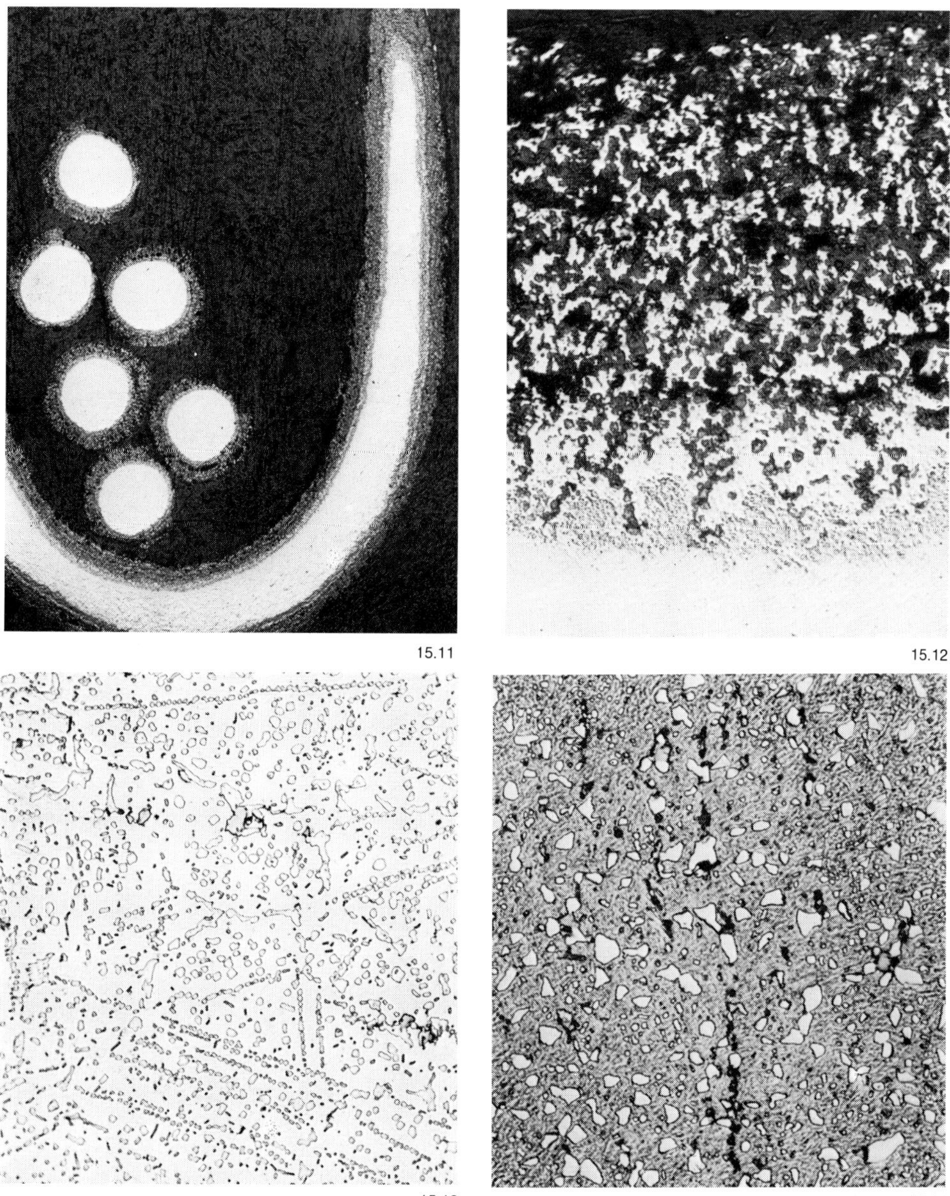

Fig. 15.11 to 15.16. Scaled and embrittled austenitic wires of a conveyor belt that was exposed to a protective atmosphere of partially combusted illuminating gas

Fig. 15.11. Unetched longitudinal and transverse sections through an embrittled and broken wire. 9 ×

Fig. 15.12. Surface structure of an embrittled wire. Transverse section, unetched. 200 ×

Fig. 15.13 and 15.14. Core structure of embrittled wire. Longitudinal section. 500 ×

Fig. 15.13. Etch: V2A-etching solution

Fig. 15.14. Etch: Aqua regia

The **wires of a conveyor belt,** that had operated in an annealing furnace at 870 to 900 °C under a protective atmosphere of partially combusted city gas, became brittle within ½ year, and the thinner wires broke frequently. The protective atmosphere contained 6.2 % CO_2, 5.4 % CO and 12.1 % H_2 but no oxygen at entry into the furnace.

This atmosphere had a slightly decarburizing effect upon the annealing charge, a heat treatable alloy steel of AISI type 5140.

The wires of the conveyor belt consisted of a heat resistant steel with 0.15 % C, 2.0 % Si., 18.4 % Cr and 10.8 % Ni, based on an analysis for a new belt.

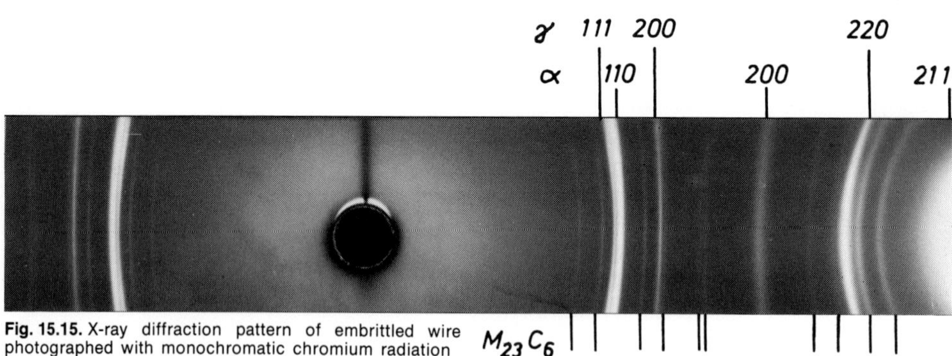

Fig. 15.15. X-ray diffraction pattern of embrittled wire photographed with monochromatic chromium radiation

Fig. 15.16. Microstructure of wire of unused conveyor belt. Longitudinal section. Etch: V2A-etchant. 200 ×

In the embrittled wires of the used belt, carbon content had risen to 0.84 %, and chromium content fallen to 16.4 %. Therefore the furnace atmosphere had a carburizing effect upon the steel of the wires and an oxidizing effect as far as chromium was concerned.

The result of chemical analysis was confirmed by metallographic examination. **Figure 15.11** shows in an overview a longitudinal section and a number of transverse sections of an embrittled and broken wire. They indicate the scaling of the surface. **Fig. 15.12** shows the structure at the surface and **Fig. 15.13** that of the core. The oxidation at the surface and the carburization of the core are clearly noticeable. Carbon has disappeared from the oxidized surface and accumulated in the interior of the wire. The basic microstructure of the core material had become martensitic by carburization and chromium loss of the austenitic matrix. This could be seen after deeper etching with aqua regia **(Fig. 15.14).** No -phase could be found either by special etching or by X-ray diffraction analysis. However, the diffraction pattern clearly showed lines of α-iron and of chromium carbide $Cr_{23}C_6$ in addition to those of γ-iron **(Fig. 15.15)** The austentic matrix of a wire of the unused conveyor belt shown in **Fig. 15.16** should be compared with Fig. 15.13. This wire proved to be completely incapable of being magnetized in contrast to the embrittled one which had become strongly magnetic.

Gas-heated annealing furnaces are sometimes operated with an excess of gas for the protection of the charge against oxidation and decarburization. This may lead to the generation of atmospheres – especially if they are rich in sulfur – that form sulfides with the metal and prevent the formation of a protective layer in the heat resistant furnace parts, as is cited in the following example.

Transport rollers of a sheet metal annealing furnace hearth oxidized almost completely after three months of operation at approximately 950 °C. They consisted of a heat resistant cast steel which contained approx. 0.4 % C, 1.75 % Si, 26 % Cr and 14 % Ni. According to the German handbook for irons and steels,* this alloy is said to be stable up to about 1150 °C in air. The furnace was operated with crude generator gas with a substantial excess of gas as expressed by the following exhaust gas analysis:

CO_2 %	H_2O %	CO %	H_2 %
12.4	9.7	8.8	4.7

Chemical analysis of base metal and scale resulted in the data in **Table 1.** Particularly noticeable is the high carbon content of the base metal and the high sulfur content of the scale. The heating gas therefore was sulfur-rich. At this overabundance in gas it may be assumed that sulfur was present at least in part in the form of hydrogen sulfide.

Table 1. Chemical Composition of Base Metal and Scale of Oxidized Transport Rollers

	Fe %	C %	Si %	Mn %	P %	S %	Cr %	Ni %	Weight gain %
Base metal	52.9	2.68	1.84	1.09	0.16	0.13	26.3	16.5	–
Oxide scale	39.3	0.46	1.17	0.53	0.13	2.76	21.5	0.16	6.56

* **Werkstoff-Handbuch Stahl und Eisen,** 4th Ed., Stahleisen, Düsseldorf, 1965, Sheet O 91

If the sum of metallic constituents is assumed to be one hundred in the composition according to Table 1, and if the data are compared to an analysis of the charge, the values summarized in **Table 2** would result. Accordingly, iron and chromium contents are considerably higher in the oxide scale and correspondingly lower in the remaining base metal. However, only minor amounts of nickel are present in the scale, whereas the remaining metal is rich in nickel.

Table 2. Comparison of Composition of the Charge with Metallic Constituents of Base Metal and Scale Adjusted to 100 % for Oxidized Transport Rollers

	Fe %	Si %	Mn %	Cr %	Ni %	Sum %
Melt	55.60	1.80	0.70	27.30	14.60	100
Base Metal	53.65	1.87	1.12	26.65	16.71	100
Scale	62.65	1.87	0.85	34.37	0.26	100

Metallographic examination verified that the roller's original 10 mm wall thickness had been reduced to only a minor amount of a metallic substance **(Fig. 15.17).** Scaling led to a material erosion on the outside, while on the inside it led to a selective attack along the interstices and microcavities of the dendritic cast structure **(Fig. 15.18).** The oxide scale consisted of two type of crystals, light grey and dark grey **(Fig. 15.19),** and contained homogeneous metallic dispersions

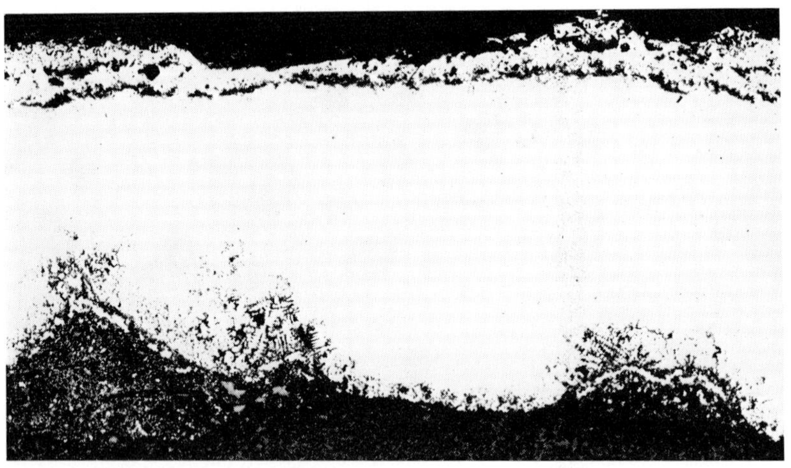

15.17

Fig. 15.17 to 15.21. Scaled transport rollers of annealing furnace operated with an excess of sulfur-rich crude generator gas

Fig. 15.17. Unetched transverse section through remaining part of wall. 10 ×

Fig. 15.18 and 15.19. Microstructure at internal surface, unetched. 200 ×

Fig. 15.18. Transition zone

Fig. 15.19. Scale

Fig. 15.20 and 15.21. Core structure. Etch: V2A-etching solution

Fig. 15.20. 100 ×

Fig. 15.21. 500 ×

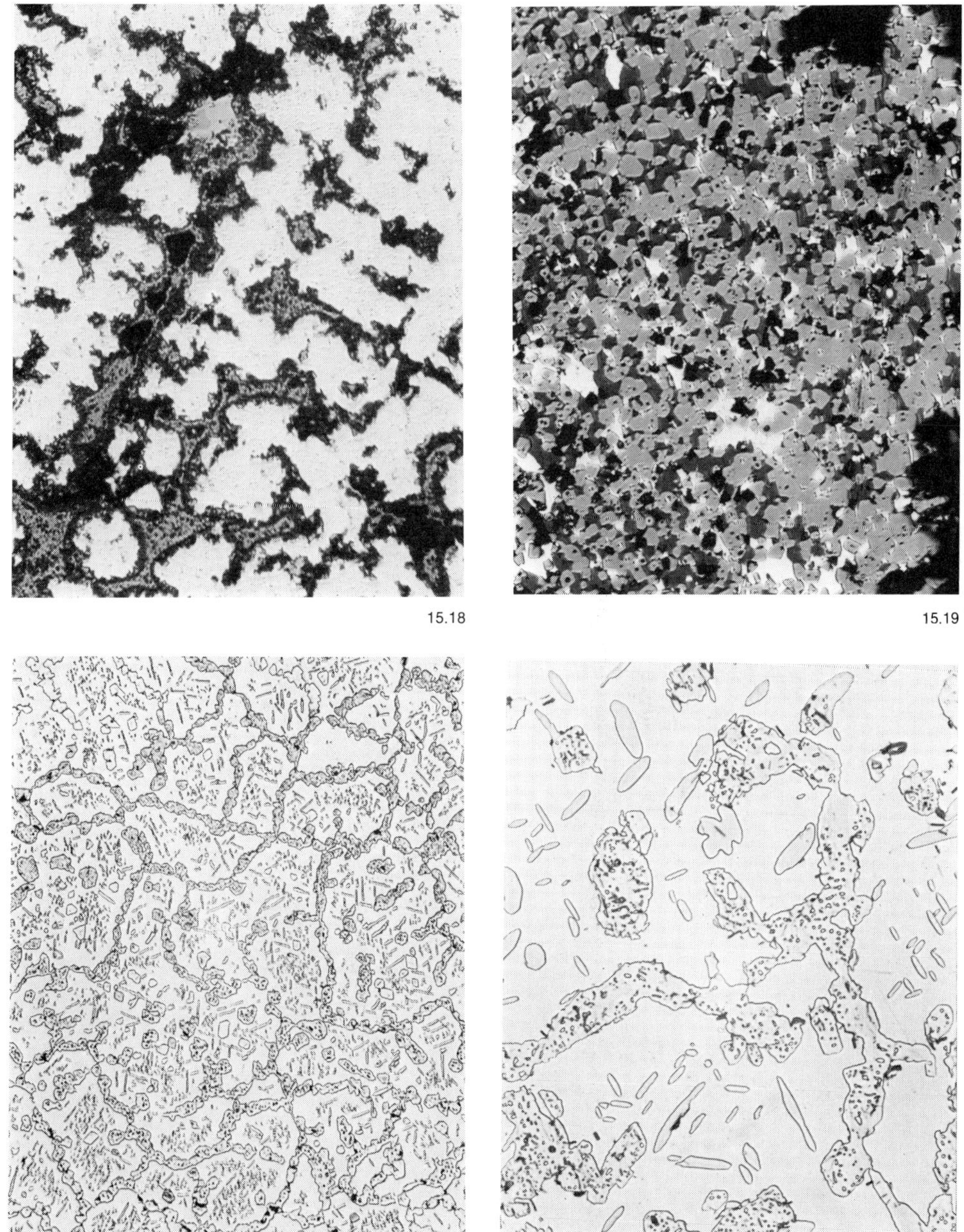

15.18

15.19

15.20

15.21

15.22

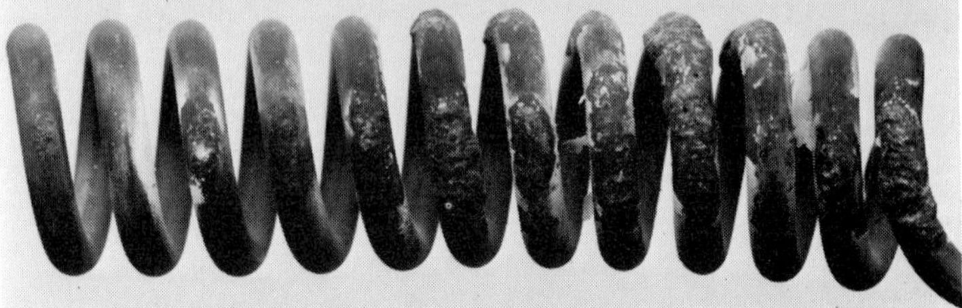

15.23

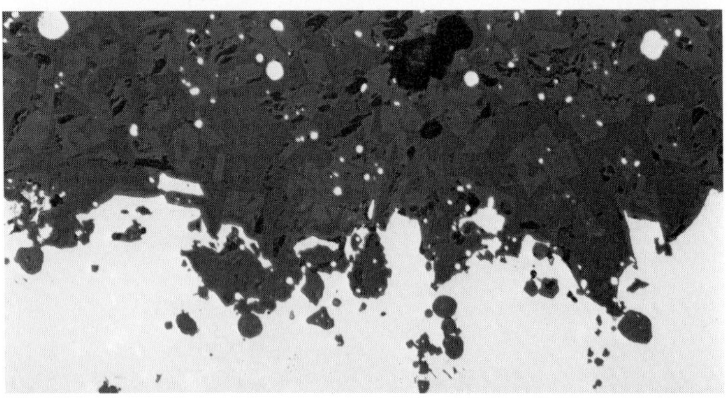

15.24

Fig. 15.22 to 15.27. Scaled heating coils of heat resistant chromium-aluminum steel that were exposed to weakly oxidizing atmosphere

Fig. 15.22. Section of lower furnace part

Fig. 15.23. Locally attacked coil. 1 ×

Fig. 15.24. Surface structure of scorious wire, unetched. 500 ×

Fig. 15.25. Microstructure of wire at an unattacked place. Transverse section. Etch: V2A-etchant. 200 ×

Fig. 15.26. Cavity in melted region. Transverse section. Etch: V2A-etchant. 200 ×

Fig. 15.27. Precipitates of aluminum nitride at an attacked spot. Unetched transverse section. 500 ×

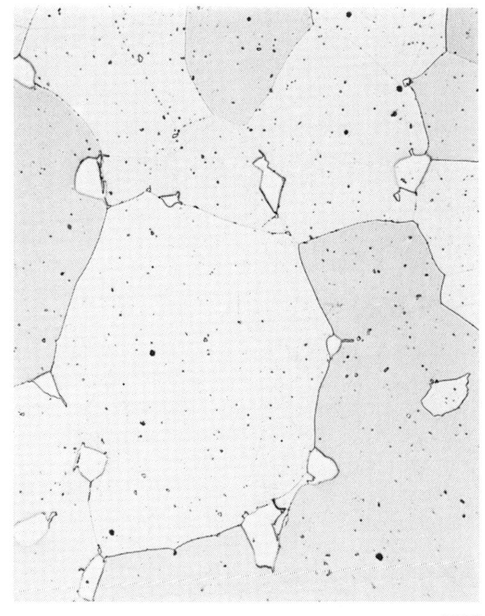

15.25

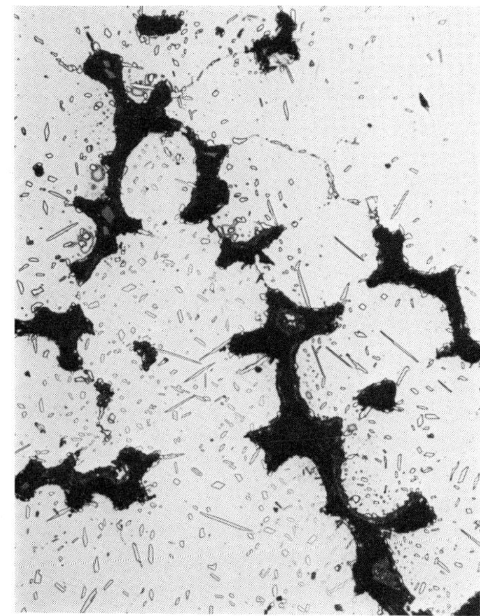

15.26

15.27

that had no discernible structure even after etching with nital and V2A-etchant. They may possibly have consisted of nickel or a nickel-rich solid solution. The matrix showed precipitates in the grain and on the grain boundaries over the entire remaining cross section, that probably consisted predominantly of carbides **(Fig. 15.20 and 15.21).**

If the excessive gas flow could not have been eliminated in view of the material to be annealed, a semi-ferritic cast or rolled steel that is lower in nickel, e.g. one containing about 0.4 % C, 27 % Cr and 4 % Ni, should have been used for the rollers. This alloy composition would offer a greater corrosion resistance to sulfur-containing gases.

Steels alloyed with aluminum are especially sensitive against weakly oxidizing and especially nitrogen-containing gases. These steels are highly stable in strongly oxidizing atmospheres, but may be attacked locally if there is a lack of oxygen[7]). The following failure analysis may serve as an example.

An industrial furnace used to anneal farm machinery parts was put out of service because the **heating coils** burned through. The wires which rested freely in the heating chamber consisted of a steel with 0.07 % C, 21.3 % Cr and 5.7 % Al. An alcoholic carburizing agent or xylene had been added to the furnace atmosphere for the absorption of oxygen. Nevertheless, the annealed parts were slightly oxidized.

The coils displayed all types of degradation. Long stretches were unattacked, but locally they showed signs of incipient melting and slag formation – partly at the bottom, partly on top – where they may have been in contact with the furnace brickwork **(Fig. 15.22).** In the vicinity of such spots, a white deposit could be seen that apparently consisted of aluminum oxide. In part the coil had completely melted under formation of a black slag through the reaction with the brickwork. Strongly attacked and completely unscathed areas were located immediately adjacent to one another in the same winding **(Fig. 15.23).**

The slag structure consisted of oxide crystals with a eutectic matrix in which metallic drops were dispersed **(Fig. 15.24).** The wire that in the unattacked regions consisted of a coarse grained ferritic structure with chromium carbide **(Fig. 15.25)** had been heated to melting in the oxidized areas through an increase in resistance due to a reduction in cross section **(Fig. 15.26).** In such places precipitates of aluminum nitride could also be found in the microstructure **(Fig. 15.27).** These contribute to a decrease in oxidation resistance as a result of nitrogen combining with the aluminum.

As a remedial step the furnace design should be altered in such a way that the coils are no longer in contact with the furnace atmosphere. If this is not possible, the steel used as the heating element should be replaced by an austenitic chromium-nickel or iron-chromium-nickel alloy that is more corrosion resistant than chromium-aluminum steel under similar conditions, provided the furnace temperature does not exceed 1100 °C.

The next failure case is an example for the lowering of stability through a reaction of the scale **with mineral deposits.** A pipe of heat resistant cast steel containing approx. 0.4 % C, 2 % Si, 17 % Cr and 37 % Ni, in use as **reaction retort in a natural gas cracking plant,** had been heated from the outside by a gas flame, whereby a maximum temperature of 1010 °C was said to have been reached at the hottest point. The pipe consisted of two sections joined together with a circumferential weld. The sections had been produced by centrifugal casting. The mold had been lined with sand.

The pipe was considerably elongated and in part warped out of shape when compared to its original state. Its wall was externally eroded on one side by more than half as a result of scaling. Remarkably the girth weld seam between the sections and the beveled edges of the V-groove were hardly oxidized at all.

According to chemical analysis, the scale contained only 0.006 % S, but 37.5 % SiO_2.

After sectioning, metallographic examination disclosed a change in the macroscopic structure **(Fig. 15.28)** on the outside below the oxide scale which had been removed before. The austenite could be seen under the microscope to be slightly carburized in this zone and permeated by sand particles **(Fig. 15.29).** The particles had reacted in many cases at their edges with the surrounding iron and scale **(Fig. 15.30 and 15.31),** and in part had formed a eutectic with the iron silicate fayalite (2FeO . SiO_2) as a component **(Fig. 15.32).** At the inner surface the microstructure was not markedly changed when compared with a new comparison pipe received from the factory **(Fig. 15.33).**

Heating tests with specimens cut from the unused pipe showed that comparable microstructural changes occurred only at temperatures above 1100 °C. Most likely, therefore, the maximum temperature stated must have been substantially exceeded. The steel was overloaded both with respect to its scale resistance and its high temperature strength; however, scale formation was no doubt accelerated by the presence of the sand particles. This became obvious from the fact that the weld seam and the machined areas of the pipe sections were less oxidized than the unmachined outer surface.

The final conclusion was that the pipes had either to be machined henceforth or that sand lining of the molds had to be eliminated. The mold maker chose the latter course.

An example for **green rot** mentioned at the beginning of this chapter is demonstrated by the following failure analysis:

Heating elements made from a 40 mm wide and 2 mm thick strip of 80-20 Ni-Cr type resistance heating alloy that is stable against oxidation in air up to 1200 °C, were rendered unusable after a few months' service in a continuous annealing furnace by strong oxidation; the operating temperature of 1050 °C had supposedly not been exceeded[8]). In this furnace generator and transformer sheets were annealed in a decarburizing atmosphere of moist hydrogen.

The elements displayed surface bulges or protrusions preferentially near the edges and the bends **(Fig. 15.34),** and internal gaps in the fracture plane **(Fig. 15.35).** The fractures were colored green by chromium oxide except for a narrow, bright metallic strip in the core of the ribbon.

Transverse and longitudinal sections showed that the regions under the surface protrusions had been heavily oxidized **(Fig. 15.36).** The attack had started from isolated points where the protective layer had apparently been punctured and from there advanced concentrically. Adjacent to the protrusions the surface of the ribbon was often completely scale-free. If scaling had started from two opposite ends of the ribbon, voids were formed in the middle between them which in part gaped wide open.

Figure 15.37 illustrates the microstructure of the scale layer. Oxidation apparently proceeded initally on the austenitic grain boundaries. Two explanations are possible for the formation of the voids. They may either be sites of incipient melting or concentrations of vacancies caused by the escape of chromium. Since no traces of a solidification eutectic structure could be found, the second explanation is the more plausible one. The voids may have been enlarged by the penetration of atomic hydrogen from the reaction of water vapor with the metal and its molecular precipitation under pressure.

As a preventive measure prior to restoring the furnace to service, heating the coil under strongly oxidizing conditions in air – a treatment to be repeated from time to time between the hydrogen treatments – came under consideration. But permanent elimination of failures could only be expected by substituting a less sensitive iron-nickel-chromium alloy for the nickel-chromium alloy used.

15.28

15.29

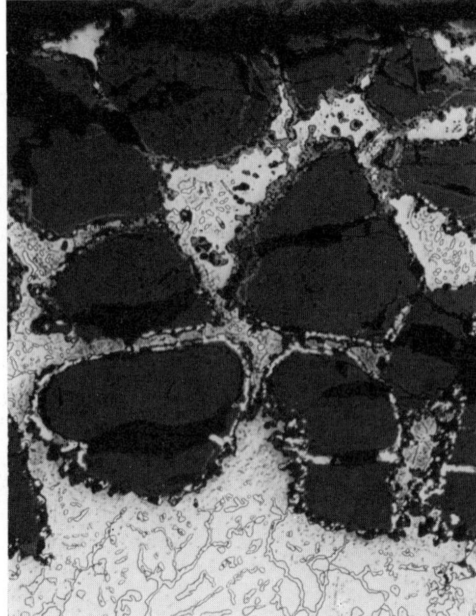

15.30

Fig. 15.28 to 15.33 Scaled and warped centrifugally cast pipe of a natural gas cracking plant

Fig. 15.28. Transverse section. Etched with aqua regia. 0.5 ×

Fig. 15.29 and 15.30 Microstructure below outside surface. Transverse section. Etch: V2A-etching solution. 100 ×

Fig. 15.31. Reaction products at edge of sand grain. Etch: V2A-etchant. 500 ×

Fig. 15.32. Fayalite eutectic below the outside surface. Etch: V2A-etchant. 200 ×

Fig. 15.33. Microstructure at inside surface. Transverse section. Etch: V2A-etching solution. 100 ×

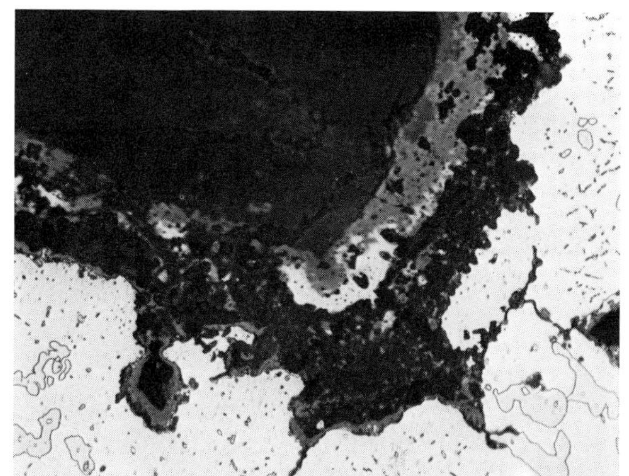

15.31

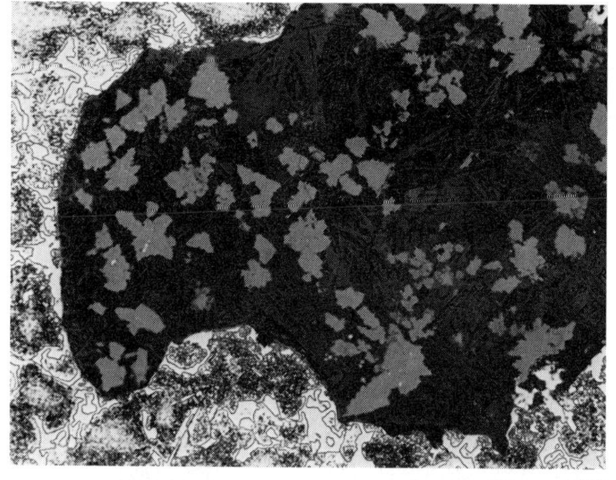

15.32

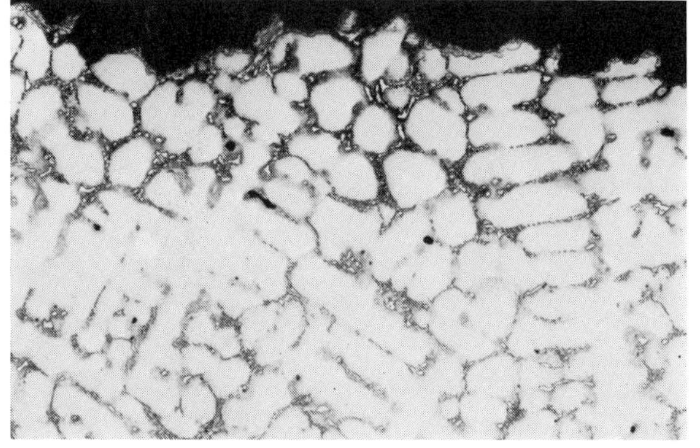

15.33

15.34

15.35

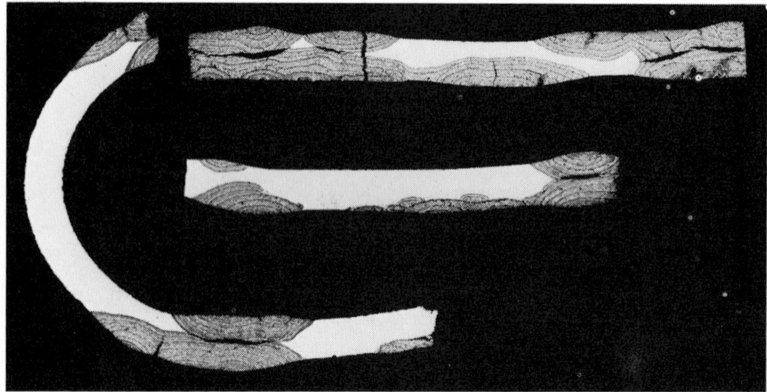

15.36

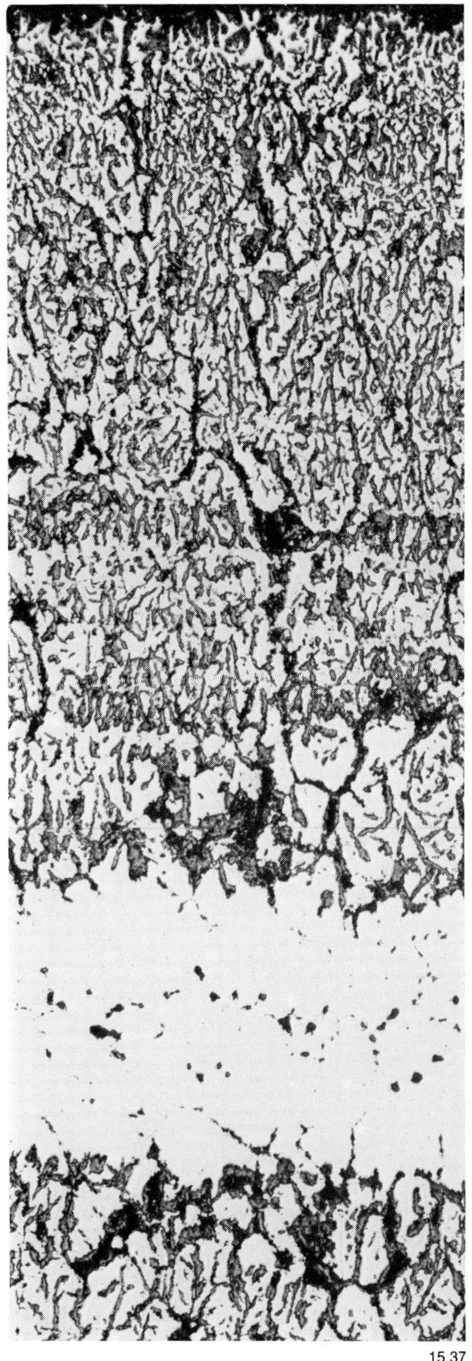

15.37

Fig. 15.34 to 15.37. 80–20 Ni-Cr resistance heating element oxidized in moist hydrogen atmosphere of an annealing furnace.

Fig. 15.34. Protrusions on the surface of heating element. 0.7 ×

Fig. 15.35. Fractures with internal gaps. 1 ×

Fig. 15.36. Unetched longitudinal sections. 3 ×

Fig. 15.37. Structure of scale layer and intergranular voids in core of heating element. Unetched transverse section. 100 ×

15.38

15.39

Fig. 15.38 to 15.43. Ruptured distribution manifold from cooling unit of ammonia synthesis plant

Fig. 15.38. Side view. 0.4 ×

Fig. 15.39. Fracture. 0.6 ×

Fig. 15.40. Intergranular incipient cracks of attacked zone. Etch: Nital. 10 ×

Fig. 15.41. Structure at edge of longitudinal boring showing hydrogen crack (horizontal in Fig. 15.39). Etch: Picral. 500 ×

Fig. 15.42 and 15.43. Traces of hydrogen attack in unetched section. 500 ×

Fig. 15.42. 11 mm below surface

Fig. 15.43. 14 mm below surface p 404

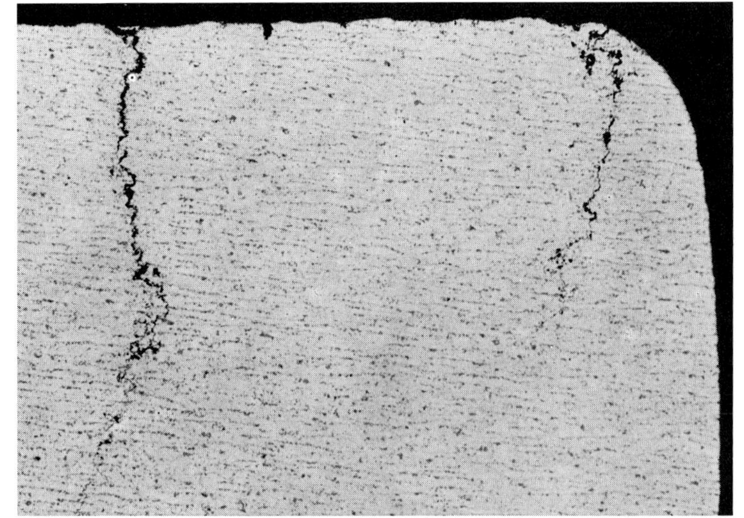

15.40

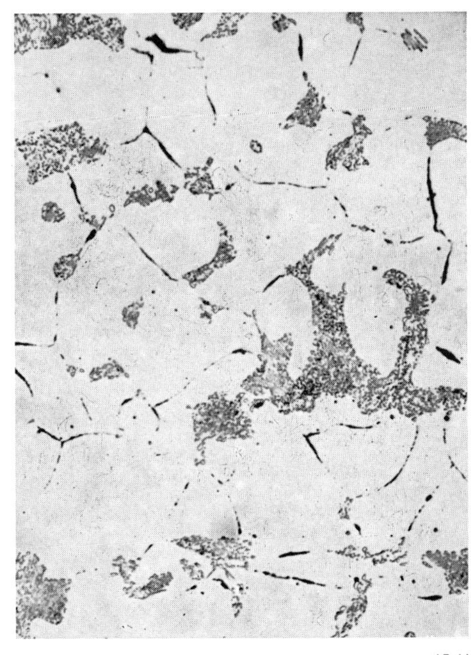

15.41

15.2 **Failures Through Decarburization, Hydrogen Damage**

Failures caused by surface decarburization were already reported in section 6. This type of decarburization is caused by oxygen or oxygen-containing gases such as water vapor and carbon dioxide. The reaction product is carbon monoxide. The process takes place at the steel surface. Its rate is determined by the diffusion rate of carbon if there is no oxide scale layer present to inhibit removal of the reaction product. This type of decarburization lowers hardness, wear resistance, and hardenability of the surface zone, as well as bend- and torsion-fatigue strength, but deformability would increase or at least not decrease.

In the following, another type of decarburization will be discussed, namely that by hydrogen. Dry hydrogen at atmospheric pressure normally has a decarburizing effect above 1000 °C[9]). On the other hand, at the high pressures that occur in technical hydriding processes, such as in ammonia synthesis and others, decarburization may already take place between 200 and 300 °C[10])[11])[12]). The reaction product is methane. Decarburization is not limited exclusively to the surface, but is present wherever methane can be formed and precipitated under pressure, such as, for instance, at the grain boundaries. Since hydrogen is very mobile in steel and since the carbon diffusion paths are short, the attack rate is determined by the rate of reaction of the decarburization which in turn depends upon the strength of the carbon bond or the stability of the carbides. The methane precipitated under pressure tears open the grain boundaries. Metallographically, hydrogen decarburization or hydrogen attack differs significantly from surface decarburization by oxygen or water vapor. The reaction with a small amount of carbon suffices to induce crack formation. At low temperatures and correspondingly slow carbon diffusion, the attack may occur before decarburization becomes metallographically discernible. The damage then can be proved by a carefully polished unetched mount because etching, particularly grain boundary etching with nitric acid, might cover up the fine cracks.

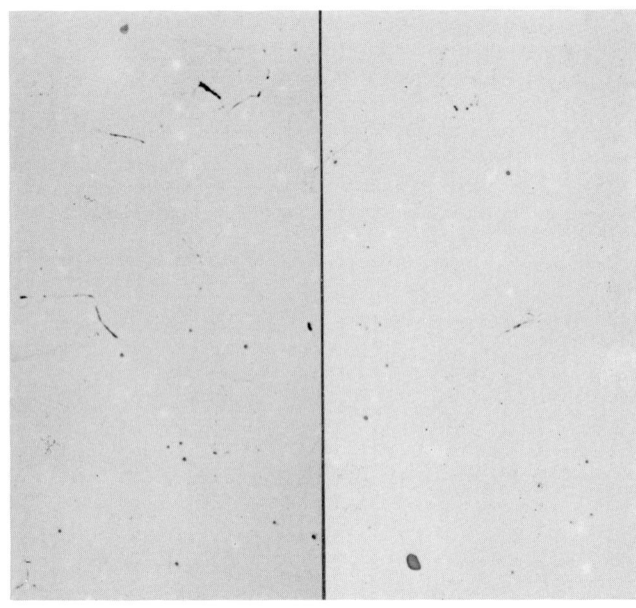

15.42 15.43

Visually hydrogen damage can sometimes be discerned by blister formation, similar to attack by pickling agents (see section 8.1). The blisters, however, are not filled with hydrogen, but methane.

The fracture surface of parts that have been damaged by hydrogen has a characteristic dull grey appearance. Strength and particularly all deformation characteristics are decisively reduced by hydrogen attack.

Conditions leading to hydrogen damage may be present too during electrolyis of molten salts containing crystal water, and during the reaction of superheated steam with steel. In the latter reaction, magnetite (Fe_3O_4) and atomic hydrogen are formed on the iron according to the equation

$$3\ Fe + 4\ H_2O \rightarrow Fe_3O_4 + 8\ H$$

if the process takes place below 570 °C.

Some examples are cited below. An **ammonia synthesis plant cooler** was destroyed after three years' service due to the rupture of a distribution manifold[13]). The cooler consisted of a row of upper distribution manifolds and lower collection manifolds with a system of pipes between them. Synthesis gas under a pressure of 800 atm and at about 300 °C, consisting of approximately 68 % hydrogen, 22 % nitrogen and 10 % NH_3, was watercooled externally to room temperature. The site and the structure of the fracture are shown in **Fig. 15.38 and 15.39,** respectively. The fracture originated from inside the cross-boring for the gas inlet tube and had the typical mat-grey fibrous structure of a material destroyed by hydrogen. The structure was permeated by intergranular cracks in the transverse boring as well as in the longitudinal boring **(Fig. 15.40).** Decarburization was so minor due to the low operating temperature that it could not be positively verified by metallography **(Fig. 15.41),** but the grain boundaries were torn open in part. Damage had penetrated to a depth of 16 mm. Careful preparation of the metallographic section and good attention by the examiner are required to recognize the last traces of the attack. These features are demonstrated in **Fig. 15.42 and 15.43.** In the damaged zone the notch toughness, which was 48 J (35 ft lb) in the unattacked part of the cross section, had dropped to 7 J (5 ft lb). Hence the operating conditions were such that a hydrogen-resistant steel should have been used.

A 110 mm O.D. and 70 mm I.D. **high pressure pipe** made of unalloyed steel that was exposed to hydrogen under the maximum conditions of 260 atm and 280 °C started to leak through longitudinal cracks after five years. An internal view of the pipe is reproduced in **Fig. 15.44.** After breaking open, the crack surface displayed the typical mat-grey fracture. The crack was predominantly intergranular as is apparent from **Fig. 15.45.** The steel was but slightly decarburized as would be expected from the low temperature. Hydrogen attack was far advanced. In the center of the pipe wall the microstructure was substantially destroyed **(Fig. 15.46),** and even 3 mm under the outer surface, torn-open grain boundaries could still be recognized **(Fig. 15.47).**

Hydrogen damage may be prevented by **alloying the steel** with elements that bind the carbon firmly. Such elements are chromium, molybdenum, vanadium, and titanium. Complete stability against hydrogen takes place only when the unstable iron carbide has been entirely eliminated from the microstructure, and the steel contains only the stable special carbides[14]). The respective amounts of metal and carbon necessary for this may be gleaned from the applicable phase diagrams. Naturally, one should make sure that the equilibrium carbide is actually present. If after fast cooling of the steel, e.g. after welding, an unstable transition carbide is formed, the steel will be attacked by hydrogen, even though it has the correct composition. Such a case has been already reported in section 7 (Fig. 7.40 to 7.42). Subsequent annealing to reinstitute the equilibrium is then absolutely essential.

Steels with 2–6 % Cr that are used in hydrogen installations also possess a certain resistance against the effect of hydrogen sulfide, which is often present in hydriding gases[1]). Additions of molybdenum and vanadium improve high temperature strength as well as hydrogen stability[15]).

Finally an example for an attack by **high pressure steam** must be cited. **Tubes of a backwall of a combustion chamber** of a 10-MPa (100-atm) boiler showed pits or trench-like erosion in a circumferential weld joint after 4½ years of service[16]). According to the plant operator, the

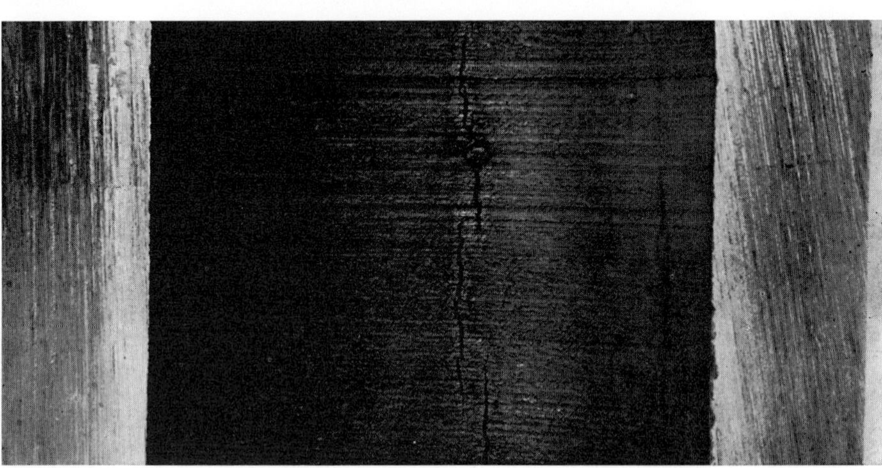

15.44

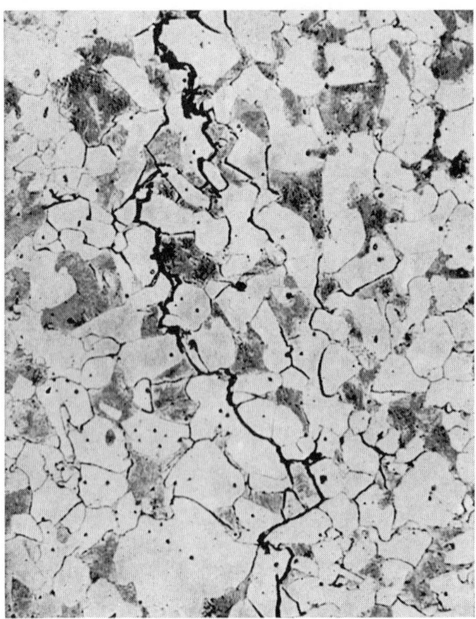

15.45

temperature in the tube may have risen to 400 °C. A bluish-black corrosion product adhered to the pits that could be distinguished clearly from the red coating of the remaining inner surface. Longitudinal and transverse sections through the corroded region are shown in **Fig. 15.48 and 15.49.** The weld seam was strongly reduced in cross section from the inside and was covered with a thick, crumbling coating of magnetite (Fe_3O_4) which was much thicker than the oxide scale at the outer surface of the pipe **(Fig. 15.50).** The attack also extended to the adjacent zone that had been overheated during welding. It ended fairly precisely at the point where the normal grain size of the tube material prevailed. Oxidation had penetrated deeper into the steel at the boundaries of the very coarse austenitic grain **(Fig. 15.51).** Moreover, the inner zone was decarburized and permeated with fine cracks on the grain boundaries and in preferred lattice planes. In the longitudinal section, the corrosive attack became less severe with increasing distance from the weld seam; similarly the corrosion severity decreased in the transverse section with distance from the pit formation, This is expressed clearly in **Fig. 15.52a to c,** which show the typical characteristics of hydrogen damage. The strong local attack in the weld seam and in the coarse-grained zone adjacent to the seam probably can be explained by the fact that formation of a firmly adhering protective coating may have been prevented by the weld scale flakes and the weld pool turbulence at the protruding ridges of the weld joint. Fresh surfaces for oxidation were created by the crack formation as a consequence of hydrogen attack; thus the scaling reaction was accelerated and deepened. A second weld seam that had been considerably less attacked had a finer grain structure. Therefore, overheating and flash formation during welding should be avoided, particularly in tubes for high pressure steam boilers.

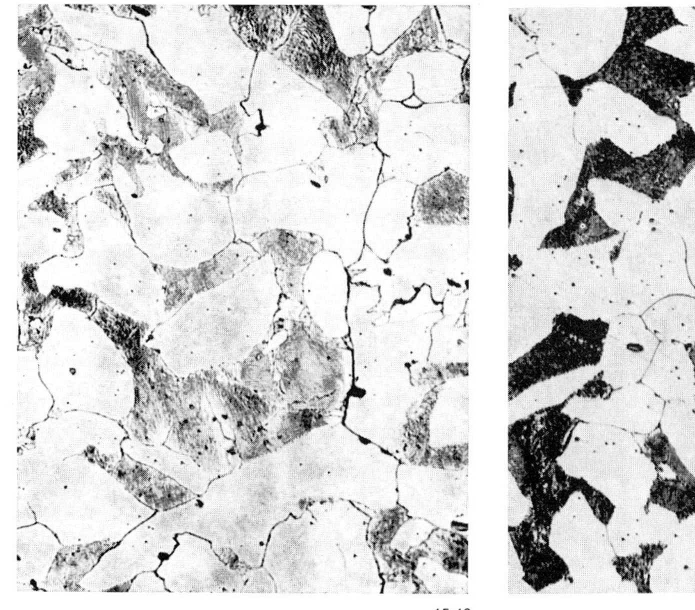

15.46

15.47

Fig. 15.44 to 15.47. High pressure pipe damaged by hydrogen attack

Fig. 15.44. View from inside. 1 ×

Fig. 15.45 to 15.47. Structure in pipe wall. Transverse section. Etch: Picral

Fig. 15.45. Intergranular crack. 100 ×

Fig. 15.46. In middle of pipe wall. 200 ×

Fig. 15.47. 3 mm away from outside surface. 200 ×

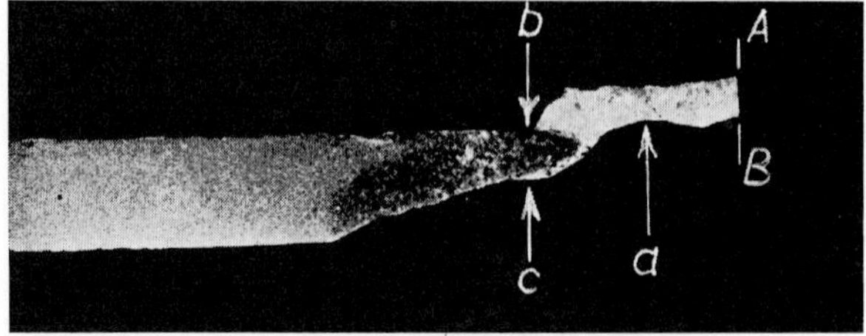

15.48

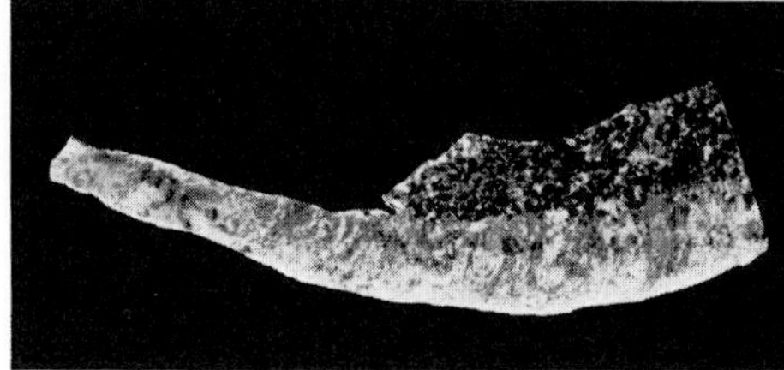

15.49

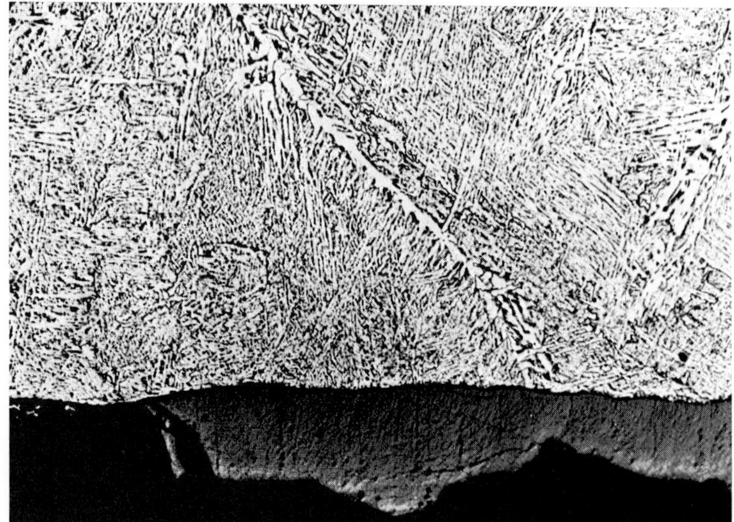

15.50

Fig. 15.48 to 15.52. Combustion chamber of 10-MPa (100-atm) boiler attacked by hot steam

Fig. 15.48 and 15.49. Section through weld. Etch: Picral. 2 ×

Fig. 15.48. Longitudinal section: Bottom: Inside. Right: Weld

Fig. 15.49. Transverse section A–B according to Fig. 15.48. Top: Inside. Left: Pitted zone. 100 ×

Fig. 15.50. Structure at inner edge of weld seam (spot a in Fig. 15.48) with heavy, spalling Fe_3O_4-coating. Longitudinal section. Etch: Picral. 100 ×

Fig. 15.51. Structure at inside boundary of transition zone (location c in Fig. 15.46) with selective oxidation and hydrogen attack on grain boundaries and lattice planes. Longitudinal section. Etch: Picral. 100 ×

Fig. 15.52.a b c Hydrogen attack in inner part of pipe wall that diminished with increasing distance from pitted zone (a–b–c). Transverse section. Etch: Picral. 200 ×

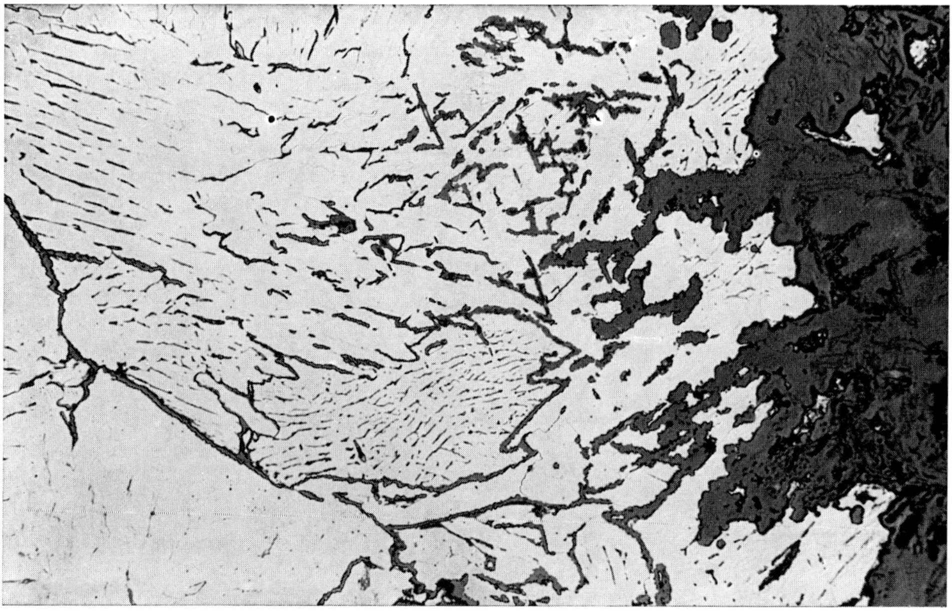

15.51

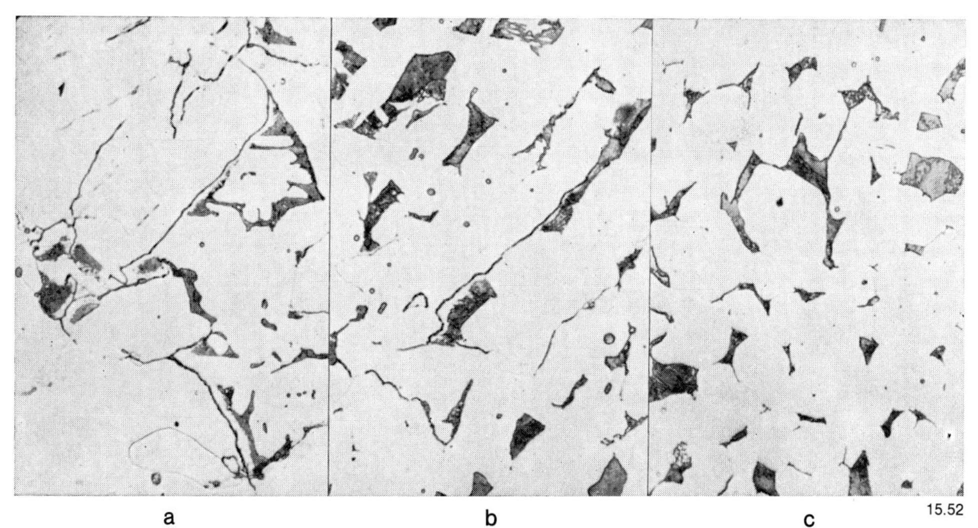

a b c

15.52

15.3 **Corrosion in Electrolytes**

Corrosion in acids and aqueous solutions is an electrolytic process. The metal goes into solution as an ion, and iron as the twice positively charged ion Fe^{2+}. In this case iron has to give off two negative charges, i.e. electrons:

$$Fe \rightarrow Fe^{2+} + 2\ominus$$

This is the anodic partial reaction, an oxidation. The two freed electrons are used to reduce a component of the electrolyte. During dissolution in hydrochloric acid, these are, for instance, the positively charged hydrogen ions that are reduced to atoms; they then combine into molecules and are discharged as a gas:

$$2\,H^+ + 2\ominus \rightarrow 2\,H \rightarrow H_2$$

This is the cathodic partial reaction.

Fig. 15.53. Recuperator pipe which developed leaks by contact corrosion. 1 × Left: Pipe of 6 % chromium steel. Center: Weld seam with austenitic-martensitic microstructure. Right: Pipe of unalloyed mild steel

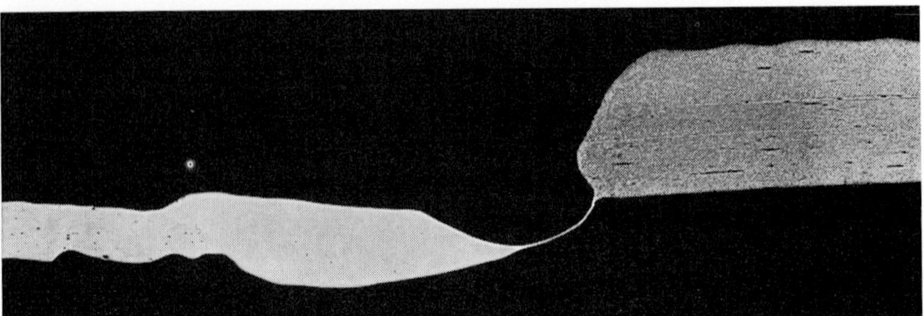

Fig. 15.54. Longitudinal section through weld connection according to Fig. 15.53. Etch: Nital. 8 ×

During oxidation of the iron the anodic process is the same, but the cathodic partial reaction proceeds differently. Here the oxygen that is dissolved in water is reduced together with the water to hydroxyl ions as follows:

$$2H_2O + O_2 + 4 \ominus \rightarrow 4\ (OH)^-$$

Corrosions of this type cannot proceed without dissolved oxygen. This reaction results in ferrous hydroxide that is immediately further oxidized to ferric hydroxide FeO(OH).

The "corrosion current" (also known as Tafel current) flows between anode and cathode. The dissolution of the iron can be aggrevated by anodic "polarization" (e.g. by external currents – vagabond currents in the ground) and may be prevented by cathodic shielding.

15.3.1 **Surface Area Corrosion**

Under normal circumstances anodes and cathodes are evenly distributed over the metal surface and are not bound to specific areas. Corrosion then leads to a uniform metal regression. It may be described in $g/m^2 \cdot h$ or in mm/yr and approximated beforehand. Surface area corrosion may have considerable economic consequences, but does not pose any particular danger. The application of metallographic testing methods is not essential for proving it.

15.3.2 **Localized Corrosion**

The so-called localized corrosion presents a different case altogether. It occurs, when two metals with different electrochemical potential, i.e. one "more noble" and one "less noble", make contact in the electrolyte, or if one metal is in contact with two solutions of different kinds or concentration. The metal with the less noble potential, or the less noble area of the metal at which it is in touch with the two solutions, then becomes the anode of a galvanic corrosion cell and is dissolved by the corrosion current, while the more noble ion is protected cathodically. Corrosion is more concentrated in the contact area, the lower the conductivity of the electrolyte. The same holds true for cathodic shielding.

15.3.2.1. **Contact Corrosion**

A process that takes place during contact of two metals with different potentials in an electrolyte, is designated as contact corrosion. It also appears in porous regions of electronically conducting coatings – including some paints[17]) – or at crevices in oxide layers that have a more noble potential than the bright metal. Various kinds of selective corrosion, which will be discussed later, are accelerated by the formation of galvanic cells.

The following failure analysis offers an example for the effect of contact corrosion[18]). **Pipes of a recuperator** for the preheating of combustion air in a rolling mill furnace failed after a relatively short service period because of leakage in the colder part. The unalloyed mild steel pipes were joined by welding with austenitic electrodes to smaller diameter 6 % Cr-alloy steel pipes used for the warmer part. Such a leaky joint is shown in **Fig. 15.53.** The weld seam had cracked in the thicker mild steel pipe side. At this transition to the unalloyed pipe, it was deeply eroded in a trench-like fashion over the entire perimeter. A longitudinal section through this location is reproduced in **Fig. 15.54.** Corrosion appeared in close contact with the mild steel pipe an had consumed almost the entire wall thickness. In contrast, the chromium steel pipe on the other side next to the weld seam was only slightly corroded over a short distance which roughly corresponded to the heat-affected zone. The fine-grained ferritic-pearlitic microstructure of the mild steel pipe had become coarse-grained and acicular by overheating (Widmannstätten structure). The weld seam microstructure was not austenitic but hat become predominantly martensitic as a result of mixing the weld metal with the fused pipe steel. The alloy steel pipe had a microstructure of ferrite with finely dispersed carbides, while that of the transition zone to the weld had been transformed to martensite or bainite. The highest difference in potential therefore must have been present between the alloyed weld seam with the martensitic structure and the unalloyed pipe, whereby the weld seam had the less noble potential. The same weld seam located further to the inside toward the warmer part of the recuperator remained unattacked because no condensate could form there.

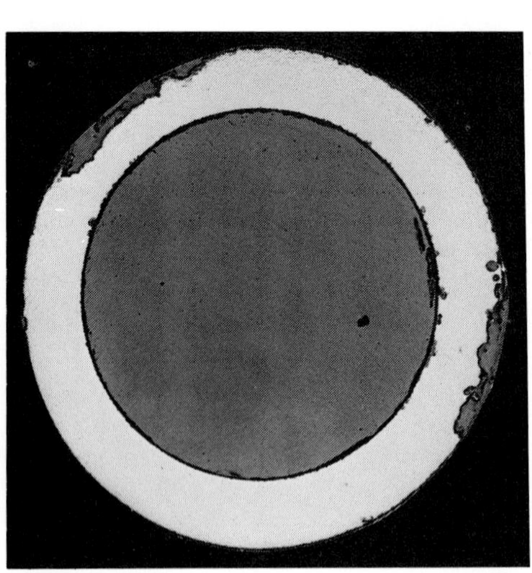

15.55

15.56

Fig. 15.55 and 15.56. Contact corrosion in lead-coated condensor tubes

Fig. 15.55. Corrosion pits in unetched transverse section. 3 ×

Fig. 15.56 Condensor tubes after etching of surface with sulfurous potassium-ferricyanide solution. Approx. 0.75 ×. Left: Unused tube. Right: Corroded tube

The effect of porous metal coatings is described in the following example: **Condenser tubes of a tubular radiator** were leaded inside and outside to facilitate soldering by dipping into a lead bath. They began to leak after only one year due to pitting corrosion, while uncoated tubes that could be welded into the bottoms of larger condensers remained unattacked. The tubes in the condensers were cooled by water at about 12 °C inlet temperature and were surrounded on the outside by ammonia vapors at 120 °C. The pitting corrosion extended over the entire circumference. A transverse section through the pitted region is shown in **Fig. 15.55;** no lead coating could be observed. Examination of an unused pipe showed that the coating was very thin right from the start. After etching of this virgin tube with a solution of 10 g potassium ferricyanide in 1000 ml 2 % sulfuric acid[19]), numerous blue spots could be seen at the inner surface, while in the used tube the surface was colored completely blue **(Fig. 15.56).** The lead coating therefore had been initially porous and had disappeared completely in the corroded tube.

Surfaces with cracked scale are also attacked by selective corrosion. In this case the oxide scale has the more noble potential. Since the rate of corrosion is determined by the cathodic partial reaction and since the scale-covered surface as a rule is much larger than the iron surface bared by the cracks, a high density corrosion current may be formed and consequently pitting and trench-like erosions, that penetrate much deeper than would be the case in uniform regression.

Crack formation in the oxide scale was the cause of corrosion in a **quadrant pipe bend.** The pipe alternately passed hot condensate and cooling water and began to leak after nine months. During an on-site inspection the spalled-off scale pattern from the outer surface was observed to correspond to the deformation lines; this damage inside the bend is shown in **Fig. 15.57.** In part, pitting corrosion extended through the wall. From the oblique location of the corrosion pits, erosion was assumed to play a part in their formation (see section 13.5). Since the damage sites were located at the outside of the bend, the erosion may have been caused by the oblique impact of water drops in the steam (droplet impact). The internal surface was covered by a strongly adherent thin scale. The scale layer had cracked locally by erosion or more probably by subsequent cold bending. The contributory effect of the erosion is expressed by the flat ground surface and the oblique position of the deep pits in the transverse and longitudinal sections of **Fig. 15.58 and 15.59.** The presumed cold deformation caused by the subsequent bend formation was confirmed by the slip line appearance **(Fig. 15.60).**

Iron becomes passive in highly alkaline solutions by the formation of a surface oxide layer that is but a few lattice parameters thick, i.e. it assumes a noble potential in these solutions. That is the reason why, for instance, reinforcement steel in concrete does not oxidize; this assumes that the latter is non-porous and the rods are covered by the concrete to a sufficient depth. Stainless chromium-nickel and chromium steels, too, owe their resistance against atmospheric corrosion and oxidizing acids to the formation of a passive layer. If this layer cannot form completely or is penetrated locally, selective corrosion may also occur.

As an example, a section through a pit in a high strength steel used in a **prestressed concrete structure** is represented in **Fig. 15.61** (see also section 15.3.4.). Here, the passive layer was destroyed locally by chlorine ions from the calcium chloride that had been added to the concrete for faster setting. Sulfides, too, have this type of effect, as is illustrated in **Fig. 15.62,** where a high strength steel which had been removed from a sulfur-rich concrete is shown (see also section 15.3.4.).

The effect of chlorides is also one of the principal causes for pitting in stainless steels, and this type of corrosion therefore is also designated as **chloride corrosion.** The passive layer is locally penetrated in this process. Sulfide inclusions – particularly manganese sulfide – that lie on the surface, facilitate the nucleation of pits[20]), but are not necessarily essential for such nucleation. The pits initiate at tiny dots and are usually extremely small. Therefore this pitting corrosion is also called **needlepoint corrosion.**

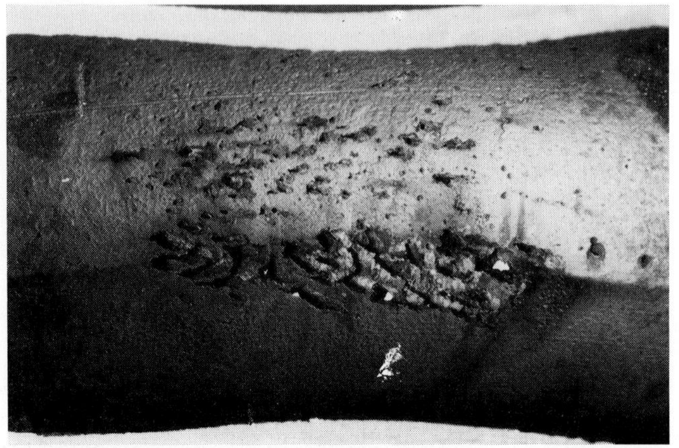

15.57

15.58

15.59

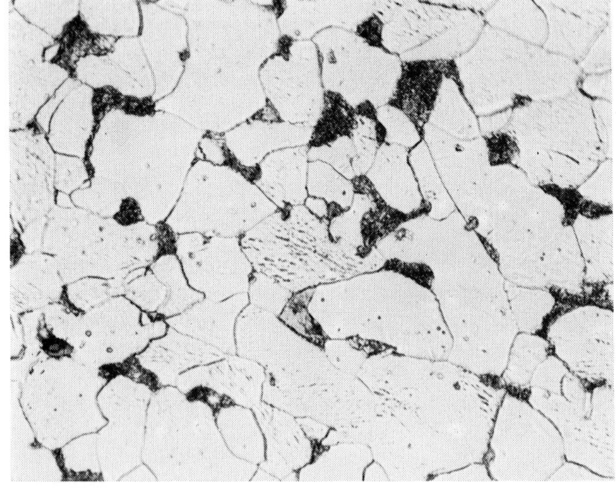

15.60

Fig. 15.57 to 15.60. Localized corrosion in a quadrant pipe bend caused by crack formation in the scale

Fig. 15.57. Inner surface of pipe bend with pitting corrosion and erosion. 1 ×

Fig. 15.58 and 15.59. Unetched sections through corrosion area. 3 ×

Fig. 15.58. Transverse section

Fig. 15.59. Longitudinal section

Fig. 15.60. Microstructure of bend with slip lines. Transverse section. Etch: Nital. 500 ×

An example, the bottom of a **bleaching kier** of a cotton mill, is shown in **Fig. 15.63.** The corroded side of the vat, made of an austenitic chromium-nickel-molybdenum steel, had been in contact with a bleaching solution containing 1.8 g/l active chlorine for one year. The vat had to be removed from service because the bottom had a brown discoloration, which appeared to be a deposit of rust. Only during observation under a magnifying glass did the corrosion show pitting characteristics **(Fig. 15.64).** The depth of the holes was still pretty shallow **(Fig. 15.65).** Since another kier made of the same steel had been in service for eight years without being attacked, the prevailing conditions must have brought the steel to the limit of its corrosion resistance; the failure could not have been decisively affected by minor differences in the composition of the steel or the operating conditions. As a remedy, a recommendation was made that either a 10 % nitric acid be used from time to time for repeated passivation, or else the application of cathodic protection through aluminum or magnesium anodes, which are claimed to have been very successful in such cases.

In a **dished head of a pressure vessel,** made of boiler plate that was clad with a sheet of AISI type 405 stainless steel, corrosion pits were found during inspection on the clad surface after open air storage. They could hardly be seen with the naked eye. The pits are shown in ten-fold magnification in **Fig. 15.66.** In a section through the holes **(Fig. 15.67),** the corrosion could be seen to have proceeded selectively in the active regions along the grain boundaries. As can be seen from **Fig. 15.68,** the latter were covered with chromium containing carbides. Section 15.3.3 will discuss selective corrosion of this and other types.

The higher the molybdenum contents, the more resistant the stainless chromium-nickel steels are to pitting corrosion by chlorides.

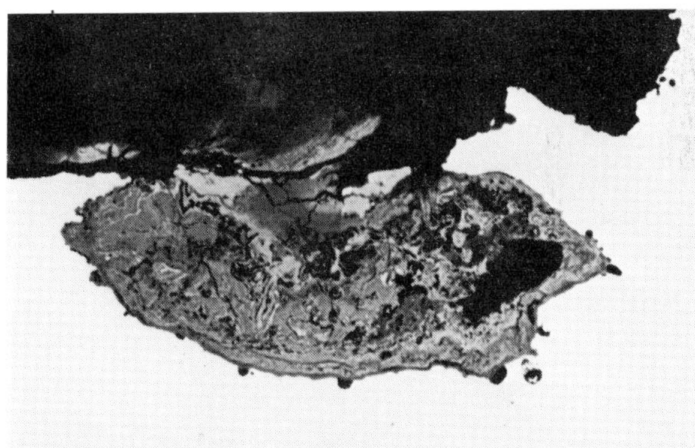

Fig. 15.61. Unetched section through pit of high strength stressed steel. 100 ×

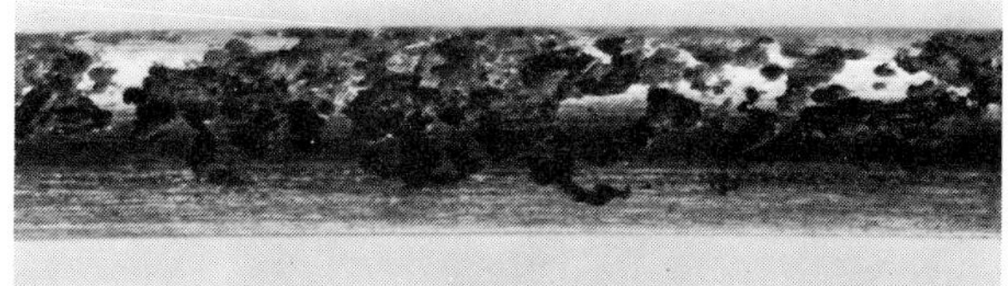

Fig. 15.62. Pitting in high strength stressed steel caused by hydrogen sulfide. 10 ×

15.63

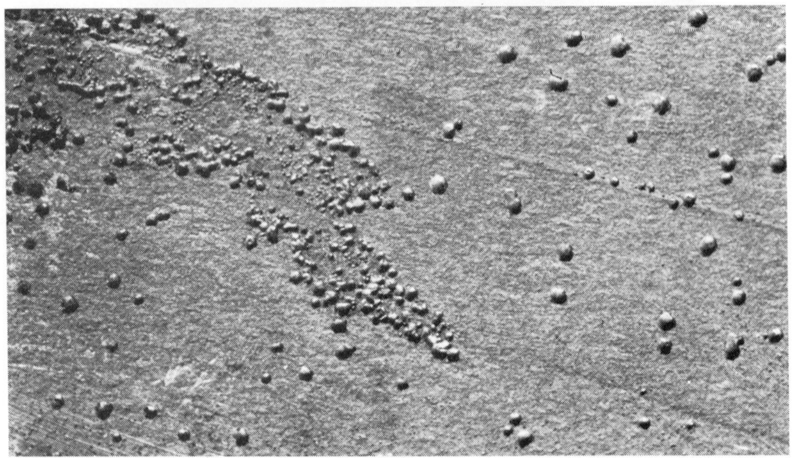

15.64

15.65

Fig. 15.63 to 15.65. Rusted bottom of stainless steel bleaching kier

Fig. 15.63. View

Fig. 15.64. Needlepoint corrosion in bottom of kier. 7 ×

Fig. 15.65. Unetched section through corrosion pits. 100 ×

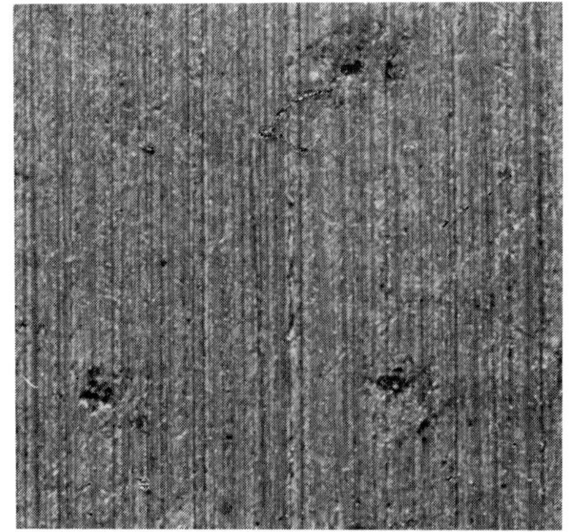

15.66

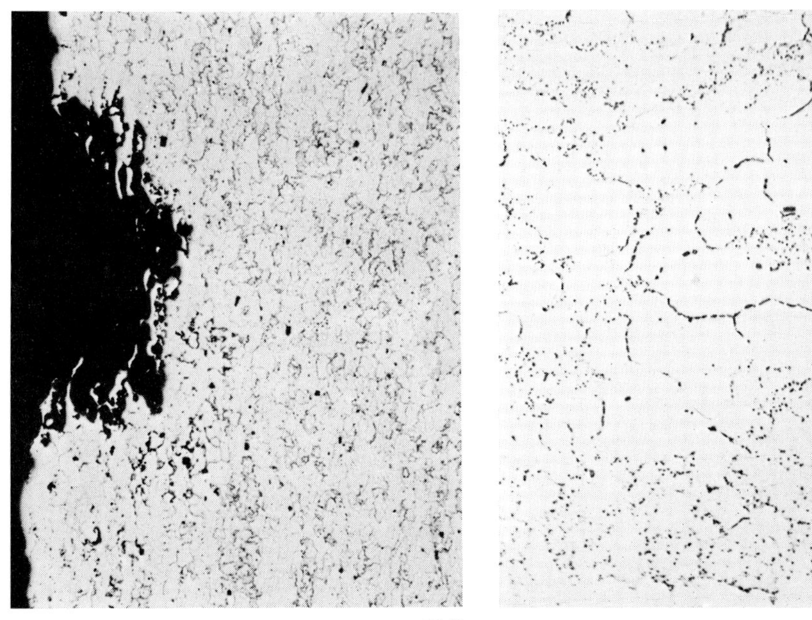

15.67

15.68

Fig. 15.66 to 15.68. Pitting corrosion in chromium steel cladding of a dished head of a pressure vessel

Fig. 15.66. Cleaned surface of the bottom. 10 ×

Fig. 15.67. Pitting with selective corrosion. Transverse section. Etch: Nital. 100 ×

Fig. 15.68. Base structure with carbide precipitates on grain boundaries. Transverse section. Carbide etch with ammonium hydroxide 2 V. 500 ×

15.3.2.2 **Concentration Cells**

Corrosion cells may also be formed if a metal is in contact with two electrolytes or two solutions of different concentration, as was mentioned before.

A corrosion-current caused by such "concentration cells" was primarily responsible for the following failure. **Prestressed concrete roofs** which consisted of three slabs each, contained three coaxially prestressed bundles of ribbed wires. They were cemented by troweling a fast hardening chloride-containing concrete at the butt joints. In many instances, the stressed wires broke at the butt joints. The condition of a wire bundle after freeing one of the butt areas is illustrated in **Fig. 15.69.** The jacket tube in which the wires were stretched had rusted through

15.69

15.70

Fig. 15.69 and 15.70. Enhanced corrosion by formation of concentration cells between two concrete types in a bundle of prestressed wires

Fig. 15.69. View of butt joint between high and low chloride-containing concrete. 1 ×. Jacket tube rusted through from outside

Fig. 15.70. Transverse sections of wires according to Fig. 15.69. 5 ×

from the outside and the wires were considerably eroded by localized corrosion, as can be seen from the transverse sections in **Fig. 15.70.** This failure was apparently caused by the combined effect of chlorine ions and the polarization of the corrosion-current flowing between the concretes of different compositions.

15.71

15.72

Fig. 15.71 and 15.72. Corrosion in bottom tank of passenger steamer

Fig. 15.71. Crevice corrosion in overlapped zone of rivet seam. 0.25 ×

Fig. 15.72. Section of Fig. 15.71. 1 ×

15.73

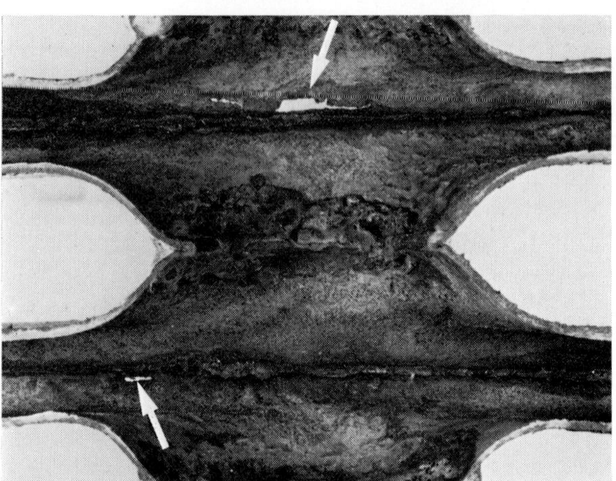

15.74

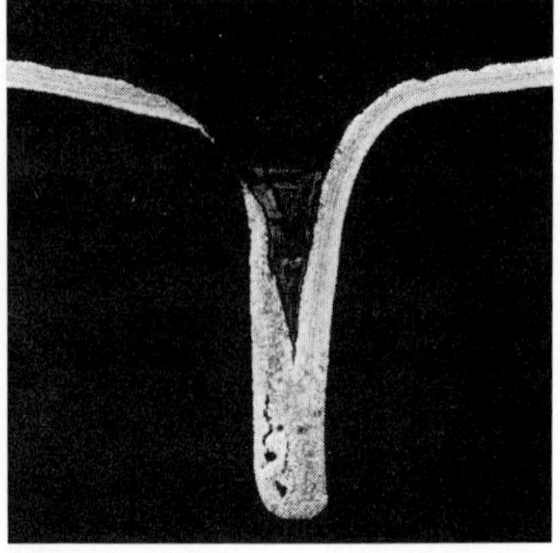

15.75

15.76

A special case of a **concentration cell** is the differential-aeration or **Evans cell.** The cell formation in this case is caused by the differences in oxygen concentration of the electrolyte. For instance, such differences can occur at the edge of a water drop, directly below the surface of stagnant water, in narrow gaps, or at rough-machined surfaces, where concentration equalization is hindered by diffusion and convection. The metal part immersed in the more oxygen deficient region of the electrolyte becomes the anode in this element and goes into solution.

The formation of Evans cells in drops is the cause for the particular aggressivity of water condensates. Experience has shown that such condensation corrosion has caused many failures during shipping of densely packed parts (see also section 10).

A case of **crevice corrosion** has already been reported in section 3.1 (Fig. 3.15 to 3.19). The corrosion in this case was favored by the faulty design of screen bars with too narrowly sinuous loops.

A **sheet from a passenger steamer bottom tank** is illustrated in **Fig. 15.71 and 15.72.** These tanks are filled alternately with oil and seawater during each trip. The sheet had been attacked preferentially by crevice corrosion in an overlapped riveted joint zone.

Crevice corrosion must also be regarded as one of the causes for **steel sheet radiator** corrosion in heating plants. These radiators sometimes become leaky in the lower part next to the rolled seam through which the sheet metal strips are welded shut. Such a leaky place ist reproduced from the outside in **Fig. 15.73,** while **Fig. 15.74** shows similar spots from the inside after severing the lower end. The leaky zone was always located in that area where a narrow crevice had formed in the transition from the bottom into the rolled seam **(Fig. 15.75).** The failure did not occur as long as the radiator was filled completely with water. An aeration cell can be formed only when the water level has sunk so low that it stands immediately above the bottom. This typically can occur when the heating plant is shut off. It was confirmed in this instance by the **water line corrosion** at the discharge level **(Fig. 15.76).** Therefore this was a typical case of **stagnant (splash zone) corrosion** which is the cause of many failures.

Characteristics of contact crevice corrosion could also be seen in five **crankshaft bearing rings.** The corrosion occurred in a narrow ring-shaped zone where the bearings had been sealed against the shaft housing with silicone rings. The seal rings had to be replaced because they started leaking after a service life between 94 and 1325 h; an example of this defective zone is shown in **Fig. 15.77.** Adjacent to a brightly rubbed circumferential strip, dark corrosion pits were located obliquely against the circumferential direction at intervals of approximately 1½ mm. As can be seen from **Fig. 15.78,** this is very definitely pitting corrosion. In the same zone there were narrow, obliquely running ribs pressed into the lip of the seal rings; these corresponded exactly to the direction and spacing of the observed corrosion pits **(Fig. 15.79).** Accordingly, the cause of failure was stagnant crevice corrosion.

Contact or crevice corrosion may also take place if nonmetallic evaporation residues from solutions or dust particles from gases are deposited on metal surfaces. A failure of this type that occurred in **facade sheets of stainless steel** on a newly constructed building has already been described in section 3.3 (Fig. 3.77). A section of this facing sheet that was oxidized locally by a lime dust deposit during construction is reproduced in **Fig. 15.80.** Stainless steel is corrosion resistant only under oxidizing conditions which subsequently guarantee the formation and maintenance of passivity. This is the case in air and aerated water. In the crevice between the dust particles and the steel surface, the steel becomes locally activated. Precipitation of dust during construction must therefore be prevented right from the start, and dust deposited later must be removed by regular cleaning.

Fig. 15.73 to 15.76. Rusted sheet steel radiators

Fig. 15.73. Rusted-through spot in bottom. 0.66 ×. Patch weld in adjoining member

Fig. 15.74. Interiour of bottom part after removal of corrosion product. 0.66 ×

Fig. 15.75. Transverse section through rolled seam with pitting by crevice corrosion. 3 ×

Fig. 15.76. Waterline corrosion in lower part of sidewall. 0.33 ×

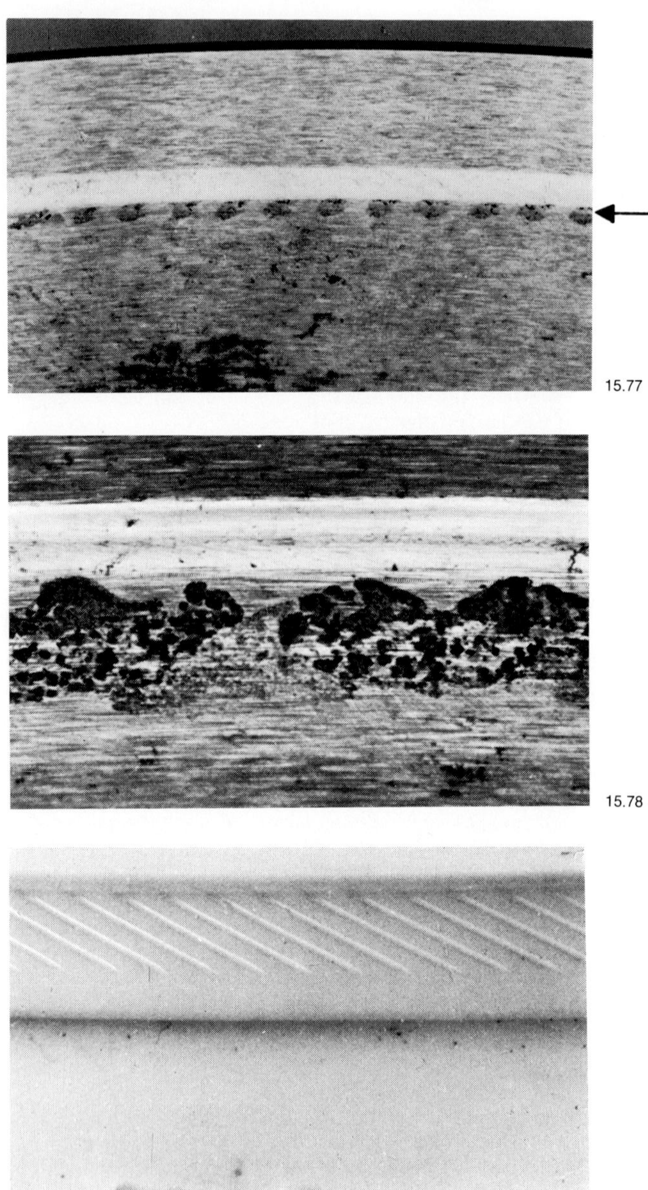

15.77

15.78

15.79

Fig. 15.77. to 15.79. Crankshaft bearing ring whose oxidation caused the crankshaft housing to leak after 125 h operation

Fig. 15.77. Crevice corrosion spots (←) on bearing surface of ring. 5 ×

Fig. 15.78. As Fig. 15.77. 15 ×

Fig. 15.79. Internal surface of silicone seal ring with fishbone pattern at lip. 5 ×

Evaporator heater tubes that were welded longitudinally and made of titanium-stabilized, molybdenum-enriched stainless steel, containing approx. 0.1 % C, 18 % Cr, 12 % Ni, 2 % Mo and 0.5 % Ti, had begun to leak after only 2 months. The tubes were used in a three stage vacuum evaporator unit in which glucose extract was heated to about 70 °C by 130 °C steam. The weakly acidic juice contained in the dry substance (approx. 35 %) about 2400 ppm cloride. The tubes of 29 mm I. D. and 32 mm O. D. showed in some cases needlepoint-like holes at the outside. The inner surface was covered in part with a transparent glaze to which nodules of a black substance were adhering **(Fig. 15.81).** The glaze was readily soluble in cold water and therefore probably consisted of sugar. The black substance contained up to 80 % carbon. In spots the steel was corroded under these nodules by pitting **(Fig. 15.82).** The pits were located in the weld seam as well as in other areas. A pitting corrosion area in transverse section is shown in **Fig. 15.83.** The steel composition corresponded to specifications, and the weld seam was tight and free of defects. In view of the juice's chloride content, a steel alloyed with molybdenum had been used – which has a higher resistance against pitting corrosion by chloride – and this steel had proved to stand up well during prior operation of the evaporator unit; therefore the chloride had to be eliminated as failure cause. Rather this was doubtlessly a case of crevice corrosion caused by carbon-rich evaporation residues from the sugar carburization.

15.3.3 **Selective Corrosion, Grain Boundary Disintegration**

Selective corrosion is commonly understood to be the preferred dissolution of small areas with less noble potential. For instance, the dezincification of α-β-brass[21])[22]) is well known, and it will not be discussed here. So is spongiosis of cast iron, which has been discussed in section 12. In a larger sense the preferred oxidation of chromium and aluminum in heat resistant steels and alloys (see also section 15.1) may also be included here. But the particular case of the intergranular corrosion or grain boundary disintegration in stainless steels must be treated in this section; it caused many technical failures and much economic damage in the first few years after the development of austenitic steels.

By a process known as **sensitization,** the otherwise highly resistant stainless chromium and chromium-nickel steels – for instance against oxidizing acids such as nitric acid – become susceptible to intergranular corrosion, if chromium carbide (and nitride) is precipitated on their grain boundaries. This may happen in ferritic steels during cooling from a high temperature after welding; the two major factors which increase their sensitivity are the higher supersaturation of the carbon in the α-iron and the increased diffusion rate of the carbon in α-iron. Precipitates are formed immediately adjacent to the weld seam at the hottest point. The austenitic steels, in which grain boundary deterioration has played a much more important role, precipitate carbides only during tempering at temperatures of 500 to 800 °C. In this temperature range, a zone exists in welded parts that runs at some distance parallel to the weld seam. Grain boundary deterioration was first observed in welded sheet constructions. In addition, it also occurs when parts of austenitic chromium-nickel steel are annealed in the critical temperature range; this temperature range is widened and lowered with longer annealing times[23]).

So much chromium is extracted from the austenite adjacent to the grain boundaries through precipitation of the chromium-rich carbide $C_{23}C_6$, and also of the nitride Cr_2N[24]), that its chemical stability falls below the limit of resistance[25])[26]). Intergranular corrosion follows along the zones which have become unstable in this manner and may take place so rapidly – accelerated through formation of cells between the active zones and the passive austenite – that the cohesion between the grains is completely destroyed in a comparatively short time.

Steels have less of a tendency for grain boundary sensitization at lower carbon contents. New melting processes permit a lowering of carbon content to less than 0.03 %. Another method of making steel more resistant to a deterioration of the grain boundaries consists of binding the carbon and nitrogen to elements such as titanium and niobium, which form hard-to-dissolve carbides and nitrides.

An example of failure by intergranular corrosion of an austenitic steel through erroneous thermal treatment has already been reported in section 6.2 (Fig. 6.80 to 6.86). Such failures may be prevented by correct treatment or selection of a grain boundary sensitization-resistant steel. If a tendency toward deterioration of the grain boundaries is present (not actual grain boundary disintegration) it may be overcome by solution annealing around 1000 °C and fast cooling (see Fig. 6.86).

The following is an example of a failure through intergranular corrosion in ferritic chromium steels. A welded **exhaust gas pipe** of such a steel, through which a gas with 92.5 % N_2, 5 % O_2 and 0.5 % NO passed, began to leak after an operating period of 2 weeks. The sheet consisted of a steel containing 0.125 % C, 19.2 % Cr and 1.02 % Nb + Ta. Therefore the carbon content exceeded that specified for these ferritic steels, and the combined chromium and tantalum contents did not suffice to bind the total amount of carbon (specification $Nb > 12 \times \% C$), especially if tantalum was present in large amounts. An etched section through a weld is reproduced in **Fig. 15.84.** The sheet had been attacked at both sides of the seam by intergranular corrosion and was permeated by cracks. The attacked region in this case was located in the zone that became hottest immediately adjacent to the weld seam, which is in contrast to intergranular corrosion in welded austenitic steel sheets. One of the wide cracks (a) and its vicinity, which was permeated by many grain boundary cracks, are presented at higher magnification in **Fig. 15.85.** This case might also have been cited under "stress corrosion" (see section 15.3.4.1), since weld and operating stresses played a part in the crack formation. In ferritic steels the tendency toward grain boundary deterioration is removed by soft annealing below 850 °C rather than by solution treatment as in austenitic steels.

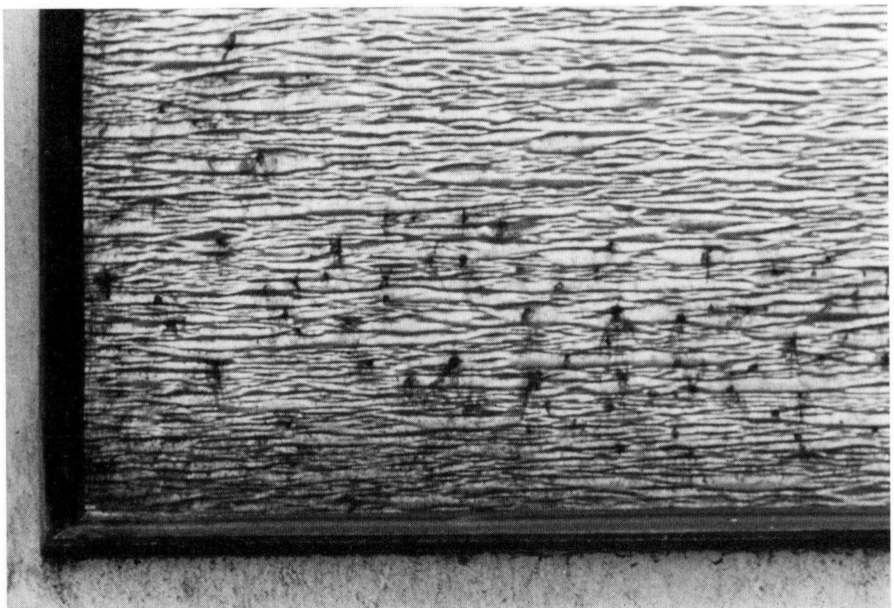

Fig. 15.80. Stainless steel facing sheet oxidized by dust deposits. Approx. 0.125 ×

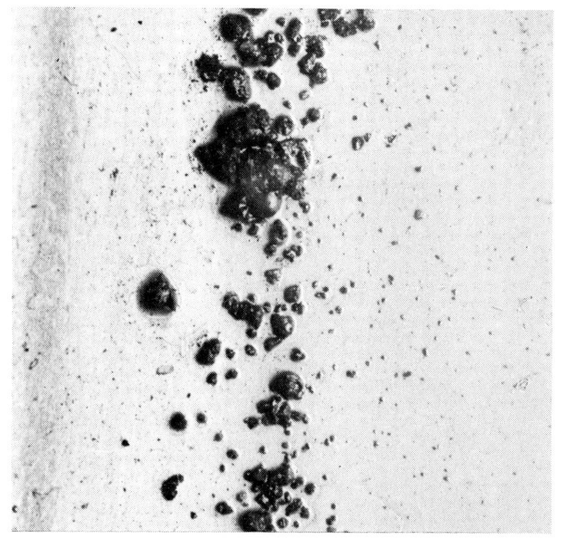

15.81

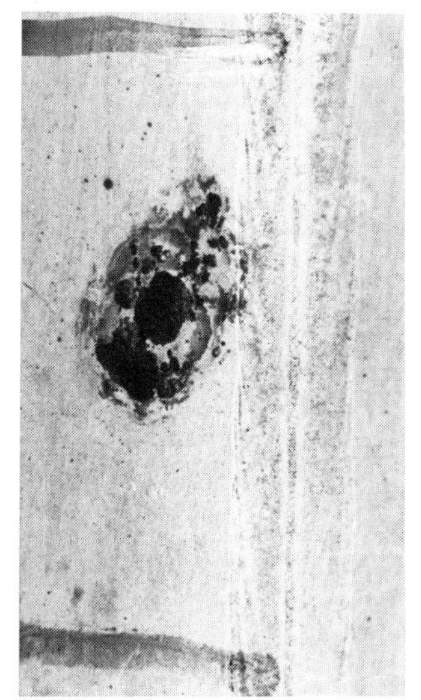

15.82

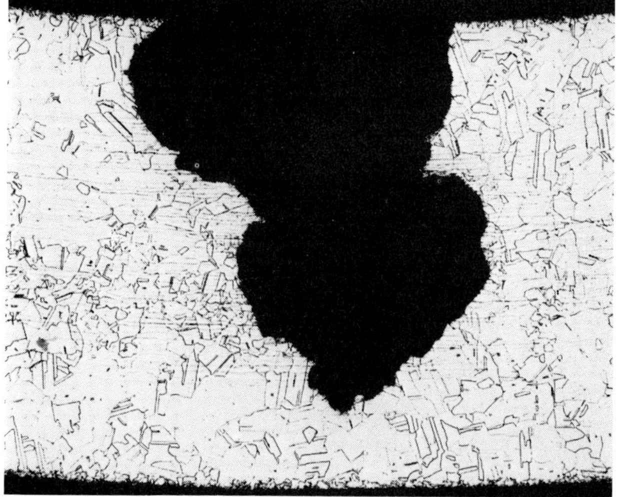

15.83

Fig. 15.81 to 15.83. Stainless steel pipe of an evaporation hood which sprung leaks by pitting corrosion beneath carbon deposits

Fig. 15.81. View from inside. 5 ×

Fig. 15.82. Pithole below a carbon nodule. 5 ×

Fig. 15.83. Pitting corrosion in transverse section. Etch: V2A-etchant. 50 ×

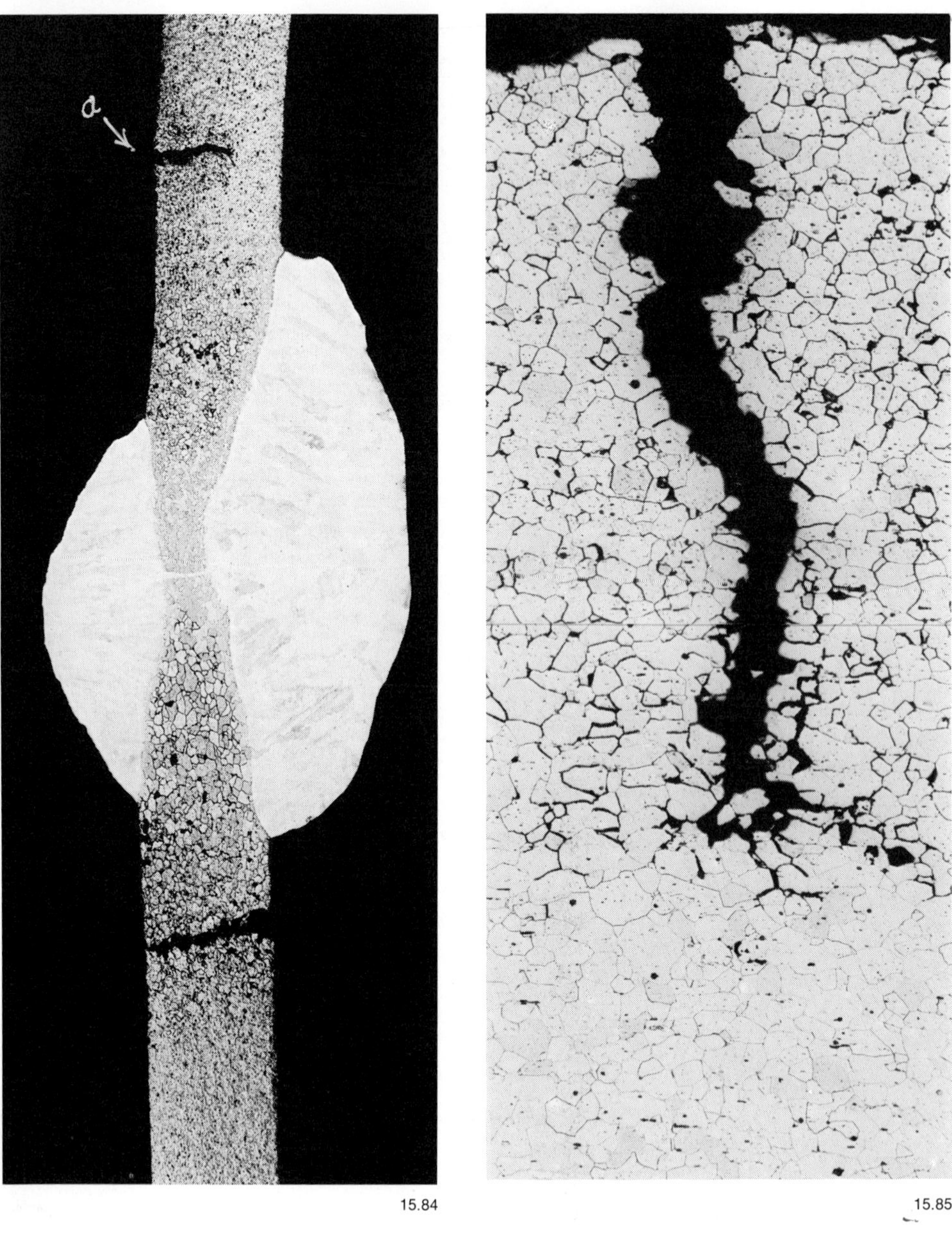

Fig. 15.84 and 15.85. Stainless 19 % Cr-steel exhaust gas pipe that sprang leaks in the weld seam due to intergranular corrosion. Etch: V2A-etching solution

Fig. 15.84. 8 ×

Fig. 15.85. Crack a according to Fig. 15.84. 80 ×

15.3.4 **Stress-Corrosion Cracking**

A further group of corrosion phenomena is characterized by the fact that stres tions play a role in addition to a specific corrosive agent. Therefore they are de comprehensive term stress-corrosion, or, because they lead to crack formation, sion cracking, even though they may vary greatly in type. Corrosion may attack soft or hard, unalloyed or alloyed, or ferritic or austenitic steels. The cracks may propagate in a transgranular or intergranular manner, or may be partially intergranular or transgranular. General metal regression is usually minor. The braze-induced fracture sensitivity treated in section 6.1.2, i.e. the effect of certain iron-soluble metals in the liquid state on steel that is under tensile stress, may also be classified as a form of stress-corrosion[27]).

There is no single interpretation for the causes and phenomena of stress-corrosion cracking. The prerequisite for actual stress-corrosion as originally designated is generally the common characteristic that the metal surface must be coated by the electrolyte with a cover or passive layer that is penetrated at particular spots, grain boundaries, or intercepts of slip bands[28]). Thus a corrosion-current of high density is concentrated from a large cathodic surface on a very much smaller anodic attack area. With the formation of the first incipient crack, a notch of ideal sharpness is formed and the corrosion-current is now directed onto a small region at the crack tip after passivation of the crack walls. Activity is maintained by continuous slip. This interpretation is applicable, for instance, to intergranular stress-corrosion cracking in soft iron and low carbon steels, and to transgranular attack of austenitic steels.

Another mechanism is responsible for such processes as the cathodic attack through hydrides or the crack formation during or subsequent to corrosion of high strength, stressed steels, also frequently designated as stress-corrosion cracking. The mechanism in this case is subject to dispute, but in the one of hydrogen sulfide it is essentially known.

In the following a classification will be made between intergranular and transgranular stress-corrosion, even though the boundaries are flexible; within these groups, corrosion of soft and high strength steels will be separately treated.

15.3.4.1 **Intergranular Stress-Corrosion Cracking**

Intergranular stress-corrosion cracking of unalloyed boiler plates has been known for a long time. It is caused by the concentration of caustic soda in the rivet seams of steam boilers as a consequence of evaporation of the feed water and therefore was originally called **caustic embrittlement.** But since the phenomenon does not cause embrittlement, but crack formation, this designation is erroneous. The term caustic crack sensitivity would correspond better to the facts, if the word caustic is to be retained. But as a matter of fact, this type of intergranular corrosion is caused by numeorus alkaline and weakly acidic oxidizing solutions. In this sense, a boiling 60 % calcium nitrate solution has an especially strong effect; it therefore is also being used as test reagent for caustic crack sensitivity*. The less noble grain boundaries in contrast to the passive grain surfaces constitute the points of origin for crack formation. The attack is accelerated by anodic polarization and inhibited cathodically.

Rimmed low carbon steels with high nitrogen contents which are sensitive to aging, are even more sensitive to caustic crack formation (see section 4.3.1). Intergranular stress-corrosion cracking is facilitated[29]) by precipitation of tertiary cementite onto the grain boundaries of α-iron. It had been reported by A. Fry in 1926[30]), that steel can be desensitized against caustic crack formation by extensive deoxidation with nitrogen-binding elements such as aluminum. This has been confirmed recently by P. Drodten and K. Forch[31]). Other special nitride and carbide forming alloying elements, such as titanium and niobium, have the same effect in this sense.

---

* **Stahl-Eisen Prüfblatt** 1860

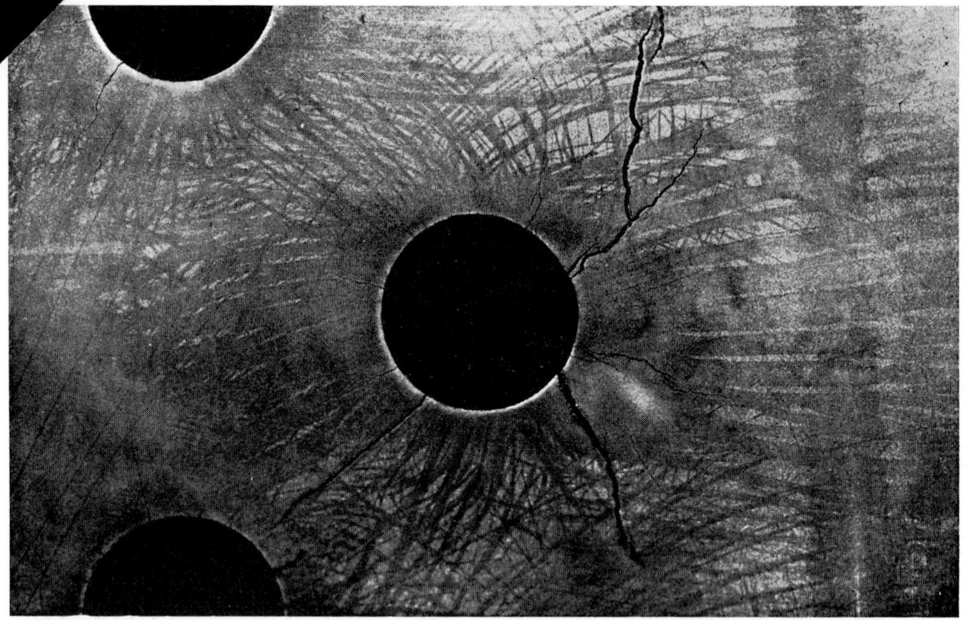

15.86

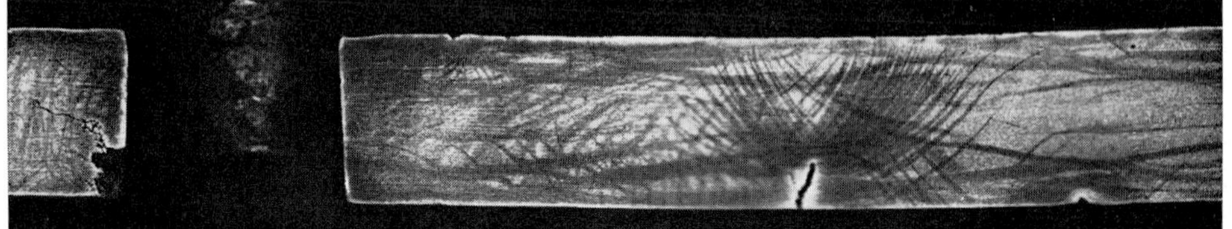

15.87

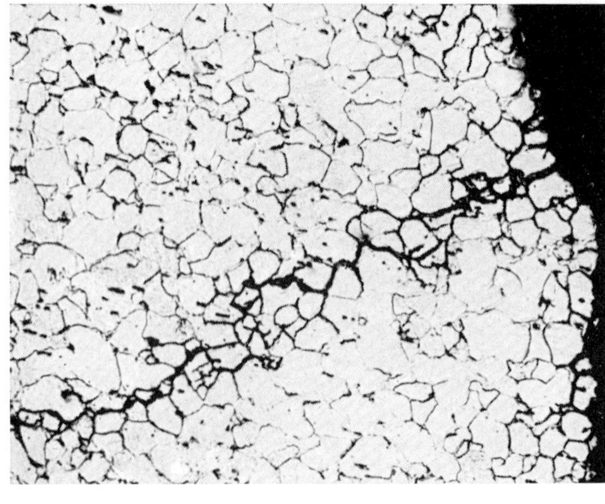

15.88

As already mentioned, caustic corrosion cracks were first observed in the rivet seams of steam boilers[30]). An example are the caustic corrosion cracks at the rivet hole of an old **incline-tube boiler with two headers,** dating from the year 1913[32])[33]). It is shown in **Fig. 15.86 and 15.87** in surface appearance and cross section, respectively, after etching with Fry's reagent. The 20 mm thick boiler plate consisted of a rimmed steel with a 380 MPa (55 ksi) strength level and good toughness. The steel had been aged after cold deformation in the zones around the rivet holes as proved by the Lüders bands. The strong aging in the zone cold deformed by calking is noticeable, just like in Fig. 15.86. The cracks are exclusively intergranular, as demonstrated in **Fig. 15.88.**

In a riveted **tank** of 12 mm thick Armco iron plate[34]), in which 45 % **caustic soda** was stored at 40–60 °C, the rivet heads on the inside split here and there and the plate cracked from rivet hole to rivet hole; such a crack on the outer surface of the tank is illustrated in **Fig. 15.89.** After detaching the rivet joints, the plate surfaces which had been face to face, showed more cracks diverging radially from the rivet hole edges into the plate **(Fig. 15.90).** All the cracks propagated from the opposing contact surfaces of the sheets **(Fig. 15.91).** The narrow crevice present here offers favorable corrosion conditions in every respect. The rivets also had cracked in numerous cases at or near the shaft head. All the cracks were found to be intergranular in character **(Fig. 15.92),** and were evidently caustic corrosion cracks from the beginning.

The following failure though was less certain[35]). After conversion of a steel plant to oil burning, the shells of the bricked-up **sheet metal chimneys** showed cracks especially in places where condensed water could collect. The shells were welded together out of rimmed and semikilled steel plate of varied qualities. The cracks were associated with circular seams or with fillet-welded ribs **(Fig. 15.93 and 15.94).** Some ran parallel to, and some ran across the welds. All cracks originated from the inside of the shells and were intergranular in nature like caustic corrosion cracks **(Fig. 15.95).** The interrelation between crack formation and the conversion to oil heating gave rise to the conclusion that the corrosive agent was a component of the oil combustion gases dissolved in the condensed water. Sulfuric or sulfurous acid or their salts, or even nitric acids or nitrates may be responsible. Up to 0.25 % organically bound nitrogen has been found in heating oil, and 80 to 92 mg/m^3 of nitric oxide was reported in combustion gases; in addition, more than 1 % calcium nitrate was found in the bricks of demolished chimneys[36]). Nitrate solutions have already been mentioned as one of the most dangerous instigators of intergranular stress-corrosion cracking. R. Flossmann and others[37]) proved that up to 0.2 % NO may be formed at high combustion temperatures in blast furnace recuperators according to $N_2 + O_2 = NO$ (Nernst). Under these contitions the weld seams are attacked in an intergranular manner after oxidation to nitrate. According to E. Bühler and others[38]), such an attack is substantially reinforced in the presence of SO_3 ions. This explains why the failures of the sheet chimneys first occurred after the conversion of the steel plant to oil.

The **fan wheel of a gas exhaust pump,** used to move coke oven gas, flew apart after one and one-half years of service at 5000 r.p.m. The cause was the cracking of a number of rivets that served to secure the blades. The appearance of a cracked rivet, along with a longitudinal section, are reproduced in **Fig. 15.96 and 15.97,** respectively. The rivet had incipient cracks in numerous places. The cracks had an intergranular pattern. In this case an aqueous solution of ammonia – always present in coke oven gas – may have served as corrosive agent.

Fig. 15.86 to 15.88. Rivet hole cracks in steam boiler

Fig. 15.86. Flat ground face. Etch according to Fry. 1 ×

Fig. 15.87. Transverse section through a rivet hole. Etch according to Fry. 1 ×

Fig. 15.88. Micrograph of intergranular rivet hole crack. Etch: Nital. 100 ×

In the **cover plate of an exhaust fan** that had operated under the same or similar conditions, cracks were found at a place that was cold worked by hammerblows **(Fig. 15.98).** Apparently deformation stresses were contributing to their formation.

These examples have shown how a variety of conditions may lead to this type of intergranular stress corrosion cracking. But if preventive measures are to be recommended, the recognition of the factors that had a bearing on the genesis of the failure are vital. The following example shows that sometimes the tracking and inquiry methods of a detective are required.

Leaks occurred at the weld seams of a **plant evaporation hood** in which calcium chloride solutions were concentrated from 15 to 40 % by heating with steam that was at 3 atm pressure and 260 °C. Magnetic particle inspection found cracks that were oriented predominantly transversely to the seam **(Fig. 15.99),** but sometimes also ran parallel to a weld undercut. The cracks orginated at the inside **(Fig. 15.100)** and took an intergranular course **(Fig. 15.101).** Accordingly they should have been caustic cracks. But no propensity for such cracks was to be expected in

15.89

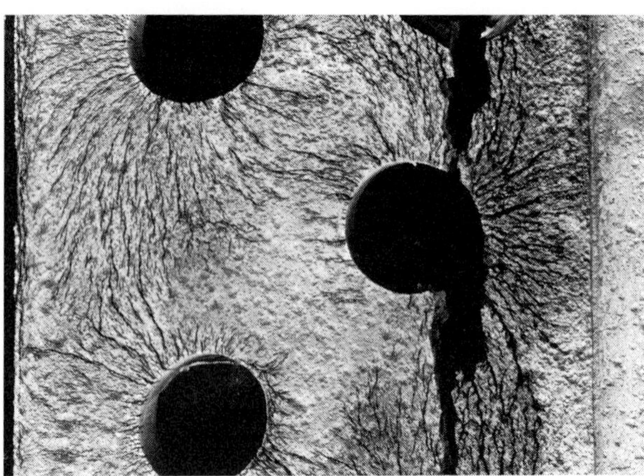

15.90

view of the stated conditions. Through intensive questioning of the service personnel, the fact that the pipes of the heating chamber had been boiled with caustic soda for 24 to 48 hours at 126 °C every two weeks for the removal of crust deposits was finally established. Previously the crusts had been removed by reaming the pipes. Sometime later the steel pipes were replaced by copper tubes and the crusts were removed in the manner previously described. This new procedure had caused the damage of the evaporation hood sheets. A variety of remedies were recommended, 1) to return to the former mechanical procedure for crust removal, 2) the selection of a suitable solvent, or 3) the production of evaporation hoods from a caustic crack-resistant steel thoroughly killed with aluminum.

Finally, a rare example should be mentioned of intergranular stress-corrosion cracking that occurred in the construction industry. On a farm, a perforated tile **roof of a stable** collapsed 26 years after building. It had been reinforced with 14 and 16 mm round rods. (This case differs in principle from others of prestressed concrete beams made of aluminum oxide cement that were found in Bavarian cow stables. See also the following section 15.3.4.2). In the tile fragments, 0.0071 % ammonia and 0.93 % nitrate were found. Four rod sections of the reinforcement were

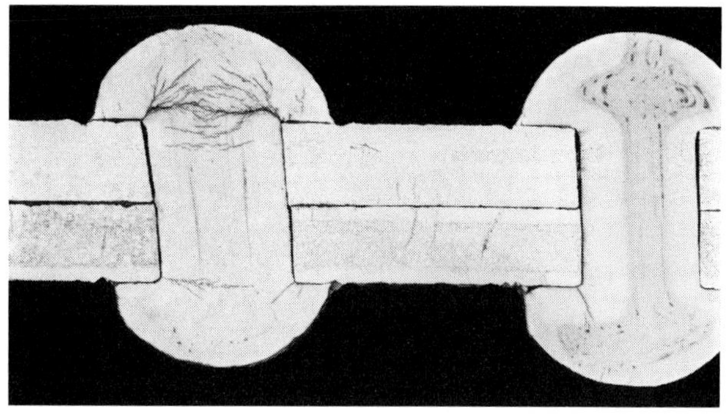

15.91

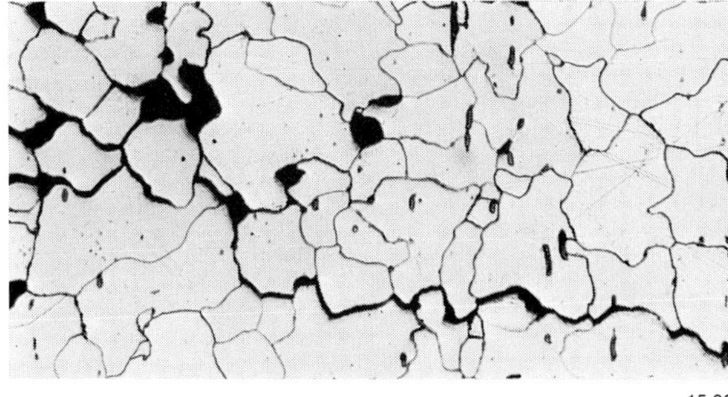

15.92

Fig. 15.89 to 15.92. Cracked rivet seam of leaching vat

Fig. 15.89. Outer surface. 0.66 ×

Fig. 15.90. Inner surface of overlap zone. 0.66 ×

Fig. 15.91. Longitudinal section through rivet row. Etch according to Heyn. 0.66 ×

Fig. 15.92. Intergranular crack. Etch: Nital. 100 ×

examined. Except for one, they consisted of a rimmed steel with 0.05 % C. All rods had incipient transverse cracks **(Fig. 15.102).** Still more cracks were made visible by pickling with hot hydrochloric acid. During tensile tests the rods sometimes cracked prematurely after minor deformation, even though an attempt had been made to select crack-free parts. Only the rod that consisted of a killed steel with 0.05 % C, attained normal values for strength and deformation at fracture. Lüders lines were found in longitudinal sections of the prematurely broken rods after etching according to Fry; therefore the rods had been strain aged. During microscopic examination, all cracks were found to be intergranular in nature **(Fig. 15.103).** Therefore these were caustic cracks. Nitrate ions may have been the corrosive agent. They may have been formed by oxidation of ammonia vapors in the stable environment.

15.93

15.94

Fig. 15.93 to 15.95. Crack damage in bricked-up sheet metal chimney

Fig. 15.93. Cracks parallel and vertical to circular seam. 1 ×

Fig. 15.94. Crack in welded-on rib. 1 ×

Fig. 15.95. Crack path in microstructure. Etch: Nital. 100 ×

In **high strength steels** the transgranular type of stress corrosion is predominant, perhaps because in that case the mechanical component of the crack formation process overshadows the chemical one. But sometimes in the initial stage of crack formation an intergranular course is found in these steels, too. Some exemples of this will be cited in the following paragraph. Another ist described in section 15.3.4.2 (Fig. 15.140 and 15.141). The intergranular initiation of hydrogen cracks (Fig. 2.13) observed under the scanning electron microscope may also be cited in this context. The fracture of a **centrifuge drum lockring** of chromium manganese steel that was heat treated to a strength level of 1030 MPa (150 ksi) is shown in **Fig. 15.104**[39]). The centrifuge had operated at 4500 r.p.m. in the processing of fruit juices for about one year when it flew apart due to the rupture of the ring. The fracture had originated at the crevice corrosion pit at the base of the trapezoidal thread **(Fig. 15.105).** Fine incipient cracks with a preferred intergranular path could be seen under these pits in the metallographic section **(Fig. 15.106).**

In this context the crack of a **drum cover** of a similar centrifuge should also be mentioned[40]), in which starch was processed. The cover consisted of a heat treated stainless steel of AISI type 431 with a tensile strength of 1030 MPa (150 ksi). The fracture propagated from a corrosion pit at the root of a radially milled groove of the cover. As can be seen from **Fig. 15.107,** its path was preferentially transgranular into the interior but was occasionally deflected by ferrite streaks whereby the ferrite was etched out, while the carbides at the periphery of the streaks remained. In the branches an intergranular path of the cracks could be found in some places, and corrosion had advanced onto certain lattice planes of the heat treated structure **(Fig. 15.108).** Therefore in this case, too, a tendency toward a selective progression of corrosion was present.

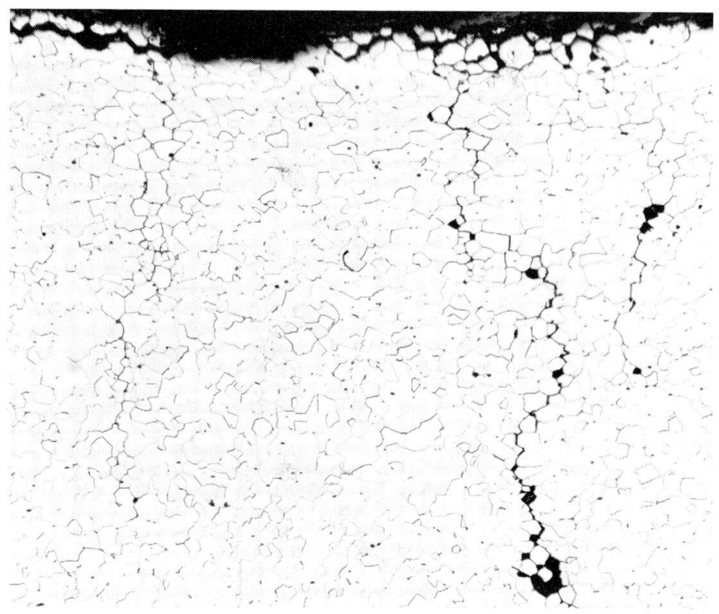

15.95

15.96

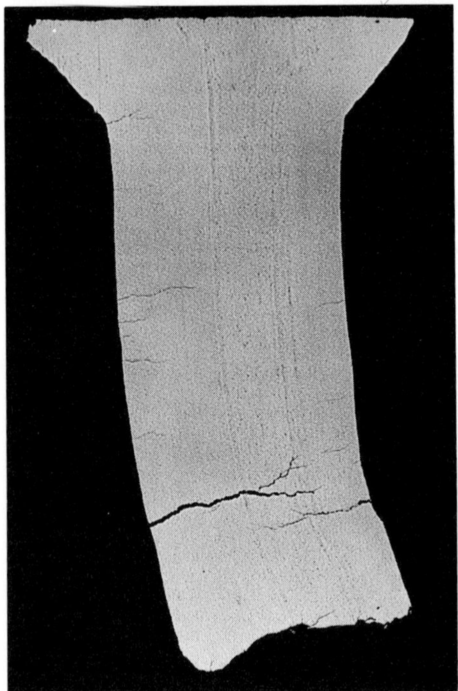

15.97

15.98

Fig. 15.96 and 15.97 Cracked rivet of gas exhaust fan

Fig. 15.96. Surface. 6 ×
Fig. 15.97. Unetched longitudinal section. 4 ×

Fig. 15.98. Section of cover plate of exhaust fan with stress corrosion cracking at hammer blow markings (arrows). 1 ×

Fig. 15.99 to 15.101. Evaporation hood that developed leaks through longitudinal cracks

Fig. 15.99. Cracks on outside, made visible by magnetic particle inspection. 0.66 ×
Fig. 15.100. Section taken parallel to weld seam. Etch according to Heyn. 1 ×. Top: Inside

Fig. 15.101. Small crack with intergranular path. Section. Etch: Picral. 200 ×

15.102

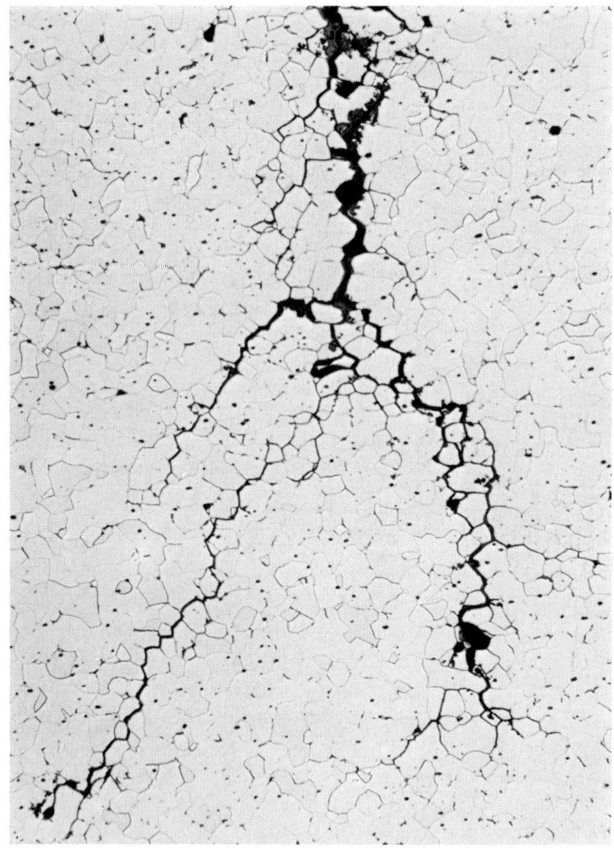

15.103

Fig. 15.102 and 15.103. Caustic cracking of reinforcement rods causing collapse of perforated tile roof of animal stable

Fig. 15.102. Reinforcement rod with incipient cracks. 1 ×
Fig. 15.103. Intergranular stress corrosion crack. Longitudinal section. Etch: Picral. 90 ×

Fig. 15.104 to 15.106. Broken centrifuge drum lock ring that had a 1030 MPa (150 ksi) tensile strength

Fig. 15.104. Fracture. Approx. 1 ×. → Circumferential fracture in thread G = uppermost turn

Fig. 15.105. Crevice corrosion in base of trapezoidal thread. 3 ×

Fig. 15.106. Micrograph of crack propagation from pitting. Etch: Nital. 100 ×

Intergranular stress-corrosion cracking may also appear in sensitized austenitic steels that are subject to grain boundary deterioration, if they are under stress while in alkaline or chloride solutions such as seawater. Especially susceptible in this sense are manganese-containing austenites **(Fig. 15.109 and 15.110).** Failures occurred in **welded ship's plates** of chromium-nickel-manganese steel that had been strengthened by cold working and had been destroyed by the combined effect of residual stresses and seawater corrosion. The phenomenon on which this failure is based could also be perceived as intergranular corrosion[41]), which had been accelerated by carbide precipitation as a consequence of cold working. However, such an interpretation is not possible in the following case.

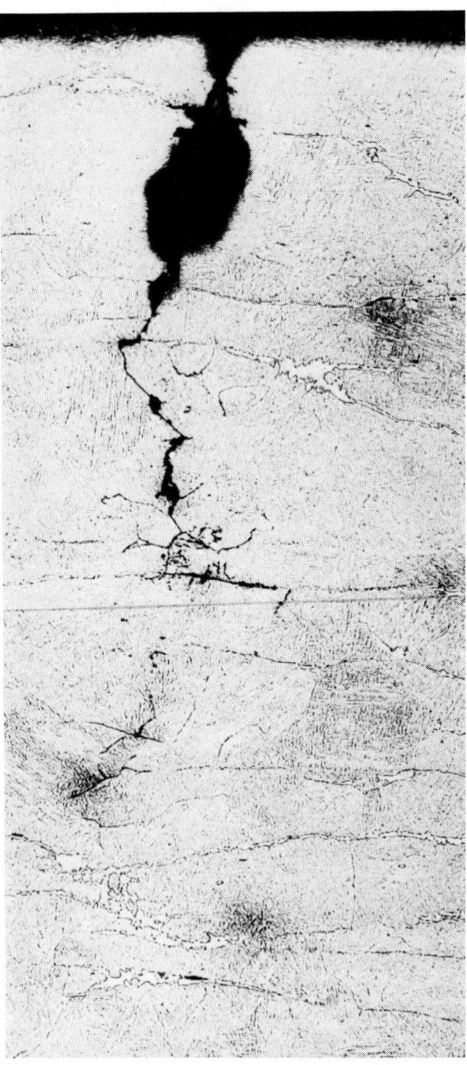

Fig. 15.107. Stress Corrosion cracking in centrifuge drum cover of heat treated stainless steel. Transverse section. Etch: V2A-etchant. 100 ×

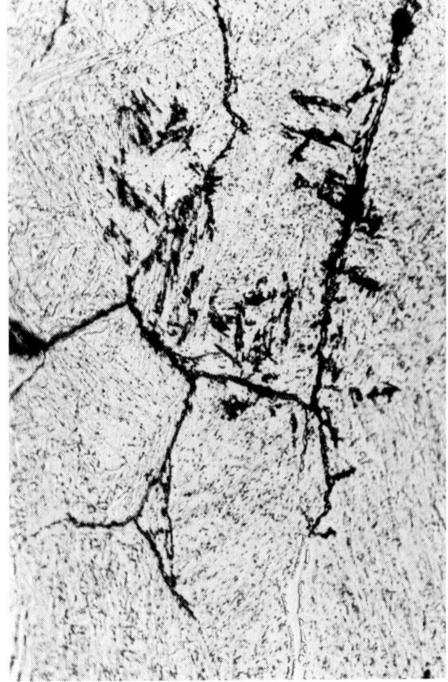

Fig. 15.108. Crack branch from Fig. 15.107. 500 ×

High pressure Nimonic 75 tubes in the waste gas boiler of an ammonia synthesis plant had been cooled from the outside with water at 100 °C and steam of 28 atm pressure. Intergranular stress corrosion cracks were observed in the steam-cooled part, even though no grain boundary precipitates were in evidence **(Fig. 15.111 to 15.114).** Yet demineralized water at 350 °C would already have a corrosive effect in this sense on alloys with more than 70 % Ni at stresses just above the yield strength[42]).

But such cases of intergranular stress corrosion are relatively rare. Furthermore, in austenitic steels, the transgranular type of stress-corrosion cracking predominates. Examples of this are cited in the following section.

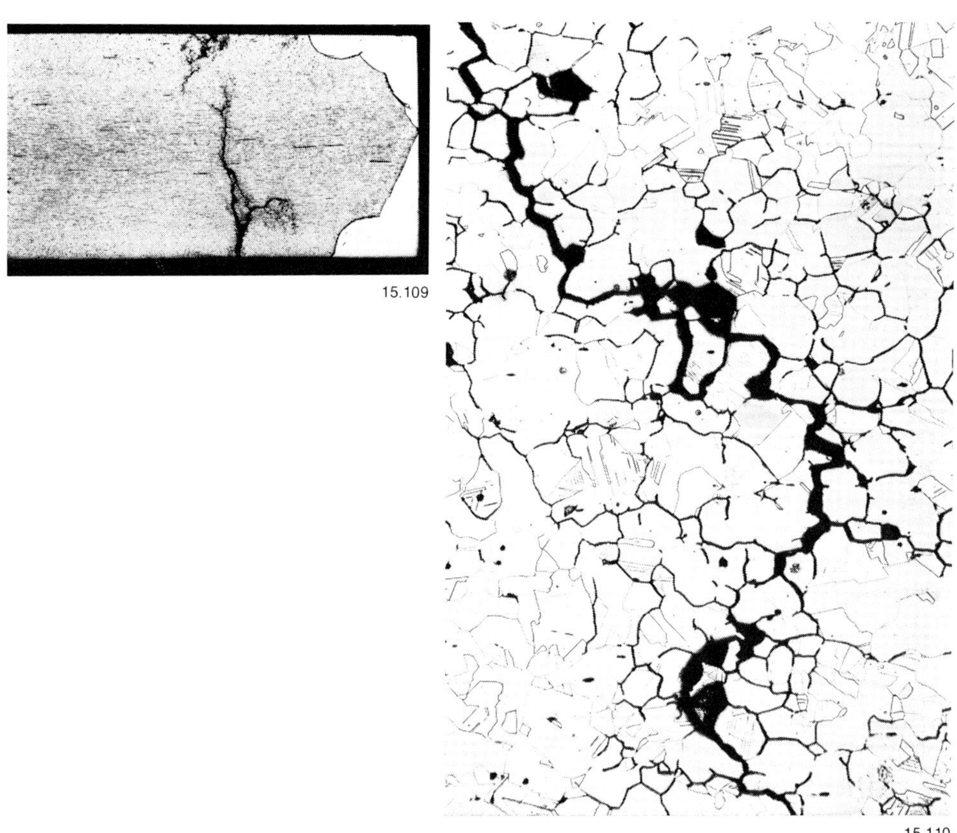

15.109

15.110

Fig. 15.109 and 15.110. Intergranular stress corrosion in cold rolled ship's plate of austenitic chromium-nickel-manganese steel

Fig. 15.109. Cracks in critical zone next to weld seam. Etch: V2A-etchant. 2 ×

Fig. 15.110. Microstructure in critical zone with grain boundary deterioration and stress corrosion crack. Transverse section. Etch: V2A-etchant. 100 ×

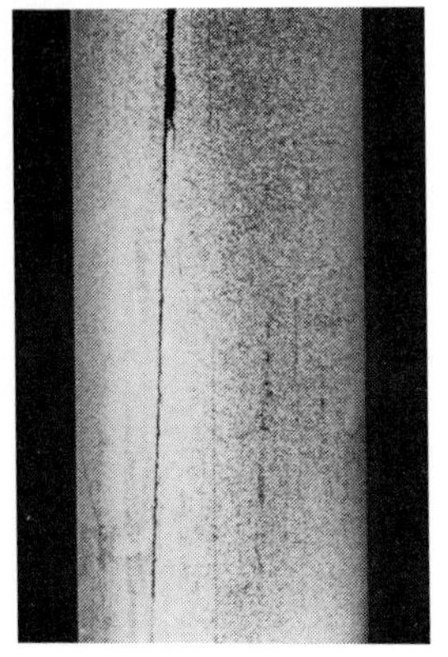

15.111

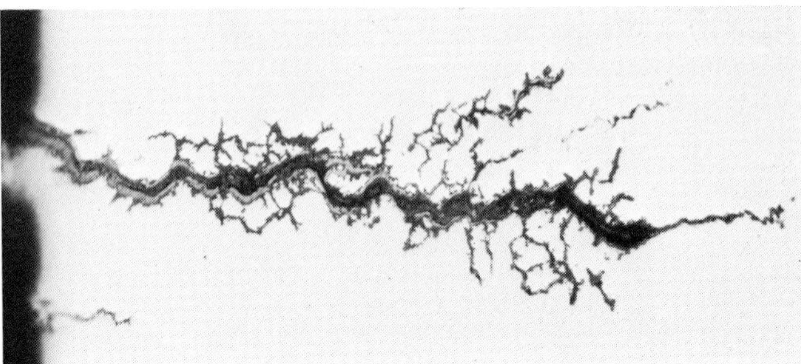

15.112

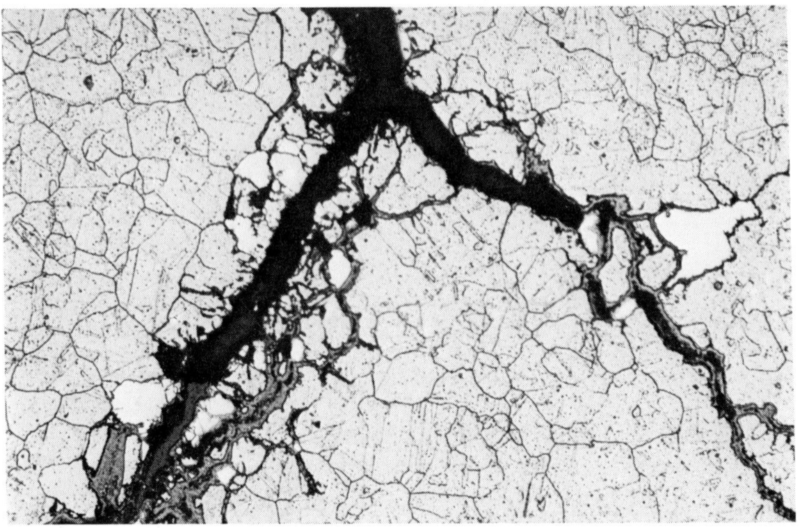

15.113

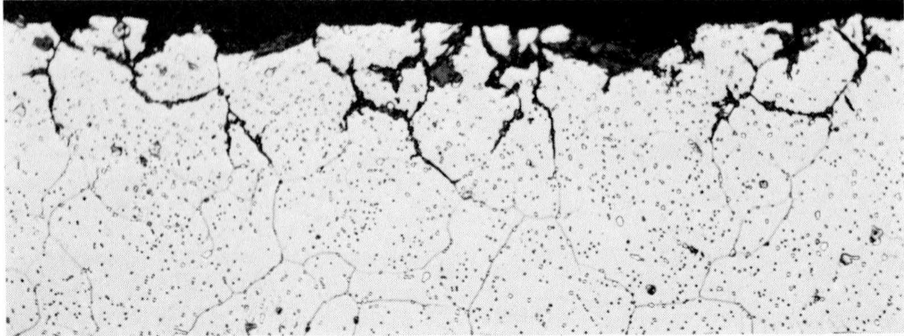

15.114

15.3.4.2 **Transgranular Stress-Corrosion Cracking**

The only case of transgranular stress-corrosion cracking in unalloyed and low alloy steels on record was for a long time a case of crack formation in pressurized bottles, as observed by H. Buchholtz an R. Pusch[43]). The bottles were used for storage of illuminating gas. Corrosion was caused by the prussic acid content of the gas. Failure in this case was not due to corrosion by anodic iron dissolution, as is normally the case, but to cathodic formation and penetration of hydrogen. Accordingly this phenomenon is designated as **hydrogen embrittlement.** Even though in this case it does not correspond correctly to the definition, it will be treated together with the analogous hydrogen sulfide attack and the crack formation in high strength steel wires in pre-stressed concrete.

During the last 25 years, substantial failures have resulted from **hydrogen sulfide corrosion** of tools used in oil drilling[44]), oil and gas pipelines[45])[46])[47]), and construction elements of reinforced concrete[48]). During tests of the effect of hydrogen sulfide in aqueous solution upon steel, F. K. Naumann and W. Carius[49]) came to the conclusion that the following process occurs:

Anodic partial reaction:	$Fe \longrightarrow Fe^{2+} + 2\ominus$
Cathodic partial reaction:	$H_2S + 2\ominus \longrightarrow S^{2-} + 2H$
or:	$2 \cdot (H_2S + 1\ominus \longrightarrow (SH)^- + H)$
Total reaction:	$Fe + H_2S \longrightarrow FeS + 2H$

Surface material is removed by the dissolution of iron, a process that slows down with the formation of a surface layer of iron sulfide which increases in thickness. The atomic hydrogen that is formed cathodically can penetrate the steel uninhibited and then is precipitated molecularly under a pressure at imperfections such as inclusions, grain boundaries, or dislocations. This brings the steel into a high state of tri-axial stress. Pressures of the precipitated hydrogen were calculated to be as high as 10^5 MPa (10^6 atm) or 3×10^4 MPa (0.3×10^6 atm), resp., from cathodic overloading and from the results of diffusion experiments[50]).

The life of stressed steels decreases logarithmically with the logarithm of concentration or the partial pressure of hydrogen sulfide[47]). However, even very low partial pressures around 10^{-3} to 10^{-4} atm may lead to failures in a comparatively short time. Hydrogen sulfide attack also decreases with increasing alkalinity of the solution and stops completely at pH values of about 9 or 10[47])[49]). In dry gases electrolysis cannot take place. Therefore, in the author's opinion, drying in the presence of hydrogen sulfide containing gases is the surest and cheapest way to avoid failures[47]). Desiccation must be extensive enough that no condensate can form even at the lowest anticipated temperatures.

Fig. 15.111 to 15.114. Nimonic 75 tube torn open by intergranular stress-corrosion cracking

Fig. 15.111. Outer surface. 1 ×. Crack accentuated by dye-penetrant application

Fig. 15.112. Crack branches in unetched transverse section. 50 ×

Fig. 15.113. As Fig. 15.112. Etched electrolytically with diluted nitric acid at 1.5 V. 200 ×

Fig. 15.114. Intergranular corrosion below surface. Longitudinal section. Etched electrolytically in diluted nitric acid at 1.5 V. 500 ×

15.115

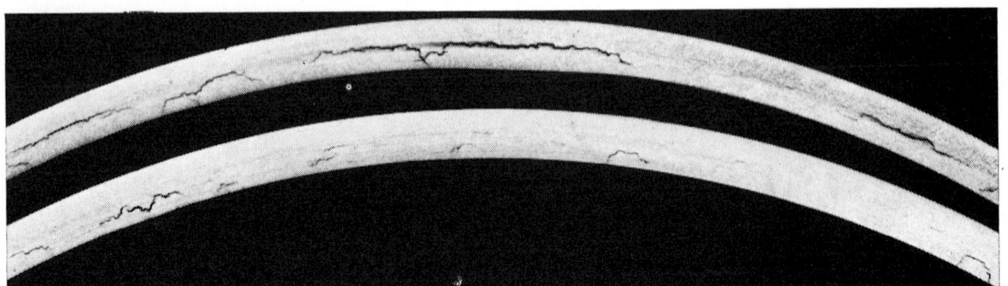

15.116

Fig. 15.115 to 15.119. Natural gas line of 400 mm diameter attacked by hydrogen sulfide

Fig. 15.115. Inside surface, colored black and covered with blisters
Fig. 15.116. Transverse section through pipe. Etch according to Heyn. 0.7 ×

Fig. 15.117 and 15.118. Unetched longitudinal section

Fig. 15.117. Separation in a slag inclusion stringer. 100 ×
Fig. 15.118. Crack propagation from one inclusion stringer to the next. 500 ×

Fig. 15.119. Pipe after burst test. Cross-hatched areas indicate cracks after ultrasonic inspection

15.117

15.118

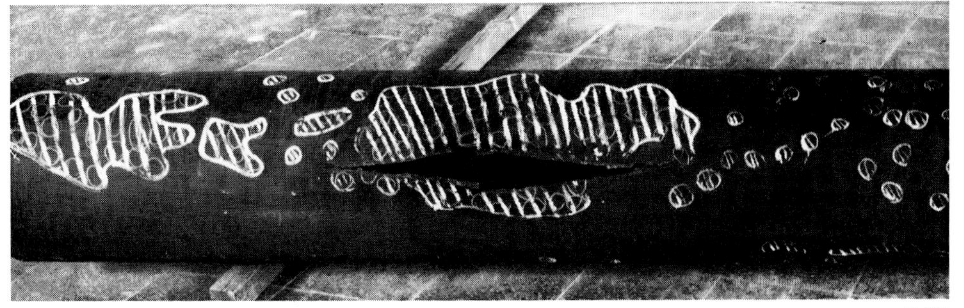

15.119

15.120

15.121

Fig. 15.120 to 15.126. Couplings for bore rods from natural gas drilling, cracked by hydrogen sulfide attack

Fig. 15.120. Socket with longitudinal crack and transverse fracture originating from it. 0.3 ×

Fig. 15.121. Coupling with longitudinal cracks. 0.3 ×

Fig. 15.122 and 15.123. Examples of fractures that originated in regions calculated as not being highly stressed. Points of origin designated by arrows

Fig. 15.122. Socket fracture. Left: With deposit. Right: Pickled. 1 ×

Fig. 15.123. Coupling fracture. 1 ×

Fig. 15.124. Baumann sulfur print of a transverse fracture. Reproduction. Approx. 0.5 ×

Fig. 15.125 and 15.126. Cracks with origin in interior. Transverse sections. Etch: Nital

Fig. 15.125. Shear stress crack in cold deformed zone below plier mark. 65 ×

Fig. 15.126. Internal stress crack. 60 ×

p 446

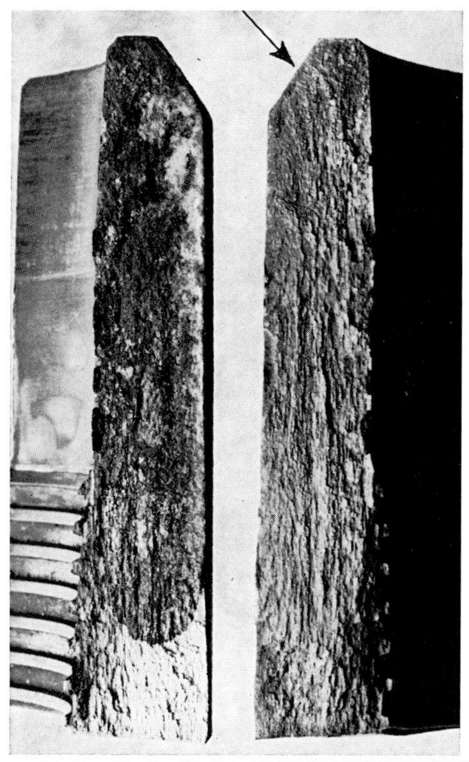

15.122

15.123

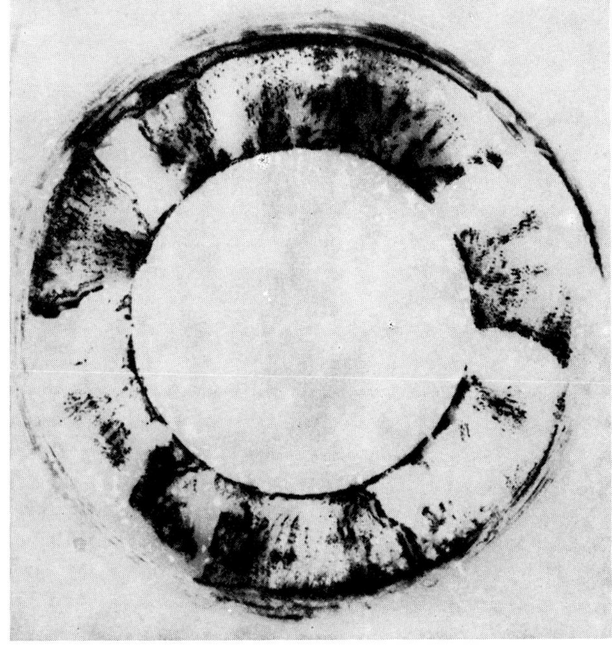

15.124

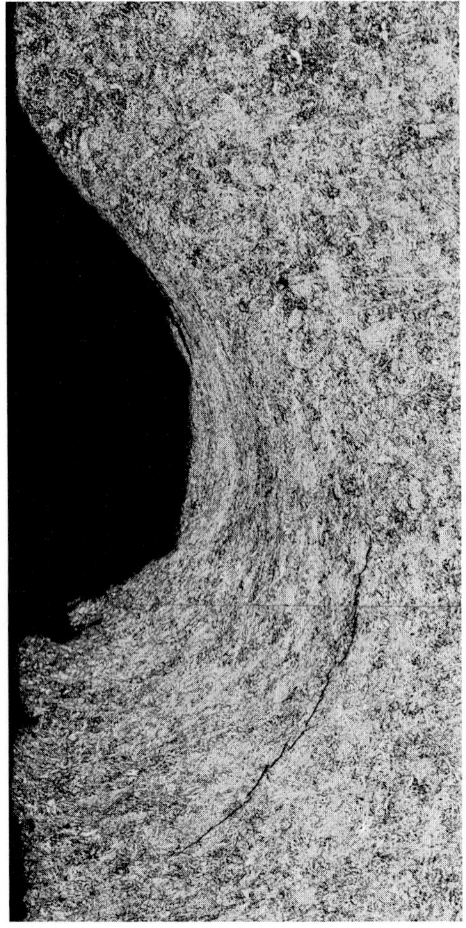

15.125

15.126

In soft, well deformable steels, damage is signified by the formation of blisters by the expansion at inclusions close to the surface. These need not be of unusual size. In hard steels, cracks may be formed or assisted by oriented tensile or shear stresses – external or residual. The steel is more sensitive to crack formation, the higher its strength, i.e. the lower its capacity to relieve stresses by deformation[51]). Even passive chromium and chromium-nickel steels are attacked depending upon their susceptibility to pitting. Therefore this problem cannot be solved from the steel selection point of view.

Failures in low carbon steel will be demonstrated for a **gas pipeline** of 400 mm diameter, through which hydrogen sulfide-containing natural gas was passed under a pressure of 60 atm[52]). The interior of the pipeline showed some hand-size blisters **(Fig. 15.115)** after removal from service for other reasons. Where no scaling had taken place at the inner surface, there were adhesions of a black deposit. This developed hydrogen sulfide when sulfuric acid was dropped onto it, and also blackened bromide paper saturated with sulfuric acid. The deposit therefore was iron sulfide. When drilled open under acidic water, the blisters gave off a gas that was under considerable pressure. The gas contained 90 % hydrogen – or 97 % if the air which penetrated during capture is disregarded. In transverse sections, fissures were not only found located directly under the surface, but also had split open the entire pipe wall **(Fig. 15.116).** They generally followed the fiber orientation parallel to the surface **(Fig. 15. 117),** but occasionally also approached the interior surface propagating from fiber to fiber **(Fig. 15.118).** During a bursting test of one of the pipes by internal water pressure, failure occurred by local bulging in a zone which was strongly permeated by fissures, as was shown by ultrasonic inspection **(Fig. 15.119).** A tangential fracture stress of 330 MPa (48 ksi) was reached which corresponded almost exactly to the yield point of the steel.

This failure could possibly have been prevented by drying of the gas.

A counterpart to this, also taken from the petroleum industry, but one that concerns a **steel heat treated to higher strength,** is represented by the following failure[44])[53]). **Bore rod couplings** of chromium-molybdenum steel, heat treated to a 880 to 1030 MPa (127.5–150 ksi) strength level, fractured consistently during natural gas drilling. The transverse fractures were induced by longitudinal cracks in the socket and coupling parts **(Fig. 15.120 and 15.121).** They often originated not at the points of highest stress calculated to be at the inner edge, but from some other point at the end faces or perimeter **(Fig. 15.122 and 15.123).** All fractures were tarnished blue. Sulfur prints established that the blue appearing deposit consisted of a thin layer of iron sulfide **(Fig. 15.124).** Metallographic examination showed that shear stresses under impression marks caused by pliers used for tightening of the screws partly contributed to crack formation **(Fig. 15.125).** Additional cracks propagated from other points below the surface without apparent reason **(Fig. 15.126).** The cracks resembled the flakes described in section 4.3.3 and were due to the same cause, namely the penetration and precipitation of hydrogen. Hydrogen sulfide was present in aqueous solution. The natural gas still contained 5–6 % of it after passing through the rinsing water. Therefore a remedy was more difficult to propose. Raising the alkalinity of the scrubbing solution would help. In addition, inocculation with inhibitors or agents forming coatings such as chromates or phosphates that have proven valuable in keeping these corrosion failures in check, should be undertaken as a remedial step.

Hydrogen sulfide may also attack **steel inserts in concrete** if the cement contains sulfur in larger amounts and if the mortar simultaneously loses its alkalinity – needed to protect the steel by passivation of the surface against corrosion – due to outside influences (carbonate formation) or through structural transformation[54]). This structural transformation occurs under certain conditions, for instance, under the influence of a hot-humid atmosphere on special concrete made from aluminous cement[48]). Considerable failures have occured through hydrogen sulfide attack of construction elements of this type of concrete that otherwise possess excellent properties. High strength steels used for prestressed concrete reinforcements are particularly sensitive to this type of attack.

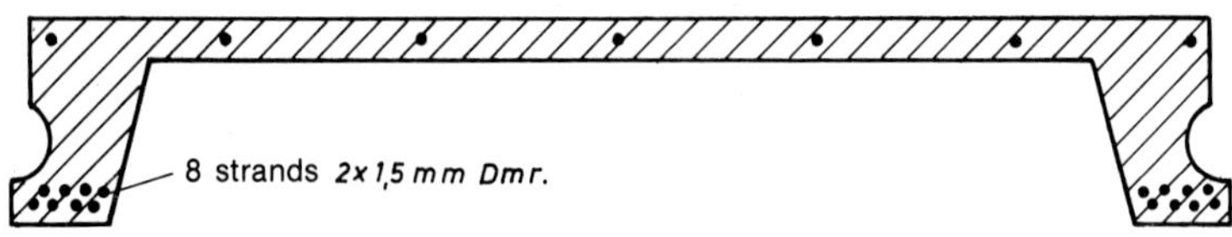

15.127

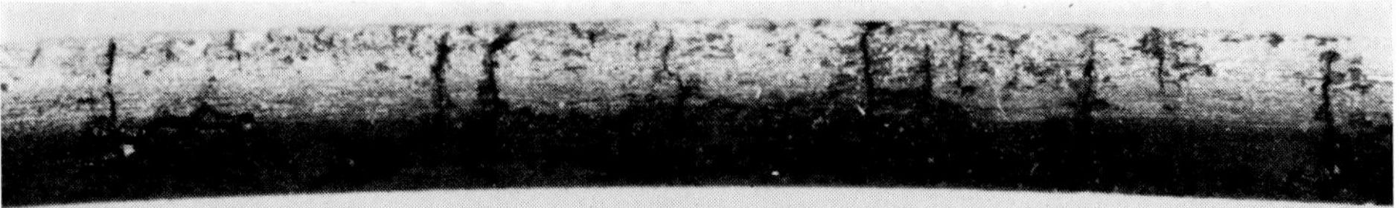

15.128

15.129

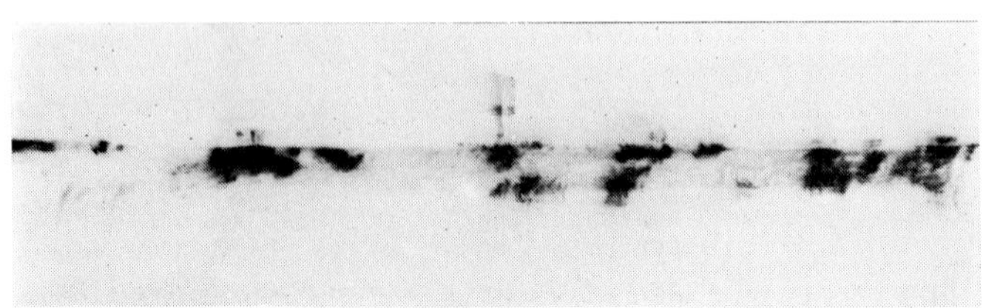

15.130

This may be visualized by an example. In the pouring bay of an iron foundry, eight **prestressed roof plates** collapsed. They were located above the tap of a cupola furnace. The roof was weighted down at this point by a deposit of flue dust. Moisture could remain in it for a long time, so that a hot-humid atmosphere prevailed. The ribbed plates of 5.30 m length and 0.60 m width were produced as finished slabs from aluminous cement type conrete. They had been prestressed in the longitudinal ribs with eight strands; these strands consisted of two 1.5 mm wires that had been cold drawn to a 1960 MPa (285 ksi) strength level. This is sketched in **Fig. 15.127.** The slabs were rejected by the foundry because the wires were claimed to be completely brittle.

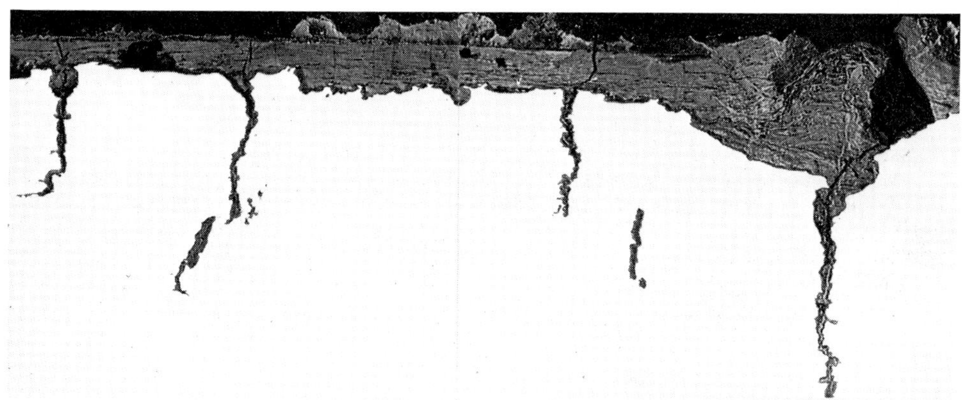

15.131

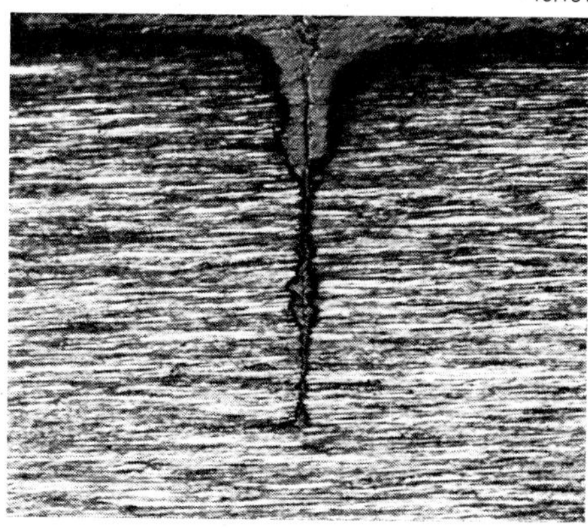

15.132

Fig. 15.127 to 15.133. Wires of prestressed concrete roof plates attacked by hydrogen sulfide

Fig. 15.127. Cross section through a prestressed concrete plate. Approx. 0.17 ×
Fig. 15.128. Hydrogen sulfide cracks in reinforcement wire. 10 ×. Surface cleaned with diluted citric acid
Fig. 15.129. Broken-open cracks in reinforcement wires. 10 ×
Fig. 15.130. Baumann sulfur print of surface of pitted reinforcement wire. Reproduced 1 ×

Fig. 15.131. Hydrogen sulfide corrosion and cracks in reinforcement wire. Unetched longitudinal section. 70 ×
Fig. 15.132. Hydrogen sulfide crack in drawn fiber structure. Longitudinal section. Etch: Picral. 100 ×

Fig. 15.133. Reverse bend tests with attacked and satisfactory 1.5 mm wires that have a 1960 MPa (286 ksi) tensile strength p 450

Site inspection indicated that many wires had broken. Others could be broken by bending with little effort. They did not break because they were brittle, but because they had incipient cracks in many places **(Fig. 15.128).** The fractures were composed of a brittle part which ran perpendicular to the surface, and a deformation type that was deflected in the direction of the fiber **(Fig. 15.129).** The brittle fractures were discolored black by iron sulfide, and therefore were hydrogen sulfide cracks. The deformation fractures, on the other hand, were in part bright, and therefore had occurred later during collapsing, dismantling or bending. Furthermore, the wire surface was heavily corroded by pitting (see also Fig. 15.62). The pits were stained black and reacted strongly with silver bromide paper dipped into sulfuric acid during the preparation of a Baumann sulfur print **(Fig. 15.130).** A crack region in a longitudinal section is shown in **Fig. 15.131.** After etching, it could be seen that the cracks ran in a transgranular path **(Fig. 15.132).** The failure then was not caused by the brittleness of the wires, but by the use of an unsuitable cement in conjunction with unfavorable environmental conditions.

Similar failures were also found in roof slabs and beams in animal stables under hay lofts, and in ceilings above laundry rooms, pickling plants and bath houses. Examination always showed the same results.

Structural changes in concrete are easily recognizable in fresh fractures. The originally grainy structure becomes sandy, and the color changes from bluish-grey to yellow-grey. The depth of change – also that due to carbonate formation – can be determined quickly by spraying the new fracture with phenolphthalein solution; this produces a contrast between the violet-colored core and the unstained region that had undergone structural changes. For fast testing of hydrogen sulfide cracks in wires, reverse bend tests on a sufficiently large number of cut-off end specimens have proven valuable. The frequency curves in **Fig. 15.133** show the results of such a test series with wires of a satisfactory and an attacked roof slab.

Since these failures and their causes have become known, the use of aluminous cement for prestressed concrete has been discontinued. But quite some time will be required before all suspected parts have been dismantled.

Oxygen corrosion, too, may endanger steel reinforcements in concrete and particularly in prestressed concrete. Failure may occur if the steel rusts prior to use which could occur during

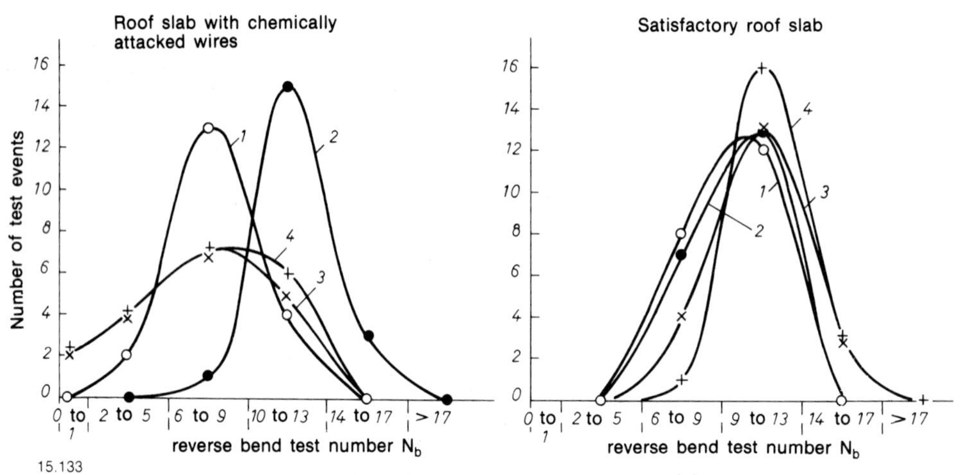

15.133

shipping or at the construction site (Fig. 1.14 and 1.15). Rust pits act as sharp notches, as has been mentioned already. W. Jäniche and coworkers[55]) have determined that high-stressed, sharply notched bars may break after a short time in air with a high humidity content. The stressed steel therefore must be carefully protected from corrosion prior to use. The conditions for inducing stress-corrosion are present if the steel is exposed to rusting for a while after stressing, but before pressing the mortar into the channels; this may sometimes happen if a delay, such as caused by a sudden freeze, is encountered. Inside the concrete the steel is made passive by the pore water of the mortar that corresponds to a saturated solution of calcium hydroxide with a pH value of 12.6, and is thus protected from corrosion. But this only holds true if the concrete is free of large pores in which the protective, highly alkaline solution cannot be retained; and if the concrete does not contain any halogen-bearing additions, such as calcium chloride for accelerated setting, which may prevent passivation or destroy the passive layer. Corrosion passivity also holds true only if the steel reinforcements are located deeply enough under the concrete surface, so that they cannot be reached by the formation of carbonate. The reaction of the concrete with carbon dioxide in the air leads to a decrease of free lime content and thus to a lower pH value.

The higher the strength, the lower the elongation, and the more severely stressed the steel is, the more sensitive it is to the formation and spreading of such stress-corrosion cracks. The cracks caused in this way resemble in their appearance those caused by cathodic corrosion and may also be due to the absorption of hydrogen into steel[56]). This phenomenon accordingly is also designated as hydrogen-induced stress-corrosion.

An example of this type of stress-corrosion cracking had been cited already during the treatment of concentration cells (Fig. 15.69 to 15.70). Another example follows. After six years a cable ruptured in a **lift cable frame for a hall roof.** The cable consisted of a stranded bundle of forty corrugated oval wires (1600 MPa (232 ksi) ultimate strength) encased in a jacket tube which was filled and compacted with a mortar of Portland cement. The cable fell down onto the hall roof even though the tension on the wires has only been about 550 MPa (80 ksi), i.e. little more than one-third of their ultimate strength. Further investigation revealed that the bundle had been under stress for months prior to filling of the jacket tube with mortar. All wires had broken – mostly without any noticeable deformation – within a region of 500 mm length around the rupture of the jacket tube **(Fig. 15.134).** During subsequent opening of the tube and removal of the mortar, many broken wires were found even outside the actual fracture site. Some wires had broken at several places. Some fractures were covered with mortar, and therefore had occurred prior to the filling of the channel. The jacket tube was in part rusted on the inside, and especially so at the sleeves. In other places the tube showed signs of contact corrosion where one of the wires had touched the covering **(Fig. 15.135).** At such contact areas, the wires frequently rusted **(Fig. 15.136).** Contact corrosion sometimes also caused the replication of the fin of one wire on the surface of an adjacent one **(Fig. 15.137).** Corrosion usually leads to the formation of deep pits **(Fig. 15.139).** The fractures originated in such pits **(Fig. 15.138).** They had formed in stages, but were not really fatigue type cracks. During microscopic examination, fine, predominantly intergranular cracks were found on occasion under the rust pits **(Fig. 15.140 and 15.141).** This type of stress-corrosion cracking therefore could also be designated as intergranular. But since by far the greatest part of the fracture was transgranular, these failures were incorporated in this section just like the hydrogen-induced types which are also occasionally intergranular at their origins (see Fig. 2.13).

Failures therefore occurred because the stressed steels were kept so long in the open channel that they had already cracked in part prior to filling with mortar. Whether it makes sense or is advisable to provide such reinforcing wires with a protective paint or coating with a view toward contingencies of this kind, is subject to dispute. Such corrosion protection in any case must not affect either the adhesive strength or the stability of the steel reinforcements in the concrete or mortar.

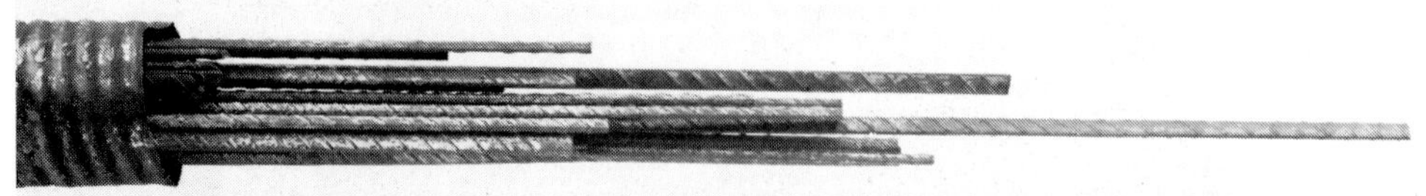

15.134

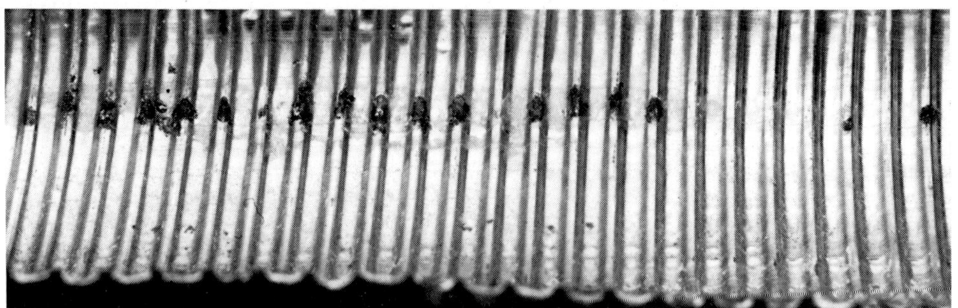

15.135

15.136

15.137

Fig. 15.134 to 15.141. Cracked lift wire suspension structure of hall roof

Fig. 15.134. Fracture. 0.2 ×

Fig. 15.135 and 15.136. Contact corrosion between jacket tube and wire

Fig. 15.135. Inner surface of jacket tube. 0.66 ×

Fig. 15.136. Wire whose upper edge was in contact with jacket tube. 1 ×

Fig. 15.137. Contact corrosion between wire and another wire. 1 ×

Fig. 15.138 and 15.139. Fracture origin from rust pit. 5 ×

Fig. 15.138. Fracture

Fig. 15.139. Surface at fracture (left)

Fig. 15.140 and 15.141. Intergranular incipient crack under corrosion pit. Longitudinal section

Fig. 15.140. Unetched. 200 ×

Fig. 15.141. Etched with diluted picric acid and Agepon wetting agent. 500 ×

15.138

15.139

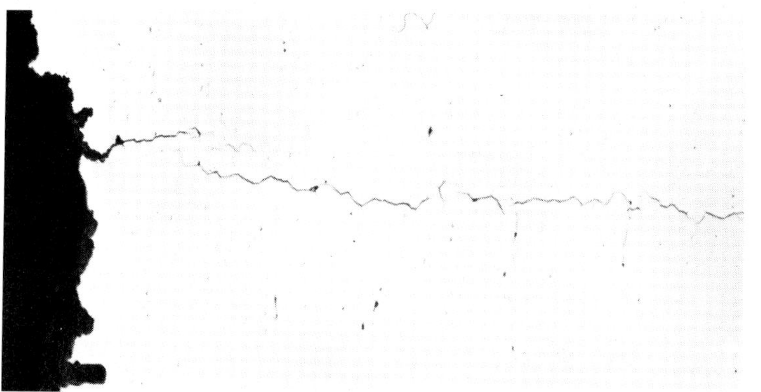

15.140

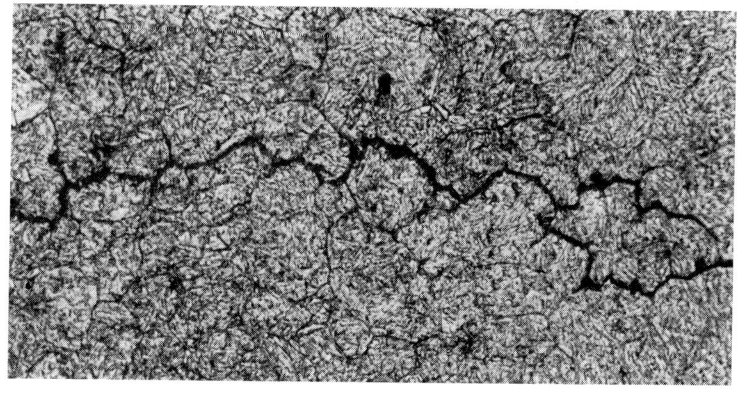

15.141

In the following, some failures caused by incomplete imbedding of the stressed steel rods or wires into the mortar are described. During inspection of **gasoline station roofs**[54]) – from which the wire bundle shown in Fig. 15.69 was taken – it was determined that the wires were heavily rusted in some areas (not only at joints as a result of chloride-containing mortar, but also in-between). The rust spots were always situated at the upward pointed side of the jacket tubes **(Fig. 15.142),** and corrosion had affected only the uppermost or the top wires **(Fig. 15.143).** The reason was that the jacket tubes were not completely filled with mortar. The smooth surface of the mortar level can still be recognized in Fig. 15.142. The tubes and the wires had corroded from these unfilled hollow spaces. A number of wires had also already started to crack. In order to determine how corrosion affects strength and elongation of the wires in static tensile tests, we dismantled a bundle with thirteen wires. The wires were cut into lengths of 1 m, and the specimens were used for tensile tests. In all rusted wires the fractures propagated from rust pits. The results, in the form of frequency curves and arranged according to rusted and rust-free specimens, are represented in **Fig. 15.144.** Strength which was around 1600 MPa (232 ksi) for the rust-free specimens, was widely scattered for the oxidized specimens down to values below 1000 MPa (145 ksi). The elongation maximum which was between 4 and 5 % for the unoxidized specimens, was displaced to between 0 and 1 % for those with rust. Accordingly, the notch effect of the corrosion pits had a particularly strong influence upon deformability. The corrosion pits would have a still greater effect in the case of tensile creep stresses[55]).

15.142

15.143

Fig. 15.142 to 15.144. Reinforced concrete roof wires rusted and broken due to incomplete imbedding in mortar

Fig. 15.142. Jacket tube rusted-through from interior hollow space at upper face. 1 ×

Fig. 15.143. Reinforcement wire bundles of an incompletely filled channel. 1 ×. Only the uppermost wire which stuck out from the mortar is corroded

Fig. 15.144. Effect of corrosion upon mechanical properties of reinforcement wires

Three **anchors of stranded guy wires** located at a Rhine harbor cracked within a few years after their installation. They served to secure sheet pilings against a concrete block at the river bank. The cable anchorages which consisted of 22 8-mm diameter wires each, were made of heat treated steel that had an ultimate strength of approx. 1500 MPa (218 ksi); these wires had been imbedded in a thin layer of mortar and secured firmly in the ground at a depth of 5–6 m, i.e. mostly in ground water. The harbor administration assumed that earth movement provoked by mining may have caused the failure. The wires were heavily rusted and therefore certainly not encased by dense concrete. Fine transverse cracks propagated from the corrosion pits **(Fig. 15.145);** the cracks subsequently had in part corroded **(Fig. 15.146).** Therefore this was a case of stress-corrosion cracking. Longitudinal sections illustrating fresh and corroded cracks are reproduced in **Fig. 15.147 and 15.148.** These probably ran predominantly in a transgranular manner, but this could not be positively established in the fine-grained heat treated structure. Bend stresses originating from earth movement or from vibration transmitted through the earth may have contributed to the cracks and fractures, but the actual cause of fracture was corrosion as a consequence of an absence of a thick mortar case.

Another failure due to a similar cause occurred through the rupture of a high strength steel rod – 10.2 mm in diameter with an ultimate strength of approx. 1500 MPa (218 ksi) – in a 14 cm thick concrete **runway of an airport** barely one year after being put into service. The fracture **(Fig. 15.149)** had originated at a corrosion pit **(Fig. 15.150).** The cause was solely a large air bubble located at the point of failure. In such a large hollow space, no saturated lime solution, which passivates the steel and is in equilibrium with the surrounding mortar, can remain permanently unchanged.

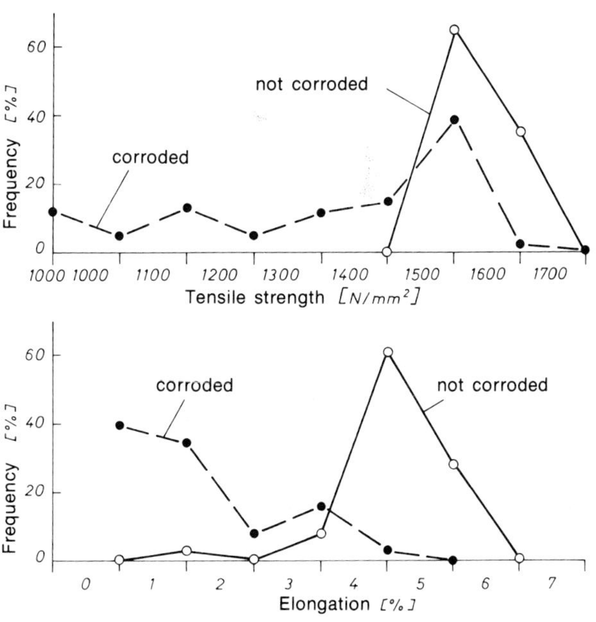

15.144

15.145

15.146

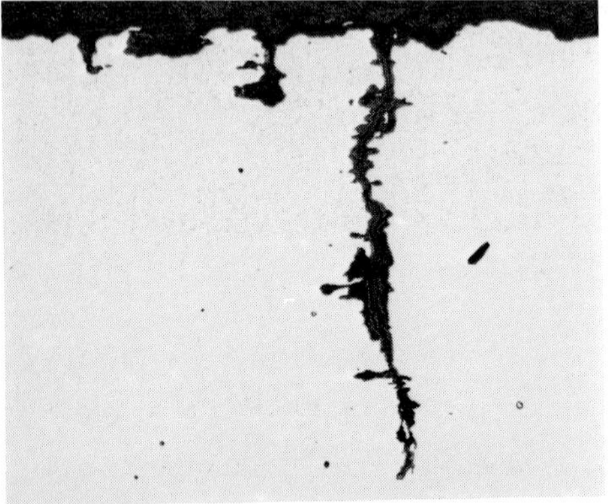

15.147

15.148

Finally an example will be cited for stress-corrosion cracking caused by the effect of chlorides upon the high strength steel in reinforced concrete. This concerns cylindrical **water tanks** of approximately 30 m in diameter that were prestressed by 4.1 mm diameter wire cold drawn through a revolving die ("Preload Process"). The wire that had an original tensile strength of 1600 to 1630 MPa (232–236 ksi) attained a strength of 1670 to 1720 MPa (242–250 ksi) after cold drawing and eventually had a preload of 1030 ± 70 MPa (150 ± 10 ksi).

After prestressing of the ring-shaped concrete core, the wires were pneumatically covered with mortar, whereupon another layer of wire was stretched out in the lower part of the tank. Onto this layer, still another layer of mortar was sprayed. After two tanks had been in service for over three months, the stressed wires broke in numerous places in two others that were under construction. Since the stressed wire in the tanks had been obtained from another supplier than the wire for the first group, the contractor assumed that the cause of failure was a lower quality product. However, examination immediately showed that only those wires had broken that had been in contact with the mortar of the intermediate layer. This was confirmed by an on-site inspection. Therefore the cause of failure had to be sought in the properties of the intermediate layer mortar and in particular in its choride content.

15.149

15.150

Fig. 15.149 and 15.150. Reinforcement rod of 11.2 mm diameter that corroded and cracked in an air bubble in a concrete runway

Fig. 15.149. Fracture

Fig. 15.150. Surface with corrosion pits

Fig. 15.145 to 15.148. Stranded anchor wires with incipient cracks caused by stress-corrosion and corrosion-fatigue

Fig. 15.145. New cracks. 10 ×

Fig. 15.146. Corroded cracks. 10 ×

Fig. 15.147 and 15.148. Unetched longitudinal sections through cracks. 100 ×

Fig. 15.147. New crack

Fig. 15.148. Corroded cracks

15.151

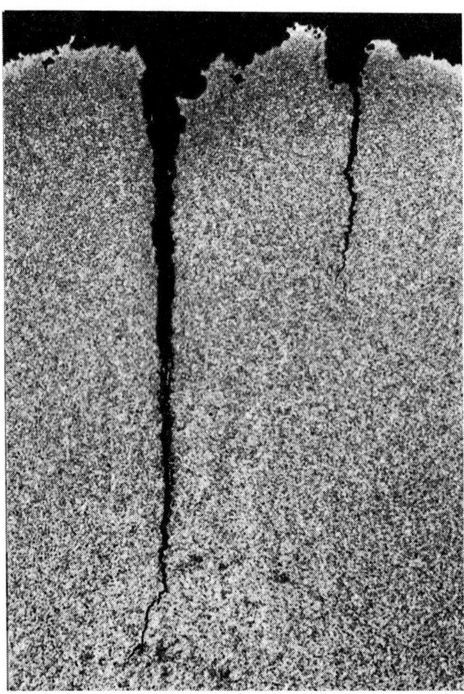

15.152

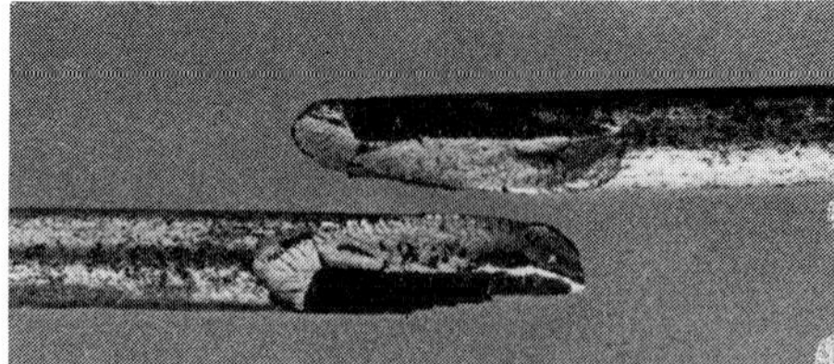

15.153

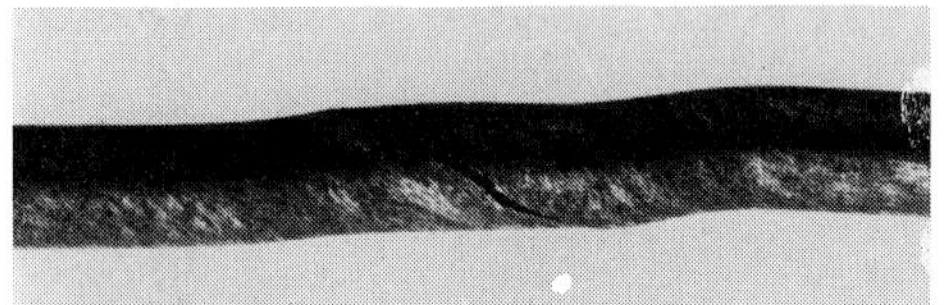

15.154

Fig. 15.151 to 15.154. Reinforcement wires of water tank, prestressed in accordance with "Preload Process", that cracked chlorine ion corrosion under the effect of tensile and torsion stresses

Fig. 15.151. Corrosion pits and crack in wire. 4 ×
Fig. 15.152. Transverse section through region with cracks. Etch: Picral. 50 ×

Fig. 15.153 and 15.154. Reverse torsion test results. 2 ×

Fig. 15.153. Wire with cracks. Broken after 180° twist
Fig. 15.154. New wire. Incipient crack after 5 × 360° twists

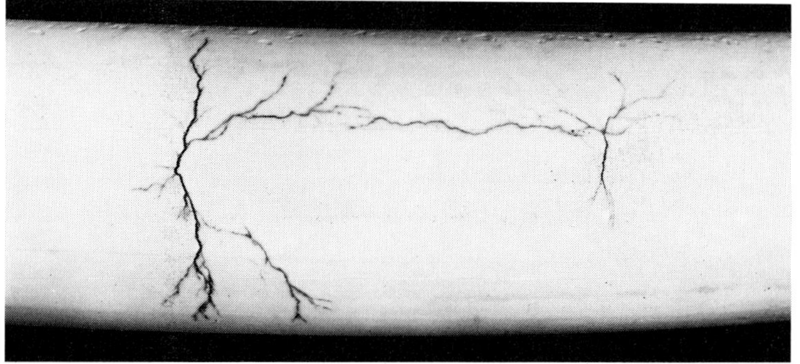

15.155

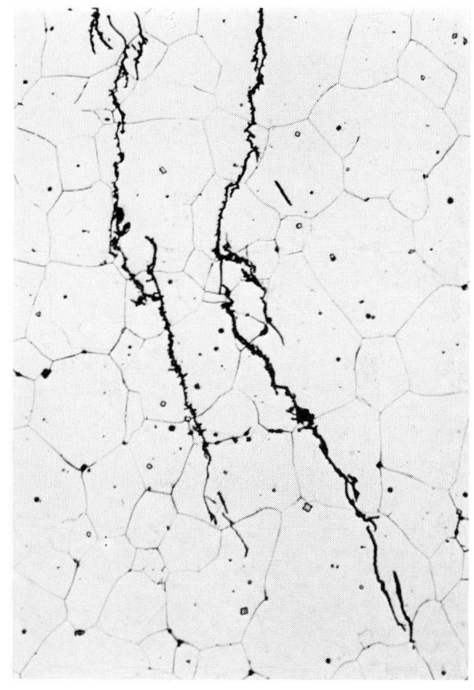

15.156

Fig. 15.155 and 15.156. Heating coil of evaporation vessel for calcium chloride solution that developed a leak by stress corrosion

Fig. 15.155. External surface. Cracks made more visible after dye-penetrant application. 0.5 ×
Fig. 15.156. Microstructure in transverse section. Etch: 50 % diluted nitric acid, 2 V. 200 ×

The investigation established that the wires were heavily rusted in those places where they were in contact with the intermediate layer, i.e. those internally located at the outside, and those externally located at the inside. Many wires had incipient cracks propagating longitudinally from the corrosion pits **(Fig. 15.151).** Often the cracks also were deflected obliquely from the wire axis or had changed into transverse fractures. Therefore, considering the type of stress applied, the wires may have come under torsional loads as well. Transverse sections showed that the cracks had taken a straight path **(Fig. 15.152).** Tensile tests resulted in a minor loss of strength, but in a major reduction in toughness due to the corrosion. The rusted wires broke prematurely at the corrosion pits or cracks, while unused comparison wires had good elongation and reduction in area considering their high strength. In reverse torsion tests, which are a severe method for determining longitudinal defects, the corroded specimens fractured in the first turn **(Fig. 15.153),** whereas the new wires could be twisted back and forth several times before they showed incipient cracks **(Fig. 15.154).**

An examination showed that the corrosive mortar of the intermediate layer contained 0.01 % S and 0.13 % Cl, while the mortar of one of the tanks whose wires were not corroded contained but 0.03 % Cl in addition to traces of S. In both cases the cement was low in chlorides. The source of the mortar's chloride content could finally be determined by a site inspection. It came from the sand that had been dredged from the nearby sea, or more correctly, from the seawater that was carried along with the sand. The chloride content of the mortar of the tank constructed first was lower, because construction coincided with the wintery rainy season. The salt content of the ocean sand stored on the harbor pier had been washed out by the rain. After the cause of failure was recognized, the sand used for new construction was taken from a sand pit located further inland.

Stress-corrosion cracking of this type can be avoided only by the above mentioned precautionary measures during construction. Prevention cannot be expected from the material's point of view considering the necessity of employing high strength steels with correspondingly higher notch sensitivity for the prestressed elements.

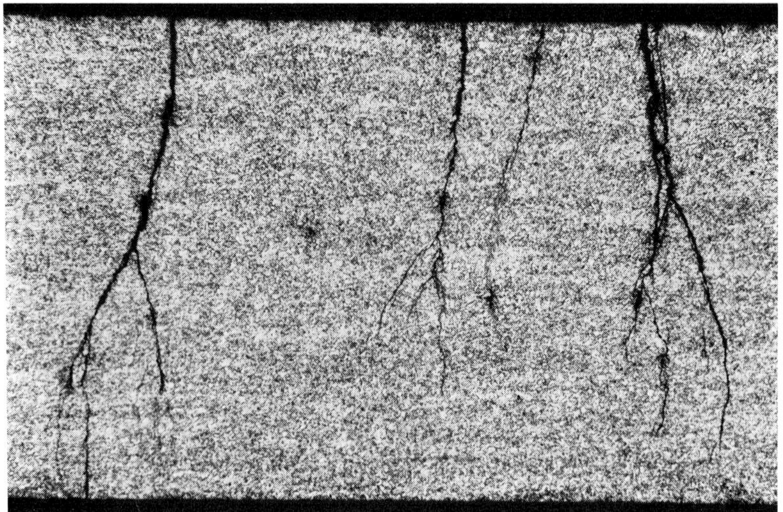

Fig. 15.157. Stress-corrosion cracks in sheet of a welded diphenyl heater. Section cut perpendicular to weld seam. Etch: V2A-etchant. 40 ×

Austenitic chromium-nickel steels suffer transgranular stress-corrosion cracking in the true sense of the definition by the effect of chlorine or hydroxyl ions[57]). The stresses may be either external or residual, such as deformation or weld stresses. Crack formation starts with local penetration of the passive layer at slip planes. The process is strongly dependent upon temperature. At room temperature stress-corrosion cracking will occur only after a long time, if at all.

H. Keller and G. Petrich[58]) have reported a metallographic explanation of failures caused by stress-corrosion cracking.

A failure due to seawater corrosion, where high stresses had occurred through the improper fastening of a ring onto the bushing of a **safety valve,** has been reported already in section 1 (Fig. 1.3 to 1.5).

The following failure was also caused by chlorides[59]).

A solution containing 50 to 70 % calcium chloride, with a pH value of 7.5 to 8.5, was concentrated by evaporation in a brick-lined vessel. Heating took place by passing steam of 1.5 MPa (15 atm) through a system of **coils** made of an austenitic stainless steel containing approx. 0.1 % C, 18 % Cr, 12 % Ni, 2 % Mo and 0.5 % Ti. The final temperature of the solution was about 170 °C. After five months one of the coils consisting of tubes having an outside diameter of 68 mm and a wall thickness of 3.4 mm developed a leak. No removal of material due to corrosion could be detected either on the inside or outside of the tube. However, indications of tightly closed cracks could be seen on the outer surface which had been in contact with the chloride solution. Numerous, multiple-branched cracks were revealed by dye-penetrant inspection **(Fig. 15.155).** The cracks had originated from the outside, partly penetrated the entire wall, and propagated mainly across the grains **(Fig. 15.156).** The cause of failure unequivocally was stress-corrosion cracking caused by the hot calcium chloride solution. The coil should have been annealed in order to remove residual stresses.

A **heat exchanger** of welded sheet metal, in which a mixture of diphenyl and diphenyl oxide containing 20 ppm chlorine was heated to 350 °C, began to leak adjacent to a weld after a few days. A tightly adhering thin, dark deposit could be seen on the sheet at the point of failure which was located close to the inlet opening. No chlorine could be found in it. A section through this defective spot showed that the leak was caused by stress-corrosion cracking **(Fig. 15.157).** The cracks originated at the side of the diphenyl and propagated transgranularly in the austenitic microstructure.

Weld stresses also contributed to the strong corrosion of a **vacuum** still made of the same Ti-stabilized, 2 % Mo-containing austenitic stainless steel used for the previously cited evaporation vessel coils. The problem concerns a vessel with a longitudinally welded shell made of a 6 mm thick sheet with welded-on dished ends. A weak sulfuric acid solution of quinidine bisulfate with toluene or isopropyl alcohol with a pH value of 2 to 3 was suctioned up through a pipe nipple whose location coincided with the longitudinal seam. The solution was heated under 6666 Pa (50 torr) pressure to approximately 100 °C through the bottom by steam, concentrated to two-thirds of the original volume after several hours, and subsequently discharged through a nipple at the bottom. Corrosion occurred only in a narrow sector of the vessel in which the suction nipple, the longitudinal weld seam and also the inlet pipe nipple for the hot steam were located. The damage extended to the circumferential seam of the bottom **(Fig. 15.158)** and the lower part of the longitudinal seam. The remaining part of the sheet metal shell had not been attacked. In some areas a group of short, radially oriented cracks could be observed in the bottom **(Fig. 15.159).** The sheets of the shell and bottom were eroded in a narrow strip parallel to the weld seam. Deep furrows or cracks ended in these trenches. They were oriented essentially perpendicular to the seam; in addition, they also ran obliquely to it in bunches or in a concentric arc around it. These were probably lines of equal stress. The furrows also permeated the weld bead which moreover was attacked laterally from the trenches, but remained standing in a web-like formation **(Fig. 15.160 and 15.161).** Both the sheet and weld bead had almost the same

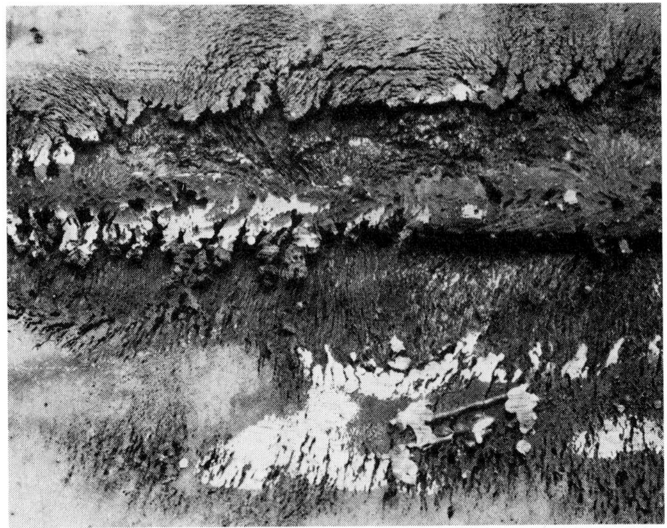

15.158

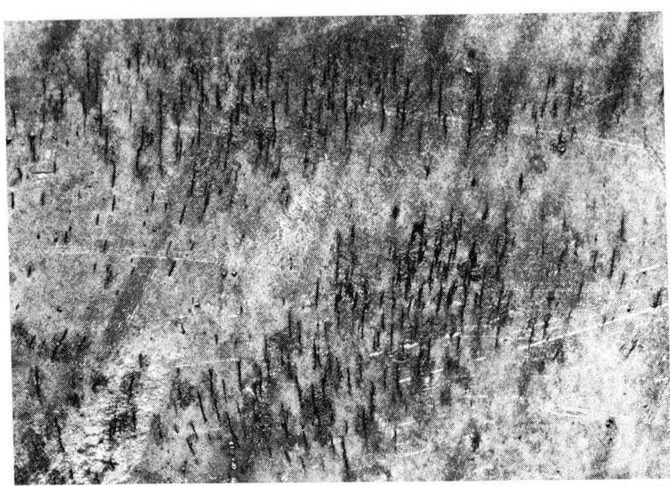

15.159

Fig. 15.158 to 15.164. Stress-corrosion in a vacuum still

Fig. 15.158 and 15.159. Internal surface. 1 ×

Fig. 15.158. Region adjacent to circumferential seam. Top: Shell. Below: Bottom
Fig. 15.159. Stress-corrosion in bottom

Fig. 15.160. Section across bottom seam (↓), unetched. 1 ×
Fig. 15.161. Corrosion originating from boundary zone weld/sheet. Etch: V2A-etchant. 5 ×

Fig. 15.162 to 15.164. Transverse section through cracks in bottom
Fig. 15.162. Unetched. 1 ×
Fig. 15.163. Unetched. 15 ×
Fig. 15.164. Etched with V2A-etchant. 200 × p 464

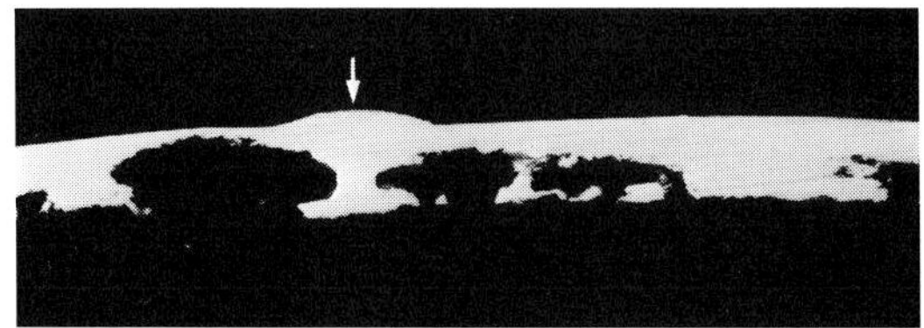

15.160

15.161

15.162

15.163

composition except that the sheet metal had been stabilized with titanium and the weld metal with niobium. That ruled out contact corrosion in this case. The weld seam was flawless in the uncorroded part. As always, it contained a little more ferrite than the sheets, which may explain the lower susceptibility to stress-corrosion cracking. The microstructure of the adjacent sheet metal was not noticeably changed by the welding heat. The short cracks in the bottom sheet showed in transverse section the profile illustrated in the **Fig. 15.162 and 15.163.** They had started from a streak-like area at which, the passive layer probably had been torn open by transversely directed stresses and then had spread out to the inside in a roof-like pattern. Corrosion apparently was impeded by ferrite streaks and deflected in the direction of the fibers **(Fig. 15.164).** The corrosive agent that was at fault could not be determined. It probably was sulfuric acid that may have concentrated in the solution below the inlet nipple during filling or while at rest. According to statements by the sender, no chlorine or hydroxyl ions were present. The stresses required for corrosion had been caused on the one hand by welding and on the other hand by the cold deformation of the bottom.

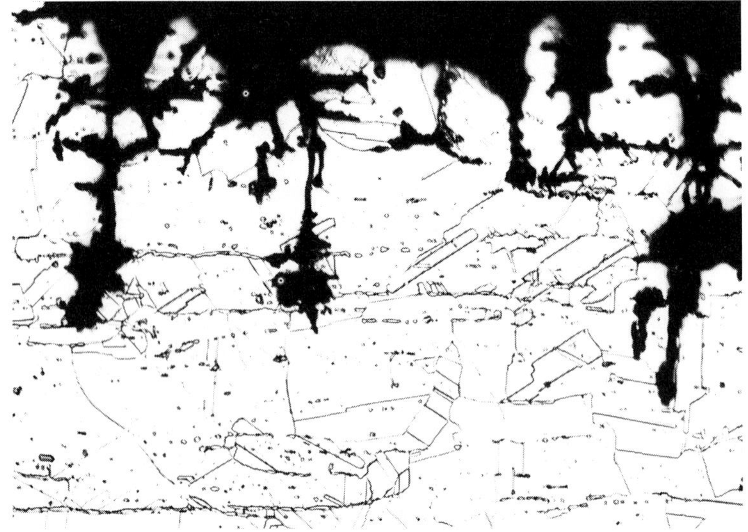

15.164

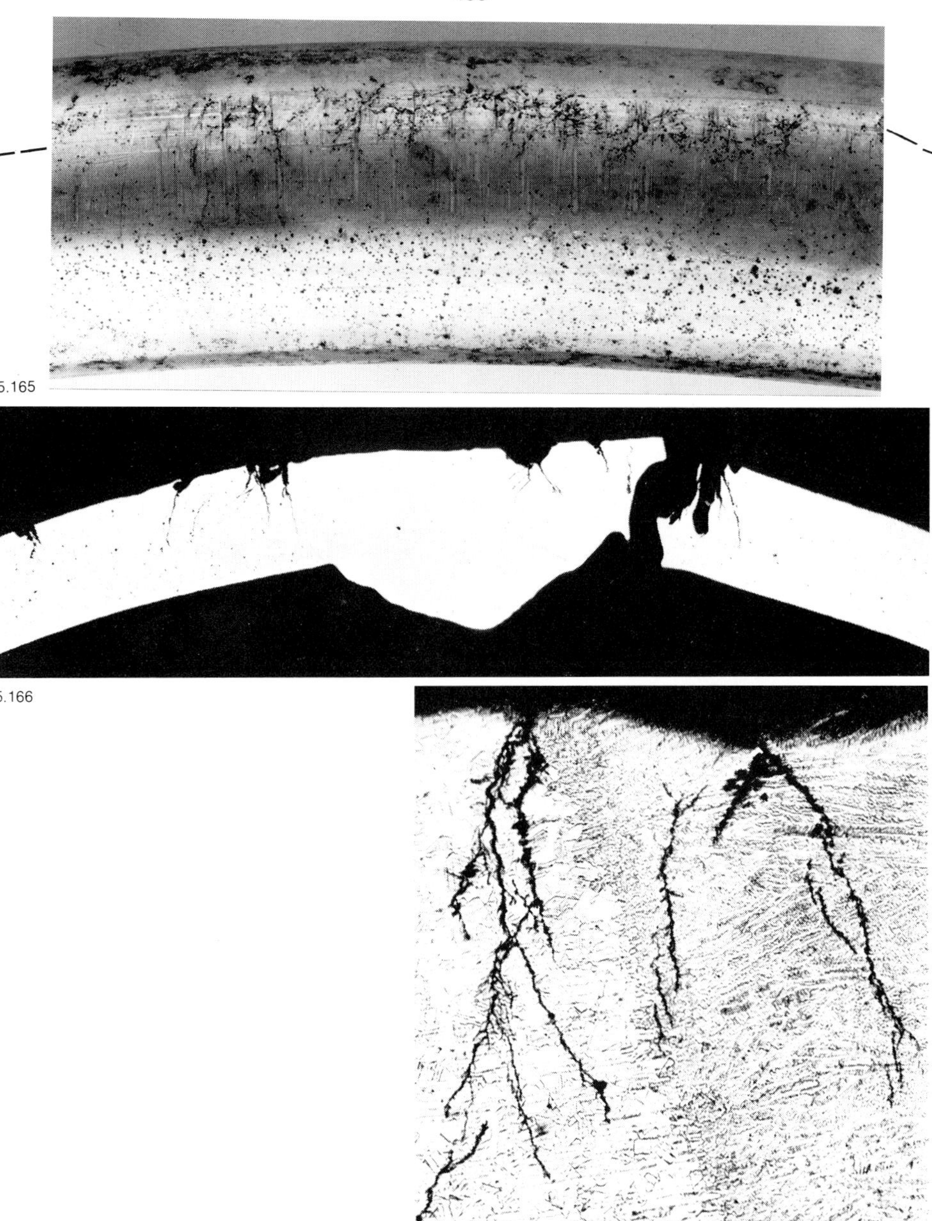

Fig. 15.165 to 15.167. Heating coil of dye vat with stress-corrosion cracks in and on both sides of weld seam and pitting

Fig. 15.165. External surface with weld seam zone (––). 1 ×

Fig. 15.166. Transverse section through weld seam. Unetched. 10 ×

Fig. 15.167. Transition from tube metal (left) to weld seam. Transverse section. Etch: V2A-etchant. 100 ×

The direct connection between stress-corrosion and weld stresses became apparent also in the corrosion of a **heating coil of a dye vat,** which normally operated with an alkaline dye liquor of a pH value of 12 to 13 at 100 °C maximum; however, between dyeing runs hydrogen chloride bleach had been passed through the coils. The tubes consisted of an austenitic stainless steel whose molybdenum content was too low, while its nickel content was at the lower limit of specifications for the titanium-stabilized steel with 18 % Cr, 12 % Ni and 2 % Mo, that is commonly used in Germany. Corrosion was demonstrated by pitting and crack formation **(Fig. 15.165).** The crack zone was located on both sides of the longitudinal weld seam **(Fig. 15.166),** and corrosion cracks had attacked the weld bead as well as the tube metal **(Fig. 15.167).** In this case (OH)' as well as Cl' ions had been present that may have acted as corrosive agents. Corrosion may have been facilitated by the low molybdenum and nickel contents.

G. Herbsleb[60]) has reported on stress-corrosion cracking caused by residual tensile stresses in the surface of coarsely ground pipes and plates of austenitic chromium-nickel and chromium-nickel-molybdenum steels.

Ferritic chromium-molybdenum steels are not susceptible to this type of stress corrosion. They can also be worked easily and are weldable if they have a very low carbon content (superferrites) and contain additions of titanium or niobium.

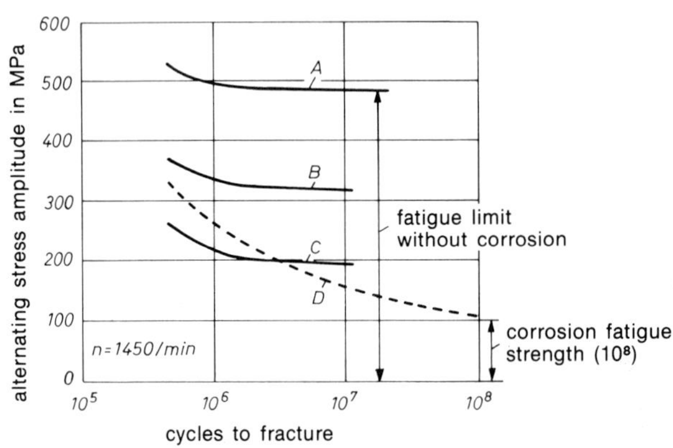

Curve A to C:	Fatigue curves from tests in air
Curve A:	Specimens not wetted
Curve B:	Specimens wetted previously by tap water for 10 days
Curve C:	Specimens wetted previously with tap water under simultaneous loading with $\sigma_a = 80$ MPa (11 600 psi)
Curve D:	Fatigue curve determined under simultaneous wetting with tap water

Fig. 15.168. Effect of corrosion through tap water upon bend strength of a chromium-nickel steel with 960 MPa (140 ksi) tensile strength. From Werkstoff-Handbuch Stahl & Eisen (according to McAdam)

15.3.5 **Corrosion-Fatigue Cracking**

As has been reported already in section 2.1 (Fig. 2.26), corrosion also decreases strength substantially. Corrosion pits act as sharp notches **(Fig. 15.168,** curve B) under subsequent fatigue type alternating stresses or cyclic loading. Corrosion pits originating under low alternating stresses that have not lead to fracture (curve C, Fig. 15.168) exert an even greater notch effect. If a material is stressed to fracture by alternating loads under a steady corrosion effect, no constant finite value of the fracture stress is discernible within a reasonable time (curve D, Fig. 15.168). The process that takes place under this type of corrosion is therefore to be considered the dynamic opposite or a special case of stress-corrosion, and is designated accordingly corrosion-fatigue cracking. The crack path is predominantly transgranular.

At high stresses and frequencies, that lead to fracture in a short time, the mechanical effect of the stress predominates. Accordingly the corrosion resistance of the steel is not of decisive significance. At low stresses the influence of time upon the corrosion plays a more important role. The number of load cycles that can be withstood then becomes greater with higher frequencies. The strength of the steel is but of minor importance. A corrosion-fatigue strenght of 100 to 200 MPa (15–30 ksi) for 10^7 load cycles may be figured on for a steel with a 300 to 1700 MPa (45–245 ksi) tensile strength that is wetted by tap water (see curve f in Fig. 2.26).

As a first example, **Figure 15.169** shows corrosion-fatigue cracks that originated at pits on the outer surface of a **high pressure pipe** rolled into the bottom **of an ammonia refrigerator**[61]). Such cracks had finally led to the rupture of the pipe. This was induced by several fatigue fractures at the outer surfaces, as illustrated in **Fig. 15.170.** There was no noticeable deformation. The cracks in the ferritic-pearlitic microstructure were predominantly transgranular **(Fig. 15.171).** Corrosion in this case was reinforced by crevice formation between pipe and refrigerator bottom.

The following failure on a **ground anchorage** could also have occurred due to faulty planning. In order to anchor a piling that served to secure a railway embankment which had been half dug out lengthwise, bundles of six stressed 12.2 mm diameter rods, each made of a heat treated steel (1370 MPa (200 ksi) ultimate strength) were used. The anchors rested directly in the ground without corrosion protection. This might have worked out well for steel of low strength, i.e. higher elongation, in rods of correspondingly heavier cross section during the limited time allowed for construction. But the rods broke 95 to 106 days after the respective installation date at an applied tensile stress of 595 MPa (86 ksi), i.e. 48 % of the measured yield point. Except for a few pits, none of the rods were heavily rusted. All the fractures originated at a point on the surface that could easily be determined **(Fig. 15.172a).** They were bend type failures with a deformation-deficient incipient fracture. Fine transverse cracks also were noticeable on the same perimeter line from which the fracture had started **(Fig. 15.172b).** In longitudinal sections these appeared as short, predominantly transgranular incipient cracks that originated in corrosion pits **(Fig. 15.173).** Metallographic examination, too, pointed to a single line on the rod perimeter where the transverse cracks from which the fractures had propagated were in evidence. This confirmed that the cracks had been formed under participating bending stresses. All observations indicate that this was a case of corrosion-fatigue cracking leading to failure. Their probable cause was a superimposition of cyclic bending loads from the operation of the railroad onto the static tensile loading. The dynamic stresses were transmitted onto the rods through the hard frozen earth at the time the fractures had occurred.

15.169

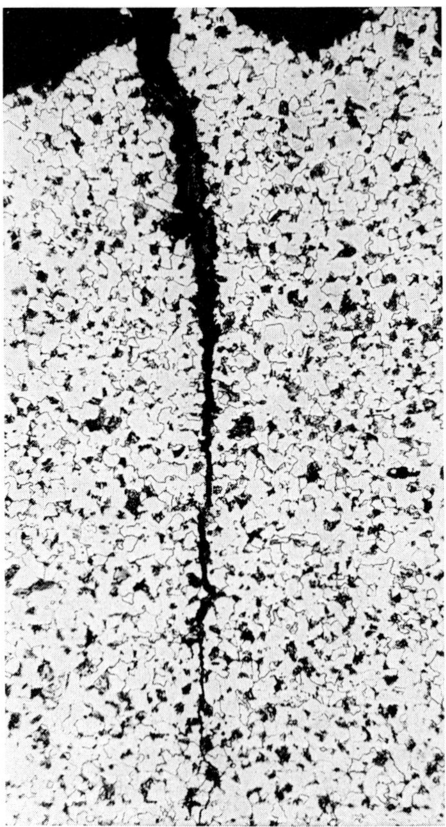

15.171

15.170

Fig. 15.169 to 15.171. Ammonia refrigerator pipe cracked due to cyclic stress-corrosion fatigue

Fig. 15.169. Corrosion-fatigue cracks originating in pits on exterior surface of pipe. 3 ×. Surface pickled with ammonium citrate solution

Fig. 15.170. Fracture originating in several incipient cracks. 5 ×. Light spot left is remaining fracture

Fig. 15.171. Transgranular fatigue crack originating in pit. Longitudinal section. Etch: Picral. 100 ×

The rupture of a **drum closing ring of a centrifuge** in which apple juice was processed was also caused by corrosion-fatigue cracking. The ring consisted of a heat treated steel which had an 880 to 980 MPa (128–142 ksi) tensile strength and high toughness. Failures occurred approximately eight years after delivery of the centrifuge. The fracture had originated at the upper thread, propagated along this for some distance and then ended with the residual fracture in the lower thread **(Fig. 15.174).** The incipient fracture consisted of a number of fatigue cracks that all had their origin in deep corrosion pits in the upper thread **(Fig. 15.175).** The profile in **Fig. 15.176** shows the depth of these pits. Corrosion was probably caused by a condensate of a sulfurous acid used for processing the apple juice; it had been enhanced by crevice formation. Since the fatigue fractures were comparatively small in relation to the entire fracture surface, the stress apparently had been very high. In order to prevent this as well as repeated similar failures, measures should be taken to eliminate corrosion entirely; these include changes in design, alloying of the steel in conformity with the respective purpose, or lining of the drum with a corrosion-resistant steel.

Very costly failures also occurred by corrosion-fatigue fractures in cooling channels of internally cooled machine parts such as piston rods of large Diesel engines. The fatigue fractures that originated in the longitudinal bore of a highly stressed hardened **chromium steel roll** (illustrated in **Fig. 15.177),** also were believed to have been caused by the participatory corrosive attack of the cooling water. In such cases an inhibitor such as an oil emulsion should be added to the cooling water or the bore should be provided with a protective coating.

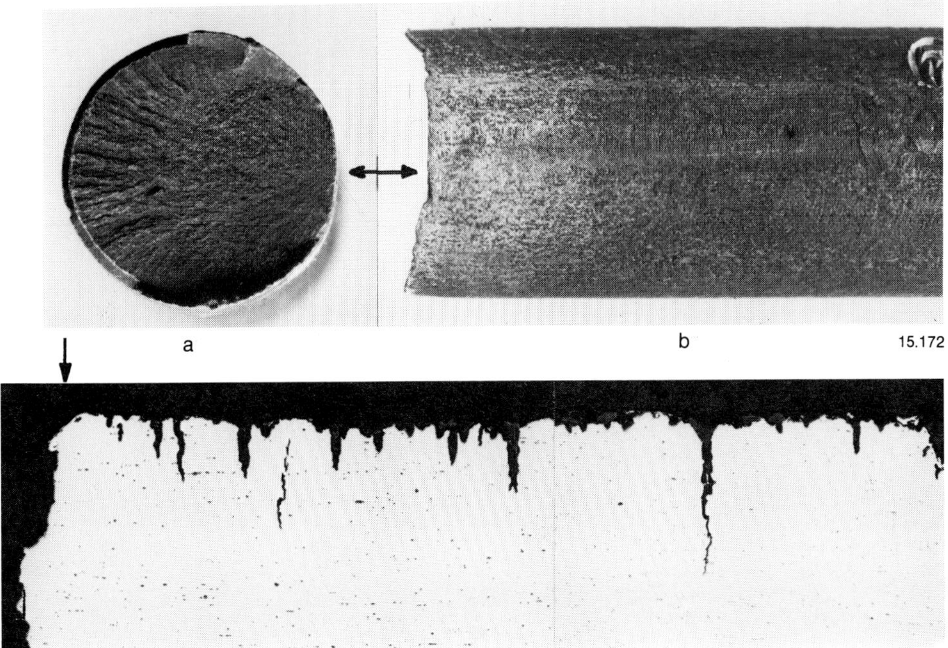

Fig. 15.172 and 15.173. Corrosion-fatigue cracks and fracture in stressed reinforcement rods of a ground anchorage

Fig. 15.172a. Fracture. 3 ×

Fig. 15.172b. Surface at fracture. 3 ×

Fig. 15.173. Longitudinal section through fracture origin (↓). Unetched. 100 ×

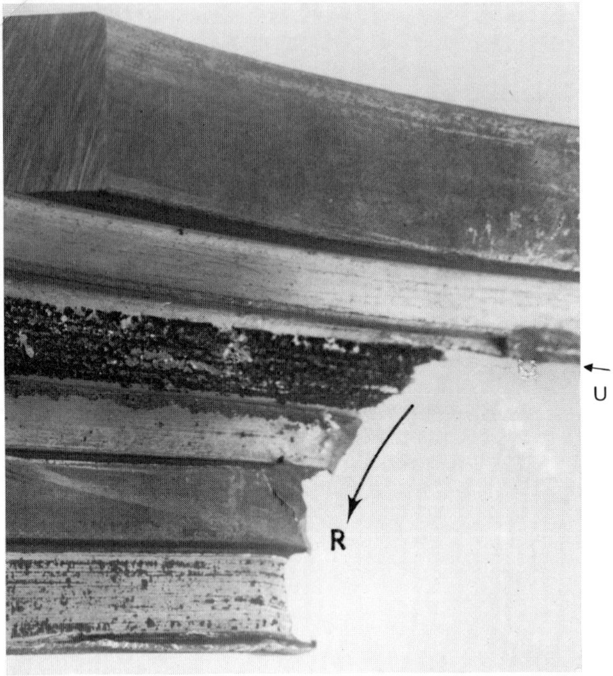

U

R

15.174

15.176

15.175

Fig. 15.174 to 15.176. Drum closure ring of centrifuge for apple juice processing plant which ruptured due to corrosion-fatigue

Fig. 15.174. Section through ring with remaining fracture. 1 ×
U = circumferential fracture (see Fig. 15.175). R = final fracture
Fig. 15.175. Circumferential fracture with incipient fatigue fractures. 1 ×

Fig. 15.176. Unetched transverse section. 1 ×

An example taken from experience in shipbuilding should be cited. They **hull of a suction dredge** had sprung leaks in several places in the vicinity of the Diesel engine as well as adjacent to the pump. Cracks ran parallel to and across the longitudinal axis of the vessel. They were located near the weld seams, but also in part midway between the seams. Cracks in the bottom that were situated parallel to the weld at 15 mm distance, as seen from the water side, are shown in **Fig. 15.178.** They seemed enlarged by corrosion. In addition to the penetrating cracks, groups of fine incipient cracks also were visible upon closer metallographic examination **(Fig. 15.179 and 15.180).** When opened, these had the characteristics of fatigue cracks **(Fig. 15.181).** Therefore they had to be interpreted as corrosion-fatigue cracks. Their location evidently coincides preferentially with the vibration antinodes of the ship bottom plates. Weld stresses may also have contributed.

Finally, a wartime example should be cited in which the connection between fatigue fracture and corrosion in expressed particularly well. Numerous crack failures occurred in **steel boxes** that were substituted for copper fire-boxes in locomotives. Such a crack in the lower part of a fire-box of strain age-resistant steel with 0.14 to 0.18 % C is shown in **Fig. 15.182.** The crack appeared at the water side of the region where the box was riveted to the bottom flange and where a maximum stress was to be expected at the transition between the sheet and the upper rim of the bottom flange. When the crack branch located at the upper left of the illustration was broken open, a number of fatigue cracks, which ultimately linked together, were observed on the fracture surface **(Fig. 15.183).** In addition to the principal crack, many small incipient cracks were present. Sections across such cracks revealed that they were filled with corrosion products **(Fig. 15.184).** Therefore this was a case of corrosion-fatigue cracking. The vibrations may have been caused by motion or temperature changes. When steel was substituted for copper, it should have been provided with corrosion protection.

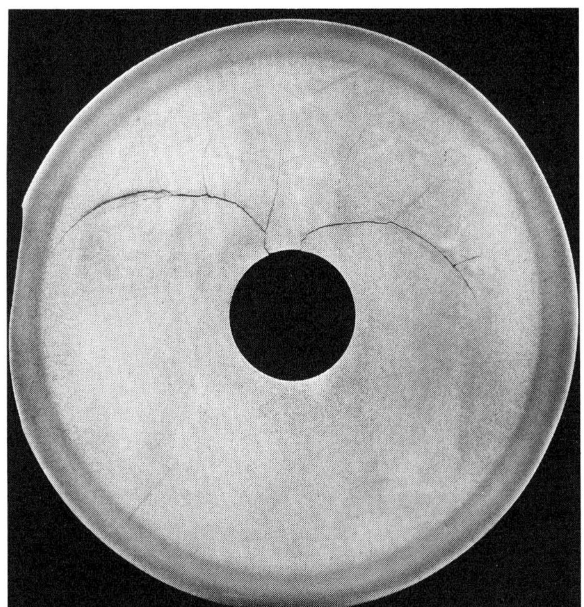

Fig. 15.177. Fatigue fractures originating in water-cooled bore of hardened chromium-steel roll. Transverse section. Etch according to Heyn. Approx. 0.25 ×

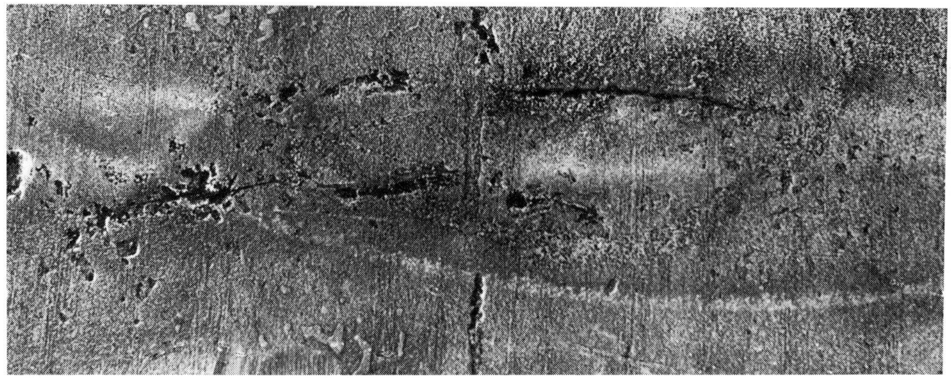

15.178

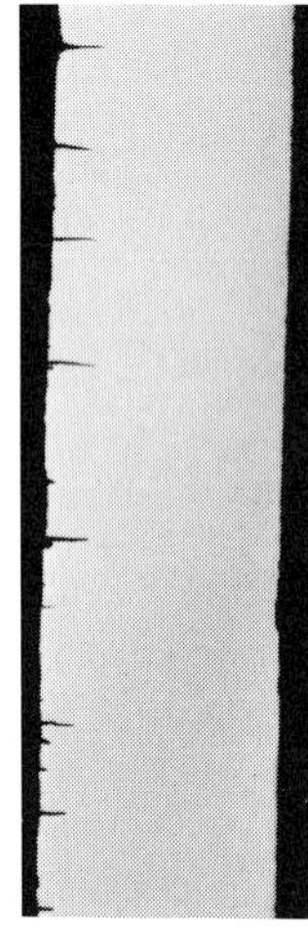

15.179

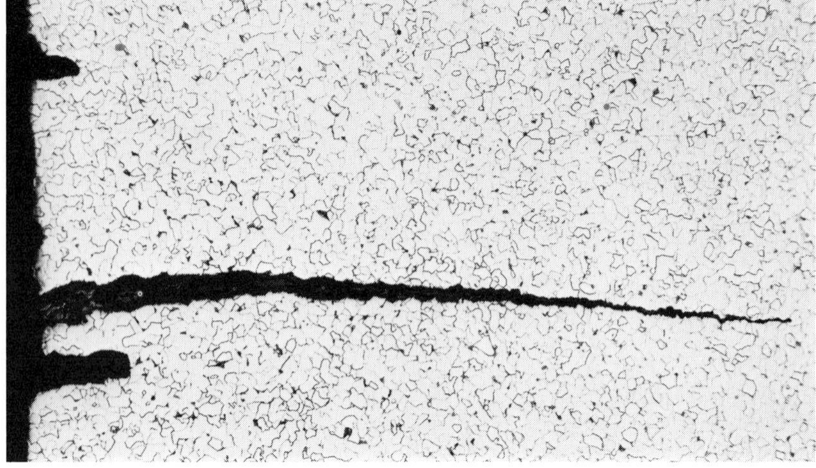

15.180

15.181

Fig. 15.178 to 15.181. Roll of a suction dredge that sprang leaks caused by corrosion-fatigue cracks

Fig. 15.178. Corrosion-fatigue cracks in ship's bottom on water side. 0.6 ×

Fig. 15.179 and 15.180. Fine cracks in ship's bottom. Transverse section. Left: water side

Fig. 15.179. Unetched. 3.5 ×

Fig. 15.180. Etch: Nital. 100 ×

Fig. 15.181. Opened crack with fatigue fracture. 3 ×

15.182

15.183

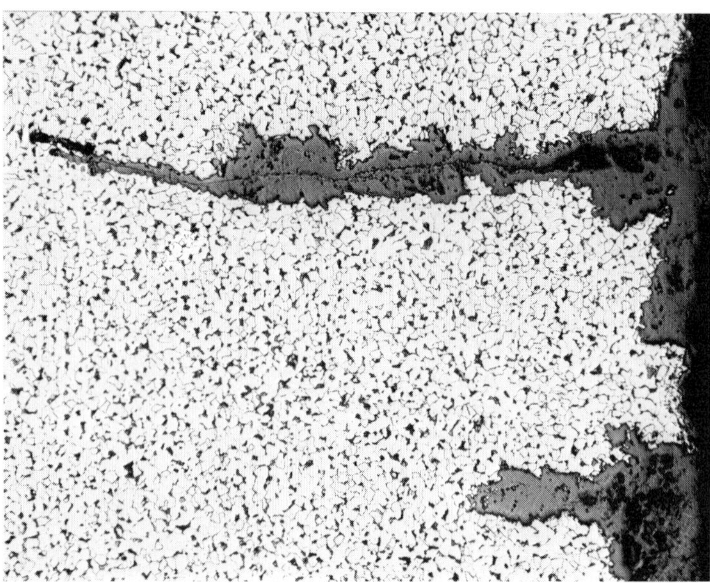

15.184

Literature Section 15

15.1

1) F. K. Naumann: Stähle für Treibstoffgewinnungsanlagen. Chem. Fabrik 11 (1938) S. 77/87
2) A. Rahmel: Korrosion durch Alkalisulfate und Schwefeltrioxid. Arch. Eisenhüttenwes. 31 (1960) S. 59/65. Mitt. Ver. Großkesselbes. H. 74 (Okt. 1961)
3) A. Rahmel: Einfluß von Calciumsulfat und Magnesiumsulfat auf die Hochtemperatur-Oxydation austenitischer Chrom-Nickel-Stähle in schwefeltrioxidhaltigen Atmosphären bei Vorhandensein von Oberflächenbelägen aus Alkalisulfaten. Mannesmann-Forsch.-Ber. 236/1963
4) H. Köhler: Hochtemperaturkorrosion an einem Wärmeaustauscher. Stahl u. Eisen 91 (1971) S. 275/81
5) H.-J. Pohle u. H. Machold: Erfahrungen beim Betrieb metallischer Rekuperatoren für Tieföfen. Stahl u. Eisen 89 (1969) S. 487/96
6) F. K. Naumann u. F. Spies: Verzunderte Rohre aus einer Generatorgasanlage. Prakt. Metallographie 13 (1976) S. 635/36
7) G. Bandel: Gefüge- und Eigenschaftsänderungen in hitzebeständigen Chrom-Aluminium- und Chrom-Silcium-Stählen durch Stickstoffaufnahme. Arch. Eisenhüttenwes. 11 (1937/38) S. 139/44
8) F. K. Naumann u. F. Spies: Verzundertes Heizbad aus einem Durchlaufglühofen. Prakt. Metallographie 5 (1968) S. 457/62

15.2

9) W. Oelsen, K.-H. Sauer u. G. Naumann: Zur Entkohlung von Stählen in zundernden und nichtzundernden Gasen. Jahrb. 1969 d. Landesamtes f. Forsch. d. Landes Nordrhein-Westfalen. Westd. Verlag, Köln u. Opladen, S. 411/67
10) F. K. Naumann: Einwirkung von Wasserstoff unter hohem Druck auf unlegierten Stahl. Stahl u. Eisen 57 (1937) S. 889/900
11) I. Class: Stand der Kenntnisse über die Eigenschaften druckwasserstoffbeständiger Stähle. Stahl u. Eisen 80 (1960) S. 1117/35
12) F. K. Naumann u. F. Spies: Entkohlung. Prakt. Metallographie 8 (1971) S. 375/84
13) Dieselben: Untersuchung von Verteilerbalken aus dem Kühler einer Ammoniak-Syntheseanlage. Prakt. Metallographie 8 (1971) S. 621/26
14) F. K. Naumann: Der Einfluß von Legierungszusätzen auf die Beständigkeit von Stahl gegen Wasserstoff unter hohem Druck. Stahl u. Eisen 58 (1938) S. 1239/50. Techn. Mitt. Krupp Forsch.-Ber. 1 (1938) S. 223/43
15) H. Kiessler: Druckwasserstoffbeständige Stähle. Werkstoff-Handbuch Stahl u. Eisen. 4. Aufl. 1965, Blatt O 85
16) F. K. Naumann u. F. Spies: Untersuchung korrodierter Kesselrohre. Prakt. Metallographie 8 (1971) S. 729/34

15.3

17) K. Meyer u. W. Schwenk: Korrosionsgefährdung an Verletzungen von Schutzanstrichen durch Elementbildung in salzreichen Wässern. Schiff u. Hafen 26 (1974) H. 11. Mannesmann-Forsch.-Ber. 662/74
18) F. K. Naumann u. F. Spies: Ausfall eines Rekuperators mit austentisch geschweißten Rohren. Prakt. Metallographie 10 (1973) S. 50/53
19) J. Billiter: Prinzipien der Galvanotechnik. Berlin 1934, S. 270
20) G. Herbsleb: Der Einfluß von Eisen- und Mangansulfideinschlüssen auf die Lochkeimbildung bei der Lochkorrosion von Stahl. Stahl u. Eisen 93 (1973) S. 837/39
21) E. Kauczor: Zerstörung von Messingteilen durch Entzinkung von Messing. Prakt. Metallographie 5 (1968) S. 353/56
22) Derselbe: Zerstörung von Messing-Hartlot durch Entzinkung. Prakt. Metallographie 7 (1970) S. 587/90
23) H.-J. Wiester u. G. Piel: Untersuchungen über die interkristalline Korrosion austenitischer Chrom-Nickel-Stähle nach langandauernder Beanspruchung zwischen 450 und 800° C Arch. Eisenhüttenwes. 30 (1959) S. 293/97
24) G. Grützner: Kornzerfallsanfälligkeit stickstoff-legierter austenitischer Chrom-Nickel-Stähle durch Chromnitridausscheidungen. Stahl u. Eisen 93 (1973) S. 9/18
25) B. Strauß, H. Schottky u. J. Hinnüber: Die Karbidausscheidung beim Glühen von nichtrostendem unmagnetischem Chrom-Nickel-Stahl Z. anorg. allgem. Chemie 188 (1930) S. 309/24
26) H.-J. Wiester, H.-J. Schüller u. P. Schwab: Ausscheidungsvorgänge in austenitischen Chrom-Nickel-Stählen und ihr Einfluß auf die Neigung zur interkristallinen Korrosion. Arch. Eisenhüttenwes. 30 (1959) S. 299/309
27) W. Friebe u. A. Hankel: Stand der Kenntnisse über die Entstehung interkristalliner Risse in Verzinkungskesseln. Stahl u. Eisen 94 (1974) S. 299/303
28) H.-J. Engell u. M. O. Speidel: Ursachen und Mechanismen der Spannungsrißkorrosion. Korrosion 22. Bericht über die Diskussionstagung 1968 der DGM, 35. Veranstaltung der Europäischen Föderation Korrosion – Spannungsrißkorrosion – in Frankfurt/Main vom 25. und 26. April 1968. Herausg. H. Kaesche. Verlag Chemie GmbH., Weinheim/Bergstr. S. 1/20
29) H. Gräfen: Derzeitiger Stand der Kenntnisse über die Spannungsrißkorrosion unlegierter und schwachlegierter Stähle. Wie 28) S. 25/33
30) A. Fry: Das Verhalten der Kesselbaustoffe im Betrieb. Kruppsche Monatsh. 7 (1926) S. 185/96
31) P. Drodten u. K. Forch: Interkristalline Spannungsrißkorrosion bei niedriglegierten Stählen. Arch. Eisenhüttenwes. 44 (1973) S. 893/98
32) A. Pomp u. P. Bardenheuer: Schäden an Dampfkesselelementen. Mitt. Kais.-Wilh.-Inst. Eisenforsch. 11 (1929) Abh. 128. S. 185/91

Fig. 15.182 to 15.184. Steel firebox cracked due to corrosion-fatigue

Fig. 15.182. Water side of firebox wall with corrosion-fatigue cracks. Approx. 0.5 ×
Fig. 15.183. Lateral crack branch after breaking open and pickling. Approx. 3 ×
Top: Water side. Bottom: Fire side
Fig. 15.184. Micrograph of corrosion-fatigue cracks. Etch: Nital. 100 ×

33) F. K. Naumann u. F. Spies: Nietlochrisse in einem Dampfkessel. Prakt. Metallographie 7 (1970) S. 459/63

34) Dieselben: Schadhafte Nietnaht eines Laugenbehälters. Prakt. Metallographie 7 (1970) S. 527/30

35) Dieselben: Schäden an geschweißten Rohren und Blechen. Prakt. Metallographie 7 (1970) S. 637/43

36) Z. Verö u. B. Zorkoczy: W. Rädeker: Spannungskorrosion an geschweißten Blechkaminen von Siemens-Martin-Stahlwerksöfen. Schweißen u. Schneiden 18 (1966) S. 83

37) R. Floßmann, K. P. Roeder u. G. Schnegelsberg: Spannungsrißkorrosion an Winderhitzern für hohe Heißwindtemperaturen Stahl u. Eisen 94 (1974) S. 84/86

38) H. E. Bühler, G. Robusch u. G. Lennartz: Arten der interkristallinen Spannungskorrosion an unlegierten und niedriglegierten Stählen unter besonderer Berücksichtigung der Spannungskorrosion an modernen Großhochöfen. Betriebliche Erfahrungen und Untersuchungen. Stahl u. Eisen 95 (1975) S. 797/802

39) F. K. Naumann u. F. Spies. Gebrochener Trommelverschlußring. Prakt. Metallographie 12 (1975) S. 381/83

40) F. K. Naumann u. F. Spies: Gerissener Trommeldeckel einer Zentrifuge. Prakt. Metallographie 12 (1975) S. 209/15

41) V. Čihal u. J. Kubelka: Influence of Cold Work on Sensitising of Stainless Steels to Intergranular Corrosion. Prakt. Metallographie 12 (1975) S. 148/55

42) H. Coriou, L. Grall, P. Olivier u. H. Willermoz: Influence of carbon and nickel content on stress corrosion cracking of austenitic stainless alloys in pure or chlorinated water at 350° C. Proceedings of Conference Fundamental Aspects of Stress Corrosion Cracking 1967, Ohio 1969 S. 352/59

43) H. Buchholtz u. R. Pusch: Beitrag zur transkristallinen Spannungsrißkorrosion von Stahl. Stahl u. Eisen 62 (1942) S. 21/30

44) F. K. Naumann: Schäden an Hughes-Schrumpfverbindern in einer schwefelwasserstofführenden Bohrung. Erdöl Zschr. 73 (1957) S. 4/14

45) W. Dahl, H. Stoffels, H. Hengstenberg u. C. Cüren: Untersuchungen über die Schädigung von Stählen unter Einfluß von feuchtem Schwefelwasserstoff. Stahl u. Eisen 87 (1967) S. 125/36

46) H. Schenck, E. Schmidtmann u. H. F. Klärner: Standzeitverhalten von Zugproben aus Rohrstählen mit einer Mindeststreckgrenze von 30 kg/mm^2 in schwefelwasserstoffhaltigen Lösungen und Gasen. Stahl u. Eisen 87 (1967) S. 136/46

47) F. K. Naumann: Standzeitversuche an Stählen hoher Zugfestigkeit in schwefelwasserstoffhaltigen Lösungen und Gasen. Stahl u. Eisen 87 (1967) S. 146/51

48) F. K. Naumann u. A. Bäumel: Bruchschäden an Spanndrähten durch Wasserstoffaufnahme in Tonerdezementbeton. Arch. Eisenhüttenwes. 32 (1961) S. 89/94

49) F. K. Naumann u. W. Carius: Die Bedeutung der Korrosionsvorgänge in wäßrigen Schwefelwasserstoff-Lösungen für die Bruchbildung an Stählen. Arch. Eisenhüttenwes. 30 (1959) S. 283/92

50) Dieselben: Die Aufnahme von Wasserstoff aus wäßrigen Schwefelwasserstoff-Lösungen durch Stahl und die Bruchbildung an gespannten Stählen. Arch. Eisenhüttenwes. 30 (1959) S. 361/68

51) Dieselben: Bruchbildung an Stählen bei Einwirkung von Schwefelwasserstoffwasser. Arch. Eisenhüttenwes. 30 (1959) S. 233/38

52) F. K. Naumann u. F. Spies: Untersuchung einer blasigen und rissigen Erdgasleitung. Prakt. Metallographie 10 (1973) S. 475/80

53) Dieselben: Schäden an Gestängeverbindern in einer schwefelwasserstofführenden Erdgasbohrung. Prakt. Metallographie 10 (1973) S. 100/08

54) F. K. Naumann: Korrosionsschäden an gespannten Stählen. Beton- und Stahlbetonbau 64 (1969) S. 10/17

55) W. Jäniche, W. Puzicha u. H. Litzke: Zum zeitabhängigen Bruch vergüteter hochfester Stähle. Arch. Eisenhüttenwes. 36 (1965) S. 887/96

56) H.-J. Engell: Korrosionserscheinungen und Werkstofffragen bei Stahl- und Spannbetonbauwerken. Stahl u. Eisen 98 (1978) S. 637/41

57) M. Ternes: Die Spannungsrißkorrosion von Eisenlegierungen unter besonderer Berücksichtigung nichtrostender austenitischer Stähle. Werkstoffe u. Korrosion 14 (1963) S. 729/39

58) H. Keller u. M. Petrich: Metallographische Aufklärung von Spannungskorrosions-Schäden an ferritischen und austenitischen Stählen. Prakt. Metallographie 1 (1964) S. 66/72

59) F. K. Naumann u. F. Spies: Undichte Heizschlange aus einem austenitischen Chrom-Nickel-Molybdän-Stahl. Prakt. Metallographie 5 (1968) S. 529/31

60) G. Herbsleb: Einfluß der Oberflächenbeschaffenheit auf die Beständigkeit nichtrostender, austenitischer Chrom-Nickel-Stähle gegen transkristalline Spannungsrißkorrosion. Werkstoffe u. Korrosion 24 (1973) S. 867/72

61) F. K. Naumann u. F. Spies: Gerissenes Rohr aus einem Ammoniakkühler. Prakt. Metallographie 10 (1973) S. 414/16

APPENDIX I

Glossary of Special or Unusual Terms Used in This Book

arrest lines. Profile of a fatigue crack front which is coarser than fatigue fracture features of the type known as striations. The lines indicate the position of the crack front after each succeeding cycle of stress.

Berlin blue. An iron blue pigment that is a complex ferriferrocyanide produced by oxidation of a ferrous ferrocyanide precipitate from a soluble ferrocyanide and ferrous sulfate.

boil-over skull pieces. Portion of liquid metal that solidifies and remains on the side of a mold. During tapping or pouring the melt, these undissolved fragments become entrapped in the ingot.

braze cracking. Cracks which are the result of low melting braze metals, such as copper, zinc, or tin, penetrating along steel grain boundaries.

butterflies. Hard microstructural alterations that resist etching and frequently contain fine cracks. They generally form at nonmetallic inclusions directly below a highly stressed surface and are oriented in the direction of principal stress.

flakes. Short discontinuous internal fissures of round or elliptical shape. They are attributed to internal stresses and regions of high internal hydrogen concentrations. The fractures typically occur during cooling after the first forging or hot rolling operation.

flash. An anomaly of the surface grain size of hot galvanized steel sheet. Hydrogen that has been absorbed during pickling is evolved during galvanizing. This causes increased nucleation sites and hence islands of finer grain clusters. Excessive entrapment of hydrogen may result in gas pockets or blisters.

green rot. A form of high temperature attack of stainless steel and heat-resistant nickel-chromium and nickel-chromium-iron alloys. If these alloys are subjected to simultaneous oxidation and carburization, chromium is first precipitated as chromium carbide and the carbide particles are then oxidized. The chromium oxide has a green color.

inhibitor. An oil generally added to pickling solutions to retard the dissolution of metal and pitting while promoting a uniform removal of oxide scale from steel surfaces. German product available under the name Vogel-Sparbeize is used as an addition to V2A-etching solution.

klanken test. A non-standardized dynamic free-handed loop tensile or tear test developed and used by one German industrial manufacturer and user for bench testing of wires.

microcavity. A void left in the dendrite interstices of a cast metal ingot as a result of solidification or casting shrinkage.

penetration hole. A hole or pin hole that penetrates the entire thickness of an enamel coating of a steel sheet.

rock candy. A mixture of extremely large intergranular and cleavage fracture modes.

runner. A conduit for molten metal or a portion of the gate assembly of a casting that connects the sprue with the gate. A runner stick is a gate pin.

scab. An imperfection consisting of a flat volume of metal joined to a casting through a small area. Usually a scab is set in a depression in a casting and is separated from the metal of the casting proper by a thin layer of sand.

spongiosis. The selective leaching of grey iron by the corrosive attack of mild aqueous salt solutions, especially on pipes and fittings buried in the ground. A porous mass of graphite and phosphide is left behind. Same as graphitic corrosion.

top crust piece. A collection of metallic inclusions which solidify and float on the top of molten metal. They remain undissolved and are subsequently trapped in the solidified ingot.

undercut. A groove melted into the base metal adjacent to the toe of a weld and left unfilled.

weld crack susceptibility. A crack formation in thin steel sheets or thinwalled pipes adjacent to gas welded seams. The cracks form in the austenitic grain boundaries immediately below the solidus temperature. Steels of high sulfur content are especially sensitive to this phenomenon.

weld sensitivity. A failure tendency characterized by hardening of the weldment or the heat-affected zone. It occurs mainly in arc welding of thick steel sections.

APPENDIX II

Identification of Etchants Referred to in This Book

Etchant or Reagent	Composition	Procedure*	Purpose	Result
		* Sequential steps to be followed as listed		
A. Macroscopic Examination				
1. Heyn's reagent	120 ml H_2O 10 g $CuCl_2 \cdot 2NH_4Cl$	Immerse finish-ground face of specimen in etchant for 2–10 minutes; wipe off Cu-deposit with soft cloth under H_2O rinse	Proof of P-segregations, fiber orientation and cracks	P-and C-rich regions are etched darker; primary structure (dendrites, fiber direction), cracks and surface decarburization become visible
2. Oberhoffer's reagent	500 ml H_2O 500 ml C_2H_5OH 0.5 g $SnCl_2$ 1 g $CuCl_2$ 30 g $FeCl_3$ 42 ml HCl	Swab etchant onto polished specimen surface for a few seconds	Proof of P-segregations and fiber orientation	Regions low in P are etched, segregation zones not; primary structure (dendrites, streaks, stringers) and gas bubble segregation become visible
3. Fry's reagent	100 ml H_2O 120 ml HCl 90 g $CuCl_2$	Rub finish-ground surface for 2–20 minutes; after etching rinse first with conc. HCl or acidic C_2H_5OH, then with H_2O	Proof of strain aging	Cold worked regions in low-carbon and nitrogen-rich steels are etched dark after natural or artificial (tempered to 200–300 °C) strain aging
4. Adler's reagent	25 ml H_2O 3 g $CuCl_2 \cdot 2NH_4Cl$ 50 ml HCl 15 g $FeCl_3$	Swab etchant onto finish-ground surface of specimen for a few seconds to 1 minute	Bring out structure of weld seams	Weld seams and their layers from different passes become clearly visible
5. Baumann's sulfur print	100 ml H_2O 5 ml H_2SO_4	Ordinary sensitized photographic silver bromide paper with semimatte finish is soaked in solution, excess dripped off; emulsion side pressed for 1 to 5 minutes onto finish-ground surface; rinse with H_2O, fix in thiosulfate solution, wash thoroughly, dry	Proof of S-segregations	FeS and MnS inclusions develop H_2S which reacts with the AgBr to produce Ag_2S (etched darker)

Etchant or Reagent	Composition	Procedure*	Purpose	Result
		* Sequential steps to be followed as listed		
B. Microscopic Examination				
1. Nital	100 ml C_2H_5OH 1 ml HNO_3	Etch polished surface for a few seconds up to 1 minute	Development of microstructure	Ferrite grains are differently etched according to orientation; grain boundaries become distinct; carbides and nitrides are not attacked
2. Picral	100 ml C_2H_5OH 4 g $C_6H_2OH\ (NO_2)_3$	Etched polished surface for a few seconds to 1 minute; used solution reacts faster	Development of microstructure	Orientation differences appear less pronounced than for nital. Hence etched structure shows less contrast
3. Alkaline sodium picrate	75 ml H_2O 25 g NaOH 2 g $C_6H_2OH\ (NO_2)_3$	After removal of scale from outer surface, immerse polished surface in etchant kept at room temperature or up to 50 °C; etch time depends on temperature	Proof of cementite; differentiation of Fe_3C, Fe_2N and Fe_2P	Fe_3C is stained fast, Fe_2N more slowly; ferrite and Fe_2P, as well as many special and solid solution carbides (except Fe_3-W_2-C) are not stained
4. Murakami's reagent	100 ml H_2O 10 g KOH 10 g $K_3Fe\ (CN)_6$	Use fresh solution; immerse or swab polished surface at room temperature for 15–60 seconds or immerse at 50 °C up to 2 minutes; rinse with H_2O, C_2H_5OH, dry	Differentiation of special and solid solution carbides as well as Fe_2P from pure Fe_3C	Special and solid solution carbides and Fe_2P are stained faster than Fe_3C
5. Ammonium hydroxide electrolyte	Diluted NH_4OH	Immerse polished surface in electrolyte at 1.5 to 2 v for a fews seconds to 1 minute	Attack of special carbides	Special and solid solution carbides are stained, but not Fe_3C
6. Caustic soda electrolyte	10N NaOH	Immerse polished surface in electrolyte at 1.5 to 2 v for a few seconds to 2 minutes	Attack of sigma-phase	Sigma-phase is stained

Etchant or Reagent	Composition	Procedure*	Purpose	Result
		* Sequential steps to be followed as listed		
C. For Steels With High Chromium Content				
V2A-etchant	100 ml H_2O 100 ml HCl 10 ml HNO_3 0.3 ml inhibitor* * see Appendix I	Use fresh solution daily; immerse or swab polished surface at room temperature or immerse at 50 °C up to 1 minute	Development of microstructure of stainless and heat resistant steels	Grain faces are etched; distinct marking of twin boundaries in austenitic steels
D. For Copper Alloys				
Copper ammonium chloride	120 ml H_2O 10 g $CuCl_2$ NH_4OH	Add NH_4OH until precipitate is dissolved; immerse polished surface of specimen; wash thoroughly	Development of Phases	Dark staining of beta in alpha-beta brass

INDEX

T

U

V